Fundamental Mathematics

FIFTH EDITION

Marvin L. Bittinger

Indiana University Purdue University Indianapolis

PEARSON

Addison
Wesley

Boston   Columbus   Indianapolis   New York   San Francisco   Upper Saddle River   Amsterdam   Cape Town
Dubai   London   Madrid   Milan   Munich   Paris   Montréal   Toronto   Delhi   Mexico City
São Paulo   Sydney   Hong Kong   Seoul   Singapore   Taipei   Tokyo

Editorial Director	Christine Hoag
Editor in Chief	Maureen O'Connor
Executive Editor	Cathy Cantin
Executive Project Manager	Kari Heen
Associate Editors	Joanna Doxey and Christine Whitlock
Editorial Assistant	Jonathan Wooding
Production Manager	Ron Hampton
Composition	Pre-Press PMG
Production Services	Sally Lifland/Lifland et al., Bookmakers
Editorial Services	Martha K. Morong/Quadrata, Inc.
Art Editor and Photo Researcher	Geri Davis/The Davis Group, Inc.
Senior Media Producer	Ceci Fleming
Associate Producer	Nathaniel Koven
Content Development Managers	Eric Gregg (MathXL), Mary Durnwald (TestGen)
Executive Marketing Manager	Michelle Renda
Marketing Assistants	Margaret Wheeler and Alicia Frankel
Prepress Supervisor	Caroline Fell
Manufacturing Manager	Evelyn Beaton
Senior Manufacturing Buyer	Carol Melville
Senior Media Buyer	Ginny Michaud
Text Designer	Geri Davis/The Davis Group, Inc.
Senior Designer/Cover Design	Beth Paquin
Cover Photograph	© Martin Harvey/Photographers Choice/AGE FotoStock

Photo Credits

Photo credits appear on page A-12.

Library of Congress Cataloging-in-Publication Data
Bittinger, Marvin L.
  Fundamental college mathematics. — 5th ed. / by Marvin L. Bittinger.
      p. cm.
  Includes index.
  1. Mathematics—Textbooks.   I. Bittinger, Marvin L.
      Fundamental mathematics. II. Title.
  QA39.3.B57 2010
  513—dc22   2009020668

1 2 3 4 5 6 7 8 9 10—CRK—14 13 12 11 10 09

© 2010, 2007, 2003, 1999. Pearson Education, Inc.

Addison-Wesley
is an imprint of

www.pearsonhighered.com

ISBN-13: 978-0-321-61342-4
ISBN-10: 0-321-61342-2

Contents

1 Whole Numbers   1

4 Decimal Notation 215

5 Ratio and Proportion 295

6 Percent Notation 347

Index of Applications

Author's Note to Students

Welcome to *Fundamental College Mathematics*. Having a solid grasp of the mathematical skills taught in this book will enrich your life in many ways, both personally and professionally, including increasing your earning power and enabling you to make wise decisions about your personal finances.

As I wrote this text, I was guided by the desire to do everything possible to help you learn its concepts and skills. The material in this book has been developed and refined with feedback from users of the four previous editions so that you can benefit from their class-tested strategies for success. Regardless of your past experiences in mathematics courses, I encourage you to consider this course as a fresh start and to approach it with a positive attitude.

One of the most important things you can do to ensure your success in this course is to allow enough time for it. This includes time spent in class and time spent out of class studying and doing homework. To help you derive the greatest benefit from this textbook, from your study time, and from the many other learning resources available to you, I have included an organizer card at the front of the book. This card serves as a handy reference for contact information for your instructor, fellow students, and campus learning resources, as well as a weekly planner. It also includes a list of the Study Tips that appear throughout the text. You might find it helpful to read all of these tips as you begin your course work.

Knowing that your time is both valuable and limited, I have designed this objective-based text to help you learn quickly and efficiently. You are led through the development of each concept, then presented with one or more examples of the corresponding skills, and finally given the opportunity to use these skills by doing the interactive margin exercises that appear on the page beside the examples. For quick assessment of your understanding, you can check your answers with the answers placed at the bottom of the page. This innovative feature, along with illustrations designed to help you visualize mathematical concepts and the extensive exercise sets keyed to section objectives, gives you the support and reinforcement you need to be successful in your math course.

To help apply and retain your knowledge, take advantage of the new Skill to Review exercises when they appear at the beginning of a section and the comprehensive mid-chapter reviews, summary and reviews, and cumulative reviews. Read through the list of supplementary material available to students that appears in the preface to make sure you get the most out of your learning experience, and investigate other learning resources that may be available to you.

Give yourself the best opportunity to succeed by spending the time required to learn. I hope you enjoy learning this material and that you will find it of benefit.

Best wishes for success!
Marv Bittinger

Related Bittinger Paperback Titles

- Bittinger: *Basic College Mathematics,* 11th Edition
- Bittinger/Penna: *Basic College Mathematics with Early Integers,* 2nd Edition
- Bittinger: *Introductory Algebra,* 11th Edition
- Bittinger: *Intermediate Algebra,* 11th Edition
- Bittinger/Beecher: *Introductory and Intermediate Algebra,* 4th Edition

Accuracy

Students rely on accurate textbooks, and my users value the Bittinger reputation for accuracy. All Bittinger titles go through an exhaustive checking process to ensure accuracy in the problem sets, mathematical art, and accompanying supplements.

Preface

New in This Edition

To maximize retention of the concepts and skills presented, five highly effective review features are included in the 5th edition. Student success is increased when review is integrated throughout each chapter.

Five Types of Integrated Review

Skill to Review exercises, found at the beginning of most sections, link to a section objective. These exercises offer a just-in-time review of a previously presented skill that relates to new material in the section. For convenient studying, section and objective references are followed by two practice exercises for immediate review and reinforcement. Exercise answers are given at the bottom of the page for immediate feedback.

Skill Maintenance Exercises, found in each exercise set, review concepts from other sections in the text to prepare students for their final examination. Section and objective references appear next to each Skill Maintenance exercise. All Skill Maintenance answers are included in the text.

A **Mid-Chapter Review** reinforces understanding of the mathematical concepts and skills just covered before students move on to new material. Section and objective references are included. Exercise types include Concept Reinforcement, Guided Solutions, Mixed Review, and Understanding Through Discussion and Writing. Answers to all exercises in the Mid-Chapter Review are given at the back of the book.

The **Chapter Summary and Review** at the end of each chapter is expanded to provide more comprehensive in-text practice and review.

- **Key Terms, Properties, and Formulas** are highlighted, with page references for convenient review.
- **Concept Reinforcement** offers true/false questions to enhance students' understanding of mathematical concepts.
- **Important Concepts** are listed by section objectives, followed by *worked-out examples* for reference and review and *similar practice exercises* for students to solve.
- **Review Exercises**, including Synthesis exercises and two new multiple-choice exercises, are organized by objective and cover the whole chapter.
- **Understanding Through Discussion and Writing** exercises strengthen understanding by giving students a chance to express their thoughts in spoken or written form.

Section and objective references for all exercises are included. Answers to all exercises in the Summary and Review are given at the back of the book.

Chapter Tests, including Synthesis questions and a new multiple-choice question, allow students to review and test their comprehension of chapter skills prior to taking an instructor's exam. Answers to all questions in the Chapter Tests are given at the back of the book. Section and objective references for each question are included with the answers.

A Cumulative Review after every chapter starting with Chapter 3 revisits skills and concepts from all preceding chapters to help students recall previously learned material and prepare for exams. Answers to all Cumulative Review exercises are coded by section and objective at the back of the book to help students identify areas where additional practice is needed.

Other New Elements

A new design enhances the Bittinger guided-learning approach. Margin exercises are now located next to examples for easier navigation, and answers for those exercises are given at the bottom of the page for immediate feedback.

Content changes include streamlined coverage in Sections 1.3 and 1.5 and also in 2.4, where two objectives are combined. Coverage of rounding and estimating, formerly found in Section 1.4, is now presented in Section 1.6, where new material on estimating quotients has been added. Material relating to credit card interest rates has moved from Section 6.8 to Section 6.7 in this edition. In addition, we have added over 130 new examples and over 900 new exercises.

Hallmark Features

Revised! The **Bittinger Student Organizer** card at the front of the text helps students keep track of important contacts and dates and provides a weekly planner to help schedule time for classes, studying, and homework. A helpful list of study tips found in each chapter is also included.

New! **Chapter Openers** feature motivating real-world applications that are revisited later in the chapters. This feature engages students and prepares them for the upcoming chapter material. (See pages 1, 147, and 215.)

New! **Real-Data Applications** encourage students to see and interpret the mathematics that appears every day in the world around them. (See pages 75, 298, and 377.) Almost 500 applications are new to this edition, and many are drawn from the fields of business and economics, life and physical sciences, social sciences, medicine, and areas of general interest such as sports and daily life.

Study Tips appear throughout the text to give students pointers on how to develop good study habits as they progress through the course, encouraging them to get involved in the learning process. (See pages 9, 113, and 217.) For easy reference, a list of Study Tips by chapter, section, and page number is included in the Bittinger Student Organizer.

Caution Boxes are found at relevant points throughout the text. The heading "*Caution!*" alerts students to coverage of a common misconception or an error often made in performing a particular mathematics operation or skill. (See pages 120, 188, and 222.)

Revised! Optional **Calculator Corners** are located where appropriate throughout the text. These streamlined Calculator Corners are written to be accessible to students and to represent current calculators. A calculator icon indicates exercises suitable for calculator use. (See pages 192, 228, and 355.)

Immediate Practice and Assessment in Each Section

OBJECTIVES ➔ SKILL TO REVIEW ➔ EXPOSITION ➔ EXAMPLES WITH DETAILED ANNOTATIONS AND VISUAL ART PIECES ➔ MARGIN EXERCISES ➔ EXERCISE SETS

Objective Boxes begin each section. A boxed list of objectives is keyed by letter not only to section subheadings, but also to the section exercise sets and the Mid-Chapter Review and the Summary and Review exercises, as well as to the answers to the questions in the Chapter Tests and Cumulative Reviews. This correlation enables students to easily find appropriate review material if they need help with a particular exercise or skill at the objective level. (See pages 88, 162, and 216.)

New! **Skill to Review** exercises, found at the beginning of most sections, link to a section objective and offer students a just-in-time review of a previously presented skill that relates to new material in the section. For convenient studying, objective references are followed by two practice exercises for immediate review and reinforcement. Answers to these exercises are given at the bottom of the page for immediate feedback. (See pages 19, 234, and 363.)

Revised! **Annotated Examples** provide annotations and color highlighting to lead students through the structured steps of the examples. The level of detail in these annotations is a significant reason for students' success with this book. This edition contains over 130 new examples. (See pages 126, 190, and 320.)

Revised! The **art and photo program** is designed to help students visualize mathematical concepts and real-data applications. Many applications include source lines and feature graphs and drawings similar to those students see in the media. Color is used consistently in geometric shapes and fractional parts to convey meaning. For example, a red border emphasizes perimeter, a black outline with a yellow fill emphasizes area, three-dimensional figures illustrating volume are blue, and fractional parts are shown in purple. (See pages 59, 103, 182, 264, and 388.)

Revised! **Margin Exercises**, now located next to examples for easier navigation, accompany examples throughout the text and give students the opportunity to work similar problems for immediate practice and reinforcement of the concept just learned. Answers are now available at the bottom of the page. (See pages 97, 258, and 356.)

Exercise Sets

To give students ample opportunity to practice what they have learned, each section is followed by an extensive exercise set *keyed by letter to the section objectives* for easy review and remediation. In addition, students also have the opportunity to synthesize the objectives from the current section with those from preceding sections. **For Extra Help** icons, shown at the beginning of each exercise set, indicate supplementary learning resources that students may need.

- **Skill Maintenance Exercises**, found in each exercise set, review concepts from other sections in the text to prepare students for their final examination. Section and objective codes appear next to each Skill Maintenance exercise for easy reference. All Skill Maintenance answers are included in the text. (See pages 116, 169, and 354.)
- **Vocabulary Reinforcement Exercises** provide an integrated review of key terms that students must know to communicate effectively in the language of mathematics. These appear once per chapter in the Skill Maintenance portion of an exercise set. (See pages 34, 139, and 269.)
- **Synthesis Exercises** help build critical-thinking skills by requiring students to use what they know to synthesize, or combine, learning objectives from the current section with those from previous sections. These are available in most exercise sets. (See pages 79, 154, and 337.)

Mid-Chapter Review

New! A **Mid-Chapter Review** gives students the opportunity to reinforce their understanding of the mathematical skills and concepts just covered before they move on to new material. Section and objective references are included for convenient studying, and answers to all the Mid-Chapter Review exercises are included in the text. The types of exercises are as follows:

- **Concept Reinforcement** are true/false questions that enhance students' understanding of mathematical concepts. These are also available in the Summary and Review at the end of the chapter. (See pages 35, 124, and 253.)
- **Guided Solutions** present worked-out problems with blanks for students to fill in the correct expressions to complete the solution. (See pages 35, 124, and 253.)

- **Mixed Review** provides free-response exercises, similar to those in the preceding sections in the chapter, reinforcing mastery of skills and concepts. (See pages 35, 124, and 253.)
- **Understanding Through Discussion and Writing** lets students demonstrate their understanding of mathematical concepts by expressing their thoughts in spoken and written form. This type of exercise is also found in each Chapter Summary and Review. (See pages 36, 125, and 254.)

Matching Feature

Translating for Success problem sets give extra practice with the important "Translate" step of the process for solving word problems. After translating each of ten problems into its appropriate equation or inequality, students are asked to choose from fifteen possible translations, encouraging them to comprehend the problem before matching. (See pages 136, 277, and 324.)

End-of-Chapter Material

Revised!   The **Chapter Summary and Review** at the end of each chapter is expanded to provide more comprehensive in-text practice and review. Section and objective references and answers to all the Chapter Summary and Review exercises are included in the text. (See pages 80, 140, and 285.)

- **Key Terms, Properties, and Formulas** are highlighted, with page references for convenient review. (See pages 80, 140, and 285.)
- **Concept Reinforcement** offers true/false questions to enhance student understanding of mathematical concepts. (See pages 80, 140, and 285.)
- New! **Important Concepts** are listed by section objectives, followed by *a worked-out example* for reference and review and *a similar practice exercise* for students to solve. (See pages 80–82, 140–142, and 285–287.)
- **Review Exercises,** including Synthesis exercises and two new multiple-choice exercises, covering the whole chapter are organized by objective. (See pages 82–84, 142–144, and 287–289.)
- **Understanding Through Discussion and Writing** exercises strengthen understanding by giving students a chance to express their thoughts in spoken or written form. (See pages 84, 144, and 289.)

Chapter Tests, including Synthesis questions and a new multiple-choice question, allow students to review and test their comprehension of chapter skills prior to taking an instructor's exam. Answers to all questions in the Chapter Test are given at the back of the book. Section and objective references for each question are included with the answers. (See pages 145, 343, and 415.)

New!   **A Cumulative Review** now follows every chapter starting with Chapter 3; this review revisits skills and concepts from all preceding chapters to help students recall previously learned material and prepare for exams. Answers to all Cumulative Review exercises are coded by section and objective at the back of the book to help students identify areas where additional practice is needed. (See pages 213, 293, and 345.)

For Extra Help

Student Supplements

New! Worksheets for Classroom or Lab Practice
(ISBN: 978-0-321-62728-5)

These classroom- and lab-friendly workbooks offer the following resources for every section of the text: a list of learning objectives, vocabulary practice problems, and extra practice exercises with ample work space.

Student's Solutions Manual (ISBN: 978-0-321-62691-2)
By Judith Penna

Contains completely worked-out annotated solutions for all the odd-numbered exercises in the text. Also includes fully worked-out annotated solutions for all the exercises (odd- and even-numbered) in the Mid-Chapter Reviews, the Summary and Reviews, the Chapter Tests, and the Cumulative Reviews.

Chapter Test Prep Videos

Chapter Tests can serve as practice tests to help you study. Watch instructors work through step-by-step solutions to all the Chapter Test exercises from the textbook. Chapter Test Prep videos are available on YouTube (search using Bittinger-BasicMath) and in MyMathLab. They are also included on the Video Resources on DVD described below and available for purchase at www.MyPearsonStore.com.

Video Resources on DVD Featuring Chapter Test Prep Videos
(ISBN: 978-0-321-62729-2)

- Complete set of lectures covering every objective of every section in the textbook
- Complete set of Chapter Test Prep videos (see above)
- All videos include optional English and Spanish subtitles.
- Ideal for distance learning or supplemental instruction
- DVD-ROM format for student use at home or on campus

InterAct Math Tutorial Website (www.interactmath.com)

Get practice and tutorial help online! This interactive tutorial website provides algorithmically generated practice exercises that correlate directly to the exercises in the textbook. Students can retry an exercise as many times as they like with new values each time for unlimited practice and mastery. Every exercise is accompanied by an interactive guided solution that provides helpful feedback for incorrect answers, and students can also view a worked-out sample problem that steps them through an exercise similar to the one they're working on.

MathXL® Tutorials on CD (ISBN: 978-0-321-62290-7)

This interactive tutorial CD-ROM provides algorithmically generated practice exercises that are correlated at the objective level to the exercises in the textbook. Every practice exercise is accompanied by an example and a guided solution designed to involve students in the solution process. Selected exercises may also include a video clip to help students visualize concepts. The software provides helpful feedback for incorrect answers and can generate printed summaries of students' progress.

Instructor Supplements

Annotated Instructor's Edition (ISBN: 978-0-321-62286-0)

Includes answers to all exercises printed in blue on the same page as the exercises. Also includes the student answer section, for easy reference.

Instructor's Solutions Manual (ISBN: 978-0-321-62288-4)
By Judith Penna

Contains brief solutions to the even-numbered exercises in the exercise sets. Also includes fully worked-out annotated solutions for all the exercises (odd- and even-numbered) in the Mid-Chapter Reviews, the Summary and Reviews, the Chapter Tests, and the Cumulative Reviews.

Printed Test Forms (ISBN: 978-0-321-62287-7)
By Laurie Hurley

- Contains one diagnostic test and one pretest for each chapter.
- **New!** Includes two versions of a short mid-chapter quiz.
- Provides eight test forms for every chapter and eight test forms for the final exam.
- For the chapter tests, four free-response tests are modeled after the chapter tests in the main text, two tests are designed for 50-minute class periods and organized so that each objective in the chapter is covered on one of the tests, and two tests consist of multiple-choice questions. Chapter tests also include more challenging Synthesis questions.
- For the final exam, three test forms are organized by chapter, three forms are organized by question type, and two forms are multiple-choice tests.

Instructor Resource Manual
(ISBN: 978-0-321-62692-9)

- Features resources and teaching tips designed to help both new and adjunct faculty with course preparation and classroom management.
- **New!** Includes a mini-lecture for each section of the text with objectives, key examples, and teaching tips.
- Additional resources include general first-time advice, sample syllabi, teaching tips, collaborative learning activities, correlation guide, video index, and transparency masters.

Additional Media Supplements

MyMathLab MyMathLab® Online Course (access code required)

MyMathLab is a series of text-specific, easily customizable online courses for Pearson Education's textbooks in mathematics and statistics. Powered by CourseCompass™ (our online teaching and learning environment) and MathXL® (our online homework, tutorial, and assessment system), MyMathLab gives instructors the tools they need to deliver all or a portion of their course online, whether their students are in a lab setting or working from home. MyMathLab provides a rich and flexible set of course materials, featuring free-response exercises that are algorithmically generated for unlimited practice and mastery. Students can also use online tools, such as video lectures, animations, interactive math games, and a multimedia textbook, to independently improve their understanding and performance. Instructors can use MyMathLab's homework and test managers to select and assign online exercises correlated directly to the textbook, and they can also create and assign their own online exercises and import TestGen tests for added flexibility. MyMathLab's online gradebook—designed specifically for mathematics and statistics—automatically tracks students' homework and test results and gives the instructor control over how to calculate final grades. Instructors can also add offline (paper-and-pencil) grades to the gradebook. MyMathLab also includes access to the **Pearson Tutor Center** (www.pearsontutorservices.com). The Tutor Center is staffed by qualified mathematics instructors who provide textbook-specific tutoring for students via toll-free phone, fax, email, and interactive Web sessions. MyMathLab is available to qualified adopters. For more information, visit our website at www.mymathlab.com or contact your sales representative.

MathXL MathXL® Online Course (access code required)

MathXL® is a powerful online homework, tutorial, and assessment system that accompanies Pearson Education's textbooks in mathematics or statistics.

With MathXL, instructors can

- create, edit, and assign online homework and tests using algorithmically generated exercises correlated at the objective level to the textbook.
- create and assign their own online exercises and import TestGen tests for added flexibility.
- maintain records of all student work tracked in MathXL's online gradebook.

With MathXL, students can

- take chapter tests in MathXL and receive personalized study plans based on their test results.
- use the study plan to link directly to tutorial exercises for the objectives they need to study and retest.
- access supplemental animations and video clips directly from selected exercises.

MathXL is available to qualified adopters. For more information, visit our website at www.mathxl.com, or contact your Pearson sales representative.

TestGen® (www.pearsoned.com/testgen) enables instructors to build, edit, and print tests using a computerized bank of questions developed to cover all the objectives of the text. TestGen is algorithmically based, allowing instructors to create multiple but equivalent versions of the same question or test with the click of a button. Instructors can also modify test bank questions or add new questions. The software and test bank are available for download from Pearson Education's online catalog.

PowerPoint® Lecture Slides present key concepts and definitions from the text. Slides are available to download from within MyMathLab and from Pearson Education's online catalog.

Pearson Math Adjunct Support Center (http://www.pearsontutorservices.com/math-adjunct.html) is staffed by qualified instructors with more than 100 years of combined experience at both the community college and university levels. Assistance is provided for faculty in the following areas: suggested syllabus consultation, tips on using materials packed with your book, book-specific content assistance, and teaching suggestions, including advice on classroom strategies.

Acknowledgments

Our deepest appreciation to all of you who helped to shape this edition by reviewing and spending time with us on your campuses. In particular, we would like to thank the following reviewers:

Kimberly Caldwell, *Volunteer State Community College*
Amelia Canavan, *Century College*
Alison Cater, *California State University, Hayward*
Julane B. Crabtrere, *Johnson County Community College*
Kelly Groginski, *Onondaga Community College*
Karen Hale, *Onondaga Community College*
Cassandra V. Johnson, *Robeson Community College*
Joe Jordan, *John Tyler Community College*
David Kedrowski, *Mid Michigan Community College*
Lynette J. King, *Gadsten State Community College*
Edward Kinley, *Pennsylvania Culinary Institute*
Dottie Lapre, *Nash Community College*
Edith Lester, *Volunteer State Community College*
Lynn Marecek, *Santa Ana College*
Shauna Mullins, *Murray State University*
Ruela Overby, *Richland College*
Brian Peterman, *Century College*
Nimisha Raval, *Central Georgia Technical College–Macon Campus*
Ronald Rushing, *Southwest Georgia Technical College*
Robert Thompson, *Lane Community College*
Stephen Toner, *Victory Valley Community College*
S. Paige Wood, *Kilgore College*

The endless hours of hard work by Martha Morong, Sally Lifland, Jane Hoover, Geri Davis, and Judy Beecher have led to products of which we are immensely proud. We also want to thank Judy Penna for writing the Student's and Instructor's Solutions Manuals and for her strong leadership in the preparation of the printed supplements and video lectures with Barbara Johnson. Other strong support has come from Laurie Hurley for the Printed Test Bank and for accuracy checking, along with checkers Holly Martinez and Barbara Johnson and proofreader Patty LaGree. Michelle Lanosga assisted with applications research. We also wish to recognize Tom Atwater and Margaret Donlan, who wrote video scripts.

In addition, a number of people at Pearson have contributed in special ways to the development and production of this textbook including the Developmental Math team: Vice President, Executive Director of Development Carol Trueheart, Senior Development Editor Dawn Nuttall, Production Manager Ron Hampton, Senior Designer Beth Paquin, Associate Editors Joanna Doxey and Christine Whitlock, Editorial Assistant Jonathan Wooding, Associate Media Producer Nathaniel Koven, and Media Producer Ceci Fleming. Executive Editor Cathy Cantin and Executive Marketing Manager Michelle Renda encouraged our vision and provided marketing insight. Kari Heen, Executive Project Manager, deserves special recognition for overseeing every phase of the project and keeping it moving.

Whole Numbers

Real-World Application

When construction is completed on the slender 150-story Chicago Spire, it will be
the tallest building in the United States, rising to a height of 2000 ft. The height of
Chicago's Willis Tower, formerly known as the Sears Tower, is 1450 ft. How much
taller will the Chicago Spire be?

Sources: Chicago Historical Society; *USA Today* research

This problem appears as Exercise 1 in Section 1.8.

1.1

Standard Notation

OBJECTIVES

a Give the meaning of digits in standard notation.

b Convert from standard notation to expanded notation.

c Convert between standard notation and word names.

TO THE STUDENT

At the front of the text, you will find the Bittinger Student Organizer card. This pullout card will help you keep track of important dates and useful contact information. You can also use it to plan time for class, study, work, and relaxation. By managing your time wisely, you will provide yourself the best possible opportunity to be successful in this course.

We study mathematics in order to be able to solve problems. In this section, we study how numbers are named. We begin with the concept of place value.

a Place Value

Consider the numbers in the following table.

Three Most Populous Countries in the World

COUNTRY	POPULATION 2008
China	1,331,074,881
India	1,150,101,887
United States	304,849,937

SOURCE: msnu.edu

A **digit** is a number 0, 1, 2, 3, 4, 5, 6, 7, 8, or 9 that names a place-value location. For large numbers, digits are separated by commas into groups of three, called **periods**. Each period has a name: *ones, thousands, millions, billions, trillions,* and so on. To understand the population of India in the table above, we can use a **place-value chart**, as shown below.

PLACE-VALUE CHART														
Periods →	Trillions			Billions			Millions			Thousands			Ones		
						1	1	5	0	1	0	1	8	8	7
	Hundreds	Tens	Ones	Hundreds	Tens	Ones	Hundreds	Tens	Ones	Hundreds	Tens	Ones	Hundreds	Tens	Ones

1 billion   150 millions   101 thousands   887 ones

What does the digit 2 mean in each number?

1. 526,555

2. 265,789

3. 42,789,654

4. 24,789,654

5. 8924

6. 5,643,201

EXAMPLES In each of the following numbers, what does the digit 8 mean?

1. 278,342   8 thousands
2. 872,342   8 hundred thousands
3. 28,343,399,223   8 billions
4. 1,023,850   8 hundreds
5. 98,413,099   8 millions
6. 6328   8 ones

Do Margin Exercises 1–6 (in the margin at left).

Answers

1. 2 ten thousands   2. 2 hundred thousands
3. 2 millions   4. 2 ten millions   5. 2 tens
6. 2 hundreds

EXAMPLE 7 *Hurricane Relief.* Private donations for relief for Hurricanes Katrina and Rita, which struck the Gulf Coast of the United States in 2005, totaled $3,378,185,879. What does each digit name?

Source: The Center on Philanthropy at Indiana University

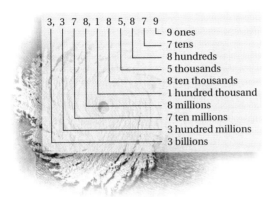

3, 3 7 8, 1 8 5, 8 7 9
- 9 ones
- 7 tens
- 8 hundreds
- 5 thousands
- 8 ten thousands
- 1 hundred thousand
- 8 millions
- 7 ten millions
- 3 hundred millions
- 3 billions

Do Exercise 7.

b | Converting from Standard Notation to Expanded Notation

To answer questions such as "How many?", "How much?", and "How tall?", we often use whole numbers. The set, or collection, of **whole numbers** is

0, 1, 2, 3, 4, 5, 6, 7, 8, 9, 10, 11, 12, ...

The set goes on indefinitely. There is no largest whole number, and the smallest whole number is 0. Each whole number can be named using various notations. The set 1, 2, 3, 4, 5, . . . , without 0, is called the set of **natural numbers**.

Consider the data in the table below showing the Advanced Placement exams taken most frequently by the class of 2007.

EXAM	NUMBER TAKEN
U.S. History	333,561
English Language and Composition	282,230
Calculus AB	211,693
Biology	144,796

SOURCE: The College Board

The number of Biology exams taken was 144,796. This number is expressed in **standard notation**. We write **expanded notation** for 144,796 as follows:

144,796 = 1 hundred thousand + 4 ten thousands
+ 4 thousands + 7 hundreds
+ 9 tens + 6 ones.

7. Federal Payroll. In December 2006, the payroll for civilian employees of the federal government was $13,896,346,626. What does each digit name?

Source: U.S. Census Bureau

Answer

7. 1 ten billion; 3 billions; 8 hundred millions; 9 ten millions; 6 millions; 3 hundred thousands; 4 ten thousands; 6 thousands; 6 hundreds; 2 tens; 6 ones

Write expanded notation.

8. 1895

9. 333,561, the number of Advanced Placement U.S. History exams taken by the class of 2007

10. 1670 ft, the height of the Taipei 101 Tower in Taiwan

11. 2718 mi, the length of the Congo River in Africa

12. 104,094 square miles, the area of Colorado

EXAMPLE 8 Write expanded notation for 1815 ft, the height of the CN Tower in Toronto, Canada.

$$1815 = 1 \text{ thousand} + 8 \text{ hundreds} + 1 \text{ ten} + 5 \text{ ones}$$

EXAMPLE 9 Write expanded notation for 3400.

$$3400 = 3 \text{ thousands} + 4 \text{ hundreds} + 0 \text{ tens} + 0 \text{ ones}, \quad \text{or}$$
$$3 \text{ thousands} + 4 \text{ hundreds}$$

EXAMPLE 10 Write expanded notation for 211,693, the number of Advanced Placement Calculus exams taken by the class of 2007.

$$211,693 = 2 \text{ hundred thousands} + 1 \text{ ten thousand}$$
$$+ 1 \text{ thousand} + 6 \text{ hundreds} + 9 \text{ tens} + 3 \text{ ones}$$

Do Exercises 8–12.

c Converting Between Standard Notation and Word Names

We often use **word names** for numbers. When we pronounce a number, we are speaking its word name. Russia won 72 medals in the 2008 Summer Olympics in Beijing, China. A word name for 72 is "seventy-two." Word names for some two-digit numbers like 36, 51, and 72 use hyphens. Others like 17 use only one word, "seventeen."

2008 Summer Olympics Medal Count

COUNTRY	GOLD	SILVER	BRONZE	TOTAL
United States of America	36	38	36	110
People's Republic of China	51	21	28	100
Russia	23	21	28	72
Great Britain	19	13	15	47
Australia	14	15	17	46

SOURCE: beijing2008.cn

Answers

8. 1 thousand + 8 hundreds + 9 tens + 5 ones
9. 3 hundred thousands + 3 ten thousands + 3 thousands + 5 hundreds + 6 tens + 1 one
10. 1 thousand + 6 hundreds + 7 tens + 0 ones, or 1 thousand + 6 hundreds + 7 tens
11. 2 thousands + 7 hundreds + 1 ten + 8 ones
12. 1 hundred thousand + 0 ten thousands + 4 thousands + 0 hundreds + 9 tens + 4 ones, or 1 hundred thousand + 4 thousands + 9 tens + 4 ones

EXAMPLES Write a word name.

11. 36, the number of gold medals won by the United States

Thirty-six

12. 15, the number of silver medals won by Australia

Fifteen

13. 100, the total number of medals won by the People's Republic of China

One hundred

Do Exercises 13–15.

For word names for larger numbers, we begin at the left with the largest period. The number named in the period is followed by the name of the period; then a comma is written and the next number and period are named. Note that the name of the ones period is not included in the word name for a whole number.

EXAMPLE 14 Write a word name for 46,605,314,732.

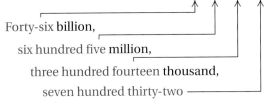

Forty-six **billion,**
six hundred five **million,**
three hundred fourteen **thousand,**
seven hundred thirty-two

The word "and" *should not* appear in word names for whole numbers. Although we commonly hear such expressions as "two hundred *and* one," the use of "and" is not, strictly speaking, correct in word names for whole numbers. For decimal notation, it is appropriate to use "and" for the decimal point. For example, 317.4 is read as "three hundred seventeen *and* four tenths."

Do Exercises 16–19.

EXAMPLE 15 Write standard notation.

Five hundred six **million,**
three hundred forty-five **thousand,**
two hundred twelve

Standard notation is 506,345,212.

Do Exercise 20.

Answers

13. Forty-six **14.** Thirteen **15.** Twenty-eight **16.** Two hundred four **17.** Fifty-two thousand, one hundred seventy-seven **18.** One million, eight hundred seventy-nine thousand, two hundred four **19.** Six billion, six hundred seventy-seven million, six hundred two thousand, two hundred ninety-two **20.** 213,105,329

a What does the digit 5 mean in each number?

1. 235,888   **2.** 253,777   **3.** 1,488,526   **4.** 500,736

Movie Receipts. Box-office receipts on the opening weekend of *Shrek the Third* were $121,629,270.
Source: Box Office Mojo

What digit names the number of:

5. thousands?   **6.** ten millions?

7. tens?   **8.** hundred thousands?

b Write expanded notation.

9. 5702   **10.** 3097   **11.** 93,986   **12.** 38,453

Stair-Climbing Races. The figure below shows the number of stairs in four buildings in which stair-climbing races are held. In Exercises 13–16, write expanded notation for the number of stairs in each race.

Stair-Climbing Races

2058   1776   1576   1081

International Towerthon, Kuala Lumpur, Malaysia   CN Tower Stair Climb, Toronto, Ontario, Canada   Empire State Building Run-Up, New York   Skytower Vertical Challenge, Auckland, New Zealand

SOURCE: towerrunning.com

13. 2058 steps in the International Towerthon, Kuala Lumpur, Malaysia

14. 1776 steps in the CN Tower Stair Climb, Toronto, Ontario, Canada

15. 1576 steps in the Empire State Building Run-Up, New York City, New York

16. 1081 steps in the Skytower Vertical Challenge, Auckland, New Zealand

Population Projections. The table below shows the projected population in 2050 for six countries. In Exercises 17–22, write expanded notation for the population of the given country.

PROJECTED POPULATION IN 2050

COUNTRY	POPULATION
China	1,424,161,948
Hong Kong	6,172,725
Japan	99,886,568
Monaco	32,964
Surinam	617,249
United States	420,080,587

SOURCES: International Programs Center, U.S. Census Bureau, U.S. Department of Commerce

17. 1,424,161,948 in China

18. 6,172,725 in Hong Kong

19. 99,886,568 in Japan

20. 32,964 in Monaco

21. 617,249 in Surinam

22. 420,080,587 in the United States

 Write a word name.

23. 85

24. 48

25. 88,000

26. 45,987

27. 123,765

28. 111,013

29. 7,754,211,577

30. 43,550,651,808

Write a word name for the number in each sentence.

31. *Wilderness Areas.* In 2009, in the most sweeping land-protection law passed in fifteen years, all development was banned on huge swaths of land in nine states. The largest was in California, where 700,634 acres were newly designated as wilderness.
Source: The Wilderness Society

32. *Busiest Airport.* In 2006, the world's busiest airport, Hartsfield in Atlanta, had 84,846,639 passengers.
Source: Airports Council International World Headquarters, Geneva, Switzerland

33. *Auto Racing.* Helio Castroneves, winner of the 2009 Indianapolis 500 auto race, won a record prize of $3,048,005.
Source: ESPN

34. *College Football.* The 2008 Rose Bowl game was attended by 93,923 fans.
Source: Fox Sports

Write standard notation.

35. Two million, two hundred thirty-three thousand, eight hundred twelve

36. Three hundred fifty-four thousand, seven hundred two

37. Eight billion

38. Seven hundred million

39. Fifty thousand, three hundred twenty-four

40. Twenty-six billion

41. Six hundred thirty-two thousand, eight hundred ninety-six

42. Seventeen thousand, one hundred twelve

Write standard notation for the number in each sentence.

43. *Ice Cream Purchases.* Americans buy one billion, six hundred million gallons of ice cream and frozen desserts each year.
Source: International Dairy Foods Association

44. *Learning a Language.* There are two hundred million Chinese children studying English.
Source: U.S. Department of Education

45. *Pacific Ocean.* The area of the Pacific Ocean is sixty-four million, one hundred eighty-six thousand square miles.

46. The average distance from the sun to Neptune is two billion, seven hundred ninety-three million miles.

Synthesis

To the student and the instructor: The Synthesis exercises found at the end of every exercise set challenge students to combine concepts or skills studied in that section or in preceding parts of the text. Exercises marked with a ▦ symbol are meant to be solved using a calculator.

47. How many whole numbers between 100 and 400 contain the digit 2 in their standard notation?

48. ▦ What is the largest number that you can name on your calculator? How many digits does that number have? How many periods?

1.2 Addition

a Addition of Whole Numbers

Addition of whole numbers corresponds to combining or putting things together.

We combine two sets.   This is the resulting set.

3 PDAs   4 PDAs   7 PDAs

We say that the **sum** of 3 and 4 is 7. The numbers added are called **addends**. The addition that corresponds to the figure above is

$$3 \quad + \quad 4 \quad = \quad 7.$$

Addend   Addend   Sum

To add whole numbers, we add the ones digits first, then the tens, then the hundreds, then the thousands, and so on.

EXAMPLE 1 Add: 6878 + 4995.

Place values are lined up in columns.

$$\begin{array}{r} \overset{\overset{1}{}}{6\;8\;7\;8} \\ +\;4\;9\;9\;5 \\ \hline 3 \end{array}$$

Add ones. We get 13 ones, or 1 ten + 3 ones. Write 3 in the ones column and 1 above the tens. This is called *carrying,* or *regrouping.*

$$\begin{array}{r} \overset{\overset{1}{}\overset{1}{}}{6\;8\;7\;8} \\ +\;4\;9\;9\;5 \\ \hline 7\;3 \end{array}$$

Add tens. We get 17 tens, so we have 10 tens + 7 tens. This is also 1 hundred + 7 tens. Write 7 in the tens column and 1 above the hundreds.

$$\begin{array}{r} \overset{\overset{1}{}\overset{1}{}\overset{1}{}}{6\;8\;7\;8} \\ +\;4\;9\;9\;5 \\ \hline 8\;7\;3 \end{array}$$

Add hundreds. We get 18 hundreds, or 10 hundreds + 8 hundreds. This is also 1 thousand + 8 hundreds. Write 8 in the hundreds column and 1 above the thousands.

$$\begin{array}{r} \overset{\overset{1}{}\overset{1}{}\overset{1}{}}{6\;8\;7\;8} \\ +\;4\;9\;9\;5 \\ \hline 1\;1\;8\;7\;3 \end{array}$$

Add thousands. We get 11 thousands.

We show you these steps for explanation. You need write only this.

$$\begin{array}{r} \overset{1\;1\;1}{6\;8\;7\;8} \\ +\;4\;9\;9\;5 \\ \hline 1\;1\;8\;7\;3 \end{array}$$ ← Addends

← Sum

OBJECTIVES

a Add whole numbers.

b Use addition in finding perimeter.

STUDY TIPS

Throughout this textbook, you will find a feature called "Study Tips." One of the most important ways in which to improve your math study skills is to learn the proper use of the textbook. Here we highlight a few points that we consider most helpful.

USING THIS TEXTBOOK

Be sure to note the symbols a , b , c , and so on, that correspond to the objectives you are to master in each section.   The first time you see them is in the margin at the beginning of the section; the second time is in the subheadings of each section; and the third time is in the exercise set for the section. You will also find symbols like [1.1a] or [1.2c] next to the skill maintenance exercises in each exercise set, in the mid-chapter review, and in the review exercises at the end of the chapter, as well as in the answers to the chapter tests and the cumulative reviews. These objective symbols allow you to refer to the appropriate place in the text when you need to review a topic.

EXAMPLE 2   Add: $391 + 276 + 789 + 498$.

$$
\begin{array}{r}
\overset{2}{} \\
3\ 9\ 1 \\
2\ 7\ 6 \\
7\ 8\ 9 \\
+\ 4\ 9\ 8 \\
\hline
4
\end{array}
$$

Add ones. We get 24, so we have 2 tens + 4 ones. Write 4 in the ones column and 2 above the tens.

$$
\begin{array}{r}
\overset{3}{}\overset{2}{} \\
3\ 9\ 1 \\
2\ 7\ 6 \\
7\ 8\ 9 \\
+\ 4\ 9\ 8 \\
\hline
5\ 4
\end{array}
$$

Add tens. We get 35 tens, so we have 30 tens + 5 tens. This is also 3 hundreds + 5 tens. Write 5 in the tens column and 3 above the hundreds.

$$
\begin{array}{r}
\overset{3}{}\overset{2}{} \\
3\ 9\ 1 \\
2\ 7\ 6 \\
7\ 8\ 9 \\
+\ 4\ 9\ 8 \\
\hline
1\ 9\ 5\ 4
\end{array}
$$

Add hundreds. We get 19 hundreds.

Do Exercises 1–4.

Add.

1. $6203 + 3542$

2.
$$
\begin{array}{r}
7\ 9\ 6\ 8 \\
+\ 5\ 4\ 9\ 7 \\
\hline
\end{array}
$$

3.
$$
\begin{array}{r}
9\ 8\ 0\ 4 \\
+\ 6\ 3\ 7\ 8 \\
\hline
\end{array}
$$

4.
$$
\begin{array}{r}
1\ 9\ 3\ 2 \\
6\ 7\ 2\ 3 \\
9\ 8\ 7\ 8 \\
+\ 8\ 9\ 4\ 1 \\
\hline
\end{array}
$$

How do we perform an addition of three numbers, like $2 + 3 + 6$? We could do it by adding 3 and 6, and then 2. We can show this with parentheses:

$$2 + (3 + 6) = 2 + 9 = 11. \qquad \text{Parentheses tell what to do first.}$$

We could also add 2 and 3, and then 6:

$$(2 + 3) + 6 = 5 + 6 = 11$$

Either way the result is 11. It does not matter how we group the numbers. This illustrates the **associative law of addition**, $a + (b + c) = (a + b) + c$. We can also add whole numbers in any order. That is, $2 + 3 = 3 + 2$. This illustrates the **commutative law of addition**, $a + b = b + a$. Together, the commutative and associative laws tell us that to add more than two numbers, we can use any order and grouping we wish. Adding 0 to a number does not change the number: $a + 0 = 0 + a = a$. That is, $6 + 0 = 0 + 6 = 6$, or $198 + 0 = 0 + 198 = 198$. We say that 0 is the **additive identity**.

Answers

1. 9745   2. 13,465   3. 16,182
4. 27,474

Calculator Corner

Adding Whole Numbers   This is the first of a series of *optional* discussions on using a calculator. A calculator is *not* a requirement for this textbook. Check with your instructor about whether you are allowed to use a calculator in the course.

There are many kinds of calculators and different instructions for their usage. We have included instructions here for a low-cost calculator. Be sure to consult your user's manual as well.

To add whole numbers on a calculator, we use the $+$ and $=$ keys. For example, to find $314 + 259 + 478$, we press 3 1 4 $+$ 2 5 9 $+$ 4 7 8 $=$. The display reads 1051, so $314 + 259 + 478 = 1051$.

Exercises: Use a calculator to find each sum.

1. $73 + 48$

2. $925 + 677$

3. $826 + 415 + 691$

4. $253 + 490 + 121$

b Finding Perimeter

Addition can be used when finding perimeter.

> **PERIMETER**
>
> The distance around an object is its **perimeter**.

EXAMPLE 3 Find the perimeter of the figure.

We add the lengths of the sides:

Perimeter = 29 in. + 19 in. + 37 in. + 44 in.

We carry out the addition as follows.

$$
\begin{array}{r}
\overset{2}{2}\,9 \\
1\,9 \\
3\,7 \\
+\ 4\,4 \\
\hline
1\,2\,9
\end{array}
$$

The perimeter of the figure is 129 in.

> Do Exercises 5 and 6.

EXAMPLE 4 Find the perimeter of the octagonal (eight-sided) resort swimming pool.

Perimeter = 13 yd + 6 yd + 6 yd + 12 yd + 8 yd
+ 12 yd + 13 yd + 16 yd

The perimeter of the pool is 86 yd.

> Do Exercise 7.

Find the perimeter of each figure.

5.

6.

Solve.

7. Index Cards. Two standard sizes for index cards are 3 in. (inches) by 5 in. and 5 in. by 8 in. Find the perimeter of each type of card.

Answers

5. 29 in. **6.** 62 ft **7.** 16 in.; 26 in.

xercise Set

For Extra Help

MyMathLab
 PRACTICE
 WATCH
 DOWNLOAD
 READ
 REVIEW

1. 3 6 4
  + 2 3

2. 1 5 2 1
  + 3 4 8

3. 8 6
  + 7 8

4. 7 3
  + 6 9

5. 1 7 1 6
  + 3 4 8 2

6. 7 5 0 3
  + 2 6 8 3

7. 9 9
  + 1

8. 9 9 9
  + 1 1

9. 8113 + 390

10. 271 + 3338

11. 356 + 4910

12. 280 + 34,902

13. 3870 + 92 + 7 + 497

14. 10,120 + 12,989 + 5738

15. 4 8 2 5
  + 1 7 8 3

16. 3 6 5 4
  + 2 7 0 0

17. 2 3,4 4 3
  + 1 0,9 8 9

18. 4 5,8 7 9
  + 2 1,7 8 6

19. 7 7,5 4 3
  + 2 3,7 6 7

20. 9 9,9 9 9
  + 1 1 2

21. 4 5
  2 5
  3 6
  4 4
  + 8 0

22. 3 8
  2 7
  3 2
  1 4
  + 7 6

23. 1 2,0 7 0
  2,9 5 4
  + 3,4 0 0

24. 4 2,4 8 7
  8 3,1 4 1
  + 3 6,7 1 2

25. 4 8 3 5
  7 2 9
  9 2 0 4
  8 9 8 6
  + 7 9 3 1

26. 9 8 9
  5 6 6
  8 3 4
  9 2 0
  + 7 0 3

b Find the perimeter of each figure.

27.

50 yd
23 yd
19 yd
40 yd

28.

14 mi
13 mi
8 mi
22 mi
10 mi
47 mi

29.

402 ft
298 ft
196 ft
212 ft
100 ft
453 ft

30.

62 yd
39 yd
28 yd
46 yd
54 yd

31. Find the perimeter of a standard hockey rink.

200 ft
85 ft

32. In Major League Baseball, how far does a batter travel in circling the bases when a home run has been hit?

90 ft
90 ft

Skill Maintenance

The exercises that follow begin an important feature called *Skill Maintenance exercises.* These exercises provide an ongoing review of topics previously covered in the book. You will see them in virtually every exercise set. It has been found that this kind of continuing review can significantly improve your performance on a final examination.

33. What does the digit 8 mean in 486,205? [1.1a]

34. Write a word name for the number in the following sentence: [1.1c]

The population of the world is projected to be 9,346,399,468 in 2050.

Source: United States Census Bureau

Synthesis

35. A fast way to add all the numbers from 1 to 10 inclusive is to pair 1 with 9, 2 with 8, and so on. Use a similar approach to add all numbers from 1 to 100 inclusive.

1.3

Subtraction

OBJECTIVE

a Subtract whole numbers.

a Subtraction of Whole Numbers

Subtraction is finding the difference of two numbers. Suppose you pick 6 pints of blueberries and give your neighbor 2 pints.

The subtraction that represents this is

$$6 \quad - \quad 2 \quad = \quad 4.$$

Minuend Subtrahend Difference

The **minuend** is the number from which another number is being subtracted. The **subtrahend** is the number being subtracted. The **difference** is the result of subtracting the subtrahend from the minuend.

In the subtraction above, note that the difference, 4, is the number we add to 2 to get 6. This illustrates the relationship between addition and subtraction and leads us to the following definition of subtraction.

STUDY TIPS

USING THIS TEXTBOOK

Read and study each step of each example. The examples include important side comments that explain each step. These examples and annotations have been carefully chosen so that you will be fully prepared to do the exercises.

Stop and do the margin exercises as you study a section. This gives you immediate reinforcement of each concept as it is introduced and is one of the most effective ways to master the mathematical skills in this text. Don't deprive yourself of this benefit!

Note the icons listed at the top of each exercise set. These refer to the distinctive multimedia study aids that accompany the book.

> **SUBTRACTION**
>
> The difference $a - b$ is that unique whole number c for which $a = c + b$.

We see that $6 - 2 = 4$ because $4 + 2 = 6$.

To subtract numbers, we subtract the ones digits first, then the tens digits, then the hundreds, then the thousands, and so on.

EXAMPLE 1 Subtract: $9768 - 4320$.

$$
\begin{array}{r}
9\ 7\ 6\ \mathbf{8} \\
-\ 4\ 3\ 2\ \mathbf{0} \\
\hline
\mathbf{8}
\end{array}
$$
Subtract ones.

$$
\begin{array}{r}
9\ 7\ \mathbf{6}\ 8 \\
-\ 4\ 3\ \mathbf{2}\ 0 \\
\hline
\mathbf{4}\ 8
\end{array}
$$
Subtract tens.

> We show these steps for explanation. You need write only this.

$$
\begin{array}{r}
9\ \mathbf{7}\ 6\ 8 \\
-\ 4\ \mathbf{3}\ 2\ 0 \\
\hline
\mathbf{4}\ 4\ 8
\end{array}
$$
Subtract hundreds.

$$
\begin{array}{r}
\mathbf{9}\ 7\ 6\ 8 \\
-\ \mathbf{4}\ 3\ 2\ 0 \\
\hline
\mathbf{5}\ 4\ 4\ 8
\end{array}
$$
Subtract thousands.

$$
\begin{array}{r}
9\ 7\ 6\ 8 \\
-\ 4\ 3\ 2\ 0 \\
\hline
5\ 4\ 4\ 8
\end{array}
$$

Because subtraction is defined in terms of addition, we can use addition to *check* subtraction.

Subtraction:

```
    9  7  6  8
 -  4  3  2  0
    5  4  4  8
```

?

Check by Addition:

```
    5  4  4  8
 +  4  3  2  0
    9  7  6  8
```

Do Exercise 1.

Do Exercise 1.

EXAMPLE 2 Subtract: 348 − 165.

We have

$$
\begin{aligned}
&\ \ 3 \text{ hundreds} + 4 \text{ tens} + 8 \text{ ones} = \ \ \ 2 \text{ hundreds} + 14 \text{ tens} + 8 \text{ ones} \\
&- 1 \text{ hundred} \ \ \ - 6 \text{ tens} - 5 \text{ ones} = -1 \text{ hundred} \ \ \ - \ \ 6 \text{ tens} - 5 \text{ ones} \\
&\phantom{- 1 \text{ hundred} \ \ \ - 6 \text{ tens} - 5 \text{ ones}} = \ \ \ 1 \text{ hundred} \ \ + \ \ 8 \text{ tens} + 3 \text{ ones} \\
&\phantom{- 1 \text{ hundred} \ \ \ - 6 \text{ tens} - 5 \text{ ones}} = \ \ \ 183.
\end{aligned}
$$

First, we subtract the ones.

```
    3  4  8
 -  1  6  5
          3
```

Subtract ones.

We cannot subtract the tens because there is no whole number that when added to 6 gives 4. To complete the subtraction, we must *borrow* 1 hundred from 3 hundreds and regroup it with the 4 tens. Then we can do the subtraction 14 tens − 6 tens = 8 tens.

```
    2  14
    3̶  4̶  8
 -  1  6  5
          3
```

Borrow one hundred. That is, 1 hundred = 10 tens, and 10 tens + 4 tens = 14 tens. Write 2 above the hundreds column and 14 above the tens.

```
    2  14
    3̶  4̶  8
 -  1  6  5
    1  8  3
```

Subtract tens; subtract hundreds.

This is what you should write.

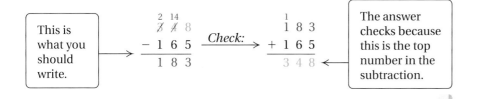

Check:

The answer checks because this is the top number in the subtraction.

---

1. Subtract. Check by adding.

```
    7  8  9  3
 -  4  0  9  2
```

Calculator Corner

Subtracting Whole Numbers To subtract whole numbers on a calculator, we use the ⎓ and = keys. For example, to find 63 − 47, we press [6] [3] [−] [4] [7] [=].
The calculator displays [16], so 63 − 47 = 16. We can check this result by adding the subtrahend, 47, and the difference, 16. To do this, we press [1] [6] [+] [4] [7] [=]. The sum is the minuend, 63, so the subtraction is correct.

Exercises: Use a calculator to perform each subtraction. Check by adding.

1. 57 − 29

2. 81 − 34

3. 145 − 78

4. 612 − 493

5.
```
    4  9  7  6
 -  2  8  4  8
```

6.
```
 1  2,  4  0  6
 -  9   8  1  3
```

EXAMPLE 3 Subtract: 6246 − 1879.

$$
\begin{array}{r}
{\scriptstyle 3\ 16} \\
6\ 2\ \cancel{4}\ \cancel{6} \\
-\ 1\ 8\ 7\ 9 \\
\hline
7
\end{array}
$$

We cannot subtract 9 ones from 6 ones, but we can subtract 9 ones from 16 ones. We borrow 1 ten to get 16 ones.

$$
\begin{array}{r}
{\scriptstyle 13} \\
{\scriptstyle 1\ \cancel{3}\ 16} \\
6\ 2\ \cancel{4}\ \cancel{6} \\
-\ 1\ 8\ 7\ 9 \\
\hline
6\ 7
\end{array}
$$

We cannot subtract 7 tens from 3 tens, but we can subtract 7 tens from 13 tens. We borrow 1 hundred to get 13 tens.

$$
\begin{array}{r}
{\scriptstyle 11\ 13} \\
{\scriptstyle 5\ \cancel{1}\ \cancel{3}\ 16} \\
\cancel{6}\ 2\ \cancel{4}\ \cancel{6} \\
-\ 1\ 8\ 7\ 9 \\
\hline
4\ 3\ 6\ 7
\end{array}
$$

We cannot subtract 8 hundreds from 1 hundred, but we can subtract 8 hundreds from 11 hundreds. We borrow 1 thousand to get 11 hundreds. Finally, we subtract the thousands.

This is what you should write.

$$
\begin{array}{r}
{\scriptstyle 11\ 13} \\
{\scriptstyle 5\ \cancel{1}\ \cancel{3}\ 16} \\
\cancel{6}\ \cancel{2}\ \cancel{4}\ \cancel{6} \\
-\ 1\ 8\ 7\ 9 \\
\hline
4\ 3\ 6\ 7
\end{array}
$$

Check:

$$
\begin{array}{r}
{\scriptstyle 1\ 1\ 1} \\
4\ 3\ 6\ 7 \\
+\ 1\ 8\ 7\ 9 \\
\hline
6\ 2\ 4\ 6
\end{array}
$$

The answer checks because this is the top number in the subtraction.

Do Exercises 2 and 3.

EXAMPLE 4 Subtract: 902 − 477.

$$
\begin{array}{r}
{\scriptstyle 8\ 9\ 12} \\
9\!\!\!\!\diagdown\ 0\ 2 \\
-\ 4\ 7\ 7 \\
\hline
4\ 2\ 5
\end{array}
$$

We cannot subtract 7 ones from 2 ones. We have 9 hundreds, or 90 tens. We borrow 1 ten to get 12 ones. We then have 89 tens.

Do Exercises 4 and 5.

EXAMPLE 5 Subtract: 8003 − 3667.

$$
\begin{array}{r}
{\scriptstyle 7\ 9\ 9\ 13} \\
8\!\!\!\!\diagdown\ 0\ 0\ 3 \\
-\ 3\ 6\ 6\ 7 \\
\hline
4\ 3\ 3\ 6
\end{array}
$$

We have 8 thousands, or 800 tens. We borrow 1 ten to get 13 ones. We then have 799 tens.

EXAMPLES

6. Subtract: 6000 − 3762.

$$
\begin{array}{r}
{\scriptstyle 5\ 9\ 9\ 10} \\
6\!\!\!\!\diagdown\ 0\ 0\ \cancel{0} \\
-\ 3\ 7\ 6\ 2 \\
\hline
2\ 3\ 3\ 8
\end{array}
$$

7. Subtract: 6024 − 2968.

$$
\begin{array}{r}
{\scriptstyle 11} \\
{\scriptstyle 5\ 9\ \cancel{1}\ 14} \\
6\ 0\ 2\ \cancel{4} \\
-\ 2\ 9\ 6\ 8 \\
\hline
3\ 0\ 5\ 6
\end{array}
$$

Do Exercises 6–9.

Subtract. Check by adding.

2.
$$
\begin{array}{r}
8\ 6\ 8\ 6 \\
-\ 2\ 3\ 5\ 8 \\
\hline
\end{array}
$$

3.
$$
\begin{array}{r}
7\ 1\ 4\ 5 \\
-\ 2\ 3\ 9\ 8 \\
\hline
\end{array}
$$

Subtract.

4.
$$
\begin{array}{r}
7\ 0 \\
-\ 1\ 4 \\
\hline
\end{array}
$$

5.
$$
\begin{array}{r}
5\ 0\ 3 \\
-\ 2\ 9\ 8 \\
\hline
\end{array}
$$

Subtract.

6.
$$
\begin{array}{r}
7\ 0\ 0\ 7 \\
-\ 6\ 3\ 4\ 9 \\
\hline
\end{array}
$$

7.
$$
\begin{array}{r}
6\ 0\ 0\ 0 \\
-\ 3\ 1\ 4\ 9 \\
\hline
\end{array}
$$

8.
$$
\begin{array}{r}
9\ 0\ 3\ 5 \\
-\ 7\ 4\ 8\ 9 \\
\hline
\end{array}
$$

9.
$$
\begin{array}{r}
2\ 0\ 0\ 1 \\
-\ \ \ 1\ 2\ 4 \\
\hline
\end{array}
$$

Answers

2. 6328 **3.** 4747 **4.** 56 **5.** 205
6. 658 **7.** 2851 **8.** 1546 **9.** 1877

1.3 | Exercise Set

For Extra Help

MyMathLab

Math XL
PRACTICE

WATCH

DOWNLOAD

33

a   Subtract. Check by adding.

1.   6 5
       − 2 1

2.   8 7
       − 3 4

3.   8 6 6
       − 3 3 3

4.   5 2 6
       − 3 2 3

5. 86 − 47

6. 73 − 28

7. 51 − 37

8. 64 − 19

9.   5 6 3
       − 1 9 4

10.   7 9 5
        − 3 9 8

11.   3 9 1
        − 3 6 5

12.   3 1 6
        − 2 4 7

13. 981 − 747

14. 887 − 698

15. 683 − 266

16. 342 − 217

17.   7 7 6 9
        − 2 3 8 7

18.   6 4 3 1
        − 2 8 9 6

19.   4 5 1 2
        − 1 7 3 4

20.   8 3 6 4
        − 5 3 7 5

21. 5318 − 2249

22. 9241 − 5643

23. 3947 − 2858

24. 7583 − 3641

25.   1 2,6 4 7
        −   4,8 9 9

26.   1 6,2 2 2
        −   5,8 8 8

27.   5 1,3 4 2
        − 4 7,1 9 8

28.   3 2,1 9 4
        − 2 9,2 3 6

29.   8 0
        − 2 4

30.   9 0
        − 7 8

31.   6 9 0
        − 2 3 6

32.   8 0 3
        − 4 1 8

$$\begin{array}{r} 7\ 6\ 4\ 0 \\ -\ 3\ 8\ 0\ 9 \\ \hline \end{array}$$

34.
$$\begin{array}{r} 5\ 2\ 8\ 0 \\ -\ 3\ 0\ 9\ 1 \\ \hline \end{array}$$

35.
$$\begin{array}{r} 6\ 8\ 0\ 8 \\ -\ 3\ 0\ 5\ 9 \\ \hline \end{array}$$

36.
$$\begin{array}{r} 9\ 4\ 0\ 5 \\ -\ \ \ 2\ 5\ 8 \\ \hline \end{array}$$

37.
$$\begin{array}{r} 2\ 3\ 0\ 0 \\ -\ \ \ 1\ 0\ 9 \\ \hline \end{array}$$

38.
$$\begin{array}{r} 7\ 5\ 0\ 0 \\ -\ 3\ 6\ 0\ 4 \\ \hline \end{array}$$

39.
$$\begin{array}{r} 6\ 0\ 0\ 7 \\ -\ 1\ 5\ 8\ 9 \\ \hline \end{array}$$

40.
$$\begin{array}{r} 8\ 0\ 0\ 3 \\ -\ \ \ 5\ 9\ 9 \\ \hline \end{array}$$

41. $90{,}237 - 47{,}209$

42. $84{,}703 - 298$

43. $101{,}734 - 5760$

44. $15{,}017 - 7809$

45.
$$\begin{array}{r} 7\ 0\ 0\ 0 \\ -\ 2\ 7\ 9\ 4 \\ \hline \end{array}$$

46.
$$\begin{array}{r} 8\ 0\ 0\ 1 \\ -\ 6\ 5\ 4\ 3 \\ \hline \end{array}$$

47.
$$\begin{array}{r} 3\ 9{,}0\ 0\ 0 \\ -\ 3\ 7{,}6\ 9\ 5 \\ \hline \end{array}$$

48.
$$\begin{array}{r} 1\ 7{,}0\ 0\ 0 \\ -\ 1\ 1{,}5\ 9\ 8 \\ \hline \end{array}$$

49. $10{,}008 - 19$

50. $40{,}006 - 147$

51. $50{,}001 - 1984$

52. $30{,}004 - 6749$

Skill Maintenance

Add. [1.2a]

53.
$$\begin{array}{r} 9\ 4\ 6 \\ +\ \ \ 7\ 8 \\ \hline \end{array}$$

54.
$$\begin{array}{r} 9\ 0\ 7\ 8 \\ +\ 3\ 6\ 5\ 4 \\ \hline \end{array}$$

55.
$$\begin{array}{r} 5\ 7{,}8\ 7\ 7 \\ +\ 3\ 2{,}4\ 0\ 6 \\ \hline \end{array}$$

56.
$$\begin{array}{r} 8\ 0\ 0\ 4 \\ 6\ 7\ 8\ 9 \\ 7\ 7\ 2\ 0 \\ +\ 6\ 8\ 5\ 1 \\ \hline \end{array}$$

57. $567 + 778$

58. $901 + 23$

59. $12{,}885 + 9807$

60. $9909 + 1011$

61. Write a word name for 6,375,602. [1.1c]

62. What does the digit 7 mean in 6,375,602? [1.1a]

Synthesis

63. Fill in the missing digits to make the subtraction true:

$9{,}\square 48{,}621 - 2{,}097{,}\square 81 = 7{,}251{,}140.$

64. ▦ Subtract: $3{,}928{,}124 - 1{,}098{,}947.$

1.4 Multiplication

a Multiplication of Whole Numbers

Repeated Addition

The multiplication 3×5 corresponds to this repeated addition.

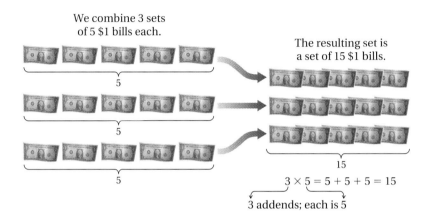

We combine 3 sets of 5 $1 bills each.

The resulting set is a set of 15 $1 bills.

$$3 \times 5 = 5 + 5 + 5 = 15$$

3 addends; each is 5

The numbers that we multiply are called **factors**. The result of the multiplication is called a **product**.

$$3 \quad \times \quad 5 \quad = \quad 15$$

Factor   Factor   Product

Rectangular Arrays

Multiplications can also be thought of as rectangular arrays. Each of the following corresponds to the multiplication 3×5.

3 rows with 5 bills in each row;
$3 \times 5 = 15$

5 columns with 3 bills in each column;
$3 \times 5 = 15$

When you write a multiplication sentence corresponding to a real-world situation, you should think of either a rectangular array or repeated addition. In some cases, it may help to think both ways.

We have used an "×" to denote multiplication. A dot " · " is also commonly used. (Use of the dot is attributed to the German mathematician Gottfried Wilhelm von Leibniz in 1698.) Parentheses are also used to denote multiplication. For example,

$$3 \times 5 = 3 \cdot 5 = (3)(5) = 3(5) = 15.$$

The product of 0 and any whole number is 0: $0 \cdot a = a \cdot 0 = 0$. For example, $0 \cdot 3 = 3 \cdot 0 = 0$. Multiplying a number by 1 does not change the number: $1 \cdot a = a \cdot 1 = a$. For example, $1 \cdot 3 = 3 \cdot 1 = 3$. We say that 1 is the **multiplicative identity**.

EXAMPLE 1 Multiply: 5×734.

We have

$$
\begin{array}{r}
7\ 3\ 4 \\
\times\quad\ \ 5 \\
\hline
2\ 0 \\
1\ 5\ 0 \\
3\ 5\ 0\ 0 \\
\hline
3\ 6\ 7\ 0
\end{array}
$$

← Multiply the 4 ones by 5: $5 \times 4 = 20$.
← Multiply the 3 tens by 5: $5 \times 30 = 150$.
← Multiply the 7 hundreds by 5: $5 \times 700 = 3500$.
← Add.

Instead of writing each product on a separate line, we can use a shorter form.

$$
\begin{array}{r}
\overset{2}{} \\
7\ \overset{2}{3}\ 4 \\
\times\quad\ \ 5 \\
\hline
0
\end{array}
$$

Multiply the ones by 5: $5 \cdot (4\ \text{ones}) = 20\ \text{ones} = 2\ \text{tens} + 0\ \text{ones}$. Write 0 in the ones column and 2 above the tens.

$$
\begin{array}{r}
\overset{1}{}\ \overset{2}{} \\
7\ 3\ 4 \\
\times\quad\ \ 5 \\
\hline
7\ 0
\end{array}
$$

Multiply the 3 tens by 5 and add 2 tens: $5 \cdot (3\ \text{tens}) = 15\ \text{tens},\ 15\ \text{tens} + 2\ \text{tens} = 17\ \text{tens} = 1\ \text{hundred} + 7\ \text{tens}$. Write 7 in the tens column and 1 above the hundreds.

$$
\begin{array}{r}
\overset{1}{}\ \overset{2}{} \\
7\ 3\ 4 \\
\times\quad\ \ 5 \\
\hline
3\ 6\ 7\ 0
\end{array}
$$

Multiply the 7 hundreds by 5 and add 1 hundred: $5 \cdot (7\ \text{hundreds}) = 35\ \text{hundreds},\ 35\ \text{hundreds} + 1\ \text{hundred} = 36\ \text{hundreds}$.

$$
\begin{array}{r}
\overset{1}{}\ \overset{2}{} \\
7\ 3\ 4 \\
\times\quad\ \ 5 \\
\hline
3\ 6\ 7\ 0
\end{array}
$$

You should write only this.

Do Exercises 1–4.

Multiplication of whole numbers is based on a property called the **distributive law**. It says that to multiply a number by a sum, $a \cdot (b + c)$, we can multiply each addend by a and then add like this: $(a \cdot b) + (a \cdot c)$. Thus, $a \cdot (b + c) = (a \cdot b) + (a \cdot c)$. For example, consider the following.

$$4 \cdot (2 + 3) = 4 \cdot 5 = 20 \qquad \text{Adding first; then multiplying}$$

$$4 \cdot (2 + 3) = (4 \cdot 2) + (4 \cdot 3) = 8 + 12 = 20 \qquad \text{Multiplying first; then adding}$$

The results are the same, so $4 \cdot (2 + 3) = (4 \cdot 2) + (4 \cdot 3)$.

Multiply.

1.
$$
\begin{array}{r}
5\ 8 \\
\times\quad 2 \\
\hline
\end{array}
$$

2.
$$
\begin{array}{r}
3\ 7 \\
\times\quad 4 \\
\hline
\end{array}
$$

3.
$$
\begin{array}{r}
8\ 2\ 3 \\
\times\quad\ \ 6 \\
\hline
\end{array}
$$

4.
$$
\begin{array}{r}
1\ 3\ 4\ 8 \\
\times\quad\ \ \ \ 5 \\
\hline
\end{array}
$$

Answers

1. 116   **2.** 148   **3.** 4938   **4.** 6740

20   CHAPTER 1   Whole Numbers

Let's find the product 51×32. Since $32 = 2 + 30$, we can think of this product as

$$51 \times 32 = 51 \times (2 + 30) = (51 \times 2) + (51 \times 30).$$

That is, we multiply 51 by 2, then we multiply 51 by 30, and finally we add. We can write our work this way.

$$
\begin{array}{r}
5\ 1 \\
\times\ 3\ 2 \\
\hline
1\ 0\ 2 \\
1\ 5\ 3\ 0 \\
\end{array}
$$

Multiplying by 2

Multiplying by 30. (We write a 0 and then multiply 51 by 3.)

> You may have learned that such a 0 need not be written. You may omit it if you wish. If you do omit it, remember, when multiplying by tens, to start writing the answer in the tens place.

We add to obtain the product.

$$
\begin{array}{r}
5\ 1 \\
\times\ 3\ 2 \\
\hline
1\ 0\ 2 \\
1\ 5\ 3\ 0 \\
\hline
1\ 6\ 3\ 2 \\
\end{array}
$$

Adding to obtain the product

EXAMPLE 2 Multiply: 457×683.

$$
\begin{array}{r}
\overset{5}{}\ \overset{2}{}\ \\
6\ 8\ 3 \\
\times\ 4\ 5\ 7 \\
\hline
4\ 7\ 8\ 1 \\
\end{array}
$$

Multiplying 683 by 7

$$
\begin{array}{r}
\overset{4}{\overset{5}{}}\ \overset{1}{\overset{2}{}}\ \\
6\ 8\ 3 \\
\times\ 4\ 5\ 7 \\
\hline
4\ 7\ 8\ 1 \\
3\ 4\ 1\ 5\ 0 \\
\end{array}
$$

Multiplying 683 by 50

$$
\begin{array}{r}
\overset{3}{\overset{4}{\overset{5}{}}}\ \overset{1}{\overset{1}{\overset{2}{}}}\ \\
6\ 8\ 3 \\
\times\ 4\ 5\ 7 \\
\hline
4\ 7\ 8\ 1 \\
3\ 4\ 1\ 5\ 0 \\
2\ 7\ 3\ 2\ 0\ 0 \\
\hline
3\ 1\ 2,1\ 3\ 1 \\
\end{array}
$$

Multiplying 683 by 400. (We write 00 and then multiply 683 by 4.)

Adding

Do Exercises 5–8.

Multiply.

5.
$$
\begin{array}{r}
4\ 5 \\
\times\ 2\ 3 \\
\end{array}
$$

6. 48×63

7.
$$
\begin{array}{r}
7\ 4\ 6 \\
\times\ \ \ 6\ 2 \\
\end{array}
$$

8. 245×837

Answers

5. 1035   **6.** 3024   **7.** 46,252   **8.** 205,065

Multiply.

9. 472
 × 306
 ___

10. 408 × 704

11. 2 3 4 4
 × 6 0 0 5
 _____

12. 1 0 0 6
 ×   7 0 3
 _____

EXAMPLE 3 Multiply: 306 × 274.

Note that 306 = 3 hundreds + 6 ones.

```
      2 7 4
    × 3 0 6      ┌ Multiplying by 6
    1 6 4 4   ←─┤ Multiplying by 3 hundreds. (We write 00
  8 2 2 0 0   ←─┘ and then multiply 274 by 3.)
  8 3,8 4 4      Adding
```

Do Exercises 9-12.

EXAMPLE 4 Multiply: 360 × 274.

Note that 360 = 3 hundreds + 6 tens.

```
      2 7 4     ┌ Multiplying by 6 tens. (We write 0 and
    × 3 6 0     │ then multiply 274 by 6.)
  1 6 4 4 0   ←─┤ Multiplying by 3 hundreds. (We write 00
  8 2 2 0 0   ←─┘ and then multiply 274 by 3.)
  9 8,6 4 0      Adding
```

Do Exercises 13-16.

Multiply.

13. 472
 × 830
 ___

14. 2 3 4 4
 × 7 4 0 0
 _____

15. 100 × 562

16. 1000 × 562

When we multiply two numbers, we can change the order of the numbers without changing their product. For example, $3 \cdot 6 = 18$ and $6 \cdot 3 = 18$. This illustrates the **commutative law of multiplication:** $a \cdot b = b \cdot a$.

Do Exercise 17.

17. a) Find $23 \cdot 47$.
 b) Find $47 \cdot 23$.
 c) Compare your answers to parts (a) and (b).

To multiply three or more numbers, we generally first group them so that we multiply two at a time. Consider $2 \cdot 3 \cdot 4$. We can group these numbers as $2 \cdot (3 \cdot 4)$ or as $(2 \cdot 3) \cdot 4$. The parentheses tell what to do first:

$2 \cdot (3 \cdot 4) = 2 \cdot (12) = 24.$ We multiply 3 and 4 and then that product by 2.

We can also multiply 2 and 3 and then that product by 4:

$(2 \cdot 3) \cdot 4 = (6) \cdot 4 = 24.$

Either way we get 24. It does not matter how we group the numbers. This illustrates the **associative law of multiplication:** $a \cdot (b \cdot c) = (a \cdot b) \cdot c$.

Multiply.
18. $5 \cdot 2 \cdot 4$

19. $4 \cdot 2 \cdot 6$

Do Exercises 18 and 19.

b Finding Area

The area of a rectangular region can be considered to be the number of square units needed to fill it. Here is a rectangle 4 cm (centimeters) long and 3 cm wide. It takes 12 square centimeters (sq cm) to fill it.

This is a square centimeter (a square unit).

In this case, we have a rectangular array of 3 rows, each of which contains 4 squares. The number of square units is given by 3 · 4, or 12. That is, $A = l \cdot w = 3\,\text{cm} \cdot 4\,\text{cm} = 12\,\text{sq cm}$.

EXAMPLE 5 *Table Tennis.* Find the area of a standard table tennis table that has dimensions of 9 ft by 5 ft.

If we think of filling the rectangle with square feet, we have a rectangular array. The length $l = 9\,\text{ft}$ and the width $w = 5\,\text{ft}$. Thus the area A is given by the formula

$$A = l \cdot w = 9\,\text{ft} \cdot 5\,\text{ft} = 45\,\text{sq ft}.$$

Do Exercise 20.

20. Professional Pool Table. The playing area of a pool table used in professional tournaments is 50 in. by 100 in. (There are 6-in. wide rails on the outside that are not included in the playing area.) Determine the playing area.

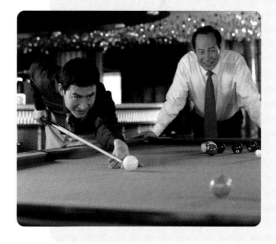

Answer
20. 5000 sq in.

a Multiply.

1. 6 5
     × 8

2. 8 7
     × 4

3. 9 4
     × 6

4. 7 6
     × 9

5. 3 · 509

6. 7 · 806

7. 7(9229)

8. 4(7867)

9. 90(53)

10. 60(78)

11. (47)(85)

12. (34)(87)

13. 8 7
      × 1 0

14. 2 3 4 0
      × 1 0 0 0

15. 9 6
      × 2 0

16. 8 0 0
      × 7 0 0

17. 6 4 3
      × 7 2

18. 7 7 7
      × 7 7

19. 4 4 4
      × 3 3

20. 5 4 9
      × 8 8

21. 5 6 4
      × 4 5 8

22. 4 3 2
      × 3 7 5

23. 8 5 3
      × 9 3 6

24. 3 4 6
      × 6 5 9

25. 6 4 2 8
      × 3 2 2 4

26. 8 9 2 8
      × 3 1 7 2

27. 3 4 8 2
      × 1 0 4

28. 6 4 0 8
      × 6 0 6 4

29. 8 7 6
      × 3 4 5

30. 3 5 5
      × 2 9 9

31. 7 8 8 9
      × 6 2 2 4

32. 6 5 2 1
      × 3 4 4 9

33. 5 6 0 8
      × 4 5 0 0
      ――――――――

34. 4 5 0 6
      × 7 8 0 0
      ――――――――

35. 5 0 0 6
      × 4 0 0 8
      ――――――――

36. 6 0
      × 2
      ―――――

b Find the area of each region.

37.

728 mi

728 mi

38. 129 yd

65 yd

39. Find the area of the region formed by the base lines on a Major League Baseball diamond.

90 ft

90 ft

40. Find the area of a standard-sized hockey rink.

200 ft

85 ft

Skill Maintenance

Add. [1.2a]

41.   4 9 0 8
        5 6 6 7
      + 2 1 1 0
      ――――――――

42.   9 8 7 6
           8 7 6
            7 6
      +   6
      ――――――――

43.   3 4 0 , 7 9 8
      +   8 6 , 6 7 9
      ―――――――――――――

44.   8 8 , 7 7 7
      + 2 2 , 3 3 3
      ―――――――――――

Subtract. [1.3a]

45.   4 9 0 8
      − 3 6 6 7
      ――――――――

46.   9 8 7 6
      −   9 8 7
      ――――――――

47.   3 4 0 , 7 9 8
      −   8 6 , 6 7 9
      ―――――――――――――

48.   8 8 , 7 7 7
      − 2 2 , 3 3 3
      ―――――――――――

Synthesis

49. 🖩 An 18-story office building is box-shaped. Each floor measures 172 ft by 84 ft with a 20-ft by 35-ft rectangular area lost to an elevator and a stairwell. How much area is available as office space?

Division

SKILL TO REVIEW
Objective 1.3a:
Subtract whole numbers.

Subtract.

1. 5 6 4
   − 3 9 7

2. 7 0 3 5
   − 2 9 4 4

a Division of Whole Numbers

Repeated Subtraction

Division of whole numbers applies to two kinds of situations. The first is repeated subtraction. Suppose we have 20 doughnuts, and we want to find out how many sets of 5 there are. One way to do this is to repeatedly subtract sets of 5 as follows.

How many sets of 5 doughnuts each?

20 doughnuts

Since there are 4 sets of 5 doughnuts each, we have

$$20 \div 5 = 4.$$

Dividend Divisor Quotient

The division $20 \div 5$ is read "20 divided by 5." The **dividend** is 20, the **divisor** is 5, and the **quotient** is 4. We divide the *dividend* by the *divisor* to get the *quotient*.

We can also express the division $20 \div 5 = 4$ as

$$\frac{20}{5} = 4 \quad \text{or} \quad 5\overline{)20}\,.$$

Rectangular Arrays

We can also think of division in terms of rectangular arrays. Consider again the 20 doughnuts and division by 5. We can arrange the doughnuts in a rectangular array with 5 rows and ask, "How many are in each row?"

We can also consider a rectangular array with 5 doughnuts in each column and ask, "How many columns are there?" The answer is still 4.

In each case, we are asking, "What do we multiply 5 by in order to get 20?"

Missing factor   Quotient

$$5 \cdot \square = 20 \qquad 20 \div 5 = \square$$

This leads us to the following definition of division.

Answers

Skill to Review:
1. 167 **2.** 4091

DIVISION

The quotient $a \div b$, where $b \neq 0$, is that unique number c for which $a = b \cdot c$.

This definition shows the relation between division and multiplication. We see, for instance, that

$20 \div 5 = 4$ because $20 = 5 \cdot 4$.

This relation allows us to use multiplication to check division.

EXAMPLE 1   Divide. Check by multiplying.

a) $16 \div 8$ **b)** $\dfrac{36}{4}$ **c)** $7\overline{)56}$

We do so as follows.

a) $16 \div 8 = 2$ *Check*: $8 \cdot 2 = 16$.

b) $\dfrac{36}{4} = 9$ *Check*: $4 \cdot 9 = 36$.

c) $\overset{8}{7\overline{)56}}$ *Check*: $7 \cdot 8 = 56$.

Do Exercises 1-3.

Divide. Check by multiplying.
1. $9\overline{)45}$

2. $27 \div 3$

3. $\dfrac{48}{6}$

Let's consider some basic properties of division.

DIVIDING BY 1

Any number divided by 1 is that same number: $a \div 1 = \dfrac{a}{1} = a$.

For example, $6 \div 1 = 6$ and $\dfrac{15}{1} = 15$.

DIVIDING A NUMBER BY ITSELF

Any nonzero number divided by itself is 1: $a \div a = \dfrac{a}{a} = 1$, $a \neq 0$.

For example, $7 \div 7 = 1$ and $\dfrac{22}{22} = 1$.

DIVIDENDS OF 0

Zero divided by any nonzero number is 0: $0 \div a = \dfrac{0}{a} = 0$, $a \neq 0$.

For example, $0 \div 14 = 0$ and $\dfrac{0}{3} = 0$.

Answers

1. 5   **2.** 9   **3.** 8

Why can't we divide by 0? Suppose the number 4 could be divided by 0. Then if $\square$ were the answer, we would have

$$4 \div 0 = \square,$$

and since 0 times any number is 0, we would have

$$4 = \square \cdot 0 = 0. \quad \text{False!}$$

Thus, the only possible number that could be divided by 0 would be 0 itself. But such a division would give us any number we wish. For instance,

$$
\left.
\begin{array}{l}
0 \div 0 = 8 \quad \text{because} \quad 0 = 8 \cdot 0; \\
0 \div 0 = 3 \quad \text{because} \quad 0 = 3 \cdot 0; \\
0 \div 0 = 7 \quad \text{because} \quad 0 = 7 \cdot 0.
\end{array}
\right\} \quad \text{All true!}
$$

We avoid the preceding difficulties by agreeing to exclude division by 0.

Do Exercises 4–7.

Division with a Remainder

Suppose we have 22 cans of soda and want to pack them in cartons of 6 cans each. We could fill 3 cartons and have 4 cans left over.

4 left over

We can think of this as the following division. The leftover cans are the **remainder**.

$$
\begin{array}{r}
3 \leftarrow \text{Quotient} \\
6 \overline{)2\ 2} \\
1\ 8 \\
\hline
4 \leftarrow \text{Remainder}
\end{array}
$$

Divide, if possible. If not possible, write "not defined."

4. $\dfrac{9}{9}$

5. $5 \div 0$

6. $0 \div 20$

7. $\dfrac{8}{1}$

Answers

4. 1 **5.** Not defined
6. 0 **7.** 8

We express the result as

$$22 \div 6 = 3 \text{ R } 4.$$

Dividend   Divisor   Quotient   Remainder

Note that

Quotient · Divisor + Remainder = Dividend.

Thus we have

$$3 \cdot 6 = 18 \qquad \text{Quotient · Divisor}$$

and $18 + 4 = 22.$ Adding the remainder. The result is the dividend.

We now show a procedure for dividing whole numbers.

EXAMPLE 2   Divide and check: $4\overline{)3\ 4\ 5\ 7}$.

First, we try to divide the first digit of the dividend, 3, by the divisor, 4. Since $3 \div 4$ is not a whole number, we consider the first *two* digits of the dividend.

$$
\begin{array}{r}
8 \\
4\overline{)3\ 4\ 5\ 7} \\
3\ 2 \\
\hline
2
\end{array}
$$

Since $4 \cdot 8 = 32$ and 32 is smaller than 34, we write an 8 in the quotient above the 4. We also write 32 below 34 and subtract.

What if we had chosen a number other than 8 for the first digit of the quotient? Suppose we had used 7 instead of 8 and subtracted $4 \cdot 7$, or 28, from 34. The result would have been $34 - 28$, or 6. Because 6 is larger than the divisor, 4, we know that there is at least one more factor of 4 in 34, and thus 7 is too small. If we had used 9 instead of 8, then we would have tried to subtract $4 \cdot 9$, or 36, from 34. That difference is not a whole number, so we know 9 is too large. When we subtract, the difference must be smaller than the divisor.

Let's continue dividing.

$$
\begin{array}{r}
8\ 6 \\
4\overline{)3\ 4\ 5\ 7} \\
3\ 2 \\
\hline
2\ 5 \\
2\ 4 \\
\hline
1
\end{array}
$$

Now bring down the 5 in the dividend and consider $25 \div 4$. Since $4 \cdot 6 = 24$ and 24 is smaller than 25, we write 6 in the quotient above the 5. We also write 24 below 25 and subtract. The difference, 1, is smaller than the divisor, so we know that 6 is the correct choice.

$$
\begin{array}{r}
8\ 6\ 4 \\
4\overline{)3\ 4\ 5\ 7} \\
3\ 2 \\
2\ 5 \\
2\ 4 \\
\hline
1\ 7 \\
1\ 6 \\
\hline
1
\end{array}
$$

Bring down the 7 and consider $17 \div 4$. Since $4 \cdot 4 = 16$ and 16 is smaller than 17, we write 4 in the quotient above the 7. We also write 16 below 17 and subtract.

$1 \leftarrow$ The remainder is 1.

Check: $864 \cdot 4 = 3456$ and $3456 + 1 = 3457.$

The answer is 864 R 1.

Do Exercises 8–10.

Divide and check.

8. $3\overline{)2\ 3\ 9}$

9. $5\overline{)5\ 8\ 6\ 4}$

10. $6\overline{)3\ 8\ 5\ 5}$

Answers
8. 79 R 2   **9.** 1172 R 4
10. 642 R 3

EXAMPLE 3 Divide: 8904 ÷ 42.

Because 42 is close to 40, we think of the divisor as 40 when we make our choices of digits in the quotient.

$$
\begin{array}{r}
2 \\
4\,2\,\overline{)\,8\,9\,0\,4} \\
8\,4 \\
\hline
5\,0
\end{array}
$$
← *Think:* 89 ÷ 40. We try 2. Multiply 42 · 2 and subtract. Then bring down the 0.

$$
\begin{array}{r}
2\;1 \\
4\,2\,\overline{)\,8\,9\,0\,4} \\
8\,4 \\
\hline
5\,0 \\
4\,2 \\
\hline
8\,4
\end{array}
$$
← *Think:* 50 ÷ 40. We try 1. Multiply 42 · 1 and subtract. Then bring down the 4.

$$
\begin{array}{r}
2\;1\;2 \\
4\,2\,\overline{)\,8\,9\,0\,4} \\
8\,4 \\
\hline
5\,0 \\
4\,2 \\
\hline
8\,4 \\
8\,4 \\
\hline
0
\end{array}
$$
← *Think:* 84 ÷ 40. We try 2. Multiply 2 · 42 and subtract.

The remainder is 0, so the answer is 212.

Do Exercises 11 and 12.

Caution!

Be careful to keep the digits lined up correctly when you divide.

Divide.

11. $4\,5\,\overline{)\,6\,0\,3\,0}$

12. $5\,2\,\overline{)\,3\,2\,8\,8}$

Calculator Corner

Dividing Whole Numbers To divide whole numbers on a calculator, we use the ÷ and = keys. For example, to divide 711 by 9, we press 7 1 1 ÷ 9 =. The display reads ▭ 79, so 711 ÷ 9 = 79.

When we enter 453 ÷ 15, the display reads ▭ 30.2. Note that the result is not a whole number. This tells us that there is a remainder. The number 30.2 is expressed in decimal notation. The symbol "." is called a decimal point. Decimal notation will be studied in Chapter 4. Although it is possible to use the number to the right of the decimal point to find the remainder, we will not do so here.

Exercises: Use a calculator to perform each division.

1. $19\,\overline{)\,532}$

2. $7\,\overline{)\,861}$

3. 9367 ÷ 29

4. 12,276 ÷ 341

Answers

11. 134 **12.** 63 R 12

Zeros in Quotients

EXAMPLE 4 Divide: 6341 ÷ 7.

```
       9
  7 ) 6 3 4 1  ←  Think: 63 ÷ 7 = 9. The first digit in the quotient
      6 3 ↓       is 9. We do not write the 0 when we find 63 − 63.
      ───         Bring down the 4.
          4
```

```
       9 0
  7 ) 6 3 4 1
      6 3 ↓
      ───
        4 1  ←  Think: 4 ÷ 7. If we subtract a group of 7's, such
                as 7, 14, 21, etc., from 4, we do not get a whole
                number, so the next digit in the quotient is 0.
                Bring down the 1.
```

```
       9 0 5
  7 ) 6 3 4 1
      6 3
      ───
        4 1  ←  Think: 41 ÷ 7. We try 5. Multiply 7 · 5 and
        3 5     subtract.
        ───
          6  ←  The remainder is 6.
```

The answer is 905 R 6.

> Do Exercises 13 and 14.

EXAMPLE 5 Divide: 8169 ÷ 34.

Because 34 is close to 30, we think of the divisor as 30 when we make our choices of digits in the quotient.

```
        2
  3 4 ) 8 1 6 9  ←  Think: 81 ÷ 30. We try 2. Multiply 34 · 2 and subtract.
        6 8 ↓         Then bring down the 6.
        ─────
        1 3 6
```

```
        2 4
  3 4 ) 8 1 6 9
        6 8 │
        ───── │
        1 3 6 │  ←  Think: 136 ÷ 30. We try 4. Multiply 34 · 4 and
        1 3 6 ↓      subtract. The difference is 0, so we do not write it.
        ─────        Bring down the 9.
              9
```

```
        2 4 0
  3 4 ) 8 1 6 9
        6 8
        ─────
        1 3 6
        1 3 6       ┌ Think: 9 ÷ 34. If we subtract a group of 34's, such as
        ─────     9 ←┤ 34 or 68, from 9, we do not get a whole number, so the
              9     └ last digit in the quotient is 0.
              0
              ─
              9  ←  The remainder is 9.
```

The answer is 240 R 9.

> Do Exercises 15 and 16.

Divide.

13. 6) 4 8 4 6

14. 7) 7 6 1 6

Divide.

15. 2 7) 9 7 2 4

16. 5 6) 4 4,8 4 7

a Divide, if possible. If not possible, write "not defined."

1. $72 \div 6$

2. $54 \div 9$

3. $\dfrac{23}{23}$

4. $\dfrac{37}{37}$

5. $22 \div 1$

6. $\dfrac{56}{1}$

7. $\dfrac{0}{7}$

8. $\dfrac{0}{32}$

9. $\dfrac{16}{0}$

10. $74 \div 0$

11. $\dfrac{48}{8}$

12. $\dfrac{20}{4}$

Divide.

13. $277 \div 5$

14. $699 \div 3$

15. $864 \div 8$

16. $869 \div 8$

17. $4\,\overline{)\,1\ 2\ 2\ 8}$

18. $3\,\overline{)\,2\ 1\ 2\ 4}$

19. $6\,\overline{)\,4\ 5\ 2\ 1}$

20. $9\,\overline{)\,9\ 1\ 1\ 0}$

21. $297 \div 4$

22. $389 \div 2$

23. $738 \div 8$

24. $881 \div 6$

25. $5\,\overline{)\,8\ 5\ 1\ 5}$

26. $3\,\overline{)\,6\ 0\ 2\ 7}$

27. $9\,\overline{)\,8\ 8\ 8\ 8}$

28. $8\,\overline{)\,4\ 1\ 3\ 9}$

29. $127{,}000 \div 10$

30. $127{,}000 \div 100$

31. $127{,}000 \div 1000$

32. $4260 \div 10$

33. $70\overline{)3692}$

34. $20\overline{)5798}$

35. $30\overline{)875}$

36. $40\overline{)987}$

37. $852 \div 21$

38. $942 \div 23$

39. $85\overline{)7672}$

40. $54\overline{)2729}$

41. $111\overline{)3219}$

42. $102\overline{)5612}$

43. $8\overline{)843}$

44. $7\overline{)749}$

45. $5\overline{)8047}$

46. $9\overline{)7273}$

47. $5\overline{)5036}$

48. $7\overline{)7074}$

49. $1058 \div 46$

50. $7242 \div 24$

51. $3425 \div 32$

52. $48\overline{)4899}$

53. $24\overline{)8880}$

54. $36\overline{)7563}$

55. $28\overline{)17,067}$

56. $36\overline{)28,929}$

57. $80\overline{)24,320}$

58. $90\overline{)88,560}$

59. $285\overline{)999,999}$

60. $306 \overline{)888{,}888}$ **61.** $456 \overline{)3{,}679{,}920}$ **62.** $803 \overline{)5{,}622{,}606}$

Skill Maintenance

In each of Exercises 63–70, fill in the blank with the correct term from the given list. Some of the choices may not be used and some may be used more than once.

63. The distance around an object is its _____. [1.2b]

64. The _____ is the number from which another number is being subtracted. [1.3a]

65. For large numbers, _____ are separated by commas into groups of three, called _____. [1.1a]

66. In the sentence $28 \div 7 = 4$, the _____ is 28. [1.5a]

67. In the sentence $10 \times 1000 = 10{,}000$, 10 and 1000 are called _____ and 10,000 is called the _____. [1.4a]

68. The number 0 is called the _____ identity. [1.2a]

69. The sentence $3 \times (6 \times 2) = (3 \times 6) \times 2$ illustrates the _____ law of multiplication. [1.4a]

70. We can use the following statement to check division:
quotient · _____ + _____ = _____. [1.5a]

associative

commutative

addends

factors

area

perimeter

minuend

subtrahend

product

digits

periods

additive

multiplicative

dividend

quotient

remainder

divisor

Synthesis

71. Complete the following table.

a	b	$a \cdot b$	$a + b$
	68	3672	
84			117
		32	12

72. Find a pair of factors whose product is 36 and:
  a) whose sum is 13.
  b) whose difference is 0.
  c) whose sum is 20.
  d) whose difference is 9.

73. A group of 1231 college students is going to take buses for a field trip. Each bus can hold 42 students. How many buses are needed?

74. ▦ Fill in the missing digits to make the equation true:

$$34{,}584{,}132 \div 76\square = 4\square{,}386.$$

Mid-Chapter Review

Concept Reinforcement

Determine whether each statement is true or false.

_____ **1.** If $a - b = c$, then $b = a + c$.   [1.3a]

_____ **2.** We can think of the multiplication 4×3 as a rectangular array containing 4 rows with 3 items in each row.   [1.4a]

_____ **3.** We can think of the multiplication 4×3 as a rectangular array containing 3 columns with 4 items in each column.   [1.4a]

_____ **4.** The product of two whole numbers is always greater than either of the factors.   [1.4a]

_____ **5.** Zero divided by any nonzero number is 0.   [1.5a]

_____ **6.** Any number divided by 1 is the number 1.   [1.5a]

Guided Solutions

Fill in each blank with the number that creates a correct statement or solution.

7. Write a word name for 95,406,237.   [1.1c]

8. Subtract: $604 - 497$.   [1.3a]

Mixed Review

In each of the following numbers what does the digit 6 mean?   [1.1a]

9. 2698   **10.** 61,204   **11.** 146,237   **12.** 586

Consider the number 306,458,129. What digit names the number of: [1.1a]

13. tens   **14.** millions   **15.** ten thousands   **16.** hundreds

Write expanded notation.   [1.1b]   Write a word name.   [1.1c]

17. 5602   **18.** 69,345   **19.** 136   **20.** 64,325

Write standard notation.   [1.1c]

21. Three hundred eight thousand, seven hundred sixteen

22. Four million, five hundred sixty-seven thousand, two hundred sixteen

Add. [1.2a]

23. 3 1 6
  + 4 8 2

24. 5 9 3
  + 4 3 7

25. 2 6 3 8
  + 5 2 8 4

26. 4 6 1 7
  2 4 3 6
  +   4 8 1

Subtract. [1.3a]

27. 7 8 6
  − 3 2 1

28. 6 2 4
  − 2 8 5

29. 3 6 0 2
  − 1 7 4 8

30. 5 0 0 4
  −   6 7 6

Multiply. [1.4a]

31. 3 6
  ×   6

32. 5 6 7
  ×   2 8

33. 4 0 7
  × 3 2 5

34. 9 4 3 5
  ×   6 0 2

Divide. [1.5a]

35. 4) 1 0 1 2

36. 3 8) 4 2 6 1

37. 6 0) 1 3 9 9

38. 5 6) 8 0 9 5

39. Find the perimeter of the figure. [1.2b]

40. Find the area of the region. [1.4b]

Understanding Through Discussion and Writing

To the student and the instructor: The Discussion and Writing exercises are meant to be answered with one or more sentences. They can be discussed and answered collaboratively by the entire class or by small groups.

41. Explain in your own words what the associative law of addition means. [1.2a]

42. Is subtraction commutative? That is, is there a commutative law of subtraction? Why or why not? [1.3a]

43. Describe a situation that corresponds to each multiplication: 4 · $150; $4 · 150. [1.4a]

44. Suppose a student asserts that "$0 \div 0 = 0$ because nothing divided by nothing is nothing." Devise an explanation to persuade the student that the assertion is false. [1.5a]

1.6 Rounding and Estimating; Order

a Rounding

We round numbers in various situations when we do not need an exact answer. For example, we might round to see if we are being charged the correct amount in a store. We might also round to check if an answer to a problem is reasonable or to check a calculation done by hand or on a calculator.

To understand how to round, we first look at some examples using the number line. The number line displays numbers at equally spaced intervals.

EXAMPLE 1 Round 47 to the nearest ten.

47 is between 40 and 50. Since 47 is closer to 50, we round up to 50.

EXAMPLE 2 Round 42 to the nearest ten.

42 is between 40 and 50. Since 42 is closer to 40, we round down to 40.

EXAMPLE 3 Round 45 to the nearest ten.

45 is halfway between 40 and 50. We could round 45 down to 40 or up to 50. We agree to round up to 50.

> When a number is halfway between rounding numbers, round up.

Do Margin Exercises 1–7.

We round whole numbers according to the following rule.

ROUNDING WHOLE NUMBERS

To round to a certain place:

a) Locate the digit in that place.

b) Consider the next digit to the right.

c) If the digit to the right is 5 or higher, round up. If the digit to the right is 4 or lower, round down.

d) Change all digits to the right of the rounding location to zeros.

Round to the nearest ten.

1. 37

2. 52   **3.** 35

4. 73   **5.** 75

6. 88   **7.** 64

Answers

Skill to Review:
1. 2   **2.** 5

Margin Exercises:
1. 40   **2.** 50   **3.** 40   **4.** 70   **5.** 80
6. 90   **7.** 60

EXAMPLE 4 Round 6485 to the nearest ten.

a) Locate the digit in the tens place, 8.

   6 4 8 5
       ↑

b) Consider the next digit to the right, 5.

   6 4 8 5
         ↑

c) Since that digit, 5, is 5 or higher, round 8 tens up to 9 tens.

d) Change all digits to the right of the tens digit to zeros.

   6 4 9 0 ← This is the answer.

Do Exercises 8-11.

Round to the nearest ten.
8. 137   **9.** 473

10. 235   **11.** 285

EXAMPLE 5 Round 6485 to the nearest hundred.

a) Locate the digit in the hundreds place, 4.

   6 4 8 5
     ↑

b) Consider the next digit to the right, 8.

   6 4 8 5
       ↑

c) Since that digit, 8, is 5 or higher, round 4 hundreds up to 5 hundreds.

d) Change all digits to the right of hundreds to zeros.

   6 5 0 0 ← This is the answer.

Do Exercises 12-15.

Round to the nearest hundred.
12. 641   **13.** 759

14. 1871   **15.** 9325

EXAMPLE 6 Round 6485 to the nearest thousand.

a) Locate the digit in the thousands place, 6.

   6 4 8 5
   ↑

b) Consider the next digit to the right, 4.

   6 4 8 5
     ↑

c) Since that digit, 4, is 4 or lower, round down, meaning that 6 thousands stays as 6 thousands.

d) Change all digits to the right of thousands to zeros.

   6 0 0 0 ← This is the answer.

Do Exercises 16-19.

Round to the nearest thousand.
16. 7896   **17.** 8459

18. 19,343   **19.** 68,500

-- *Caution!* --

7000 is not a correct answer to Example 6. It is incorrect to round from the ones digit over, as follows:

   6485 ↛ 6490 ↛ 6500 ↛ 7000.

Note that 6485 is closer to 6000 than it is to 7000.

Answers

8. 140   9. 470   10. 240   11. 290
12. 600   13. 800   14. 1900   15. 9300
16. 8000   17. 8000   18. 19,000
19. 69,000

Sometimes rounding involves changing more than one digit in a number.

EXAMPLE 7 Round 78,595 to the nearest ten.

a) Locate the digit in the tens place, 9.

7 8,5 9 5
      ↑

b) Consider the next digit to the right, 5.

7 8,5 9 5
       ↑

c) Since that digit, 5, is 5 or higher, round 9 tens to 10 tens. To carry this out, we think of 10 tens as 1 hundred + 0 tens and increase the hundreds digit by 1, to get 6 hundreds + 0 tens. We then write 6 in the hundreds place and 0 in the tens place.

d) Change the digit to the right of the tens digit to zero.

7 8,6 0 0 ← This is the answer.

Note that if we round this number to the nearest hundred, we get the same answer.

> Do Exercises 20 and 21.

(b) Estimating

Estimating can be done in many ways. In general, an estimate made by rounding to the nearest ten is more accurate than one rounded to the nearest hundred, and an estimate rounded to the nearest hundred is more accurate than one rounded to the nearest thousand, and so on.

EXAMPLE 8 Estimate this sum by first rounding to the nearest ten:

78 + 49 + 31 + 85.

We round each number to the nearest ten. Then we add.

```
  7 8        8 0
  4 9        5 0
  3 1        3 0
+ 8 5      + 9 0
           ─────
           2 5 0  ← Estimated answer
```

> Do Exercises 22 and 23.

EXAMPLE 9 Estimate the difference by first rounding to the nearest thousand: 9324 − 2849.

We have

```
  9 3 2 4      9 0 0 0
− 2 8 4 9    − 3 0 0 0
─────────    ─────────
             6 0 0 0  ← Estimated answer
```

> Do Exercises 24 and 25.

20. Round 48,968 to the nearest ten, hundred, and thousand.

21. Round 269,582 to the nearest ten, hundred, and thousand.

22. Estimate the sum by first rounding to the nearest ten. Show your work.

```
  7 4
  2 3
  3 5
+ 6 6
```

23. Estimate the sum by first rounding to the nearest hundred. Show your work.

```
  6 5 0
  6 8 5
  2 3 8
+ 1 6 8
```

24. Estimate the difference by first rounding to the nearest hundred. Show your work.

```
  9 2 8 5
− 6 7 3 9
```

25. Estimate the difference by first rounding to the nearest thousand. Show your work.

```
  2 3,2 7 8
− 1 1,6 9 8
```

Answers

20. 48,970; 49,000; 49,000
21. 269,580; 269,600; 270,000
22. 70 + 20 + 40 + 70 = 200
23. 700 + 700 + 200 + 200 = 1800
24. 9300 − 6700 = 2600
25. 23,000 − 12,000 = 11,000

EXAMPLE 10 Estimate the following product by first rounding to the nearest ten and then to the nearest hundred: 683×457.

Nearest ten

$$
\begin{array}{r}
6\ 8\ 0 \\
\times \quad 4\ 6\ 0 \\
\hline
4\ 0\ 8\ 0\ 0 \\
2\ 7\ 2\ 0\ 0\ 0 \\
\hline
3\ 1\ 2{,}8\ 0\ 0
\end{array}
$$

$683 \approx 680$
$457 \approx 460$

Nearest hundred

$$
\begin{array}{r}
7\ 0\ 0 \\
\times \quad 5\ 0\ 0 \\
\hline
3\ 5\ 0{,}0\ 0\ 0
\end{array}
$$

$683 \approx 700$
$457 \approx 500$

Exact

$$
\begin{array}{r}
6\ 8\ 3 \\
\times \quad 4\ 5\ 7 \\
\hline
4\ 7\ 8\ 1 \\
3\ 4\ 1\ 5\ 0 \\
2\ 7\ 3\ 2\ 0\ 0 \\
\hline
3\ 1\ 2{,}1\ 3\ 1
\end{array}
$$

We see that rounding to the nearest ten gives a better estimate than rounding to the nearest hundred.

> Do Exercise 26.

EXAMPLE 11 Estimate the following quotient by first rounding to the nearest ten and then to the nearest hundred: $12{,}238 \div 175$.

Nearest ten

$$
\begin{array}{r}
6\ 8 \\
1\ 8\ 0\ \overline{)\ 1\ 2{,}2\ 4\ 0} \\
1\ 0\ 8\ 0 \\
\hline
1\ 4\ 4\ 0 \\
1\ 4\ 4\ 0 \\
\hline
0
\end{array}
$$

Nearest hundred

$$
\begin{array}{r}
6\ 1 \\
2\ 0\ 0\ \overline{)\ 1\ 2{,}2\ 0\ 0} \\
1\ 2\ 0\ 0 \\
\hline
2\ 0\ 0 \\
2\ 0\ 0 \\
\hline
0
\end{array}
$$

> Do Exercise 27.

In the sentence $7 - 5 = 2$, the equals sign indicates that $7 - 5$ is the *same* as 2. When we round to make an estimate, the outcome is rarely the same as the exact result. Thus we cannot use an equals sign when we round. Instead, we use the symbol $\approx$. This symbol means "**is approximately equal to.**" In Example 9, for instance, we could write

$$9324 - 2849 \approx 6000.$$

26. Estimate the product by first rounding to the nearest ten and then to the nearest hundred. Show your work.

$$
\begin{array}{r}
8\ 3\ 7 \\
\times\ 2\ 4\ 5 \\
\hline
\end{array}
$$

27. Estimate the quotient by first rounding to the nearest hundred. Show your work.

$$64{,}534 \div 349$$

The next two examples show how estimating can be used in making a purchase.

EXAMPLE 12 *Microwave Ovens.* Ellen manages a small apartment building and is planning to purchase a new over-the-range microwave oven for each of the 12 units in the building. One model that she is considering costs $248. Estimate, by rounding to the nearest ten, the total cost of the purchase.

We have

$$\begin{array}{r} 2\ 5\ 0 \\ \times\quad 1\ 0 \\ \hline 2\ 5\ 0\ 0. \end{array}$$

The microwave ovens will cost about $2500.

Do Exercise 28.

EXAMPLE 13 *Purchasing a New Car.* Jon and Joanna are shopping for a new car. They are considering buying a Saturn ASTRA 5-Door XE. The base price of the car is $16,495. A 4-speed automatic transmission package can be added to this, as well as several other options, as shown in the chart below. Jon and Joanna want to stay within a budget of $20,000.

Estimate, by rounding to the nearest hundred, the cost of the ASTRA with the automatic transmission package and all other options and determine whether this will fit within their budget.

SATURN ASTRA 5-DOOR XE	PRICE
Base price	$16,495
4-speed automatic transmission	$1,325
16-in. twin-spoke machined alloy wheels	$350
Air conditioning	$960
Dual-panel power sunroof (Requires purchase of air conditioning)	$1,200
Heated cloth front seats	$250
StabiliTrak Stability Control	$495

SOURCE: Saturn

First, we list the base price of the car and then the cost of each of the options. We then round each number to the nearest hundred and add.

$$\begin{array}{r} 1\ 6,4\ 9\ 5 \\ 1\ 3\ 2\ 5 \\ 3\ 5\ 0 \\ 9\ 6\ 0 \\ 1\ 2\ 0\ 0 \\ 2\ 5\ 0 \\ +\quad 4\ 9\ 5 \\ \hline \end{array} \qquad \begin{array}{r} 1\ 6,5\ 0\ 0 \\ 1\ 3\ 0\ 0 \\ 4\ 0\ 0 \\ 1\ 0\ 0\ 0 \\ 1\ 2\ 0\ 0 \\ 3\ 0\ 0 \\ +\quad 5\ 0\ 0 \\ \hline 2\ 1,2\ 0\ 0 \ \leftarrow \text{Estimated cost} \end{array}$$

The estimated cost is $21,200. This exceeds Jon and Joanna's budget of $20,000, so they will have to forgo at least one option.

Do Exercises 29 and 30.

28. Microwave Ovens. Suppose Ellen chooses a smaller model of microwave oven that costs $198. Estimate, by rounding to the nearest ten, the total cost of 12 ovens.

Refer to the chart at left to do Margin Exercises 29 and 30.

29. By eliminating at least one option, determine how Jon and Joanna can buy an ASTRA and stay within their budget. Keep in mind that purchasing the sunroof also requires the purchase of air conditioning.

30. Elizabeth and C.J. are also considering buying a Saturn ASTRA 5-Door XE. Estimate, by rounding to the nearest hundred, the cost of this car with automatic transmission, air conditioning, and StabiliTrak Stability Control.

Answers

28. $2000   **29.** Eliminate either the automatic transmission or the sunroof. There are other correct answers as well.   **30.** $19,300

c Order

We know that 2 is not the same as 5. We express this by the sentence $2 \neq 5$. We also know that 2 is less than 5. We symbolize this by the expression $2 < 5$. We can see this order on the number line: 2 is to the left of 5. The number 0 is the smallest whole number.

ORDER OF WHOLE NUMBERS

For any whole numbers a and b:

1. $a < b$ (read "a is less than b") is true when a is to the left of b on the number line.
2. $a > b$ (read "a is greater than b") is true when a is to the right of b on the number line.

We call $<$ and $>$ **inequality symbols**.

EXAMPLE 14 Use $<$ or $>$ for ☐ to write a true sentence: 7 ☐ 11.

Since 7 is to the left of 11 on the number line, $7 < 11$.

EXAMPLE 15 Use $<$ or $>$ for ☐ to write a true sentence: 92 ☐ 87.

Since 92 is to the right of 87 on the number line, $92 > 87$.

A sentence like $8 + 5 = 13$ is called an **equation**. It is a *true* equation. The equation $4 + 8 = 11$ is a *false* equation. A sentence like $7 < 11$ is called an **inequality**. The sentence $7 < 11$ is a *true* inequality. The sentence $23 > 69$ is a *false* inequality.

Do Exercises 31–36.

Use $<$ or $>$ for ☐ to write a true sentence. Draw the number line if necessary.

31. 8 ☐ 12

32. 12 ☐ 8

33. 76 ☐ 64

34. 64 ☐ 76

35. 217 ☐ 345

36. 345 ☐ 217

STUDY TIPS

TEXTBOOK SUPPLEMENTS

Are you aware of all the supplements that exist for this textbook? See the preface for a description of each supplement including the *Student's Solutions Manual* and Video Resources on DVD Featuring Chapter Test Prep Videos.

Answers

31. $<$ 32. $>$ 33. $>$ 34. $<$
35. $<$ 36. $>$

1.6 Exercise Set

For Extra Help

MyMathLab

Math XL
PRACTICE

WATCH

DOWNLOAD

READ

REVIEW

a Round to the nearest ten.

1. 48

2. 532

3. 463

4. 8945

5. 731

6. 54

7. 895

8. 798

Round to the nearest hundred.

9. 146

10. 874

11. 957

12. 650

13. 9079

14. 4645

15. 32,839

16. 198,402

Round to the nearest thousand.

17. 5876

18. 4500

19. 7500

20. 2001

21. 45,340

22. 735,562

23. 373,405

24. 6,713,255

b Estimate each sum or difference by first rounding to the nearest ten. Show your work.

25.
```
    7 8
 +  9 2
```

26.
```
    6 2
    9 7
    4 6
 +  8 1
```

27.
```
  8 0 7 4
− 2 3 4 7
```

28.
```
    6 7 3
 −     2 8
```

Estimate each sum by first rounding to the nearest ten. State if the given sum seems to be incorrect when compared to the estimate.

29.
```
    4 5
    7 7
    2 5
 +  5 6
  ─────
  3 4 3
```

30.
```
    4 1
    2 1
    5 5
 +  6 0
  ─────
  1 7 7
```

31.
```
  6 2 2
    7 8
    8 1
+ 1 1 1
─────
  9 3 2
```

32.
```
  8 3 6
  3 7 4
  7 9 4
+ 9 3 8
─────
3 9 4 7
```

Estimate each sum or difference by first rounding to the nearest hundred. Show your work.

33.
```
  7 3 4 8
+ 9 2 4 7
```

34.
```
    5 6 8
    4 7 2
    9 3 8
 +  4 0 2
```

35.
```
  6 8 5 2
− 1 7 4 8
```

36.
```
  9 4 3 8
− 2 7 8 7
```

Estimate each sum by first rounding to the nearest hundred. State if the given sum seems to be incorrect when compared to the estimate.

37.	2 1 6	**38.**	4 8 1	**39.**	7 5 0	**40.**	3 2 6
	8 4		7 0 2		4 2 8		2 7 5
	7 4 5		6 2 3		6 3		7 5 8
	+ 5 9 5		+ 1 0 4 3		+ 2 0 5		+ 9 4 3
	1 6 4 0		1 8 4 9		1 4 4 6		2 3 0 2

Estimate each sum or difference by first rounding to the nearest thousand. Show your work.

41.	9 6 4 3	**42.**	7 6 4 8	**43.**	9 2,1 4 9	**44.**	8 4,8 9 0
	4 8 2 1		9 3 4 8		− 2 2,5 5 5		− 1 1,1 1 0
	8 9 4 3		7 8 4 2				
	+ 7 0 0 4		+ 2 2 2 2				

Estimate each product by first rounding to the nearest ten. Show your work.

45.	4 5	**46.**	5 1	**47.**	3 4	**48.**	6 3
	× 6 7		× 7 8		× 2 9		× 5 4

Estimate each product by first rounding to the nearest hundred. Show your work.

49.	8 7 6	**50.**	3 5 5	**51.**	4 3 2	**52.**	7 8 9
	× 3 4 5		× 2 9 9		× 1 9 9		× 4 3 4

Estimate each quotient by first rounding to the nearest ten. Show your work.

53. $347 \div 73$ **54.** $454 \div 87$ **55.** $8452 \div 46$ **56.** $1263 \div 29$

Estimate each quotient by first rounding to the nearest hundred. Show your work.

57. $1165 \div 236$ **58.** $3641 \div 571$ **59.** $8358 \div 295$ **60.** $32,854 \div 748$

Planning a Kitchen. Perfect Kitchens offers custom kitchen packages with three choices for each of four items: cabinets, countertops, appliances, and flooring. The chart below lists the price for each choice. Customers design their kitchens by making one selection from each group of items.

CABINETS	TYPE	PRICE
(a)	Oak	$7450
(b)	Cherry	8820
(c)	Painted	9630
COUNTERTOPS	TYPE	PRICE
(d)	Laminate	$1595
(e)	Solid surface	2870
(f)	Granite	3528
APPLIANCES	PRICE RANGE	PRICE
(g)	Low	$1540
(h)	Medium	3575
(i)	High	6245
FLOORING	TYPE	PRICE
(j)	Vinyl	$625
(k)	Travertine	985
(l)	Hardwood	1160

61. Estimate the cost of remodeling a kitchen with choices (a), (d), (g), and (j) by rounding to the nearest hundred dollars.

62. Estimate the cost of a kitchen with choices (c), (f), (i), and (l) by rounding to the nearest hundred dollars.

63. Sara and Ben are planning to remodel their kitchen and have a budget of $17,700. Estimate by rounding to the nearest hundred dollars the cost of their kitchen remodeling project if they choose options (b), (e), (i), and (k). Can they afford their choices?

64. The Davidsons must make a final decision on the kitchen choices for their new home. The allotted kitchen budget is $16,000. Estimate by rounding to the nearest hundred dollars the kitchen cost if they choose options (a), (f), (h), and (l). Does their budget allotment cover the cost?

65. Suppose you are planning a new kitchen and must stay within a budget of $14,500. Decide on the options you would like and estimate the cost by rounding to the nearest hundred dollars. Does your budget support your choices?

66. Suppose you are planning a new kitchen and must stay within a budget of $18,500. Decide on the options you would like and estimate the cost by rounding to the nearest hundred dollars. Does your budget support your choices?

67. Company Cars. A publishing company buys a Honda Accord LX for each of its 112 sales representatives. Each car costs $21,160 with an additional $670 per car in destination charges.

a) Estimate the total cost of the purchase by rounding the cost of each car, the destination charge, and the number of sales representatives to the nearest hundred.

b) Estimate the total cost of the purchase by rounding the cost of each car to the nearest thousand and the destination charge and the number of sales representatives to the nearest hundred.

Source: Honda

68. Airline Tickets. A travel club of 176 people decides to fly from Chicago to Seattle. The cost of a round-trip ticket is $535.

a) Estimate the total cost of the trip by rounding the cost of the airfare and the number of travelers to the nearest ten.

b) Estimate the total cost of the trip by rounding the cost of the airfare and the number of travelers to the nearest hundred.

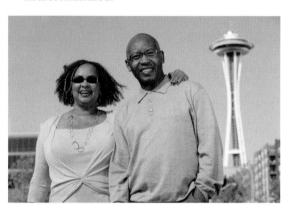

69. Banquet Attendance. Tickets to the annual awards banquet for the Riviera Swim Club cost $28 each. Ticket sales for the banquet totaled $2716. Estimate the number of people who attended the banquet by rounding the cost of a ticket to the nearest ten and the total sales to the nearest hundred.

70. School Fundraiser. For a school fundraiser, Charlotte sells trash bags at a price of $11 per box. If her sales total $2211, estimate the number of boxes sold by rounding the price per box to the nearest ten and the total sales to the nearest hundred.

c   Use < or > for ☐ to write a true sentence. Draw the number line if necessary.

71. 0 ☐ 17

72. 32 ☐ 0

73. 34 ☐ 12

74. 28 ☐ 18

75. 1000 ☐ 1001

76. 77 ☐ 117

77. 133 ☐ 132

78. 999 ☐ 997

79. 460 ☐ 17

80. 345 ☐ 456

81. 37 ☐ 11

82. 12 ☐ 32

New Book Titles. The number of new book titles published in the United States in each of three recent years is shown in the table below. Use this table to do Exercises 83 and 84.

YEAR	NEW BOOK TITLES
2004	190,078
2005	172,000
2007	276,649

SOURCE: R. R. Bowker

83. Write an inequality to compare the number of new titles published in 2004 and in 2005.

84. Write an inequality to compare the number of new titles published in 2005 and in 2007.

85. *Wind-Power Capacity.* Wind-power capacity in the United States has increased from 1694 megawatts installed in 2001 to 5249 megawatts installed in 2007. Write an inequality to compare these numbers of megawatts of wind power installed.

U. S. Wind-Power Boom

SOURCE: American Wind Energy Association

86. *Life Expectancy.* The life expectancy of a female in the United States in 2015 is predicted to be about 82 yr and that of a male about 76 yr. Write an inequality to compare these life expectancies.

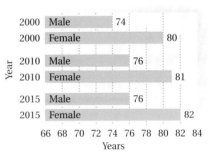

Life Expectancy in the United States

SOURCE: U.S. Census Bureau

Skill Maintenance

Add. [1.2a]

87.
$$\begin{array}{r} 67{,}789 \\ +\ 18{,}965 \\ \hline \end{array}$$

88.
$$\begin{array}{r} 9002 \\ +\ 4587 \\ \hline \end{array}$$

Subtract. [1.3a]

89.
$$\begin{array}{r} 67{,}789 \\ -\ 18{,}965 \\ \hline \end{array}$$

90.
$$\begin{array}{r} 9002 \\ -\ 4587 \\ \hline \end{array}$$

Multiply. [1.4a]

91.
$$\begin{array}{r} 46 \\ \times\ 37 \\ \hline \end{array}$$

92.
$$\begin{array}{r} 306 \\ \times\ 58 \\ \hline \end{array}$$

Divide. [1.5a]

93. $328 \div 6$

94. $4784 \div 23$

Synthesis

95.–98. ▦ Use a calculator to find the sums and the differences in each of Exercises 41–44. Then compare your answers with those found using estimation. Even when using a calculator it is possible to make an error if you press the wrong buttons, so it is a good idea to check by estimating.

1.7

Solving Equations

OBJECTIVES

a Solve simple equations by trial.

b Solve equations like $t + 28 = 54$, $28 \cdot x = 168$, and $98 \cdot 2 = y$.

Find a number that makes each sentence true.

1. $8 = 1 + \square$

2. $\square + 2 = 7$

3. Determine whether 7 is a solution of $\square + 5 = 9$.

4. Determine whether 4 is a solution of $\square + 5 = 9$.

a Solutions by Trial

Let's find a number that we can put in the blank to make this sentence true:

$$9 = 3 + \square.$$

We are asking "9 is 3 plus what number?" The answer is 6.

$$9 = 3 + 6$$

Do Exercises 1 and 2.

A sentence with = is called an **equation**. A **solution** of an equation is a number that makes the sentence true. Thus, 6 is a solution of

$$9 = 3 + \square \quad \text{because} \quad 9 = 3 + 6 \text{ is true.}$$

However, 7 is not a solution of

$$9 = 3 + \square \quad \text{because} \quad 9 = 3 + 7 \text{ is false.}$$

Do Exercises 3 and 4.

We can use a letter in an equation instead of a blank:

$$9 = 3 + n.$$

We call n a **variable** because it can represent any number. If a replacement for a variable makes an equation true, it is a **solution** of the equation.

> ## SOLUTIONS OF AN EQUATION
>
> A **solution of an equation** is a replacement for the variable that makes the equation true. When we find all the solutions, we say that we have **solved** the equation.

EXAMPLE 1 Solve $y + 12 = 27$ by trial.

We replace y with several numbers.

If we replace y with 13, we get a false equation: $13 + 12 = 27$.
If we replace y with 14, we get a false equation: $14 + 12 = 27$.
If we replace y with 15, we get a true equation: $15 + 12 = 27$.

No other replacement makes the equation true, so the solution is 15.

Solve by trial.

5. $n + 3 = 8$

6. $x - 2 = 8$

7. $45 \div 9 = y$

8. $10 + t = 32$

EXAMPLES Solve.

2. $7 + n = 22$
(7 plus what number is 22?)
The solution is 15.

3. $63 = 3 \cdot x$
(63 is 3 times what number?)
The solution is 21.

Do Exercise 5-8.

Answers

1. 7 **2.** 5 **3.** No **4.** Yes **5.** 5
6. 10 **7.** 5 **8.** 22

b Solving Equations

We now begin to develop more efficient ways to solve certain equations. When an equation has a variable alone on one side and a calculation on the other side, we can find the solution by carrying out the calculation.

EXAMPLE 4   Solve: $x = 245 \times 34$.

To solve the equation, we carry out the calculation.

$$
\begin{array}{r}
2\ 4\ 5 \\
\times\ \ \ 3\ 4 \\
\hline
9\ 8\ 0 \\
7\ 3\ 5\ 0 \\
\hline
8\ 3\ 3\ 0
\end{array}
$$

$x = 245 \times 34$
$x = 8330$

The solution is 8330.

Do Exercises 9–12.

Look at the equation

$$x + 12 = 27.$$

We can get x alone by subtracting 12 *on both sides*. Thus,

$x + 12 - 12 = 27 - 12$ Subtracting 12 on both sides

$x + 0 = 15$ Carrying out the subtraction

$x = 15.$

SOLVING $x + a = b$

To solve $x + a = b$, subtract a on both sides.

If we can get an equation in a form with the variable alone on one side, we can "see" the solution.

EXAMPLE 5   Solve: $t + 28 = 54$.

We have

$t + 28 = 54$

$t + 28 - 28 = 54 - 28$ Subtracting 28 on both sides

$t + 0 = 26$

$t = 26.$

To check the answer, we substitute 26 for t in the original equation.

Check: $\dfrac{t + 28 = 54}{26 + 28 \ ? \ 54}$

$54 \ \big|$ TRUE   Since $54 = 54$ is true, 26 checks.

The solution is 26.

Do Exercises 13 and 14.

Solve.

9. $346 \times 65 = y$

10. $x = 2347 + 6675$

11. $4560 \div 8 = t$

12. $x = 6007 - 2346$

Solve. Be sure to check.

13. $x + 9 = 17$

14. $77 = m + 32$

EXAMPLE 6 Solve: $182 = 65 + n$.

We have

$$182 = 65 + n$$
$$182 - 65 = 65 + n - 65 \qquad \text{Subtracting 65 on both sides}$$
$$117 = 0 + n \qquad \text{65 plus } n \text{ minus 65 is } 0 + n.$$
$$117 = n.$$

Check:
$$\frac{182 = 65 + n}{182 \; \overset{?}{} \; 65 + 117}$$
$$\;\; | \;\; 182 \qquad \text{TRUE}$$

The solution is 117.

Do Exercise 15.

15. Solve: $155 = t + 78$. Be sure to check.

EXAMPLE 7 Solve: $7381 + x = 8067$.

We have

$$7381 + x = 8067$$
$$7381 + x - 7381 = 8067 - 7381 \qquad \text{Subtracting 7381 on both sides}$$
$$x = 686.$$

Check:
$$\frac{7381 + x = 8067}{7381 + 686 \; \overset{?}{} \; 8067}$$
$$8067 \;\; | \qquad \text{TRUE}$$

The solution is 686.

Do Exercises 16 and 17.

Solve. Be sure to check.

16. $4566 + x = 7877$

17. $8172 = h + 2058$

We now learn to solve equations like $8 \cdot n = 96$. Look at

$$8 \cdot n = 96.$$

We can get n alone by dividing by 8 *on both sides*. Thus,

$$\frac{8 \cdot n}{8} = \frac{96}{8} \qquad \text{Dividing by 8 on both sides}$$
$$n = 12. \qquad \text{8 times } n \text{ divided by 8 is } n.$$

To check the answer, we substitute 12 for n in the original equation.

Check:
$$\frac{8 \cdot n = 96}{8 \cdot 12 \; \overset{?}{} \; 96}$$
$$96 \;\; | \qquad \text{TRUE}$$

Since $96 = 96$ is a true equation, 12 is the solution of the equation.

SOLVING $a \cdot x = b$

To solve $a \cdot x = b$, divide by a on both sides.

EXAMPLE 8   Solve: $10 \cdot x = 240$.

We have

$$10 \cdot x = 240$$

$$\frac{10 \cdot x}{10} = \frac{240}{10} \qquad \text{Dividing by 10 on both sides}$$

$$x = 24.$$

Check: $10 \cdot x = 240$
          ───────────────
          $10 \cdot 24$? 240
          240 │   TRUE

The solution is 24.

EXAMPLE 9   Solve: $5202 = 9 \cdot t$.

We have

$$5202 = 9 \cdot t$$

$$\frac{5202}{9} = \frac{9 \cdot t}{9} \qquad \text{Dividing by 9 on both sides}$$

$$578 = t.$$

Check: $5202 = 9 \cdot t$
          ───────────────
          5202 ? $9 \cdot 578$
               │ 5202 TRUE

The solution is 578.

Do Exercises 18–20.

Solve. Be sure to check.

18. $8 \cdot x = 64$

19. $144 = 9 \cdot n$

20. $5152 = 8 \cdot t$

EXAMPLE 10   Solve: $14 \cdot y = 1092$.

We have

$$14 \cdot y = 1092$$

$$\frac{14 \cdot y}{14} = \frac{1092}{14} \qquad \text{Dividing by 14 on both sides}$$

$$y = 78.$$

The check is left to the student. The solution is 78.

EXAMPLE 11   Solve: $n \cdot 56 = 4648$.

We have

$$n \cdot 56 = 4648$$

$$\frac{n \cdot 56}{56} = \frac{4648}{56} \qquad \text{Dividing by 56 on both sides}$$

$$n = 83.$$

The check is left to the student. The solution is 83.

Do Exercises 21 and 22.

Solve. Be sure to check.

21. $18 \cdot y = 1728$

22. $n \cdot 48 = 4512$

Answers

18. 8   **19.** 16   **20.** 644   **21.** 96   **22.** 94

a Solve by trial.

1. $x + 0 = 14$

2. $x - 7 = 18$

3. $y \cdot 17 = 0$

4. $56 \div m = 7$

b Solve. Be sure to check.

5. $x = 12{,}345 + 78{,}555$

6. $t = 5678 + 9034$

7. $908 - 458 = p$

8. $9007 - 5667 = m$

9. $16 \cdot 22 = y$

10. $34 \cdot 15 = z$

11. $t = 125 \div 5$

12. $w = 256 \div 16$

13. $13 + x = 42$

14. $15 + t = 22$

15. $12 = 12 + m$

16. $16 = t + 16$

17. $10 + x = 89$

18. $20 + x = 57$

19. $61 = 16 + y$

20. $53 = 17 + w$

21. $3 \cdot x = 24$

22. $6 \cdot x = 42$

23. $112 = n \cdot 8$

24. $162 = 9 \cdot m$

25. $3 \cdot m = 96$

26. $4 \cdot y = 96$

27. $715 = 5 \cdot z$

28. $741 = 3 \cdot t$

29. $8322 + 9281 = x$

30. $9281 - 8322 = y$

31. $47 + n = 84$

32. $56 + p = 92$

33. $45 \cdot 23 = x$

34. $23 \cdot 78 = y$

35. $x + 78 = 144$

36. $z + 67 = 133$

37. $6 \cdot p = 1944$ **38.** $4 \cdot w = 3404$ **39.** $5 \cdot x = 3715$ **40.** $9 \cdot x = 1269$

41. $x + 214 = 389$ **42.** $x + 221 = 333$ **43.** $567 + x = 902$ **44.** $438 + x = 807$

45. $234 \cdot 78 = y$ **46.** $10,534 \div 458 = q$ **47.** $18 \cdot x = 1872$ **48.** $19 \cdot x = 6080$

49. $40 \cdot x = 1800$ **50.** $20 \cdot x = 1500$ **51.** $2344 + y = 6400$ **52.** $9281 = 8322 + t$

53. $m = 7006 - 4159$ **54.** $n = 3004 - 1745$ **55.** $165 = 11 \cdot n$ **56.** $660 = 12 \cdot n$

57. $58 \cdot m = 11,890$ **58.** $233 \cdot x = 22,135$ **59.** $491 - 34 = y$ **60.** $512 - 63 = z$

Skill Maintenance

Divide.   [1.5a]

61. $1283 \div 9$ **62.** $1278 \div 9$ **63.** $1\,7\,\overline{)\,5\,6\,7\,8}$ **64.** $1\,7\,\overline{)\,5\,6\,8\,9}$

Use > or < for ☐ to write a true sentence.   [1.6c]

65. $123\ \square\ 789$ **66.** $342\ \square\ 339$ **67.** $688\ \square\ 0$ **68.** $0\ \square\ 11$

69. Round 6,375,602 to the nearest thousand.   [1.6a]

70. Round 6,375,602 to the nearest ten.   [1.6a]

Synthesis

Solve.

71. ▦ $23,465 \cdot x = 8,142,355$ **72.** ▦ $48,916 \cdot x = 14,332,388$

1.8

Applications and Problem Solving

OBJECTIVE

a) Solve applied problems involving addition, subtraction, multiplication, or division of whole numbers.

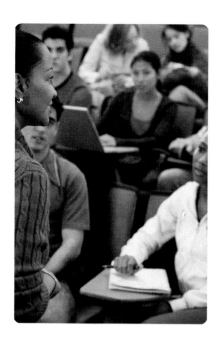

a) A Problem-Solving Strategy

One of the most important ways in which we use mathematics is as a tool in solving problems. To solve a problem, we use the following five-step strategy.

FIVE STEPS FOR PROBLEM SOLVING

1. **Familiarize** yourself with the problem situation. If the problem is presented in words, this means to read and reread it carefully until you understand what you are being asked to find. Some or all of the following can also be helpful.

   a) Make a drawing, if it makes sense to do so.

   b) Make a written list of the known facts and a list of what you wish to find out.

   c) Assign a letter, or *variable*, to the unknown.

   d) Organize the information in a chart or a table.

   e) Find further information. Look up a formula, consult a reference book or an expert in the field, or do research on the Internet.

   f) Guess or estimate the answer and check your guess or estimate.

2. **Translate** the problem to an equation using the variable.

3. **Solve** the equation.

4. **Check** to see whether your possible solution actually fits the problem situation and is thus really a solution of the problem. Although you may have solved an equation, the solution of the equation might not be a solution of the original problem.

5. **State** the answer clearly using a complete sentence and appropriate units.

The first of these five steps, becoming familiar with the problem, is probably the most important. It provides a solid foundation for translating the problem to an equation that represents the situation accurately.

EXAMPLE 1 *Community Colleges and New Jobs.* Community colleges are playing an increasingly large role in educating America's work force. The numbers of new jobs requiring a community college degree that will be created between 2004 and 2014 are shown in the table on the next page. Find the total number of new jobs for registered nurses, nursing aides and orderlies, and medical assistants.

New Jobs Created, 2004–2014

JOB	NUMBER
Registered nurse	703,000
Customer service	471,000
Nursing aide; orderly	325,000
Heavy-truck driver	223,000
Maintenance; repair	202,000
Medical assistant	202,000
Executive secretary/assistant	192,000
Sales representative	187,000
Carpenter	186,000

SOURCES: U.S. Bureau of Labor Statistics;
College Board Center for Innovative Thought

1. **Familiarize.** First, we assign a letter, or variable, to the number we wish to find. We let n = the total number of new jobs created for registered nurses, nursing aides and orderlies, and medical assistants. Since we are combining numbers, we will add.

2. **Translate.** We translate to an equation:

Registered nurses' jobs	plus	Nursing aides' and orderlies' jobs	plus	Medical assistants' jobs	is	Total number of jobs
↓	↓	↓	↓	↓	↓	↓
703,000	+	325,000	+	202,000	=	n.

3. **Solve.** We solve the equation by carrying out the addition.

$$\begin{array}{r} 703{,}000 \\ 325{,}000 \\ +\,202{,}000 \\ \hline 1{,}230{,}000 \end{array}$$

$$703{,}000 + 325{,}000 + 202{,}000 = n$$
$$1{,}230{,}000 = n$$

4. **Check.** We check 1,230,000 in the original problem. There are many ways in which this can be done. For example, we can repeat the calculation. (We leave this to the student.) Another way is to check whether the answer is reasonable. In this case, we would expect the total to be greater than the number of each individual type of new job, and it is. We can also estimate the expected result by rounding. Here we round to the nearest hundred thousand:

$$703{,}000 + 325{,}000 + 202{,}000 \approx 700{,}000 + 300{,}000 + 200{,}000$$
$$\approx 1{,}200{,}000.$$

Since $1{,}200{,}000 \approx 1{,}230{,}000$, our answer seems reasonable. If the estimate had differed greatly from the possible solution found in step (3), we would suspect that the possible solution is incorrect.

5. **State.** The total number of new jobs created for registered nurses, nursing aides and orderlies, and medical assistants between 2004 and 2014 is 1,230,000.

Do Exercises 1–3.

Refer to the table above to do Margin Exercises 1–3.

1. Find the total number of new jobs that will be created for customer-service representatives, executive secretaries/assistants, and sales representatives.

2. Find the total number of new jobs that will be created for heavy-truck drivers, maintenance and repair workers, and carpenters.

3. Find the total number of new jobs listed in the table.

Answers

1. 850,000 new jobs   2. 611,000 new jobs
3. 2,691,000 new jobs

EXAMPLE 2 *Checking Account Balance.* The balance in Francisco's checking account is $573. He uses his debit card to buy a juicer that costs $99. Find the new balance in his checking account.

1. **Familiarize.** We first make a drawing or at least visualize the situation. We let B = the new balance in Francisco's checking account. We start with $573 and take away $99.

2. **Translate.** We translate to an equation:

$$
\underbrace{\text{Money in the account}}_{573} \;\; \underbrace{\text{minus}}_{-} \;\; \underbrace{\text{Money spent}}_{99} \;\; \underbrace{\text{is}}_{=} \;\; \underbrace{\text{New balance}}_{B.}
$$

3. **Solve.** This equation tells us what to do. We subtract.

$$
\begin{array}{r}
16 \\
4\ \ \cancel{6}\ \ 13 \\
\cancel{5}\ \ \cancel{7}\ \ \cancel{3} \\
-\ \ \ 9\ \ 9 \\
\hline
4\ \ 7\ \ 4
\end{array}
\qquad
\begin{array}{l}
573 - 99 = B \\
\quad\ \ 474 = B
\end{array}
$$

4. **Check.** To check our answer of $474, we can repeat the calculation. We can also note that the answer should be less than the original amount, $573, and it is. Another way to check is to add the money spent, $99, to the new balance, $474: $99 + $474 = $573. We get the original balance, so the answer checks. We can also estimate:

$$
\$573 - \$99 \approx \$570 - \$100 \approx \$470 \approx \$474.
$$

This tells us that the answer is reasonable.

5. **State.** The new balance in Francisco's checking account is $474.

Do Exercise 4.

In the real world, problems may not be stated in written words. You must still become familiar with the situation before you can solve the problem.

EXAMPLE 3 *Travel Distance.* Abigail is driving from Indianapolis to Salt Lake City to attend a family reunion. The distance from Indianapolis to Salt Lake City is 1634 mi. In the first two days, she travels 1154 mi to Denver. How much farther must she travel?

1. **Familiarize.** We first make a drawing or at least visualize the situation. We let d = the remaining distance to Salt Lake City.

4. Checking Account Balance.
The balance in Laura's checking account is $457. She uses her debit card to buy a digital-picture frame that costs $49. Find the new balance in her checking account.

Answer

4. $408

2. Translate. We want to determine how many more miles Abigail must travel. We translate to an equation:

Distance already traveled	plus	Distance to go	is	Total distance of trip
↓	↓	↓	↓	↓
1154	+	d	=	1634.

3. Solve. To solve the equation, we subtract 1154 on both sides.

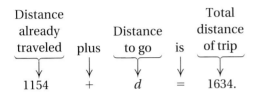

$$1154 + d = 1634$$
$$1154 + d - 1154 = 1634 - 1154$$
$$d = 480$$

$$\begin{array}{r} {\scriptstyle 5\ \ 13} \\ 1\ \cancel{6}\ \cancel{3}\ 4 \\ -\ 1\ 1\ 5\ 4 \\ \hline 4\ 8\ 0 \end{array}$$

4. Check. We check our answer of 480 mi in the original problem. This number should be less than the total distance, 1634 mi, and it is. We can add the distance traveled, 1154, and the distance left to go, 480: $1154 + 480 = 1634$. We can also estimate:

$$1634 - 1154 \approx 1600 - 1200$$
$$= 400 \approx 480.$$

The answer, 480 mi, checks.

5. State. Abigail must travel 480 mi farther to Salt Lake City.

Do Exercise 5.

5. Reading Assignment. William has been assigned 234 pages of reading for his history class. He has read 86 pages. How many more pages does he have to read?

EXAMPLE 4 *Total Cost of Chairs.* What is the total cost of 6 Adirondack chairs if each one costs $169?

1. Familiarize. We first make a drawing or at least visualize the situation. We let $C =$ the cost of 6 chairs.

$169 $169 $169 $169 $169 $169

2. Translate. We translate to an equation:

Number of chairs	times	Cost of each chair	is	Total cost
↓	↓	↓	↓	↓
6	×	$169	=	C.

3. Solve. This sentence tells us what to do. We multiply.

$$
\begin{array}{r}
1\ 6\ 9 \\
\times\ \ \ \ \ 6 \\
\hline
1\ 0\ 1\ 4
\end{array}
$$

$6 \times 169 = C$

$1014 = C$

4. Check. We have an answer, 1014, that is much greater than the cost of any individual chair, which is reasonable. We can repeat our calculation. We can also check by estimating:

$$6 \times 169 \approx 6 \times 170 \approx 1020 \approx 1014.$$

The answer checks.

5. State. The total cost of 6 chairs is $1014.

Do Exercise 6.

6. Total Cost of Gas Grills. What is the total cost of 14 gas grills, each with 520 sq in. of total cooking surface, if each one costs $398?

EXAMPLE 5 *Area of an Oriental Rug.* The dimensions of the oriental rug in the Fosters' front hallway are 42 in. by 66 in. What is the area of the rug?

1. Familiarize. We first make a drawing to visualize the situation. We let A = the area of the rug and use the formula for the area of a rectangle, $A = $ length · width $= l \cdot w$. Since we usually consider length to be larger than width, we will let $l = 66$ in. and $w = 42$ in.

42 in.

66 in.

2. Translate. We substitute in the formula:

$$A = l \cdot w = 66 \cdot 42.$$

3. Solve. We carry out the multiplication.

$$
\begin{array}{r}
6\ 6 \\
\times\ 4\ 2 \\
\hline
1\ 3\ 2 \\
2\ 6\ 4\ 0 \\
\hline
2\ 7\ 7\ 2
\end{array}
$$

$A = 66 \cdot 42$

$A = 2772$

4. Check. We can repeat the calculation. We can also round and estimate:

$$66 \times 42 \approx 70 \times 40 \approx 2800 \approx 2772.$$

The answer checks.

5. State. The area of the rug is 2772 sq in.

Do Exercise 7.

7. Bed Sheets. The dimensions of a flat sheet for a queen-size bed are 90 in. by 102 in. What is the area of the sheet?

Answers

6. $5572   **7.** 9180 sq in.

EXAMPLE 6 *Packages of Paper Towels.* A paper-products company produces 3304 rolls of paper towels. How many 12-roll packages can be filled? How many rolls will be left over?

1. **Familiarize.** We first make a drawing. We let n = the number of 12-roll packages that can be filled. The problem can be considered as repeated subtraction, taking successive sets of 12 rolls and putting them into n packages.

12-roll packages

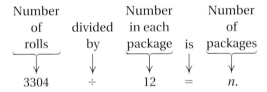

How many packages?

How many rolls are left over?

2. **Translate.** We translate to an equation:

Number of rolls	divided by	Number in each package	is	Number of packages
↓	↓	↓	↓	↓
3304	÷	12	=	n.

3. **Solve.** We solve the equation by carrying out the division.

```
        2 7 5
  1 2 ) 3 3 0 4
        2 4
        ─────
          9 0
          8 4
          ─────
            6 4
            6 0
            ─────
              4
```

$$3304 \div 12 = n$$
$$275 \, R \, 4 = n$$

4. **Check.** We can check by multiplying the number of packages by 12 and adding the remainder, 4:

$$12 \cdot 275 = 3300,$$
$$3300 + 4 = 3304.$$

5. **State.** Thus, 275 twelve-roll packages of paper towels can be filled. There will be 4 rolls left over.

Do Exercise 8.

8. **Packages of Paper Towels.** The paper-products company in Example 6 also produces 6-roll packages. How many 6-roll packages can be filled with 2269 rolls of paper towels? How many rolls will be left over?

Answer

8. 378 packages with 1 roll left over

EXAMPLE 7 *Automobile Mileage.* The 2009 Toyota Matrix gets 21 miles to the gallon (mpg) in city driving. How many gallons will it use in 3843 mi of city driving?

Source: Toyota

1. **Familiarize.** We first make a drawing. We let g = the number of gallons of gasoline used in 3843 mi of city driving.

21 mi   21 mi   21 mi   • • •   21 mi

3843 mi to drive

2. **Translate.** Repeated addition applies here. Thus the following multiplication applies to the situation.

Number of miles per gallon	times	Number of gallons used	is	Number of miles driven
21	·	g	=	3843

3. **Solve.** To solve the equation, we divide by 21 on both sides.

$$21 \cdot g = 3843$$
$$\frac{21 \cdot g}{21} = \frac{3843}{21}$$
$$g = 183$$

$$
\begin{array}{r}
1\ 8\ 3 \\
2\ 1\ \overline{)\ 3\ 8\ 4\ 3} \\
2\ 1 \\
\overline{1\ 7\ 4} \\
1\ 6\ 8 \\
\overline{6\ 3} \\
6\ 3 \\
\overline{0}
\end{array}
$$

4. **Check.** To check, we multiply 183 by 21.

$$
\begin{array}{r}
1\ 8\ 3 \\
\times\ \ 2\ 1 \\
\hline
1\ 8\ 3 \\
3\ 6\ 6\ 0 \\
\hline
3\ 8\ 4\ 3
\end{array}
$$

The answer checks.

5. **State.** The Toyota Matrix will use 183 gal of gasoline.

Do Exercise 9.

9. Automobile Mileage. The 2009 Toyota Matrix gets 29 miles to the gallon (mpg) in highway driving. How many gallons will it use in 2291 mi of highway driving?

Source: Toyota

Multistep Problems

Sometimes we must use more than one operation to solve a problem, as in the following example.

EXAMPLE 8 *Weight Loss.* To lose one pound, you must burn about 3500 calories in excess of what you already burn doing your regular daily activities. The chart on the next page shows how long a person must engage in several types of exercise in order to burn 100 calories. For how long would a person have to run at a brisk pace in order to lose one pound?

Answer

9. 79 gal

1. **Familiarize.** This is a multistep problem. We begin by visualizing the situation.

ONE POUND 3500 CALORIES			
100 cal 8 min	100 cal 8 min	...	100 cal 8 min

- Run for 8 minutes at a brisk pace, or
- Swim for 2 minutes at a brisk pace, or
- Bicycle for 15 minutes at 9 mph, or
- Do aerobic exercises for 15 minutes, or
- Golf, walking, for 20 minutes, or
- Play tennis, singles, for 11 minutes

We will first find how many hundreds are in 3500. This will tell us how many times a person must run for 8 min in order to lose one pound. Then we will multiply to find the total number of minutes required for the weight loss.

We let x = the number of hundreds in 3500.

2. **Translate.** We translate to an equation. Repeated addition applies here, so we will multiply.

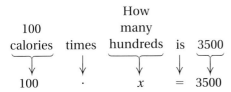

$$100 \cdot x = 3500$$

3. **Solve.** We divide by 100 on both sides of the equation.

$$100 \cdot x = 3500$$
$$\frac{100 \cdot x}{100} = \frac{3500}{100}$$
$$x = 35$$

$$\begin{array}{r} 3\ 5 \\ 1\ 0\ 0 \overline{)\ 3\ 5\ 0\ 0} \\ 3\ 0\ 0 \\ \hline 5\ 0\ 0 \\ 5\ 0\ 0 \\ \hline 0 \end{array}$$

We know that running for 8 min will burn 100 calories. This must be done 35 times in order to lose one pound. We let t = the time it takes to lose one pound. Thus we have the following.

$$t = 35 \times 8$$
$$t = 280$$

$$\begin{array}{r} 3\ 5 \\ \times\ \ \ 8 \\ \hline 2\ 8\ 0 \end{array}$$

4. **Check.** $280 \div 8 = 35$, so there are 35 8's in 280 min, and $35 \cdot 100 = 3500$, the number of calories that must be burned in order to lose one pound. The answer checks.

5. **State.** You must run for 280 min, or 4 hr 40 min, at a brisk pace in order to lose one pound.

Do Exercise 10.

10. **Weight Loss.** Use the information in Example 8 to determine how long an individual must swim at a brisk pace in order to lose one pound.

Answer

10. 70 min, or 1 hr 10 min

You will find it helpful to look for the words, phrases, and concepts in the table below as you familiarize yourself with applied problems. They will be useful when you translate the problems to equations.

KEYWORDS, PHRASES, AND CONCEPTS

ADDITION (+)	SUBTRACTION (−)	MULTIPLICATION (·)	DIVISION (÷)
add	subtract	multiply	divide
added to	subtracted from	multiplied by	divided by
sum	difference	product	quotient
total	minus	times	repeated subtraction
plus	less than	of	missing factor
more than	decreased by	repeated addition	finding equal quantities
increased by	take away	rectangular arrays	
	how much more		

The following tips will also be helpful in problem solving.

PROBLEM-SOLVING TIPS

1. Look for patterns when solving problems. Each time you study an example in the text, you may observe a pattern for problems found later in the exercise sets or in other practical situations.

2. When translating in mathematics, consider the dimensions of the variables and constants in the equation. The variables that represent length should all be in the same unit, those that represent money should all be in dollars or all in cents, and so on.

3. Make sure that units appear in the answer whenever appropriate and that you completely answer the original problem.

STUDY TIPS

SOLVING APPLIED PROBLEMS

Don't be discouraged if, at first, you find the exercises in this section to be more challenging than those in earlier sections or if you have had difficulty working applied problems in the past. Your skill will improve with each problem you solve. After you have done your homework for this section, you might want to work extra problems from the text or from the online tutorial that accompanies the text. As you gain experience solving applied problems, you will find yourself becoming comfortable with them.

Translating for Success

1. *Brick-Mason Expense.* A commercial contractor is building 30 two-unit condominiums in a retirement community. The brick-mason expense for each building is $10,860. What is the total cost of bricking the buildings?

2. *Heights.* Dean's sons are on the high school basketball team. Their heights are 73 in., 69 in., and 76 in. How much taller is the tallest son than the shortest son?

3. *Account Balance.* James has $423 in his checking account. Then he deposits $73 and uses his debit card for purchases of $76 and $69. How much is left in the account?

4. *Purchasing Camcorder.* A camcorder is on sale for $423. Jenny has only $69. How much more does she need to buy the camcorder?

5. *Purchasing Coffee Makers.* Sara purchases 8 coffee makers for the newly remodeled bed-and-breakfast hotel that she manages. If she pays $52 for each coffee maker, what is the total cost of her purchase?

The goal of these matching questions is to practice step (2), *Translate*, of the five-step problem-solving process. Translate each word problem to an equation and select a correct translation from equations A–O.

A. $8 \cdot 52 = n$

B. $69 \cdot n = 76$

C. $73 - 76 - 69 = n$

D. $423 + 73 - 76 - 69 = n$

E. $30 \cdot 10{,}860 = n$

F. $15 \cdot n = 195$

G. $69 + n = 423$

H. $n = 10{,}860 - 300$

I. $n = 423 \div 69$

J. $30 \cdot n = 10{,}860$

K. $15 \cdot 195 = n$

L. $n = 52 - 8$

M. $69 + n = 76$

N. $15 \div 195 = n$

O. $52 + n = 60$

Answers on page A-2

6. *Hourly Rate.* Miller Auto Repair charges $52 per hour for labor. Jackson Auto Care charges $60 per hour. How much more does Jackson charge than Miller?

7. *College Band.* A college band with 195 members marches in a 15-row formation in the homecoming halftime performance. How many members are in each row?

8. *Shoe Purchase.* A professional football team purchases 15 pairs of shoes at $195 a pair. What is the total cost of this purchase?

9. *Loan Payment.* Kendra's uncle loans her $10,860, interest free, to buy a car. The loan is to be paid off in 30 payments. How much is each payment?

10. *College Enrollment.* At the beginning of the fall term, the total enrollment in Lakeview Community College was 10,860. By the end of the first two weeks, 300 students had withdrawn. How many students were then enrolled?

Solve.

Chicago's Tallest Buildings. When construction is completed on the slender, 150-story Chicago Spire, it will be the tallest building in the United States, rising to a height of 2000 ft. The figure below shows the heights of the four tallest buildings in Chicago. (The Willis Tower was formerly known as the Sears Tower.) Use this information to do Exercises 1–4.

Chicago's Tallest Buildings

SOURCES: Chicago Historical Society; *USA Today* research

1. How much taller will the Chicago Spire be than the Willis Tower?

2. How much taller will the Chicago Spire be than the Trump Tower?

3. The Empire State Building in New York is 114 ft taller than the Aon Center. What is the height of the Empire State Building?

4. The Chicago Spire will be 638 ft taller than the Freedom Tower, which is under construction in New York. What is the height of the Freedom Tower?

5. *Caffeine Content.* An 8-oz serving of Red Bull energy drink contains 76 milligrams of caffeine. An 8-oz serving of brewed coffee contains 19 more milligrams of caffeine than the energy drink. How many milligrams of caffeine does the 8-oz serving of coffee contain?
Source: The Mayo Clinic

6. *Caffeine Content.* Hershey's 6-oz milk chocolate almond bar contains 25 milligrams of caffeine. A 20-oz bottle of Coca-Cola has 32 more milligrams of caffeine than the Hershey bar. How many milligrams of caffeine does the 20-oz bottle of Coca-Cola have?
Source: *National Geographic*, "Caffeine," by T. R. Reid, January 2005

7. A carpenter drills 216 holes in a rectangular array in a pegboard. There are 12 holes in each row. How many rows are there?

8. Lou arranges 504 entries on a spreadsheet in a rectangular array that has 36 rows. How many entries are in each row?

9. *TV Watching.* On average, people age 12 and older spent 1704 hr watching TV in 2008. This is 202 hr more than in 2000. Determine the number of hours people in this age group spent watching TV in 2000.

Source: U. S. Census Bureau

10. *Drilling Activity.* In 1988, there were 554 rotary rigs drilling for crude oil in the United States. This was 257 more rigs than were active in 2007. Find the number of active rotary oil rigs in 2007.

Source: Energy Information Administration

11. *Boundaries between Countries.* The boundary between mainland United States and Canada including the Great Lakes is 3987 mi long. The length of the boundary between the United States and Mexico is 1933 mi. How much longer is the Canadian border?

Source: U.S. Geological Survey

12. *Longest Rivers.* The longest river in the world is the Nile in Egypt at about 4135 mi. The longest river in the United States is the Missouri–Mississippi at about 3860 mi. How much longer is the Nile?

13. *Pixels.* A high-definition television (HDTV) screen consists of small rectangular dots called *pixels*. How many pixels are there on a screen that has 1080 rows with 1920 pixels in each row?

Pixel

14. *Crossword.* The *USA Today* crossword puzzle is a rectangle containing 15 rows with 15 squares in each row. How many squares does the puzzle have altogether?

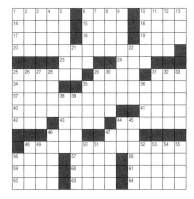

15. *Associate's Degrees.* About 697,000 associate's degrees were earned in the United States in 2005. Of these, 429,000 were earned by women. How many men earned associate's degrees in 2005?

16. *Bachelor's Degrees.* About 613,000 bachelor's degrees were earned by men in the United States in 2005. This is 215,000 fewer than the number of bachelor's degrees earned by women the same year. How many bachelor's degrees were earned by women in 2005?

17. There are 24 hr in a day and 7 days in a week. How many hours are there in a week?

18. There are 60 min in an hour and 24 hr in a day. How many minutes are there in a day?

Housing Costs.   The graph below shows the average monthly rent for a one-bedroom apartment in several cities in June 2008. Use this graph to do Exercises 19–24.

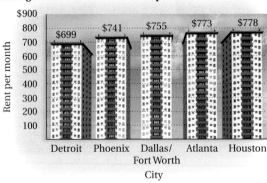

Average Rent for a One-Bedroom Apartment

SOURCE: Apartments.com

19. How much higher is the average monthly rent in Houston than in Dallas/Fort Worth?

20. How much lower is the average monthly rent in Phoenix than in Atlanta?

21. Phil, Scott, and Julio plan to rent a one-bedroom apartment in Detroit immediately after graduation, sharing the rent equally. What average monthly rent can each of them expect to pay?

22. Maria and her sister Theresa share a one-bedroom apartment in Houston, dividing the average monthly rent equally between them. How much does each pay?

23. On average, how much rent would a tenant pay for a one-bedroom apartment in Atlanta during a 12-month period?

24. On average, how much rent would a tenant pay for a one-bedroom apartment in Dallas/Fort Worth during a 6-month period?

25. *Colonial Population.*   Before the establishment of the U.S. Census in 1790, it was estimated that the Colonial population in 1780 was 2,780,400. This was an increase of 2,628,900 from the population in 1680. What was the Colonial population in 1680?

Source: *Time Almanac,* 2005

26. *Interstate Speed Limits.*   The speed limit for passenger cars on interstate highways in rural areas in Montana is 75 mph. This is 10 mph faster than the speed limit for trucks on the same roads. What is the speed limit for trucks?

27. *Motorcycle Sales.* Motorcycle sales totaled 710,000 in the United States in 2000. Sales in 2006 exceeded this total by 480,000. How many motorcycles were sold in 2006?

Source: Motorcycle Industry Council

28. *Summer Olympics.* There were 43 events in the first modern Olympic games in Athens, Greece, in 1896. There were 259 more events in the 2008 Summer Olympics in Beijing, China. How many events were there in 2008?

Sources: *USA Today* research; beijing2008.cn

29. *Yard-Sale Profit.* Ruth made $312 at her yard sale and divided the money equally among her four grandchildren. How much did each child receive?

30. *Paper Measures.* A quire of paper consists of 25 sheets, and a ream of paper consists of 500 sheets. How many quires are in a ream?

31. *Parking Rates.* The most expensive parking in the United States occurs in midtown New York City, where the average rate is $585 per month. This is $545 per month more than in the city with the least expensive rate, Bakersfield, California. What is the average monthly parking rate in Bakersfield?

Source: Colliers International

32. *Trade Balance.* In 2006, foreign visitors spent $422,594,000,000 traveling in the United States, while Americans spent $342,845,000,000 traveling abroad. How much more was spent by visitors to the United States than by Americans traveling abroad?

Source: U. S. Census Bureau

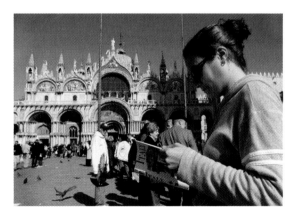

33. *Refrigerator Purchase.* Gourmet Deli has a chain of 24 restaurants. It buys a commercial refrigerator for each store at a cost of $1019 each. Determine the total cost of the purchase.

34. *Microwave Purchase.* Bridgeway College is constructing new dorms, in which each room has a small kitchen. It buys 96 microwave ovens at $88 each. Determine the total cost of the purchase.

35. *"Seinfeld."* A local television station plans to air the 177 episodes of the long-running comedy series "Seinfeld." If the station airs 5 episodes per week, how many full weeks will pass before it must begin re-airing previously shown episodes? How many unaired episodes will be shown the following week before the previously aired episodes are rerun?

36. *"Everybody Loves Raymond."* The popular television comedy series "Everybody Loves Raymond" had 208 scripted episodes and 2 additional episodes consisting of clips from previous shows. A local television station plans to air the 208 scripted episodes, showing 5 episodes per week. How many full weeks will pass before it must begin re-airing episodes? How many unaired episodes will be shown the following week before the previously aired episodes are rerun?

37. *Automobile Mileage.* The 2008 Hyundai Tucson GLS gets 25 miles to the gallon (mpg) in highway driving. How many gallons will it use in 5900 mi of highway driving?

Source: Hyundai

38. *Automobile Mileage.* The 2008 Volkswagen Jetta (5 cylinder) gets 21 miles to the gallon (mpg) in city driving. How many gallons will it use in 3465 mi of city driving?

Source: Volkswagen of America, Inc.

39. *Crossword.* The *Los Angeles Times* crossword puzzle is a rectangle containing 441 squares arranged in 21 rows. How many columns does the puzzle have?

40. *Mailing Labels.* A box of mailing labels contains 750 labels on 25 sheets. How many labels are on each sheet?

41. *High School Court.* The standard basketball court used by high school players has dimensions of 50 ft by 84 ft.
  a) What is its area?
  b) What is its perimeter?

42. *College Court.* The standard basketball court used by college players has dimensions of 50 ft by 94 ft.
  a) What is its area?
  b) What is its perimeter?
  c) How much greater is the area of a college court than a high school court? (See Exercise 41.)

43. *Loan Payments.* Dana borrows $5928 for a used car. The loan is to be paid off in 24 equal monthly payments. How much is each payment (excluding interest)?

44. *Home Improvement Loan.* The Van Reken family borrows $7824 to build a sunroom on the back of their home. The loan is to be paid off in equal monthly payments of $163 (excluding interest). How many months will it take to pay off the loan?

45. *Map Drawing.* A map has a scale of 215 mi to the inch. How far apart *in reality* are two cities that are 3 in. apart on the map? How far apart *on the map* are two cities that, in reality, are 1075 mi apart?

46. *Map Drawing.* A map has a scale of 288 mi to the inch. How far apart *on the map* are two cities that, in reality, are 2016 mi apart? How far apart *in reality* are two cities that are 8 in. apart on the map?

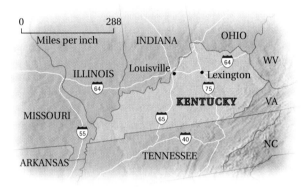

47. Copies of this book are usually shipped from the warehouse in cartons containing 24 books each. How many cartons are needed to ship 1344 books?

48. The H. J. Heinz Company ships 16-oz bottles of ketchup in cartons containing 12 bottles each. How many cartons are needed to ship 528 bottles of ketchup?

49. Elena buys 5 video games at $64 each and pays for them with $10 bills. How many $10 bills does it take?

50. Pedro buys 5 video games at $64 each and pays for them with $20 bills. How many $20 bills does it take?

51. The balance in Meg's bank account is $568. She uses her debit card for purchases of $46, $87, and $129. Then she deposits $94 in the account after returning a book. How much is left in her account?

52. The balance in Dylan's bank account is $749. He uses his debit card for purchases of $34 and $65. Then he makes a deposit of $123 from his paycheck. What is the new balance?

Refer to the information in Example 8 to do Exercises 53–56.

53. For how long must you do aerobic exercises in order to lose one pound?

54. For how long must you bicycle at 9 mph in order to lose one pound?

55. For how long must you play golf in order to lose one pound?

56. For how long must you play tennis singles in order to lose one pound?

57. *Seating Configuration.* The seats in the Boeing 737-500 airplanes in United Airlines' North American fleet are configured with 2 rows of 4 seats across in first class and 16 rows of 6 seats across in economy class. Determine the total seating capacity of the plane.
Source: United Airlines

Economy class: 16 rows of 6 seats   First class: 2 rows of 4 seats

58. *Seating Configuration.* The seats in the Airbus 320 airplanes in United Airlines' North American fleet are configured with 3 rows of 4 seats across in first class and 21 rows of 6 seats across in economy class. Determine the total seating capacity of the plane.
Source: United Airlines

Economy class: 21 rows of 6 seats   First class: 3 rows of 4 seats

59. *Bones in the Hands and Feet.* There are 27 bones in each human hand and 26 bones in each human foot. How many bones are there in all in the hands and feet?

60. An office for adjunct instructors at a community college has 6 bookshelves, each of which is 3 ft wide. The office is moved to a new location that has dimensions of 16 ft by 21 ft. Is it possible for the bookshelves to be put side by side on the 16-ft wall?

Skill Maintenance

Round 234,562 to the nearest: [1.6a]

61. Hundred.   **62.** Ten.   **63.** Thousand.

Estimate each sum or difference by rounding to the nearest thousand.   [1.6b]

64. $2783 + 4602 + 5797 + 8111$

65. $28,430 - 11,977$

66. $5800 - 2100$

67. $2100 + 5800$

Estimate each product by rounding to the nearest hundred.   [1.6b]

68. $787 \cdot 363$ **69.** $887 \cdot 799$ **70.** $10,362 \cdot 4531$

Synthesis

71. ▦ *Speed of Light.* Light travels about 186,000 miles per second (mi/sec) in a vacuum such as in outer space. In ice it travels about 142,000 mi/sec, and in glass it travels about 109,000 mi/sec. In 18 sec, how many more miles will light travel in a vacuum than in ice? than in glass?

72. Carney Community College has 1200 students. Each instructor teaches 4 classes and each student takes 5 classes. There are 30 students and 1 instructor in each classroom. How many instructors are there at Carney Community College?

1.9 Exponential Notation and Order of Operations

a Writing Exponential Notation

Consider the product $3 \cdot 3 \cdot 3 \cdot 3$. Such products occur often enough that mathematicians have found it convenient to create a shorter notation, called **exponential notation**, for them. For example,

$3 \cdot 3 \cdot 3 \cdot 3$ is shortened to $3^4 \leftarrow$ exponent

4 factors / base

We read exponential notation as follows.

NOTATION	WORD DESCRIPTION
3^4	"three to the fourth power," or "the fourth power of three"
5^3	"five-cubed," or "the cube of five," or "five to the third power," or "the third power of five"
7^2	"seven squared," or "the square of seven," or "seven to the second power," or "the second power of seven"

The wording "seven squared" for 7^2 is derived from the fact that a square with side s has area A given by $A = s^2$.

$A = s^2$, s, s

An expression like $3 \cdot 5^2$ is read "three times five squared," or "three times the square of five."

EXAMPLE 1 Write exponential notation for $10 \cdot 10 \cdot 10 \cdot 10 \cdot 10$.

Exponential notation is 10^5.

5 is the *exponent*.
10 is the *base*.

EXAMPLE 2 Write exponential notation for $2 \cdot 2 \cdot 2$.

Exponential notation is 2^3.

Do Exercises 1–4.

OBJECTIVES

a Write exponential notation for products such as $4 \cdot 4 \cdot 4$.

b Evaluate exponential notation.

c Simplify expressions using the rules for order of operations.

d Remove parentheses within parentheses.

Write exponential notation.

1. $5 \cdot 5 \cdot 5 \cdot 5$

2. $5 \cdot 5 \cdot 5 \cdot 5 \cdot 5 \cdot 5$

3. $10 \cdot 10$

4. $10 \cdot 10 \cdot 10 \cdot 10$

Answers

1. 5^4 2. 5^6 3. 10^2 4. 10^4

(b) Evaluating Exponential Notation

We evaluate exponential notation by rewriting it as a product and then computing the product.

EXAMPLE 3 Evaluate: 10^3.

$$10^3 = 10 \cdot 10 \cdot 10 = 1000$$

-------- *Caution!* --------

10^3 does not mean $10 \cdot 3$.

EXAMPLE 4 Evaluate: 5^4.

$$5^4 = 5 \cdot 5 \cdot 5 \cdot 5 = 625$$

Do Exercises 5–8.

Do Exercises 5–8.

Evaluate.

5. 10^4 **6.** 10^2

7. 8^3 **8.** 2^5

(c) Simplifying Expressions

Suppose we have a calculation like the following:

$$3 + 4 \cdot 8.$$

How do we find the answer? Do we add 3 to 4 and then multiply by 8, or do we multiply 4 by 8 and then add 3? In the first case, the answer is 56. In the second, the answer is 35. We agree to compute as in the second case:

$$3 + 4 \cdot 8 = 3 + 32 = 35.$$

Now consider the calculation

$$7 \cdot 14 - (12 + 18).$$

What do the parentheses mean? To deal with these questions, we must make some agreement regarding the order in which we perform operations. The rules are as follows.

Calculator Corner

Exponential Notation

Many calculators have a $\boxed{y^x}$ or $\boxed{\wedge}$ key for raising a base to a power. To find 16^3, for example, we press $\boxed{1}\ \boxed{6}\ \boxed{y^x}\ \boxed{3}\ \boxed{=}$ or $\boxed{1}\ \boxed{6}\ \boxed{\wedge}\ \boxed{3}\ \boxed{=}$. The result is 4096.

Exercises: Use a calculator to find each of the following.

1. 3^5

2. 5^6

3. 12^4

4. 2^{11}

RULES FOR ORDER OF OPERATIONS
1. Do all calculations within parentheses ( ), brackets [ ], or braces { } before operations outside.
2. Evaluate all exponential expressions.
3. Do all multiplications and divisions in order from left to right.
4. Do all additions and subtractions in order from left to right.

It is worth noting that these are the rules that computers and most scientific calculators use to do computations.

EXAMPLE 5 Simplify: $16 \div 8 \cdot 2$.

There are no parentheses or exponents, so we begin with the third step.

$$16 \div 8 \cdot 2 = 2 \cdot 2$$
$$= 4$$

Doing all multiplications and divisions in order from left to right

---

Answers

5. 10,000 6. 100 7. 512 8. 32

EXAMPLE 6 Simplify: $7 \cdot 14 - (12 + 18)$.

$$7 \cdot 14 - (12 + 18) = 7 \cdot 14 - 30 \qquad \text{Carrying out operations inside parentheses}$$
$$= 98 - 30 \qquad \text{Doing all multiplications and divisions}$$
$$= 68 \qquad \text{Doing all additions and subtractions}$$

Do Exercises 9–12.

Do Exercises 9–12.

Simplify.

9. $93 - 14 \cdot 3$

10. $104 \div 4 + 4$

11. $25 \cdot 26 - (56 + 10)$

12. $75 \div 5 + (83 - 14)$

EXAMPLE 7 Simplify and compare: $23 - (10 - 9)$ and $(23 - 10) - 9$.

We have

$$23 - (10 - 9) = 23 - 1 = 22;$$
$$(23 - 10) - 9 = 13 - 9 = 4.$$

We can see that $23 - (10 - 9)$ and $(23 - 10) - 9$ represent different numbers. Thus subtraction is not associative.

Do Exercises 13 and 14.

Simplify and compare.

13. $64 \div (32 \div 2)$ and $(64 \div 32) \div 2$

14. $(28 + 13) + 11$ and $28 + (13 + 11)$

EXAMPLE 8 Simplify: $7 \cdot 2 - (12 + 0) \div 3 - (5 - 2)$.

$$7 \cdot 2 - (12 + 0) \div 3 - (5 - 2)$$
$$= 7 \cdot 2 - 12 \div 3 - 3 \qquad \text{Carrying out operations inside parentheses}$$
$$= 14 - 4 - 3 \qquad \text{Doing all multiplications and divisions in order from left to right}$$
$$= 10 - 3$$
$$= 7 \qquad \text{Doing all additions and subtractions in order from left to right}$$

Do Exercise 15.

15. Simplify:

$$9 \times 4 - (20 + 4) \div 8 - (6 - 2).$$

EXAMPLE 9 Simplify: $15 \div 3 \cdot 2 \div (10 - 8)$.

$$15 \div 3 \cdot 2 \div (10 - 8)$$
$$= 15 \div 3 \cdot 2 \div 2 \qquad \text{Carrying out operations inside parentheses}$$
$$= 5 \cdot 2 \div 2$$
$$= 10 \div 2 \qquad \text{Doing all multiplications and divisions in order from left to right}$$
$$= 5$$

Do Exercises 16–18.

Simplify.

16. $5 \cdot 5 \cdot 5 + 26 \cdot 71 - (16 + 25 \cdot 3)$

17. $30 \div 5 \cdot 2 + 10 \cdot 20 + 8 \cdot 8 - 23$

18. $95 - 2 \cdot 2 \cdot 2 \cdot 5 \div (24 - 4)$

Answers

9. 51   **10.** 30   **11.** 584   **12.** 84   **13.** 4; 1
14. 52; 52   **15.** 29   **16.** 1880   **17.** 253
18. 93

EXAMPLE 10 Simplify: $4^2 \div (10 - 9 + 1)^3 \cdot 3 - 5$.

$$4^2 \div (10 - 9 + 1)^3 \cdot 3 - 5$$
$$= 4^2 \div (1 + 1)^3 \cdot 3 - 5 \qquad \text{Subtracting inside parentheses}$$
$$= 4^2 \div 2^3 \cdot 3 - 5 \qquad \text{Adding inside parentheses}$$
$$= 16 \div 8 \cdot 3 - 5 \qquad \text{Evaluating exponential expressions}$$
$$= 2 \cdot 3 - 5 \left.\vphantom{\begin{matrix}a\\b\end{matrix}}\right\} \quad \text{Doing all multiplications and}$$
$$= 6 - 5 \qquad\qquad\quad \text{divisions in order from left to right}$$
$$= 1 \qquad\qquad\qquad \text{Subtracting}$$

Do Exercises 19–21.

Simplify.
19. $5^3 + 26 \cdot 71 - (16 + 25 \cdot 3)$

20. $(1 + 3)^3 + 10 \cdot 20 + 8^2 - 23$

21. $81 - 3^2 \cdot 2 \div (12 - 9)$

EXAMPLE 11 Simplify: $2^9 \div 2^6 \cdot 2^3$.

$$2^9 \div 2^6 \cdot 2^3 = 512 \div 64 \cdot 8 \qquad \text{There are no parentheses. Evaluating exponential expressions}$$
$$= 8 \cdot 8 \left.\vphantom{\begin{matrix}a\\b\end{matrix}}\right\} \quad \text{Doing all multiplications and}$$
$$= 64 \qquad\quad \text{divisions in order from left to right}$$

Do Exercise 22.

22. Simplify: $2^3 \cdot 2^8 \div 2^9$.

Calculator Corner

Order of Operations To determine whether a calculator is programmed to follow the rules for order of operations, we can enter a simple calculation that requires using those rules. For example, we enter ⬚3⬚ ⬚+⬚ ⬚4⬚ ⬚×⬚ ⬚2⬚ ⬚=⬚. If the result is 11, we know that the rules for order of operations have been followed. That is, the multiplication $4 \times 2 = 8$ was performed first and then 3 was added to produce a result of 11. If the result is 14, we know that the calculator performs operations as they are entered rather than following the rules for order of operations. That means, in this case, that 3 and 4 were added first to get 7 and then that sum was multiplied by 2 to produce the result of 14. For such calculators, we would have to enter the operations in the order in which we want them performed. In this case, we would press ⬚4⬚ ⬚×⬚ ⬚2⬚ ⬚+⬚ ⬚3⬚ ⬚=⬚.

Many calculators have parenthesis keys that can be used to enter an expression containing parentheses. To enter $5(4 + 3)$, for example, we press ⬚5⬚ ⬚(⬚ ⬚4⬚ ⬚+⬚ ⬚3⬚ ⬚)⬚ ⬚=⬚. The result is 35.

Exercises: Simplify.

1. $84 - 5 \cdot 7$

2. $80 + 50 \div 10$

3. $3^2 + 9^2 \div 3$

4. $4^4 \div 64 - 4$

5. $15 \cdot 7 - (23 + 9)$

6. $(4 + 3)^2$

Averages

In order to find the average of a set of numbers, we use addition and then division. For example, the average of 2, 3, 6, and 9 is found as follows.

$$\text{Average} = \underbrace{\frac{2 + 3 + 6 + 9}{4}}_{\text{Divide by 4.}} = \frac{20}{4} = 5 \qquad \text{The number of addends is 4.}$$

The fraction bar acts as a pair of grouping symbols so

$$\frac{2 + 3 + 6 + 9}{4} \quad \text{is equivalent to} \quad (2 + 3 + 6 + 9) \div 4.$$

Thus we are using order of operations when we compute an average.

AVERAGE

The **average** of a set of numbers is the sum of the numbers divided by the number of addends.

EXAMPLE 12 *Average Height of Waterfalls.* The heights of the four highest waterfalls in the world are shown in the figure below. Determine the average height of the four falls.

World's Highest Waterfalls

SOURCE: World Almanac

The average is given by

$$\frac{3212 + 2425 + 2149 + 2014}{4} = \frac{9800}{4} = 2450.$$

The average height of the world's four highest waterfalls is 2450 ft.

Do Exercise 23.

d Removing Parentheses within Parentheses

When parentheses occur within parentheses, we can make them different shapes, such as [ ] (also called "brackets") and { } (also called "braces"). All of these have the same meaning. When parentheses occur within parentheses, computations in the innermost ones are to be done first.

23. Average Number of Career Hits. The numbers of career hits of five Hall of Fame baseball players are given in the bar graph below. Find the average number of career hits of all five.

Career Hits

SOURCES: Associated Press; Major League Baseball

Answer

23. 3118 hits

EXAMPLE 13 Simplify: $[25 - (4 + 3) \cdot 3] \div (11 - 7)$.

$$[25 - (4 + 3) \cdot 3] \div (11 - 7)$$

$= [25 - 7 \cdot 3] \div (11 - 7)$	Doing the calculations in the innermost parentheses first
$= [25 - 21] \div (11 - 7)$	Doing the multiplication in the brackets
$= 4 \div 4$	Subtracting
$= 1$	Dividing

EXAMPLE 14 Simplify: $16 \div 2 + \{40 - [13 - (4 + 2)]\}$.

$$16 \div 2 + \{40 - [13 - (4 + 2)]\}$$

$= 16 \div 2 + \{40 - [13 - 6]\}$	Doing the calculations in the innermost parentheses first
$= 16 \div 2 + \{40 - 7\}$	Again, doing the calculations in the innermost brackets
$= 16 \div 2 + 33$	Subtracting inside the braces
$= 8 + 33$	Doing all multiplications and divisions in order from left to right
$= 41$	Adding

Do Exercises 24 and 25.

Simplify.

24. $9 \times 5 + \{6 \div [14 - (5 + 3)]\}$

25. $[18 - (2 + 7) \div 3]$
$- (31 - 10 \times 2)$

Answers

24. 46 **25.** 4

a Write exponential notation.

1. $3 \cdot 3 \cdot 3 \cdot 3$

2. $2 \cdot 2 \cdot 2 \cdot 2 \cdot 2$

3. $5 \cdot 5$

4. $13 \cdot 13 \cdot 13$

5. $7 \cdot 7 \cdot 7 \cdot 7 \cdot 7$

6. $9 \cdot 9$

7. $10 \cdot 10 \cdot 10$

8. $1 \cdot 1 \cdot 1 \cdot 1$

b Evaluate.

9. 7^2

10. 5^3

11. 9^3

12. 8^2

13. 12^4

14. 10^5

15. 3^5

16. 2^6

c Simplify.

17. $12 + (6 + 4)$

18. $(12 + 6) + 18$

19. $52 - (40 - 8)$

20. $(52 - 40) - 8$

21. $1000 \div (100 \div 10)$

22. $(1000 \div 100) \div 10$

23. $(256 \div 64) \div 4$

24. $256 \div (64 \div 4)$

25. $(2 + 5)^2$

26. $2^2 + 5^2$

27. $(11 - 8)^2 - (18 - 16)^2$

28. $(32 - 27)^3 + (19 + 1)^3$

29. $16 \cdot 24 + 50$

30. $23 + 18 \cdot 20$

31. $83 - 7 \cdot 6$

32. $10 \cdot 7 - 4$

33. $10 \cdot 10 - 3 \cdot 4$

34. $90 - 5 \cdot 5 \cdot 2$

35. $4^3 \div 8 - 4$

36. $8^2 - 8 \cdot 2$

37. $17 \cdot 20 - (17 + 20)$

38. $1000 \div 25 - (15 + 5)$

39. $6 \cdot 10 - 4 \cdot 10$

40. $3 \cdot 8 + 5 \cdot 8$

41. $300 \div 5 + 10$

42. $144 \div 4 - 2$

43. $3 \cdot (2 + 8)^2 - 5 \cdot (4 - 3)^2$

44. $7 \cdot (10 - 3)^2 - 2 \cdot (3 + 1)^2$

45. $4^2 + 8^2 \div 2^2$

46. $6^2 - 3^4 \div 3^3$

47. $10^3 - 10 \cdot 6 - (4 + 5 \cdot 6)$

48. $7^2 + 20 \cdot 4 - (28 + 9 \cdot 2)$

49. $6 \cdot 11 - (7 + 3) \div 5 - (6 - 4)$

50. $8 \times 9 - (12 - 8) \div 4 - (10 - 7)$

51. $120 - 3^3 \cdot 4 \div (5 \cdot 6 - 6 \cdot 4)$

52. $80 - 2^4 \cdot 15 \div (7 \cdot 5 - 45 \div 3)$

53. $2^3 \cdot 2^8 \div 2^6$

54. $2^7 \div 2^5 \cdot 2^4 \div 2^2$

55. Find the average of $64, $97, and $121.

56. Find the average of four test grades of 86, 92, 80, and 78.

57. Find the average of 320, 128, 276, and 880.

58. Find the average of $1025, $775, $2062, $942, and $3721.

d Simplify.

59. $8 \times 13 + \{42 \div [18 - (6 + 5)]\}$

60. $72 \div 6 - \{2 \times [9 - (4 \times 2)]\}$

61. $[14 - (3 + 5) \div 2] - [18 \div (8 - 2)]$

62. $[92 \times (6 - 4) \div 8] + [7 \times (8 - 3)]$

63. $(82 - 14) \times [(10 + 45 \div 5) - (6 \cdot 6 - 5 \cdot 5)]$

64. $(18 \div 2) \cdot \{[(9 \cdot 9 - 1) \div 2] - [5 \cdot 20 - (7 \cdot 9 - 2)]\}$

65. $4 \times \{(200 - 50 \div 5) - [(35 \div 7) \cdot (35 \div 7) - 4 \times 3]\}$

66. $15(23 - 4 \cdot 2)^3 \div (3 \cdot 25)$

67. $\{[18 - 2 \cdot 6] - [40 \div (17 - 9)]\} + \{48 - 13 \times 3 + [(50 - 7 \cdot 5) + 2]\}$

68. $(19 - 2^4)^5 - (141 \div 47)^2$

Skill Maintenance

Solve. [1.7b]

69. $x + 341 = 793$

70. $4197 + x = 5032$

71. $7 \cdot x = 91$

72. $1554 = 42 \cdot y$

73. $3240 = y + 898$

74. $6000 = 1102 + t$

75. $25 \cdot t = 625$

76. $10,000 = 100 \cdot t$

Solve. [1.8a]

77. *Colorado.* The state of Colorado is roughly the shape of a rectangle that is 273 mi by 382 mi. What is its area?

78. On a long four-day trip, a family bought the following amounts of gasoline for their motor home:

23 gallons,   24 gallons,

26 gallons,   25 gallons.

How much gasoline did they buy in all?

Synthesis

Each of the answers in Exercises 79–81 is incorrect. First find the correct answer. Then place as many parentheses as needed in the expression in order to make the incorrect answer correct.

79. $1 + 5 \cdot 4 + 3 = 36$

80. $12 \div 4 + 2 \cdot 3 - 2 = 2$

81. $12 \div 4 + 2 \cdot 3 - 2 = 4$

82. Use one occurrence each of 1, 2, 3, 4, 5, 6, 7, 8, and 9 and any of the symbols $+$, $-$, $\cdot$, $\div$, and $(\ )$ to represent 100.

Key Terms and Properties

digit, p. 2
periods, p. 2
place-value chart, p. 2
whole numbers, p. 3
natural numbers, p. 3
sum, p. 9
addend, p. 9
additive identity, p. 10
perimeter, p. 11

minuend, p. 14
subtrahend, p. 14
difference, p. 14
factor, p. 19
product, p. 19
multiplicative identity, p. 20
dividend, p. 26
divisor, p. 26
quotient, p. 26

remainder, p. 28
rounding, p. 37
equation, p. 42
inequality, p. 42
solution of an equation, p. 48
exponential notation, p. 71
base, p. 71
exponent, p. 71

Associative Law of Addition: $\quad a + (b + c) = (a + b) + c$

Commutative Law of Addition: $\quad a + b = b + a$

Associative Law of Multiplication: $\quad a \cdot (b \cdot c) = (a \cdot b) \cdot c$

Commutative Law of Multiplication: $\quad a \cdot b = b \cdot a$

Concept Reinforcement

Determine whether each statement is true or false.

_____ **1.** $a > b$ is true when a is to the right of b on the number line.   [1.6c]

_____ **2.** Any nonzero number divided by itself is 1.   [1.5a]

_____ **3.** For any whole number a, $a \div 0 = 0$.   [1.5a]

_____ **4.** Every equation is true.   [1.7a]

_____ **5.** The rules for order of operations tell us to multiply and divide before adding and subtracting.   [1.9c]

_____ **6.** The average of three numbers is the middle number.   [1.9c]

Important Concepts

Objective 1.1a   Give the meaning of digits in standard notation.

Example   What does the digit 7 mean in 2,379,465?

   2,3 7 9,465

7 means 7 ten thousands.

Practice Exercise

  1. What does the digit 2 mean in 432,079?

Objective 1.2a   Add whole numbers.

Example   Add: 7368 + 3547.

$$
\begin{array}{r}
\overset{1\ \ 1}{7\ 3\ 6\ 8} \\
+\ 3\ 5\ 4\ 7 \\
\hline
1\ 0,9\ 1\ 5
\end{array}
$$

Practice Exercise

  2. Add: 36,047 + 29,255.

Objective 1.3a Subtract whole numbers.

Example Subtract: $8045 - 2897$.

$$
\begin{array}{r}
{}^{7}\!\!\!\not8\ {}^{9}\!\!\!\not0\ {}^{\overset{13}{\not3}}\!4\ {}^{15}\!\!\!\not5 \\
-\ 2\ 8\ 9\ 7 \\
\hline
5\ 1\ 4\ 8
\end{array}
$$

Practice Exercise

3. Subtract: $4805 - 1568$.

Objective 1.4a Multiply whole numbers.

Example Multiply: 57×315.

$$
\begin{array}{r}
{}^{1}\ {}^{\ \ 2}_{\ 3} \\
3\ 1\ 5 \\
\times\quad 5\ 7 \\
\hline
2\ 2\ 0\ 5 \leftarrow 315 \times 7 \\
1\ 5\ 7\ 5\ 0 \leftarrow 315 \times 50 \\
\hline
1\ 7{,}9\ 5\ 5
\end{array}
$$

Practice Exercise

4. Multiply: 329×684.

Objective 1.5a Divide whole numbers.

Example Divide: $6463 \div 26$.

$$
\begin{array}{r}
2\ 4\ 8 \\
26\ \overline{)\ 6\ 4\ 6\ 3} \\
5\ 2 \\
\hline
1\ 2\ 6 \\
1\ 0\ 4 \\
\hline
2\ 2\ 3 \\
2\ 0\ 8 \\
\hline
1\ 5
\end{array}
$$

The answer is 248 R 15.

Practice Exercise

5. Divide: $8519 \div 27$.

Objective 1.6a Round to the nearest ten, hundred, or thousand.

Example Round 6471 to the nearest hundred.

6 4 **7** 1
$\uparrow$

 The digit 4 is in the hundreds place. We consider the next digit to the right. Since the digit, 7, is 5 or higher, we round 4 hundreds up to 5 hundreds. Then we change all digits to the right of the hundreds digit to zeros. The answer is 6500.

Example Round to the nearest thousand.

6 **4** 7 1
$\uparrow$

 The digit 6 is in the thousands place. We consider the next digit to the right. Since the digit, 4, is 4 or lower, we round 6 thousands down, meaning that 6 thousands stays as 6 thousands. Change all digits to the right of the thousands digit to zeros. The answer is 6000.

Practice Exercises

6. Round 36,468 to the nearest hundred.

7. Round 36,468 to the nearest thousand.

Objective 1.6c   Use $<$ or $>$ for $\square$ to write a true sentence in a situation like 6 $\square$ 10.

Example   Use $<$ or $>$ for $\square$ to write a true sentence:
34 $\square$ 29.

    Since 34 is to the right of 29 on the number line,

      $34 > 29$.

Practice Exercise

8. Use $<$ or $>$ for $\square$ to write a true sentence:
78 $\square$ 81.

Objective 1.7b   Solve equations like $x + 28 = 54$, $28 \cdot x = 168$, and $98 \cdot 2 = y$.

Example   Solve: $y + 12 = 27$.

$$y + 12 = 27$$
$$y + 12 - 12 = 27 - 12$$
$$y + 0 = 15$$
$$y = 15$$

The solution is 15.

Practice Exercise

9. Solve: $24 \cdot x = 864$.

Objective 1.9b   Evaluate exponential notation.

Example   Evaluate: 5^4.

$$5^4 = 5 \cdot 5 \cdot 5 \cdot 5 = 625$$

Practice Exercise

10. Evaluate: 6^3.

Review Exercises

The review exercises that follow are for practice. Answers are given at the back of the book. If you miss an exercise, restudy the objective indicated in red next to the exercise or on the direction line that precedes it.

1. What does the digit 8 mean in 4,678,952?   [1.1a]

2. In 13,768,940, what digit tells the number of millions?   [1.1a]

Write expanded notation.   [1.1b]

3. 2793   **4.** 56,078

5. 4,007,101

Write a word name.   [1.1c]

6. 67,819   **7.** 2,781,427

8. 4,817,941, the number of vehicles sold globally by Toyota Motor Corporation in the first six months of 2008

Source: *Indianapolis Star*, July 24, 2008

Write standard notation.   [1.1c]

9. Four hundred seventy-six thousand, five hundred eighty-eight

10. *Subway Ridership.*   Ridership on the New York City Subway system totaled one billion, five hundred sixty-three million in 2007.

Source: Metropolitan Transit Authority

Add. [1.2a]

11. 7304 + 6968

12. 27,609 + 38,415

13. 2703 + 4125 + 6004 + 8956

14. $\begin{array}{r} 9\,1,4\,2\,6 \\ +\quad 7,4\,9\,5 \\ \hline \end{array}$

Subtract. [1.3a]

15. 8045 − 2897

16. 9001 − 7312

17. 6003 − 3729

18. $\begin{array}{r} 3\,7,4\,0\,5 \\ -\,1\,9,6\,4\,8 \\ \hline \end{array}$

Multiply. [1.4a]

19. 17,000 · 300

20. 7846 · 800

21. 726 · 698

22. 587 · 47

23. $\begin{array}{r} 8\,3\,0\,5 \\ \times\quad 6\,4\,2 \\ \hline \end{array}$

Divide. [1.5a]

24. 63 ÷ 5

25. 80 ÷ 16

26. $7\,\overline{)\,6\,3\,9\,4}$

27. 3073 ÷ 8

28. $6\,0\,\overline{)\,2\,8\,6}$

29. 4266 ÷ 79

30. $3\,8\,\overline{)\,1\,7,1\,7\,6}$

31. $1\,4\,\overline{)\,7\,0,1\,1\,2}$

32. 52,668 ÷ 12

Round 345,759 to the nearest: [1.6a]

33. Hundred.

34. Ten.

35. Thousand.

36. Hundred thousand.

Use < or > for ☐ to write a true sentence. [1.6c]

37. 67 ☐ 56

38. 1 ☐ 23

Estimate each sum, difference, or product by first rounding to the nearest hundred. Show your work. [1.6b]

39. 41,348 + 19,749

40. 38,652 − 24,549

41. 396 · 748

Solve. [1.7b]

42. 46 · n = 368

43. 47 + x = 92

44. 1 · y = 58

45. 24 = x + 24

46. Write exponential notation: 4 · 4 · 4. [1.9a]

Evaluate. [1.9b]

47. 10^4

48. 6^2

Simplify. [1.9c, d]

49. 8 · 6 + 17

50. 10 · 24 − (18 + 2) ÷ 4 − (9 − 7)

51. (80 ÷ 16) × [(20 − 56 ÷ 8) + (8 · 8 − 5 · 5)]

52. Find the average of 157, 170, and 168.

Solve. [1.8a]

53. *Computer Workstation.*   Natasha has $196 and wants to buy a computer workstation for $698. How much more does she need?

54. Toni has $406 in her checking account. She is paid $78 for a part-time job and deposits that in her checking account. How much is then in her account?

55. *Lincoln-Head Pennies.*   In 1909, the first Lincoln-head pennies were minted. Seventy-three years later, these pennies were first minted with a decreased copper content. In what year was the copper content reduced?

56. A beverage company packed 228 cans of soda into 12-can cartons. How many cartons did they fill?

57. An apartment builder bought 13 gas stoves at $425 each and 13 refrigerators at $620 each. What was the total cost?

58. An apple farmer keeps bees in her orchard to help pollinate the apple blossoms. The bees from an average beehive can pollinate 30 surrounding trees during one growing season. A farmer has 420 trees. How many beehives does she need to pollinate all of them?
Source: Jordan Orchards, Westminster, PA

59. *Olympic Trampoline.* Shown below is an Olympic trampoline. Determine the area and the perimeter of the trampoline. [1.2b], [1.4b]
Source: International Trampoline Industry Association, Inc.

60. A chemist has 2753 mL of alcohol. How many 20-mL beakers can be filled? How much will be left over?

61. A family budgeted $7825 for food and clothing and $2860 for entertainment. The yearly income of the family was $38,283. How much of this income remained after these two allotments?

62. Simplify: $7 + (4 + 3)^2$. [1.9c]
A. 32   B. 56
C. 151   D. 196

63. Simplify: $7 + 4^2 + 3^2$. [1.9c]
A. 32   B. 56
C. 130   D. 196

64. $[46 - (4 - 2) \cdot 5] \div 2 + 4$
A. 6   B. 20
C. 114   D. 22

Synthesis

65. ▦ Determine the missing digit d. [1.4a]

$$\begin{array}{r} 9\ d \\ \times\ \ d\ 2 \\ \hline 8\ 0\ 3\ 6 \end{array}$$

66. ▦ Determine the missing digits a and b. [1.5a]

$$\begin{array}{r} 9\ a\ 1 \\ 2\ b\ 1 \overline{)\ 2\ 3\ 6,4\ 2\ 1} \end{array}$$

67. A mining company estimates that a crew must tunnel 2000 ft into a mountain to reach a deposit of copper ore. Each day, the crew tunnels about 500 ft. Each night, about 200 ft of loose rocks roll back into the tunnel. How many days will it take the mining company to reach the copper deposit? [1.8a]

Understanding Through Discussion and Writing

1. Is subtraction associative? Why or why not? [1.3a]

2. Explain how estimating and rounding can be useful when shopping for groceries. [1.6b]

3. Write a problem for a classmate to solve. Design the problem so that the solution is "The driver still has 329 mi to travel." [1.8a]

4. Consider the expressions $9 - (4 \cdot 2)$ and $(3 \cdot 4)^2$. Are the parentheses necessary in each case? Explain. [1.9c]

CHAPTER
1

Test For Extra Help
CHAPTER
Test Prep
VIDEOS

Step-by-step test solutions are found on the Chapter Test Prep Videos available via the Video Resources on DVD, in *MyMathLab* , and on You Tube (search "BittingerBasicMath" and click on "Channels").

1. In the number 546,789, which digit tells the number of hundred thousands?

2. Write expanded notation: 8843.

3. Write a word name: 38,403,277.

Add.

4.
```
  6 8 1 1
+ 3 1 7 8
```

5.
```
  4 5,8 8 9
+ 1 7,9 0 2
```

6.
```
  1 2 3 9
    8 4 3
    3 0 1
+   7 8 2
```

7.
```
  6 2 0 3
+ 4 3 1 2
```

Subtract.

8.
```
  7 9 8 3
- 4 3 5 3
```

9.
```
  2 9 7 4
- 1 9 3 5
```

10.
```
  8 9 0 7
- 2 0 5 9
```

11.
```
  2 3,0 6 7
- 1 7,8 9 2
```

Multiply.

12.
```
  4 5 6 8
×       9
```

13.
```
  8 8 7 6
×     6 0 0
```

14.
```
    6 5
×   3 7
```

15.
```
    6 7 8
×   7 8 8
```

Divide.

16. $15 \div 4$

17. $420 \div 6$

18. $8\,9 \overline{)\, 8\,6\,3\,3}$

19. $4\,4 \overline{)\, 3\,5,4\,2\,8}$

Round 34,528 to the nearest:

20. Thousand.

21. Ten.

22. Hundred.

Estimate each sum, difference, or product by first rounding to the nearest hundred. Show your work.

23.
```
  2 3,6 4 9
+ 5 4,7 4 6
```

24.
```
  5 4,7 5 1
- 2 3,6 4 9
```

25.
```
    8 2 4
×   4 8 9
```

Use $<$ or $>$ for ☐ to write a true sentence.

26. 34 ☐ 17

27. 117 ☐ 157

Solve.

28. $28 + x = 74$

29. $169 \div 13 = n$

30. $38 \cdot y = 532$

31. $381 = 0 + a$

Solve.

32. *Calorie Content.* An 8-oz serving of whole milk contains 146 calories. This is 63 calories more than the number of calories in an 8-oz serving of skim milk. How many calories are in an 8-oz serving of skim milk?

Source: *American Journal of Clinical Nutrition*

33. A box contains 5000 staples. How many staplers can be filled from the box if each stapler holds 250 staples?

34. *Largest States.* The following table lists the five largest states in terms of their land area. Find the total land area of these states.

STATE	AREA (in square miles)
Alaska	571,951
Texas	261,797
California	155,959
Montana	145,552
New Mexico	121,356

SOURCES: U.S. Department of Commerce; U.S. Census Bureau

35. *Pool Tables.* The Bradford™ pool table made by Brunswick Billiards comes in three sizes of playing area, 50 in. by 100 in., 44 in. by 88 in., and 38 in. by 76 in.

Source: Brunswick Billiards

a) Determine the perimeter and the playing area of each table.

b) By how much does the area of the largest table exceed the area of the smallest table?

36. *Hostess Ding Dongs®.* Hostess packages its Ding Dong snack cakes in 12-packs. How many 12-packs can it fill with 22,231 cakes? How many will be left over?

37. *Office Supplies.* Morgan manages the office of a small graphics firm. He buys 3 black inkjet cartridges at $15 each and 2 photo inkjet cartridges at $25 each. How much does the purchase cost?

38. Write exponential notation: $12 \cdot 12 \cdot 12 \cdot 12$.

Evaluate.

39. 7^3

40. 10^5

Simplify.

41. $35 - 1 \cdot 28 \div 4 + 3$

42. $10^2 - 2^2 \div 2$

43. $(25 - 15) \div 5$

44. $2^4 + 24 \div 12$

45. $8 \times \{(20 - 11) \cdot [(12 + 48) \div 6 - (9 - 2)]\}$

46. Find the average of 97, 99, 87, and 89.

A. 93   **B.** 124   **C.** 186   **D.** 372

Synthesis

47. An open cardboard container is 8 in. wide, 12 in. long, and 6 in. high. How many square inches of cardboard are used?

48. Use trials to find the single-digit number a for which
$$359 - 46 + a \div 3 \times 25 - 7^2 = 339.$$

49. Cara spends $229 a month to repay her student loan. If she has already paid $9160 on the 10-yr loan, how many payments remain?

Fraction Notation: Multiplication and Division

Real-World Application

Of every 100 voters in the 2004 presidential election, 16 voters were in the 18–29 age group. In the 2008 presidential election, this number increased to 18.5 voters. What is the ratio of voters in the 18–29 age group to all voters in the 2004 presidential election? in the 2008 election?

Sources: The Center for Information & Research on Civic Learning & Engagement; www.civicyouth.org

This problem appears as Exercise 41 in Section 2.3.

2.1

Factorizations

OBJECTIVES

a Determine whether one number is a factor of another, and find the factors of a number.

b Find some multiples of a number, and determine whether a number is divisible by another.

c Given a number from 1 to 100, tell whether it is prime, composite, or neither.

d Find the prime factorization of a composite number.

SKILL TO REVIEW

Objective 1.5a: Divide whole numbers.

Divide.

1. $329 \div 8$

2. $2 \, 3 \overline{)\, 1 \, 0 \, 8 \, 1}$

In this chapter, we begin our work with fractions and fraction notation. Certain skills make such work easier. For example, in order to simplify $\frac{12}{32}$, it is important that we be able to *factor* 12 and 32, as follows:

$$\frac{12}{32} = \frac{4 \cdot 3}{4 \cdot 8}.$$

Then we "remove" a factor of 1:

$$\frac{4 \cdot 3}{4 \cdot 8} = \frac{4}{4} \cdot \frac{3}{8} = 1 \cdot \frac{3}{8} = \frac{3}{8}.$$

Thus factoring is an important skill in working with fractions.

a Factors and Factorization

In Sections 2.1 and 2.2, we consider only the **natural numbers** 1, 2, 3, and so on.

Let's look at the product $3 \cdot 4 = 12$. We say that 3 and 4 are **factors** of 12. When we divide 12 by 3, we get a remainder of 0. We say that the divisor 3 is a **factor** of the dividend 12.

FACTOR

- In the product $a \cdot b$, a and b are called **factors**.
- If we divide Q by d and get a remainder of 0, then the divisor d is a **factor** of the dividend Q.

EXAMPLE 1 Determine by long division **(a)** whether 6 is a factor of 198 and **(b)** whether 15 is a factor of 198.

a) $\begin{array}{r} 33 \\ 6\overline{)198} \\ 18 \\ \hline 18 \\ 18 \\ \hline 0 \end{array}$ ← Remainder is 0.

b) $\begin{array}{r} 13 \\ 15\overline{)198} \\ 15 \\ \hline 48 \\ 45 \\ \hline 3 \end{array}$ ← Not 0

The remainder is 0, so 6 is a factor of 198.

The remainder is not 0, so 15 is not a factor of 198.

Determine whether the second number is a factor of the first.

1. 72; 8   **2.** 2384; 28

Do Margin Exercises 1 and 2.

Consider $12 = 3 \cdot 4$. We say that $3 \cdot 4$ is a **factorization** of 12. Similarly, $6 \cdot 2$, $12 \cdot 1$, $2 \cdot 2 \cdot 3$, and $1 \cdot 3 \cdot 4$ are also factorizations of 12. Since $a = a \cdot 1$, every number has a factorization, and every number has factors. For some numbers, the factors consist of only the number itself and 1. For example, the only factorization of 17 is $17 \cdot 1$, so the only factors of 17 are 17 and 1.

Answers

Skill to Review:
1. 41 R 1   **2.** 47

Margin Exercises:
1. Yes   **2.** No

EXAMPLE 2   Find all the factors of 70.

We find as many "two-factor" factorizations as we can. We check sequentially the numbers 1, 2, 3, and so on, to see if we can form any factorizations:

70

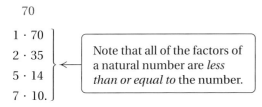

$$1 \cdot 70$$
$$2 \cdot 35$$
$$5 \cdot 14$$
$$7 \cdot 10.$$

Note that all of the factors of a natural number are *less than or equal to* the number.

Note that 3, 4, and 6 are not factors. If there are additional factors, they must be between 7 and 10. Since 8 and 9 are not factors, we are finished. The factors of 70 are 1, 2, 5, 7, 10, 14, 35, and 70.

Do Exercises 3–6.

Find all the factors of each number.

3. 10   **4.** 45

5. 62   **6.** 24

b   Multiples and Divisibility

A **multiple** of a natural number is a product of that number and some natural number. For example, some multiples of 2 are

2   (because $2 = 1 \cdot 2$);
4   (because $4 = 2 \cdot 2$);
6   (because $6 = 3 \cdot 2$);
8   (because $8 = 4 \cdot 2$).

Note that all but one of the multiples of a number are *larger* than the number.

We find multiples of 2 by counting by twos: 2, 4, 6, 8, and so on. We can find multiples of 3 by counting by threes: 3, 6, 9, 12, and so on.

EXAMPLE 3   Show that each of the numbers 8, 12, 20, and 36 is a multiple of 4.

$$8 = 2 \cdot 4 \qquad 12 = 3 \cdot 4 \qquad 20 = 5 \cdot 4 \qquad 36 = 9 \cdot 4$$

Do Exercises 7 and 8.

7. Show that each of the numbers 5, 45, and 100 is a multiple of 5.

8. Show that each of the numbers 10, 60, and 110 is a multiple of 10.

EXAMPLE 4   Multiply by 1, 2, 3, and so on, to find ten multiples of 7.

$$1 \cdot 7 = 7 \qquad 3 \cdot 7 = 21 \qquad 5 \cdot 7 = 35 \qquad 7 \cdot 7 = 49 \qquad 9 \cdot 7 = 63$$
$$2 \cdot 7 = 14 \qquad 4 \cdot 7 = 28 \qquad 6 \cdot 7 = 42 \qquad 8 \cdot 7 = 56 \qquad 10 \cdot 7 = 70$$

Do Exercise 9.

9. Multiply by 1, 2, 3, and so on, to find ten multiples of 5.

DIVISIBILITY

The number a is **divisible** by another number b if there exists a number c such that $a = b \cdot c$. The statements "a is **divisible** by b," "a is a **multiple** of b," and "b is a **factor** of a" all have the same meaning.

Thus, 27 is *divisible* by 3 because $27 = 3 \cdot 9$. We can also say that 27 is a *multiple* of 3, and 3 is a *factor* of 27.

EXAMPLE 5 Determine **(a)** whether 45 is divisible by 9 and **(b)** whether 45 is divisible by 4.

a) We divide 45 by 9 to determine whether 9 is a factor of 45.

$$\begin{array}{r} 5 \\ 9\overline{)45} \\ \underline{45} \\ 0 \end{array} \leftarrow \text{Remainder is 0.}$$

Because the remainder is 0, 45 is divisible by 9.

b) We divide 45 by 4 to determine whether 4 is a factor of 45.

$$\begin{array}{r} 11 \\ 4\overline{)45} \\ \underline{4} \\ 5 \\ \underline{4} \\ 1 \end{array} \leftarrow \text{Not 0}$$

Since the remainder is not 0, 45 is not divisible by 4.

10. Determine whether 16 is divisible by 2.

11. Determine whether 125 is divisible by 5.

12. Determine whether 125 is divisible by 6.

Do Exercises 10–12.

c **Prime and Composite Numbers**

> **PRIME AND COMPOSITE NUMBERS**
>
> - A natural number that has exactly two *different* factors, only itself and 1, is called a **prime number**.
> - The number 1 is *not* prime.
> - A natural number, other than 1, that is not prime is **composite**.

EXAMPLE 6 Determine whether the numbers 1, 2, 7, 8, 9, 11, 18, 27, 39, 43, 56, 59, and 77 are prime, composite, or neither.

The number 1 is not prime. It does not have *two* different factors.

The number 2 is prime. It has only the factors 2 and 1.

The numbers 7, 11, 43, and 59 are prime. Each has only two factors, itself and 1.

The number 8 is not prime. It has the factors 1, 2, 4, and 8 and is composite.

The numbers 9, 18, 27, 39, 56, and 77 are composite. Each has more than two factors.

Thus we have

Prime: 2, 7, 11, 43, 59;

Composite: 8, 9, 18, 27, 39, 56, 77;

Neither: 1.

The number 2 is the *only* even prime number. It is also the smallest prime number. The number 0 is neither prime nor composite, but 0 is *not* a natural number and thus is not considered here. We are considering only natural numbers.

Do Exercise 13.

13. Tell whether each number is prime, composite, or neither.

1, 2, 6, 12, 13, 19, 41, 65, 73, 99

Answers

10. Yes   **11.** Yes   **12.** No   **13.** 2, 13, 19, 41, 73 are prime; 6, 12, 65, 99 are composite; 1 is neither.

The table at right lists the prime numbers from 2 to 97. There are more extensive tables, but these prime numbers will be the most helpful to you in this text.

d Prime Factorizations

When we factor a composite number into a product of primes, we find the **prime factorization** of the number. To do this, we consider the primes

2, 3, 5, 7, 11, 13, 17, 19, 23, and so on,

and determine whether a given number is divisible by the primes.

EXAMPLE 7 Find the prime factorization of 39.

a) We divide by the first prime, 2.

$$\begin{array}{r} 19 \\ 2\overline{)39} \\ 2 \\ \overline{19} \\ 18 \\ \overline{1} \end{array}$$

Because the remainder is not 0, 2 is not a factor of 39, and 39 is not divisible by 2.

b) We divide by the next prime, 3.

$$\begin{array}{r} 13 \quad R = 0 \\ 3\overline{)39} \end{array}$$

The remainder is 0, so we know that $39 = 3 \cdot 13$. Because 13 is a prime, we are finished. The prime factorization is

$$39 = 3 \cdot 13.$$

EXAMPLE 8 Find the prime factorization of 220.

a) We divide by the first prime, 2.

$$\begin{array}{r} 110 \quad R = 0 \\ 2\overline{)220} \end{array} \qquad 220 = 2 \cdot 110$$

b) Because 110 is composite, we continue to divide, starting with 2 again.

$$\begin{array}{r} 55 \quad R = 0 \\ 2\overline{)110} \end{array} \qquad 220 = 2 \cdot 2 \cdot 55$$

c) Since 55 is composite and is not divisible by 2 or 3, we divide by the next prime, 5.

$$\begin{array}{r} 11 \quad R = 0 \\ 5\overline{)55} \end{array} \qquad 220 = 2 \cdot 2 \cdot 5 \cdot 11$$

Because 11 is prime, we are finished. The prime factorization is

$$220 = 2 \cdot 2 \cdot 5 \cdot 11.$$

We abbreviate our procedure as follows.

$$
\begin{array}{r}
11 \\
5\overline{)55} \\
2\overline{)110} \\
2\overline{)220}
\end{array}
$$

$$220 = 2 \cdot 2 \cdot 5 \cdot 11$$

Because multiplication is commutative, a factorization such as $2 \cdot 2 \cdot 5 \cdot 11$ could also be expressed as $5 \cdot 2 \cdot 2 \cdot 11$ or $2 \cdot 5 \cdot 11 \cdot 2$ (or, in exponential notation, as $2^2 \cdot 5 \cdot 11$ or $11 \cdot 2^2 \cdot 5$), but the prime factors are the same in each case. For this reason, we agree that any of these is "the" prime factorization of 220.

> Every number has just one (unique) prime factorization.

Caution!

Keep in mind the difference between finding all the factors of a number and finding the prime factorization. In Example 9, the prime factorization of 72 is $2 \cdot 2 \cdot 2 \cdot 3 \cdot 3$. The factors of 72 are 1, 2, 3, 4, 6, 8, 9, 12, 18, 24, 36, and 72.

EXAMPLE 9 Find the prime factorization of 72.

We can do divisions "up" as follows:

$$
\begin{array}{r}
3 \leftarrow \text{Prime quotient}\\
3\overline{)\ 9} \\
2\overline{)18} \\
2\overline{)36} \\
2\overline{)72} \leftarrow \text{Begin here}
\end{array}
$$

$$72 = 2 \cdot 2 \cdot 2 \cdot 3 \cdot 3$$

Another way to find the prime factorization of 72 uses a **factor tree** as follows. Begin by determining any factorization you can, and then continue factoring.

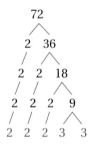

Find the prime factorization of each number.

14. 6   **15.** 12

16. 45   **17.** 98

18. 126   **19.** 144

20. 1960   **21.** 1925

EXAMPLE 10 Find the prime factorization of 189.

We can use a string of successive divisions or a factor tree. Since 189 is not divisible by 2, we begin with 3.

 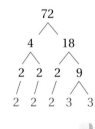

$$189 = 3 \cdot 3 \cdot 3 \cdot 7$$

Do Exercises 14–21.

Answers

14. $2 \cdot 3$ **15.** $2 \cdot 2 \cdot 3$ **16.** $3 \cdot 3 \cdot 5$
17. $2 \cdot 7 \cdot 7$ **18.** $2 \cdot 3 \cdot 3 \cdot 7$
19. $2 \cdot 2 \cdot 2 \cdot 2 \cdot 3 \cdot 3$ **20.** $2 \cdot 2 \cdot 2 \cdot 5 \cdot 7 \cdot 7$
21. $5 \cdot 5 \cdot 7 \cdot 11$

a   Determine whether the second number is a factor of the first.

1. 52; 14   **2.** 52; 13   **3.** 625; 25   **4.** 680; 16

Find all the factors of each number.

5. 18   **6.** 16   **7.** 54   **8.** 48

9. 4   **10.** 9   **11.** 1   **12.** 13

13. 98   **14.** 100   **15.** 255   **16.** 120

b   Multiply by 1, 2, 3, and so on, to find ten multiples of each number.

17. 4   **18.** 11   **19.** 20   **20.** 50

21. 3   **22.** 5   **23.** 12   **24.** 13

25. 10   **26.** 6   **27.** 9   **28.** 14

29. Determine whether 26 is divisible by 6.   **30.** Determine whether 48 is divisible by 8.

31. Determine whether 1880 is divisible by 8.   **32.** Determine whether 4227 is divisible by 3.

33. Determine whether 256 is divisible by 16.   **34.** Determine whether 102 is divisible by 4.

35. Determine whether 4227 is divisible by 9.   **36.** Determine whether 200 is divisible by 25.

37. Determine whether 8650 is divisible by 16.   **38.** Determine whether 4143 is divisible by 7.

c   Determine whether each number is prime, composite, or neither.

39. 1   **40.** 2   **41.** 9   **42.** 19

43. 11   **44.** 27   **45.** 29   **46.** 49

Find the prime factorization of each number.

47. 8   **48.** 16   **49.** 14   **50.** 15   **51.** 42

52. 32   **53.** 25   **54.** 40   **55.** 50   **56.** 62

57. 169   **58.** 140   **59.** 100   **60.** 110   **61.** 35

62. 70   **63.** 72   **64.** 86   **65.** 77   **66.** 99

67. 2884   **68.** 484   **69.** 51   **70.** 91   **71.** 1200

72. 1800   **73.** 273   **74.** 675   **75.** 1122   **76.** 6435

Skill Maintenance

Multiply. [1.4a]

77. $2 \cdot 13$ **78.** $8 \cdot 32$ **79.** $17 \cdot 25$ **80.** $25 \cdot 168$

Divide. [1.5a]

81. $0 \div 22$ **82.** $22 \div 1$ **83.** $22 \div 22$ **84.** $66 \div 22$

Solve. [1.8a]

85. *Morel Mushrooms.* During the spring, Kate's Country Market sold 43 pounds of fresh morel mushrooms. The mushrooms sold for $22 a pound. Find the total amount Kate took in from the sale of the mushrooms.

86. Sandy can type 62 words per minute. How long will it take her to type 12,462 words?

Synthesis

87. *Factors and Sums.* The top number in each column of the table below can be factored as a product of two numbers whose sum is the bottom number in the column. For example, in the first column, 56 has been factored as $7 \cdot 8$, and $7 + 8 = 15$. Fill in the blank spaces in the table.

PRODUCT	56	63	36	72	140	96		168	110			
FACTOR	7									9	24	3
FACTOR	8					8	8			10	18	
SUM	15	16	20	38	24	20	14		21			24

2.2 Divisibility

Suppose you are asked to find the simplest fraction notation for

$$\frac{117}{225}.$$

Since the numbers are quite large, you might feel that the task is difficult. However, both the numerator and the denominator are divisible by 9. If you knew this, you could factor and simplify quickly as follows:

$$\frac{117}{225} = \frac{9 \cdot 13}{9 \cdot 25} = \frac{9}{9} \cdot \frac{13}{25} = 1 \cdot \frac{13}{25} = \frac{13}{25}.$$

How did we know that both numbers have 9 as a factor? There is a simple test for determining this.

In this section, you learn quick ways to determine whether a number is divisible by 2, 3, 4, 5, 6, 8, 9, or 10. This will make simplifying fraction notation much easier.

OBJECTIVE

a Determine whether a number is divisible by 2, 3, 4, 5, 6, 8, 9, or 10.

SKILL TO REVIEW
Objective 1.2a: Add whole numbers.

Add.
  1. $9 + 8 + 2 + 1$
  2. $7 + 3 + 0 + 2 + 6$

a Rules for Divisibility

Divisibility by 2

You may already know the test for divisibility by 2.

> **BY 2**
>
> A number is **divisible by 2** (is *even*) if it has a ones digit of 0, 2, 4, 6, or 8 (that is, it has an even ones digit).

Let's see why this test works. Consider 354, which is

3 hundreds + 5 tens + 4.

Hundreds and tens are both multiples of 2. If the last digit is a multiple of 2, then the entire number is a multiple of 2.

EXAMPLES Determine whether each number is divisible by 2.

  1. 355   is *not* divisible by 2; 5 is *not* even.
  2. 4786   is divisible by 2; 6 is even.
  3. 8990   is divisible by 2; 0 is even.
  4. 4261   is *not* divisible by 2; 1 is *not* even.

Do Margin Exercises 1–4.

Determine whether each number is divisible by 2.

  1. 84   2. 59

  3. 998   4. 2225

Answers

Skill to Review:
1. 20   2. 18

Margin Exercises:
1. Yes   2. No   3. Yes   4. No

Divisibility by 3

> **BY 3**
>
> A number is **divisible by 3** if the sum of its digits is divisible by 3.

Let's illustrate why the test for divisibility by 3 works. Consider 852; since $852 = 3 \cdot 284$, 852 is divisible by 3.

$$852 = 8 \cdot 100 + 5 \cdot 10 + 2 \cdot 1$$
$$= 8(99 + 1) + 5(9 + 1) + 2(1)$$
$$= 8 \cdot 99 + 8 \cdot 1 + 5 \cdot 9 + 5 \cdot 1 + 2 \cdot 1 \qquad \text{Using the distributive law:}$$
$$a(b + c) = a \cdot b + a \cdot c$$

Since 99 and 9 are each a multiple of 3, we see that $8 \cdot 99$ and $5 \cdot 9$ are multiples of 3. This leaves $8 \cdot 1 + 5 \cdot 1 + 2 \cdot 1$, or $8 + 5 + 2$. If $8 + 5 + 2$, the sum of the digits, is divisible by 3, then 852 is divisible by 3.

EXAMPLES Determine whether each number is divisible by 3.

5. 18 $1 + 8 = 9$
6. 93 $9 + 3 = 12$ Each is divisible by 3 because the sum of its digits is divisible by 3.
7. 201 $2 + 0 + 1 = 3$

8. 256 $2 + 5 + 6 = 13$ The sum of the digits, 13, is *not* divisible by 3, so 256 is *not* divisible by 3.

Do Exercises 5–8.

Determine whether each number is divisible by 3.

5. 111   6. 1111

7. 309   8. 17,216

Divisibility by 6

A number divisible by 6 is a multiple of 6. But $6 = 2 \cdot 3$, so the number is also a multiple of 2 and 3. Thus we have the following.

> **BY 6**
>
> A number is **divisible by 6** if its ones digit is 0, 2, 4, 6, or 8 (is even) and the sum of its digits is divisible by 3.

EXAMPLES Determine whether each number is divisible by 6.

9. 720

Because 720 is even, it is divisible by 2. Also, $7 + 2 + 0 = 9$, so 720 is divisible by 3. Thus, 720 is divisible by 6.

$$720 \qquad 7 + 2 + 0 = 9$$
Even   Divisible by 3

10. 73

73 is *not* divisible by 6 because it is *not* even.

11. 256

Although 256 is even, it is *not* divisible by 6 because the sum of its digits, $2 + 5 + 6$, or 13, is *not* divisible by 3.

Do Exercises 9–12.

Determine whether each number is divisible by 6.

9. 420   10. 106

11. 321   12. 444

Answers

5. Yes   6. No   7. Yes   8. No   9. Yes
10. No   11. No   12. Yes

Divisibility by 9

The test for divisibility by 9 is similar to the test for divisibility by 3.

> **BY 9**
>
> A number is **divisible by 9** if the sum of its digits is divisible by 9.

EXAMPLES   Determine whether each number is divisible by 9.

12. 6984

Because $6 + 9 + 8 + 4 = 27$ and 27 is divisible by 9, 6984 is divisible by 9.

13. 322

Because $3 + 2 + 2 = 7$ and 7 is *not* divisible by 9, 322 is *not* divisible by 9.

Do Exercises 13–16.

Determine whether each number is divisible by 9.

13. 16   **14.** 117

15. 930   **16.** 29,223

Divisibility by 10

> **BY 10**
>
> A number is **divisible by 10** if its ones digit is 0.

We know that this test works because the product of 10 and *any* number has a ones digit of 0.

EXAMPLES   Determine whether each number is divisible by 10.

14. 3440   is divisible by 10 because the ones digit is 0.

15. 3447   is *not* divisible by 10 because the ones digit is not 0.

Do Exercises 17–20.

Determine whether each number is divisible by 10.

17. 305   **18.** 847

19. 300   **20.** 8760

Divisibility by 5

> **BY 5**
>
> A number is **divisible by 5** if its ones digit is 0 or 5.

EXAMPLES   Determine whether each number is divisible by 5.

16.   220   is divisible by 5 because the ones digit is 0.

17.   475   is divisible by 5 because the ones digit is 5.

18. 6514   is *not* divisible by 5 because the ones digit is neither 0 nor 5.

Do Exercises 21–24.

Determine whether each number is divisible by 5.

21. 5780   **22.** 3427

23. 34,678   **24.** 7775

Let's see why the test for 5 works. Consider 7830:

$$7830 = 10 \cdot 783 = 5 \cdot 2 \cdot 783.$$

Since 7830 is divisible by 10 and 5 is a factor of 10, 7830 is divisible by 5.

Answers

13. No   14. Yes   15. No   16. Yes
17. No   18. No   19. Yes   20. Yes
21. Yes   22. No   23. No   24. Yes

Consider 6734:

$$6734 = 673 \text{ tens} + 4.$$

Tens are multiples of 5, so the only number that must be checked is the ones digit. If the last digit is a multiple of 5, then the entire number is. In this case, 4 is *not* a multiple of 5, so 6734 is *not* divisible by 5.

Divisibility by 4

The test for divisibility by 4 is similar to the test for divisibility by 2.

> **BY 4**
>
> A number is **divisible by 4** if the number named by its last *two* digits is divisible by 4.

EXAMPLES Determine whether the number is divisible by 4.

19. 8212 is divisible by 4 because 12 is divisible by 4.

20. 5216 is divisible by 4 because 16 is divisible by 4.

21. 8211 is *not* divisible by 4 because 11 is *not* divisible by 4.

22. 7538 is *not* divisible by 4 because 38 is *not* divisible by 4.

Do Exercises 25–28.

To see why the test for divisibility by 4 works, consider 516:

$$516 = 5 \text{ hundreds} + 16.$$

Hundreds are multiples of 4. If the number named by the last two digits is a multiple of 4, then the entire number is a multiple of 4.

Divisibility by 8

The test for divisibility by 8 is an extension of the tests for divisibility by 2 and 4.

> **BY 8**
>
> A number is **divisible by 8** if the number named by its last *three* digits is divisible by 8.

EXAMPLES Determine whether the number is divisible by 8.

23.   5648 is divisible by 8 because 648 is divisible by 8.

24. 96,088 is divisible by 8 because 88 is divisible by 8.

25.   7324 is *not* divisible by 8 because 324 is *not* divisible by 8.

26. 13,420 is *not* divisible by 8 because 420 is *not* divisible by 8.

Do Exercises 29–32.

A Note about Divisibility by 7

There are several tests for divisibility by 7, but all of them are more complicated than simply dividing by 7. So if you want to test for divisibility by 7, simply divide by 7, either by hand or using a calculator.

Determine whether each number is divisible by 4.

25. 216   **26.** 217

27. 5862   **28.** 23,524

Determine whether each number is divisible by 8.

29. 7564   **30.** 7864

31. 17,560   **32.** 25,716

Answers

25. Yes   26. No   27. No   28. Yes
29. No   30. Yes   31. Yes   32. No

a To answer Exercises 1–8, consider the following numbers.

46	300	85	256
224	36	711	8064
19	45,270	13,251	1867
555	4444	254,765	21,568

1. Which of the above are divisible by 2?

2. Which of the above are divisible by 3?

3. Which of the above are divisible by 4?

4. Which of the above are divisible by 5?

5. Which of the above are divisible by 6?

6. Which of the above are divisible by 8?

7. Which of the above are divisible by 9?

8. Which of the above are divisible by 10?

To answer Exercises 9–16, consider the following numbers.

56	200	75	35
324	42	812	402
784	501	2345	111,111
55,555	3009	2001	1005

9. Which of the above are divisible by 3?

10. Which of the above are divisible by 2?

11. Which of the above are divisible by 5?

12. Which of the above are divisible by 4?

13. Which of the above are divisible by 8?

14. Which of the above are divisible by 6?

15. Which of the above are divisible by 10?

16. Which of the above are divisible by 9?

To answer Exercises 17–24, consider the following numbers.

305	313,332	876	64,000
1101	7624	1110	9990
13,205	111,126	5128	126,111

17. Which of the above are divisible by 2?

18. Which of the above are divisible by 3?

19. Which of the above are divisible by 6?

20. Which of the above are divisible by 5?

21. Which of the above are divisible by 9?

22. Which of the above are divisible by 8?

23. Which of the above are divisible by 10?

24. Which of the above are divisible by 4?

Skill Maintenance

Solve.   [1.7b]

25. $56 + x = 194$

26. $y + 124 = 263$

27. $3008 = x + 2134$

28. $18 \cdot t = 1008$

29. $24 \cdot m = 624$

30. $338 = a \cdot 26$

Divide.   [1.5a]

31. $2106 \div 9$

32. $4\,5\,\overline{)\,1\,8\,0,1\,3\,5}$

Solve.   [1.8a]

33. Marty's automobile has a 5-speed transmission and gets 33 mpg in city driving. How many gallons of gas will it use to travel 1485 mi of city driving?

34. There are 60 min in 1 hr. How many minutes are there in 72 hr?

Synthesis

Find the prime factorization of each number. Use divisibility tests where applicable.

35. 7800

36. 2520

37. 2772

38. 1998

39. ▦ Fill in the missing digits of the number

95, ☐ ☐ 8

so that it is divisible by 99.

40. A passenger in a taxicab asks for the driver's company number. The driver says abruptly, "Sure—you can have my number. Work it out: If you divide it by 2, 3, 4, 5, or 6, you will get a remainder of 1. If you divide it by 11, the remainder will be 0 and no driver has a company number that meets these requirements and is smaller than this one." Determine the number.

2.3   Fractions and Fraction Notation

The study of arithmetic begins with the set of whole numbers

0, 1, 2, 3, 4, 5, 6, 7, 8, 9, 10, 11, and so on.

But we also need to be able to use fractional parts of numbers such as halves, thirds, fourths, and so on. Here is an example.

The tread depth of an IRL Indy Car Series tire is $\frac{3}{32}$ in. Tires for a normal car have a tread depth of $\frac{10}{32}$ in. when new and are considered bald at $\frac{2}{32}$ in.

Sources: Indy500.com; *Consumer Reports*

OBJECTIVES

a Identify the numerator and the denominator of a fraction and write fraction notation for part of an object or part of a set of objects and as a ratio.

b Simplify fraction notation like n/n to 1, $0/n$ to 0, and $n/1$ to n.

a Fractions and the Real World

Numbers like those above and the ones below are written in **fraction notation**. The top number is called the **numerator** and the bottom number is called the **denominator**.

$$\frac{1}{2}, \frac{7}{8}, \frac{7}{4}, \frac{11}{19}, \frac{3}{32}$$

EXAMPLE 1 Identify the numerator and the denominator.

$$\frac{7}{8} \begin{array}{l} \leftarrow \text{Numerator} \\ \leftarrow \text{Denominator} \end{array}$$

Do Exercises 1–3.

Let's look at various situations that involve fractions.

Fractions as a Partition of an Object Divided into Equal Parts

Consider a candy bar divided into 5 equal sections. If you eat 2 sections, you have eaten $\frac{2}{5}$ of the candy bar.

The denominator 5 tells us the unit, $\frac{1}{5}$. The numerator 2 tells us the number of equal parts we are considering, 2.

For each fraction, identify the numerator and the denominator.

1. $\frac{3}{20}$ of people age 5 and older in Maryland speak a language other than English at home.
   Source: 2006 American Community Survey

2. In 2006, about $\frac{4}{25}$ of the motor vehicles produced in the world were produced in the United States.
   Source: Automotive News Data Center and R.L. Polk Marketing Systems GmbH

3. $\frac{19}{100}$ of the U.S. population in 2050 is projected to be foreign-born.
   Source: Pew Research Center

Answers

1. Numerator: 3; denominator: 20
2. Numerator: 4; denominator: 25
3. Numerator: 19; denominator: 100

EXAMPLE 2 What part is shaded?

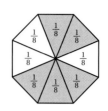

There are 8 equal parts. This tells us the unit, $\frac{1}{8}$. The *denominator* is 8. We have 5 of the units shaded. This tells us the *numerator*, 5. Thus,

$$\frac{5}{8} \quad \begin{matrix} \leftarrow 5 \text{ units are shaded.} \\ \leftarrow \text{The unit is } \frac{1}{8}. \end{matrix}$$

is shaded.

EXAMPLE 3 What part is shaded?

There are 18 equal parts. Thus the unit is $\frac{1}{18}$. The denominator is 18. We have 7 units shaded. This tells us the numerator, 7. Thus, $\frac{7}{18}$ is shaded.

Do Exercises 4–7.

The markings on a ruler use fractions.

EXAMPLE 4 What part of an inch is highlighted?

16 equal spaces

Each inch on the ruler shown above is divided into 16 equal parts. The highlighting extends to the 11th mark. Thus, $\frac{11}{16}$ of an inch is highlighted.

Do Exercise 8.

What part is shaded?

4.

5. 1 mile

6.

1 gallon

7.

8. What part of an inch is highlighted?

Answers

4. $\frac{5}{6}$ 5. $\frac{1}{3}$ 6. $\frac{3}{4}$ 7. $\frac{8}{15}$ 8. $\frac{15}{16}$

Fractions greater than or equal to 1, such as $\frac{24}{24}$, $\frac{10}{3}$, and $\frac{5}{4}$, correspond to situations like the following.

EXAMPLE 5 What part is shaded?

a)

The rectangle is divided into 24 equal parts. Thus the unit is $\frac{1}{24}$. The denominator is 24. All 24 equal parts are shaded. This tells us that the numerator is 24. Thus, $\frac{24}{24}$ is shaded.

b)

Each rectangle is divided into 6 parts. Thus the unit is $\frac{1}{6}$. The denominator is 6. We see that 11 of the equal units are shaded. This tells us that the numerator is 11. Thus, $\frac{11}{6}$ is shaded.

EXAMPLE 6 *Ice-Cream Roll-up Cake.* What part of an ice-cream roll-up cake is shaded?

3 ice cream roll-up cakes

Each cake is divided into 6 equal slices. The unit is $\frac{1}{6}$. The *denominator* is 6. We see that 13 of the slices are shaded. This tells us that the *numerator* is 13. Thus, $\frac{13}{6}$ are shaded.

> Do Exercises 9–11.

Fractions larger than or equal to 1, such as $\frac{13}{6}$ or $\frac{9}{9}$, are sometimes referred to as "improper" fractions. We will not use this terminology because notation such as $\frac{27}{8}$, $\frac{11}{3}$, and $\frac{4}{4}$ is quite "proper" and very common in algebra.

Fractions as Ratios

A **ratio** is a quotient of two quantities. We can express a ratio with fraction notation. (We will consider ratios in more detail in Chapter 5.)

What part is shaded?

9.

10.

11.

Answers

9. $\frac{15}{15}$ **10.** $\frac{10}{3}$ **11.** $\frac{7}{4}$

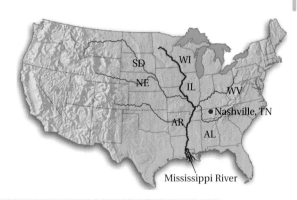

Mississippi River

12. What part of the set of states in Example 7 is north of Nashville, Tennessee?

EXAMPLE 7 *States East and West of the Mississippi River.* What part of this group of states is east of the Mississippi River? west of the Mississippi River?

Arkansas	Wisconsin
Illinois	West Virginia
Alabama	South Dakota
Nebraska	

There are 7 states in the set. We know that 4 of them, Illinois, Alabama, Wisconsin, and West Virginia, are east of the Mississippi River. Thus, 4 of 7, or $\frac{4}{7}$, are east of the Mississippi River. The 3 remaining states are west of the Mississippi River. Thus, $\frac{3}{7}$ are west of the Mississippi River.

Do Exercise 12.

EXAMPLE 8 *Baseball Standings.* The following are the final standings in the National League West for 2008, when the division was won by the Los Angeles Dodgers. Find the ratio of Dodger wins to losses, wins to total games, and losses to total games.

WEST	W	L	Pct.	GB	STRK	LAST 10	HOME	ROAD
LA Dodgers*	84	78	.519	—	L1	5-5	48-33	36-45
Arizona	82	80	.506	2	W3	6-4	48-33	34-47
Colorado	74	88	.457	10	W3	5-5	43-38	31-50
San Francisco	72	90	.444	12	W1	4-6	37-44	35-46
San Diego	63	99	.389	21	L1	5-5	35-46	28-53
*clinched division								

SOURCE: Major League Baseball

The Dodgers won 84 games and lost 78 games. They played a total of 84 + 78, or 162 games. Thus we have the following.

The ratio of wins to losses is $\frac{84}{78}$.

The ratio of wins to total games is $\frac{84}{162}$.

The ratio of losses to total games is $\frac{78}{162}$.

13. Baseball Standings. Refer to the table in Example 8. The Arizona Diamondbacks finished second in the National League West in 2008. Find the ratio of Diamondback wins to losses, wins to total games, and losses to total games.

Source: Major League Baseball

Do Exercise 13.

b Some Fraction Notation for Whole Numbers

Fraction Notation for 1

The number 1 corresponds to situations like those shown here. If we divide an object into *n* parts and take *n* of them, we get all of the object (1 whole object).

THE NUMBER 1 IN FRACTION NOTATION

$\frac{n}{n} = 1$, for any whole number *n* that is not 0.

Answers

12. $\frac{5}{7}$ 13. $\frac{82}{80}$; $\frac{82}{162}$; $\frac{80}{162}$

EXAMPLES Simplify.

9. $\dfrac{5}{5} = 1$ **10.** $\dfrac{9}{9} = 1$ **11.** $\dfrac{23}{23} = 1$

Fraction Notation for 0

Consider the fraction $\frac{0}{4}$. This corresponds to dividing an object into 4 parts and taking none of them. We get 0.

> **THE NUMBER 0 IN FRACTION NOTATION**
>
> $\dfrac{0}{n} = 0$, for any whole number n that is not 0.

EXAMPLES Simplify.

12. $\dfrac{0}{1} = 0$ **13.** $\dfrac{0}{9} = 0$ **14.** $\dfrac{0}{23} = 0$

Fraction notation with a denominator of 0, such as $n/0$, is meaningless because we cannot speak of an object being divided into *zero* parts. See also the discussion of excluding division by 0 in Section 1.5.

> **A DENOMINATOR OF 0**
>
> $\dfrac{n}{0}$ is not defined for any whole number n.

Do Exercises 20–25.

Other Whole Numbers

Consider the fraction $\frac{4}{1}$. This corresponds to taking 4 objects and dividing each into 1 part. (In other words, we do not divide them.) We have 4 objects.

> **ANY WHOLE NUMBER IN FRACTION NOTATION**
>
> Any whole number divided by 1 is the whole number. That is,
>
> $\dfrac{n}{1} = n$, for any whole number n.

EXAMPLES Simplify.

15. $\dfrac{2}{1} = 2$ **16.** $\dfrac{9}{1} = 9$ **17.** $\dfrac{34}{1} = 34$

Do Exercises 26–29.

Simplify.

14. $\dfrac{1}{1}$ **15.** $\dfrac{4}{4}$

16. $\dfrac{34}{34}$ **17.** $\dfrac{100}{100}$

18. $\dfrac{2347}{2347}$ **19.** $\dfrac{103}{103}$

Simplify, if possible.

20. $\dfrac{0}{1}$ **21.** $\dfrac{0}{8}$

22. $\dfrac{0}{107}$ **23.** $\dfrac{4-4}{567}$

24. $\dfrac{15}{0}$ **25.** $\dfrac{0}{3-3}$

Simplify.

26. $\dfrac{8}{1}$ **27.** $\dfrac{10}{1}$

28. $\dfrac{346}{1}$ **29.** $\dfrac{24-1}{23-22}$

Answers

14. 1   **15.** 1   **16.** 1   **17.** 1   **18.** 1
19. 1   **20.** 0   **21.** 0   **22.** 0   **23.** 0
24. Not defined   **25.** Not defined
26. 8   **27.** 10   **28.** 346   **29.** 23

2.3   Fractions and Fraction Notation   **105**

a Identify the numerator and the denominator.

1. $\dfrac{3}{4}$

2. $\dfrac{9}{10}$

3. $\dfrac{11}{2}$

4. $\dfrac{18}{5}$

5. $\dfrac{0}{7}$

6. $\dfrac{1}{13}$

What part of each object or set of objects is shaded?

7.

1 acre

8.

1 square inch

9.

1 yard

10.

1 mile

11.

12.

13.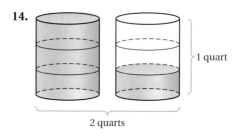

1 gold bar

2 gold bars

14.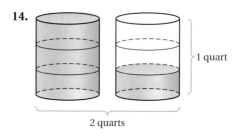

1 quart

2 quarts

15.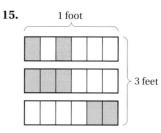

1 foot

3 feet

16.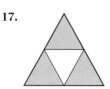

1 mile

2 miles

17.

18.

1 year

19.

20.

21.

22.

23.

24.

25.

26.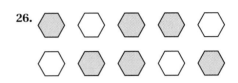

What part of an inch is highlighted?

27.

28.

29.

30.

For each of Exercises 31–34, give fraction notation for the amount of gas **(a)** in the tank and **(b)** used from a full tank.

31.

32.

33.

34.

35. For the following set of people, what is the ratio of:
  a) women to the total number of people?
  b) women to men?
  c) men to the total number of people?
  d) men to women?

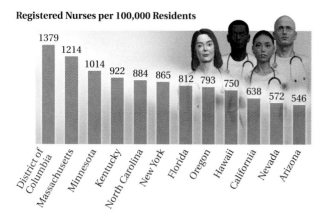

36. For the following set of nuts and bolts, what is the ratio of:
  a) nuts to bolts?
  b) bolts to nuts?
  c) nuts to the total number of elements?
  d) total number of elements to nuts?

For Exercises 37 and 38, use the following bar graph, which lists a state and the number of registered nurses per 100,000 residents in that state.

Registered Nurses per 100,000 Residents

District of Columbia 1379, Massachusetts 1214, Minnesota 1014, Kentucky 922, North Carolina 884, New York 865, Florida 812, Oregon 793, Hawaii 750, California 638, Nevada 572, Arizona 546

SOURCES: www.statehealthfacts.org; Kaiser Family Foundation, 2007

37. What is the ratio of registered nurses to 100,000 residents in the given state?
  a) Minnesota
  b) Hawaii
  c) Florida
  d) Kentucky
  e) New York
  f) District of Columbia

38. What is the ratio of registered nurses per 100,000 residents in the given state?
  a) North Carolina
  b) California
  c) Massachusetts
  d) Nevada
  e) Arizona
  f) Oregon

39. Bryce delivers car parts to auto service centers. On Thursday he had 15 deliveries scheduled. By noon he had delivered only 4 orders. What is the ratio of:

a) orders delivered to total number of orders?
b) orders delivered to orders not delivered?
c) orders not delivered to total number of orders?

40. *Gas Mileage.* A Mazda MX-5 will travel 308 mi on 11 gal of gasoline in highway driving. What is the ratio of:

a) miles driven to gasoline used?
b) gasoline used to miles driven?

Source: *Road & Track*, August 2008, p. 52

41. *18–29 Voter Age Group.* Of every 100 voters in the 2004 presidential election, 16 voters were in the 18–29 age group. In the 2008 presidential election, this number increased to 18.5 voters. What is the ratio of voters in the 18–29 age group to all voters in the 2004 presidential election? in the 2008 election?

Sources: The Center for Information & Research on Civic Learning & Engagement; www.civicyouth.org

42. *Air-Conditioned Houses.* In 1971, of every 1000 new houses in the United States, 360 houses had air conditioning. In 2006, this ratio rose to 890 houses of every 1000 new houses. What is the ratio of new houses with air conditioning to all new houses in 1971? in 2006?

Source: U.S. Census Bureau

b Simplify.

43. $\dfrac{0}{8}$

44. $\dfrac{8}{8}$

45. $\dfrac{8-1}{9-8}$

46. $\dfrac{16}{1}$

47. $\dfrac{20}{20}$

48. $\dfrac{20}{1}$

49. $\dfrac{45}{45}$

50. $\dfrac{11-1}{10-9}$

51. $\dfrac{0}{238}$

52. $\dfrac{238}{1}$

53. $\dfrac{238}{238}$

54. $\dfrac{0}{16}$

55. $\dfrac{3}{3}$

56. $\dfrac{56}{56}$

57. $\dfrac{87}{87}$

58. $\dfrac{98}{98}$

59. $\dfrac{18}{18}$ **60.** $\dfrac{0}{18}$ **61.** $\dfrac{18}{1}$ **62.** $\dfrac{8-8}{1247}$

63. $\dfrac{729}{0}$ **64.** $\dfrac{1317}{0}$ **65.** $\dfrac{5}{6-6}$ **66.** $\dfrac{13}{10-10}$

Skill Maintenance

Round 34,562 to the nearest: [1.6a]

67. Ten.   **68.** Hundred.   **69.** Thousand.   **70.** Ten thousand.

Solve.   [1.8a]

71. *New York Road Runners.* The New York Road Runners, a group that runs the New York City Marathon, had only 22,190 members in 1983. The membership increased to 46,655 in 2008. How many more members were there in 2008 than in 1983?

Source: New York Road Runners

72. *Gas Mileage.* The 2009 Chevrolet Corvette Z51 gets 26 mpg in highway driving. How many gallons will it use in 1846 mi of highway driving?

Source: *Road & Track*, August 2008, p. 52

Subtract.   [1.3a]

73. $9001 - 6798$ **74.** $2037 - 1189$ **75.** $67{,}113 - 29{,}874$ **76.** $12{,}327 - 476$

Synthesis

What part of each object is shaded?

77.   **78.**   **79.**   **80.**

Shade or mark each figure to show $\frac{3}{5}$.

81.   **82.**   **83.**   **84.**

2.4 Multiplication and Applications

a Multiplication Using Fraction Notation

Let's visualize the product of two fractions. We consider the multiplication

$$\frac{3}{5} \cdot \frac{3}{4}.$$

This is equivalent to finding $\frac{3}{5}$ of $\frac{3}{4}$. We first consider an object and take $\frac{3}{4}$ of it. We divide the object into 4 equal parts using vertical lines and take 3 of them. That is shown by the shading below.

Next, we take $\frac{3}{5}$ of the shaded area. We divide it into 5 equal parts using horizontal lines and take 3 of them. That is shown by the darker shading below.

The entire object has now been divided into 20 parts, and we have shaded 9 of them twice. Thus we see that $\frac{3}{5}$ of $\frac{3}{4}$ is $\frac{9}{20}$, or

$$\frac{3}{5} \cdot \frac{3}{4} = \frac{9}{20}.$$

The figure above shows a rectangular array inside a rectangular array. The number of pieces in the entire array is $5 \cdot 4$ (the product of the denominators). The number of pieces shaded a second time is $3 \cdot 3$ (the product of the numerators). The product is represented by 9 pieces out of a set of 20, or $\frac{9}{20}$, which is the product of the numerators over the product of the denominators. This leads us to a statement of the procedure for multiplying a fraction by a fraction.

Do Margin Exercise 1.

1. Draw a diagram like the one at left to show the multiplication $\frac{2}{3} \cdot \frac{4}{5}$.

Answers

Skill to Review:
1. 408   2. 65

Margin Exercise:
1.

We find a product such as $\frac{9}{7} \cdot \frac{3}{4}$ as follows.

To multiply a fraction by a fraction,

a) multiply the numerators to get the new numerator, and

b) multiply the denominators to get the new denominator.

$$\frac{9}{7} \cdot \frac{3}{4} = \frac{9 \cdot 3}{7 \cdot 4} = \frac{27}{28}$$

EXAMPLES Multiply.

1. $\frac{5}{6} \times \frac{7}{4} = \frac{5 \times 7}{6 \times 4} = \frac{35}{24}$

Skip writing this step whenever you can.

2. $\frac{3}{5} \cdot \frac{7}{8} = \frac{3 \cdot 7}{5 \cdot 8} = \frac{21}{40}$

3. $\frac{5}{3} \cdot \frac{4}{3} = \frac{20}{9}$

4. $\frac{1}{4} \cdot \frac{1}{3} = \frac{1}{12}$

Do Exercises 2–5.

Multiply.

2. $\frac{3}{8} \cdot \frac{5}{7}$

3. $\frac{4}{3} \times \frac{8}{5}$

4. $\frac{3}{10} \cdot \frac{1}{10}$

5. $\frac{5}{2} \cdot \frac{9}{4}$

Multiplication by a Whole Number

When multiplying a fraction by a whole number, we first express the whole number in fraction notation. We find a product such as $6 \cdot \frac{4}{5}$ as follows:

$$6 \cdot \frac{4}{5} = \frac{6}{1} \cdot \frac{4}{5} \qquad 6 = \frac{6}{1}$$

$$= \frac{6 \cdot 4}{1 \cdot 5} \qquad \text{Multiplying}$$

$$= \frac{24}{5}.$$

EXAMPLES Multiply.

5. $5 \times \frac{3}{8} = \frac{5}{1} \times \frac{3}{8} = \frac{5 \times 3}{1 \times 8} = \frac{15}{8}$

6. $\frac{2}{7} \cdot 13 = \frac{2}{7} \cdot \frac{13}{1} = \frac{2 \cdot 13}{7 \cdot 1} = \frac{26}{7}$

Do Exercises 6 and 7.

Multiply.

6. $5 \times \frac{2}{3}$

7. $\frac{3}{8} \cdot 11$

Answers

2. $\frac{15}{56}$ 3. $\frac{32}{15}$ 4. $\frac{3}{100}$

5. $\frac{45}{8}$ 6. $\frac{10}{3}$ 7. $\frac{33}{8}$

b Applications and Problem Solving

Many problems that can be solved by multiplying fractions can be thought of in terms of rectangular arrays.

EXAMPLE 7   A real estate developer owns a plot of land and plans to use $\frac{4}{5}$ of the plot for a small strip mall and parking lot. Of this, $\frac{2}{3}$ will be needed for the parking lot. What part of the plot will be used for parking?

1. **Familiarize.**   We first make a drawing to help familiarize ourselves with the problem. The land may not be rectangular, but we can think of it as a rectangle. The strip mall, including the parking lot, uses $\frac{4}{5}$ of the plot. We shade $\frac{4}{5}$ as shown on the left below. The parking lot alone uses $\frac{2}{3}$ of the part we just shaded. We shade that as shown on the right below.

2. **Translate.**   We let n = the part of the plot that is used for parking. We are taking "two-thirds of four-fifths." Recall from Section 1.8 that the word "of" corresponds to multiplication. Thus the following multiplication sentence corresponds to the situation:

$$\frac{2}{3} \cdot \frac{4}{5} = n.$$

3. **Solve.**   The number sentence tells us what to do. We multiply:

$$\frac{2}{3} \cdot \frac{4}{5} = \frac{2 \cdot 4}{3 \cdot 5} = \frac{8}{15}.$$

Thus, $\dfrac{8}{15} = n.$

4. **Check.**   We can do a partial check by noting that the answer is a fraction less than 1, which we expect since the developer is using only part of the original plot of land. Thus, $\frac{8}{15}$ is a reasonable answer. We can also check this in the figure above, where we see that 8 of 15 parts represent the parking lot.

5. **State.**   The parking lot takes up $\frac{8}{15}$ of the plot of land.

> Do Exercise 8.

8. A resort hotel uses $\frac{3}{4}$ of its extra land for recreational purposes. Of that, $\frac{1}{2}$ is used for swimming pools. What part of the land is used for swimming pools?

Answer

8. $\dfrac{3}{8}$

EXAMPLE 8 *Area of a Cornfield.* The length of a rectangular cornfield is $\frac{15}{16}$ mi. The width is $\frac{5}{8}$ mi. What is the area of the cornfield?

1. **Familiarize.** Recall that area is length times width. We make a drawing and let A = the area of the cornfield.

2. **Translate.** Then we translate:

$$
\begin{array}{ccccc}
\text{Area} & \text{is} & \text{Length} & \text{times} & \text{Width} \\
\downarrow & \downarrow & \downarrow & \downarrow & \downarrow \\
A & = & \dfrac{15}{16} & \times & \dfrac{5}{8}.
\end{array}
$$

3. **Solve.** The sentence tells us what to do. We multiply:

$$A = \frac{15}{16} \cdot \frac{5}{8} = \frac{15 \cdot 5}{16 \cdot 8} = \frac{75}{128}.$$

4. **Check.** We check by repeating the calculation. This is left to the student.

5. **State.** The area is $\frac{75}{128}$ square mile (mi²).

Do Exercise 9.

EXAMPLE 9 A recipe for oatmeal chocolate chip cookies calls for $\frac{3}{4}$ cup of brown sugar. Monica is making $\frac{1}{2}$ of the recipe. How much brown sugar should she use?

1. **Familiarize.** We first make a drawing or at least visualize the situation. We let n = the amount of brown sugar that Monica should use.

$\frac{3}{4}$ cup in the recipe $\frac{1}{2} \cdot \frac{3}{4}$ cup in $\frac{1}{2}$ the recipe

2. **Translate.** We are finding $\frac{1}{2}$ of $\frac{3}{4}$, so the multiplication sentence $\frac{1}{2} \cdot \frac{3}{4} = n$ corresponds to the situation.

3. **Solve.** We carry out the multiplication:

$$\frac{1}{2} \cdot \frac{3}{4} = \frac{1 \cdot 3}{2 \cdot 4} = \frac{3}{8}.$$

Thus, $\frac{3}{8} = n$.

4. **Check.** We check by repeating the calculation. This is left to the student.

5. **State.** Monica should use $\frac{3}{8}$ cup of brown sugar.

Do Exercise 10.

9. Area of a Ceramic Tile. The length of a rectangular ceramic tile on an inlaid ceramic counter is $\frac{4}{9}$ ft. The width is $\frac{2}{9}$ ft. What is the area of one tile?

10. Of the students at Overton Junior College, $\frac{1}{8}$ participate in sports and $\frac{3}{5}$ of these play football. What fractional part of the students play football?

Answers

9. $\frac{8}{81}$ ft²   **10.** $\frac{3}{40}$

Multiply.

1. $\dfrac{2}{5} \cdot \dfrac{2}{3}$

2. $\dfrac{3}{4} \cdot \dfrac{3}{5}$

3. $10 \cdot \dfrac{7}{9}$

4. $9 \cdot \dfrac{5}{8}$

5. $\dfrac{7}{8} \cdot \dfrac{7}{8}$

6. $\dfrac{4}{5} \cdot \dfrac{4}{5}$

7. $\dfrac{2}{3} \times \dfrac{1}{5}$

8. $\dfrac{3}{5} \times \dfrac{1}{5}$

9. $\dfrac{8}{7} \cdot \dfrac{5}{3}$

10. $\dfrac{11}{2} \cdot \dfrac{9}{8}$

11. $\dfrac{2}{5} \cdot 3$

12. $\dfrac{3}{5} \cdot 4$

13. $\dfrac{1}{2} \cdot \dfrac{1}{3}$

14. $\dfrac{1}{6} \cdot \dfrac{1}{4}$

15. $17 \times \dfrac{5}{6}$

16. $\dfrac{3}{7} \cdot 40$

17. $\dfrac{1}{10} \cdot \dfrac{7}{10}$

18. $\dfrac{3}{10} \cdot \dfrac{7}{100}$

19. $\dfrac{2}{5} \cdot 1$

20. $2 \cdot \dfrac{1}{3}$

21. $\dfrac{2}{3} \cdot \dfrac{7}{13}$

22. $\dfrac{3}{11} \cdot \dfrac{4}{5}$

23. $5 \times \dfrac{1}{8}$

24. $4 \times \dfrac{1}{5}$

25. $\dfrac{1}{4} \times \dfrac{1}{10}$

26. $\dfrac{21}{4} \cdot \dfrac{7}{5}$

27. $\dfrac{8}{3} \cdot \dfrac{20}{9}$

28. $\dfrac{1}{3} \times \dfrac{1}{10}$

29. $\dfrac{14}{15} \cdot \dfrac{13}{19}$

30. $\dfrac{12}{13} \cdot \dfrac{12}{13}$

31. $\dfrac{3}{4} \cdot \dfrac{3}{4}$

32. $\dfrac{3}{7} \cdot \dfrac{4}{5}$

33. $\dfrac{2}{11} \cdot 4$

34. $\dfrac{2}{5} \cdot 3$

b Solve.

35. *Hair Bows.* It takes $\frac{5}{3}$ yd of ribbon to make a hair bow. How much ribbon is needed to make 8 bows?

36. A gasoline can holds $\frac{5}{2}$ gal. How much will the can hold when it is $\frac{1}{2}$ full?

37. *Basketball: High School to Pro.* One of 35 high school basketball players plays college basketball. One of 75 college players plays professional basketball. What fractional part of high school basketball players play professional basketball?

Source: National Basketball Association

38. *Football: High School to Pro.* One of 42 high school football players plays college football. One of 85 college players plays professional football. What fractional part of high school football players play professional football?

Source: National Football League

39. *Serving of Cheesecake.* At the Cheesecake Factory, a piece of cheesecake is $\frac{1}{12}$ of a cheesecake. How much of the cheesecake is $\frac{1}{2}$ piece?

Source: The Cheesecake Factory

40. *Tossed Salad.* The recipe for a tossed salad calls for $\frac{3}{4}$ cup of sliced almonds. How much is needed to make $\frac{1}{2}$ of the recipe?

41. *Floor Tiling.* The floor of a room is being covered with tile. An area $\frac{3}{5}$ of the length and $\frac{3}{4}$ of the width is covered. What fraction of the floor has been tiled?

42. A rectangular table top measures $\frac{4}{5}$ m long by $\frac{3}{5}$ m wide. What is its area?

Skill Maintenance

Divide. [1.5a]

43. $7140 \div 35$

44. $32{,}200 \div 46$

45. $9\overline{)27{,}009}$

46. $35\overline{)7148}$

What does the digit 8 mean in each number? [1.1a]

47. 4,678,952

48. 8,473,901

49. 7148

50. 23,803

Simplify. [1.9c]

51. $8 \cdot 12 - (63 \div 9 + 13 \cdot 3)$

52. $(10 - 3)^4 + 10^3 \cdot 4 - 10 \div 5$

Synthesis

Multiply. Write the answer using fraction notation.

53. ▦ $\frac{341}{517} \cdot \frac{209}{349}$

54. ▦ $\left(\frac{57}{61}\right)^3$

55. ▦ $\left(\frac{2}{5}\right)^3 \left(\frac{7}{9}\right)$

56. ▦ $\left(\frac{1}{2}\right)^5 \left(\frac{3}{5}\right)$

2.5 Simplifying

a Multiplying by 1

Recall the following:

$$1 = \frac{1}{1} = \frac{2}{2} = \frac{3}{3} = \frac{4}{4} = \frac{10}{10} = \frac{45}{45} = \frac{100}{100} = \frac{n}{n}.$$

$$1 = \qquad = \qquad = \qquad =$$
$$1 \qquad\quad \frac{2}{2} \qquad\quad \frac{3}{3} \qquad\quad \frac{4}{4}$$

Any nonzero number divided by itself is 1. (See Section 1.5.)

Now recall the multiplicative identity from Section 1.4. For any whole number a, $1 \cdot a = a \cdot 1 = a$. This holds for fractions as well.

MULTIPLICATIVE IDENTITY FOR FRACTIONS

When we multiply a number by 1, we get the same number:

$$\frac{3}{5} = \frac{3}{5} \cdot 1 = \frac{3}{5} \cdot \frac{4}{4} = \frac{12}{20}.$$

Since $\frac{3}{5} = \frac{12}{20}$, we know that $\frac{3}{5}$ and $\frac{12}{20}$ are two names for the same number. We also say that $\frac{3}{5}$ and $\frac{12}{20}$ are **equivalent fractions**.

$$\frac{3}{5} = \frac{12}{20}$$

Do Margin Exercises 1–4.

Suppose we want to find a name for $\frac{2}{3}$, but one that has a denominator of 9. We can multiply by 1 to find equivalent fractions. Since $9 = 3 \cdot 3$, we choose $\frac{3}{3}$ for 1 in order to get a denominator of 9:

$$\frac{2}{3} = \frac{2}{3} \cdot 1 = \frac{2}{3} \cdot \frac{3}{3} = \frac{2 \cdot 3}{3 \cdot 3} = \frac{6}{9}.$$

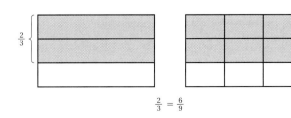

$$\frac{2}{3} = \frac{6}{9}$$

Rest of page — sidebar

OBJECTIVES

a Multiply a number by 1 to find fraction notation with a specified denominator.

b Simplify fraction notation.

c Use the test for equality to determine whether two fractions name the same number.

SKILL TO REVIEW
Objective 2.1d: Find the prime factorization of a composite number.

Find the prime factorization.
1. 84   2. 2250

Multiply.

1. $\dfrac{1}{2} \cdot \dfrac{8}{8}$ 2. $\dfrac{3}{5} \cdot \dfrac{10}{10}$

3. $\dfrac{13}{25} \cdot \dfrac{4}{4}$ 4. $\dfrac{8}{3} \cdot \dfrac{25}{25}$

Answers

Skill to Review:
1. $2 \cdot 2 \cdot 3 \cdot 7$ 2. $2 \cdot 3 \cdot 3 \cdot 5 \cdot 5 \cdot 5$

Margin Exercises:
1. $\dfrac{8}{16}$ 2. $\dfrac{30}{50}$ 3. $\dfrac{52}{100}$ 4. $\dfrac{200}{75}$

2.5 Simplifying   **117**

EXAMPLE 1 Find a name for $\frac{2}{5}$ with a denominator of 35.

Since $5 \cdot 7 = 35$, we multiply by $\frac{7}{7}$:

$$\frac{2}{5} = \frac{2}{5} \cdot \frac{7}{7} = \frac{2 \cdot 7}{5 \cdot 7} = \frac{14}{35}.$$

We say that $\frac{2}{5}$ and $\frac{14}{35}$ represent the same number. They are equivalent.

EXAMPLE 2 Find a name for $\frac{1}{4}$ with a denominator of 24.

Since $4 \cdot 6 = 24$, we multiply by $\frac{6}{6}$:

$$\frac{1}{4} = \frac{1}{4} \cdot \frac{6}{6} = \frac{1 \cdot 6}{4 \cdot 6} = \frac{6}{24}.$$

The numbers $\frac{1}{4}$ and $\frac{6}{24}$ are equivalent.

Do Exercises 5–8.

Find another name for each number, but with the denominator indicated. Use multiplying by 1.

5. $\frac{4}{3} = \frac{?}{9}$

6. $\frac{8}{7} = \frac{?}{49}$

7. $\frac{9}{10} = \frac{?}{100}$

8. $\frac{3}{15} = \frac{?}{45}$

(b) Simplifying Fraction Notation

All of the following are names for three-fourths:

$$\frac{3}{4}, \frac{6}{8}, \frac{9}{12}, \frac{12}{16}, \frac{15}{20}.$$

We say that $\frac{3}{4}$ is **simplest** because it has the smallest numerator and the smallest denominator. That is, the numerator and the denominator have no common factor other than 1.

To simplify, we reverse the process of multiplying by 1:

$$\frac{12}{18} = \frac{2 \cdot 6}{3 \cdot 6} \quad \begin{array}{l} \leftarrow \text{Factoring the numerator} \\ \leftarrow \text{Factoring the denominator} \end{array}$$

$$= \frac{2}{3} \cdot \frac{6}{6} \qquad \text{Factoring the fraction}$$

$$= \frac{2}{3} \cdot 1 \qquad \frac{6}{6} = 1$$

$$= \frac{2}{3}. \qquad \text{Removing a factor of 1: } \frac{2}{3} \cdot 1 = \frac{2}{3}$$

EXAMPLES Simplify.

3. $\dfrac{8}{20} = \dfrac{2 \cdot 4}{5 \cdot 4} = \dfrac{2}{5} \cdot \dfrac{4}{4} = \dfrac{2}{5}$

4. $\dfrac{2}{6} = \dfrac{1 \cdot 2}{3 \cdot 2} = \dfrac{1}{3} \cdot \dfrac{2}{2} = \dfrac{1}{3}$

> The number 1 allows for pairing of factors in the numerator and the denominator.

Do Exercises 9–12.

Simplify.

9. $\dfrac{2}{8}$

10. $\dfrac{24}{18}$

11. $\dfrac{10}{12}$

12. $\dfrac{15}{80}$

Answers

5. $\dfrac{12}{9}$ **6.** $\dfrac{56}{49}$ **7.** $\dfrac{90}{100}$ **8.** $\dfrac{9}{45}$ **9.** $\dfrac{1}{4}$

10. $\dfrac{4}{3}$ **11.** $\dfrac{5}{6}$ **12.** $\dfrac{3}{16}$

The use of prime factorizations can be helpful for simplifying when numerators and/or denominators are large numbers.

EXAMPLE 5 Simplify: $\dfrac{90}{84}$.

$$\dfrac{90}{84} = \dfrac{2 \cdot 3 \cdot 3 \cdot 5}{2 \cdot 2 \cdot 3 \cdot 7} \qquad \text{Factoring the numerator and the denominator into primes}$$

$$= \dfrac{2 \cdot 3 \cdot 3 \cdot 5}{2 \cdot 3 \cdot 2 \cdot 7} \qquad \text{Changing the order so that like primes are above and below each other}$$

$$= \dfrac{2}{2} \cdot \dfrac{3}{3} \cdot \dfrac{3 \cdot 5}{2 \cdot 7} \qquad \text{Factoring the fraction}$$

$$= 1 \cdot 1 \cdot \dfrac{3 \cdot 5}{2 \cdot 7}$$

$$= \dfrac{3 \cdot 5}{2 \cdot 7} \qquad \text{Removing factors of 1}$$

$$= \dfrac{15}{14}$$

The tests for divisibility (Section 2.2) are very helpful in simplifying fraction notation. We could have shortened the preceding example had we noted that 6 is a factor of both the numerator and the denominator. Then we have

$$\dfrac{90}{84} = \dfrac{6 \cdot 15}{6 \cdot 14} = \dfrac{6}{6} \cdot \dfrac{15}{14} = \dfrac{15}{14}.$$

EXAMPLE 6 Simplify: $\dfrac{603}{207}$.

At first glance this looks difficult. But note, using the test for divisibility by 9 (sum of digits is divisible by 9), that both the numerator and the denominator are divisible by 9. Thus we write both numbers with a factor of 9:

$$\dfrac{603}{207} = \dfrac{9 \cdot 67}{9 \cdot 23} = \dfrac{9}{9} \cdot \dfrac{67}{23} = \dfrac{67}{23}.$$

EXAMPLE 7 Simplify: $\dfrac{660}{1140}$.

Using the tests for divisibility, we have

$$\dfrac{660}{1140} = \dfrac{10 \cdot 66}{10 \cdot 114} = \dfrac{10}{10} \cdot \dfrac{66}{114} = \dfrac{66}{114} \qquad \text{Both 660 and 1140 are divisible by 10.}$$

$$= \dfrac{6 \cdot 11}{6 \cdot 19} = \dfrac{6}{6} \cdot \dfrac{11}{19} = \dfrac{11}{19}. \qquad \text{Both 66 and 114 are divisible by 6.}$$

Do Exercises 13–19.

Simplify.

13. $\dfrac{35}{40}$ **14.** $\dfrac{801}{702}$

15. $\dfrac{24}{21}$ **16.** $\dfrac{75}{300}$

17. $\dfrac{280}{960}$ **18.** $\dfrac{1332}{2880}$

19. Simplify each fraction in this circle graph.

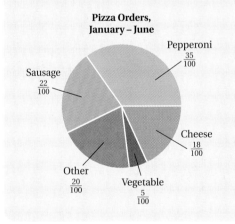

Pizza Orders, January – June

Pepperoni $\dfrac{35}{100}$

Cheese $\dfrac{18}{100}$

Vegetable $\dfrac{5}{100}$

Other $\dfrac{20}{100}$

Sausage $\dfrac{22}{100}$

Canceling

Canceling is a shortcut that you may have used for removing a factor of 1 when working with fraction notation. With *great* concern, we mention it as a possibility for speeding up your work. Canceling may be done only when removing common factors in numerators and denominators. Each such pair allows us to remove a factor of 1 in a fraction.

Our concern is that canceling be done with care and understanding. In effect, slashes are used to indicate factors of 1 that have been removed. For instance, Example 5 might have been done faster as follows:

$$\frac{90}{84} = \frac{2 \cdot 3 \cdot 3 \cdot 5}{2 \cdot 2 \cdot 3 \cdot 7} \qquad \text{Factoring the numerator and the denominator}$$

$$= \frac{2 \cdot \cancel{3} \cdot 3 \cdot 5}{2 \cdot 2 \cdot \cancel{3} \cdot 7} \qquad \begin{array}{l}\text{When a factor of 1 is noted,} \\ \text{it is "canceled" as shown: } \dfrac{2}{2} \cdot \dfrac{3}{3} = 1.\end{array}$$

$$= \frac{3 \cdot 5}{2 \cdot 7} = \frac{15}{14}.$$

-------------------------------------- *Caution!* --------------------------------------

The difficulty with canceling is that it is often applied incorrectly in situations like the following:

$$\frac{2 + 3}{2} = 3; \qquad \frac{4 + 1}{4 + 2} = \frac{1}{2}; \qquad \frac{15}{54} = \frac{1}{4}.$$

   Wrong!   Wrong!   Wrong!

The correct answers are

$$\frac{2 + 3}{2} = \frac{5}{2}; \qquad \frac{4 + 1}{4 + 2} = \frac{5}{6}; \qquad \frac{15}{54} = \frac{3 \cdot 5}{3 \cdot 18} = \frac{3}{3} \cdot \frac{5}{18} = \frac{5}{18}.$$

In each situation, the number canceled was not a factor of 1. Factors are parts of products. For example, in $2 \cdot 3$, 2 and 3 are factors, but in $2 + 3$, 2 and 3 are *not* factors. Canceling may not be done when sums or differences are in numerators or denominators, as shown here. **If you cannot factor, you cannot cancel! If in doubt, do not cancel!**

----

c A Test for Equality

When denominators are the same, we say that fractions have a **common denominator**. When fractions have a common denominator, we can compare them by comparing numerators. Suppose we want to compare $\frac{3}{6}$ and $\frac{2}{4}$. First, we find a common denominator. To do this, we multiply each fraction by 1, using the denominator of the other fraction to form the symbol for 1. We multiply $\frac{3}{6}$ by $\frac{4}{4}$ and $\frac{2}{4}$ by $\frac{6}{6}$:

$$\frac{3}{6} = \frac{3}{6} \cdot \frac{4}{4} = \frac{3 \cdot 4}{6 \cdot 4} = \frac{12}{24}; \qquad \text{Multiplying by } \frac{4}{4}$$

$$\frac{2}{4} = \frac{2}{4} \cdot \frac{6}{6} = \frac{2 \cdot 6}{4 \cdot 6} = \frac{12}{24}. \qquad \text{Multiplying by } \frac{6}{6}$$

Once we have a common denominator, 24, we compare the numerators. And since these numerators are both 12, the fractions are equal:

$$\frac{3}{6} = \frac{2}{4}.$$

Note in the preceding that if

$$\frac{3}{6} = \frac{2}{4}, \quad \text{then} \quad 3 \cdot 4 = 6 \cdot 2.$$

This tells us that we need to check only the products $3 \cdot 4$ and $6 \cdot 2$ to compare the fractions.

A TEST FOR EQUALITY

We multiply these two numbers: $3 \cdot 4$.

We multiply these two numbers: $6 \cdot 2$.

$$\frac{3}{6} \;\square\; \frac{2}{4}$$

We call $3 \cdot 4$ and $6 \cdot 2$ **cross products**. Since the cross products are the same—that is, $3 \cdot 4 = 6 \cdot 2$—we know that

$$\frac{3}{6} = \frac{2}{4}.$$

If a sentence $a = b$ is true, it means that a and b name the same number. If a sentence $a \neq b$ (read "a is not equal to b") is true, it means that a and b do *not* name the same number.

EXAMPLE 8 Use $=$ or $\neq$ for $\square$ to write a true sentence:

$$\frac{6}{7} \;\square\; \frac{7}{8}.$$

We multiply these two numbers: $6 \cdot 8 = 48$.

We multiply these two numbers: $7 \cdot 7 = 49$.

$$\frac{6}{7} \;\square\; \frac{7}{8}$$

Because $48 \neq 49$, $\frac{6}{7}$ and $\frac{7}{8}$ do not name the same number. Thus,

$$\frac{6}{7} \neq \frac{7}{8}.$$

EXAMPLE 9 Use $=$ or $\neq$ for $\square$ to write a true sentence:

$$\frac{6}{10} \;\square\; \frac{3}{5}.$$

We multiply these two numbers: $6 \cdot 5 = 30$.

We multiply these two numbers: $10 \cdot 3 = 30$.

$$\frac{6}{10} \;\square\; \frac{3}{5}.$$

Because the cross products are the same, we have

$$\frac{6}{10} = \frac{3}{5}.$$

Do Exercises 20 and 21.

Use $=$ or $\neq$ for $\square$ to write a true sentence.

20. $\frac{2}{6} \;\square\; \frac{3}{9}$ **21.** $\frac{2}{3} \;\square\; \frac{14}{20}$

Answers

20. $=$ **21.** $\neq$

a Find another name for the given number, but with the denominator indicated. Use multiplying by 1.

1. $\dfrac{1}{2} = \dfrac{?}{10}$

2. $\dfrac{1}{6} = \dfrac{?}{18}$

3. $\dfrac{5}{8} = \dfrac{?}{32}$

4. $\dfrac{2}{9} = \dfrac{?}{18}$

5. $\dfrac{9}{10} = \dfrac{?}{30}$

6. $\dfrac{5}{6} = \dfrac{?}{48}$

7. $\dfrac{7}{8} = \dfrac{?}{32}$

8. $\dfrac{2}{5} = \dfrac{?}{25}$

9. $\dfrac{5}{12} = \dfrac{?}{48}$

10. $\dfrac{3}{8} = \dfrac{?}{56}$

11. $\dfrac{17}{18} = \dfrac{?}{54}$

12. $\dfrac{11}{16} = \dfrac{?}{256}$

13. $\dfrac{5}{3} = \dfrac{?}{45}$

14. $\dfrac{11}{5} = \dfrac{?}{30}$

15. $\dfrac{7}{22} = \dfrac{?}{132}$

16. $\dfrac{10}{21} = \dfrac{?}{126}$

b Simplify.

17. $\dfrac{2}{4}$

18. $\dfrac{4}{8}$

19. $\dfrac{6}{8}$

20. $\dfrac{8}{12}$

21. $\dfrac{3}{15}$

22. $\dfrac{8}{10}$

23. $\dfrac{24}{8}$

24. $\dfrac{36}{9}$

25. $\dfrac{18}{24}$

26. $\dfrac{42}{48}$

27. $\dfrac{14}{16}$

28. $\dfrac{15}{25}$

29. $\dfrac{12}{10}$

30. $\dfrac{16}{14}$

31. $\dfrac{16}{48}$

32. $\dfrac{100}{20}$

33. $\dfrac{150}{25}$

34. $\dfrac{19}{76}$

35. $\dfrac{17}{51}$

36. $\dfrac{425}{525}$

37. $\dfrac{540}{810}$

38. $\dfrac{1000}{1080}$

39. $\dfrac{210}{2700}$

40. $\dfrac{300}{2250}$

c Use = or ≠ for ☐ to write a true sentence.

41. $\dfrac{3}{4} \,\square\, \dfrac{9}{12}$

42. $\dfrac{4}{8} \,\square\, \dfrac{3}{6}$

43. $\dfrac{1}{5} \,\square\, \dfrac{2}{9}$

44. $\dfrac{1}{4} \,\square\, \dfrac{2}{9}$

45. $\dfrac{3}{8} \,\square\, \dfrac{6}{16}$

46. $\dfrac{2}{6} \,\square\, \dfrac{6}{18}$

47. $\dfrac{2}{5} \,\square\, \dfrac{3}{7}$

48. $\dfrac{1}{3} \,\square\, \dfrac{1}{4}$

49. $\dfrac{12}{9}$ ☐ $\dfrac{8}{6}$

50. $\dfrac{16}{14}$ ☐ $\dfrac{8}{7}$

51. $\dfrac{5}{2}$ ☐ $\dfrac{17}{7}$

52. $\dfrac{3}{10}$ ☐ $\dfrac{7}{24}$

53. $\dfrac{3}{10}$ ☐ $\dfrac{30}{100}$

54. $\dfrac{700}{1000}$ ☐ $\dfrac{70}{100}$

55. $\dfrac{5}{10}$ ☐ $\dfrac{520}{1000}$

56. $\dfrac{49}{100}$ ☐ $\dfrac{50}{1000}$

Use the circle graph below to answer Exercises 57–60. What do college graduates do if they are not offered a full-time job in their first-choice field? The responses of 700 recent graduates are organized in the circle graph below. Use this graph in Exercises 57–60. Simplify the fraction representing each of the following.

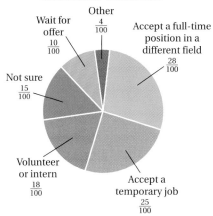

What College Graduates Will Do If They Are Not Offered a Full-Time Job in Their First-Choice Field

SOURCE: Yahoo Hotjobs survey

57. Accept a full-time position in a different field

58. Wait for offer

59. Accept a temporary job

60. Volunteer or intern

Skill Maintenance

Solve. [1.8a]

61. *Tiananmen Square.* Tiananmen Square in Beijing, China, is the largest public square in the world. The length is 963 yd and the width is 547 yd. What is its area? its perimeter?

62. A landscaper buys 13 small maple trees and 17 small oak trees for a project. A maple costs $23 and an oak costs $37. How much is spent altogether for the trees?

Subtract. [1.3a]

63. $803 - 617$

64. $50 - 18$

Solve. [1.7b]

65. $5280 = 1760 + t$

66. $10,947 = 123 \cdot y$

Synthesis

67. 🖩 On a test of 82 questions, a student got 63 correct. On another test of 100 questions, she got 77 correct. Did she get the same portion of each test correct? Why or why not?

68. *Batting Averages.* For the 2008 season, Joe Mauer, of the Minnesota Twins, won the American League batting title with 176 hits in 536 times at bat. Chipper Jones, of the Atlanta Braves, won the National League title with 160 hits in 439 times at bat. Did they have the same fraction of hits in times at bat (batting average)? Why or why not?
Source: Major League Baseball

Concept Reinforcement

Determine whether each statement is true or false.

_____ **1.** A number a is divisible by another number b if b is a factor of a.   [2.1b]

_____ **2.** If a number is not divisible by 6, then it is not divisible by 3.   [2.2a]

_____ **3.** The fraction $\frac{13}{7}$ is larger than the fraction $\frac{13}{6}$.   [2.5c]

_____ **4.** The number 1 is not prime.   [2.1c]

Guided Solutions

Fill in each blank with the number that creates a correct statement or solution.

5. $\frac{25}{\square} = 1$ [2.3b]

6. $\frac{\square}{9} = 0$ [2.3b]

7. $\frac{8}{\square} = 8$ [2.3b]

8. $\frac{6}{13} = \frac{\square}{39}$ [2.5a]

9. Simplify: $\frac{70}{225}$.   [2.5b]

$$\frac{70}{225} = \frac{2 \cdot \square \cdot 7}{\square \cdot 3 \cdot 5 \cdot \square}$$ Factoring the numerator
Factoring the denominator

$$= \frac{5}{5} \cdot \frac{\square \cdot 7}{3 \cdot \square \cdot 5}$$ Factoring the fraction

$$= \square \cdot \frac{\square}{45} \qquad \frac{5}{5} = 1$$

$$= \frac{\square}{\square} \qquad \text{Removing a factor of 1}$$

Mixed Review

To answer Exercises 10–14, consider the following numbers.   [2.2a]

84	132	594	350
300	500	120	14,850
17,576	180	1125	504
224	351	495	1632

10. Which of the above are divisible by 2 but not by 10?

11. Which of the above are divisible by 4 but not by 8?

12. Which of the above are divisible by 4 but not by 6?

13. Which of the above are divisible by 3 but not by 9?

14. Which of the above are divisible by 4, 5, and 6?

Determine whether each number is prime, composite, or neither.   [2.1c]

15. 61   **16.** 2   **17.** 91   **18.** 1

Find all the factors of each composite number. Then find the prime factorization of the number.   [2.1a], [2.1d]

19. 160
20. 222
21. 98
22. 315

What part of each object or set of objects is shaded?   [2.3a]

23.

24.

Multiply.   [2.4a]

25. $7 \cdot \dfrac{1}{9}$

26. $\dfrac{4}{15} \cdot \dfrac{2}{3}$

27. $\dfrac{5}{11} \cdot 8$

Simplify.   [2.3b], [2.5b]

28. $\dfrac{24}{60}$

29. $\dfrac{220}{60}$

30. $\dfrac{17}{17}$

31. $\dfrac{0}{23}$

32. $\dfrac{54}{186}$

33. $\dfrac{36}{20}$

34. $\dfrac{75}{630}$

35. $\dfrac{315}{435}$

36. $\dfrac{14}{0}$

Use = or ≠ for ☐ to write a true sentence.   [2.5c]

37. $\dfrac{3}{7}$ ☐ $\dfrac{48}{112}$

38. $\dfrac{19}{3}$ ☐ $\dfrac{95}{18}$

39. *Veterinary Care.*   Of every 500 households that own pet birds, 60 obtain veterinary care for their birds. What is the ratio of households that own birds that seek veterinary care for their birds to all households that own birds?   [2.3a]
Source: American Veterinary Medical Association

40. *Area of an Ice-Skating Rink.*   The length of a rectangular ice-skating rink in the atrium of a shopping mall is $\dfrac{7}{100}$ mi. The width is $\dfrac{3}{100}$ mi. What is the area of the rink? [2.4b]

Understand Through Discussion and Writing

41. Explain a method for finding a composite number that contains exactly two factors other than itself and 1.   [2.1c]

42. Which of the years from 2000 to 2020, if any, also happen to be prime numbers? Explain at least two ways in which you might go about solving this problem.   [2.2a]

43. Explain in your own words when it *is* possible to "cancel" and when it *is not* possible to "cancel."   [2.5b]

44. Can fraction notation be simplified if the numerator and the denominator are two different prime numbers? Why or why not?   [2.5b]

2.6 Multiplying, Simplifying, and Applications

OBJECTIVES

a) Multiply and simplify using fraction notation.

b) Solve applied problems involving multiplication of fractions.

SKILL TO REVIEW

Objective 2.2a: Determine whether a number is divisible by 2, 3, 4, 5, 6, 8, 9, or 10.

Determine whether each number is divisible by 9.

1. 486   2. 129

a) Multiplying and Simplifying Using Fraction Notation

It is often possible to simplify after we multiply. To make such simplifying easier, it is usually best not to carry out the products in the numerator and the denominator immediately, but to factor and simplify first. Consider the product

$$\frac{3}{8} \cdot \frac{4}{9}.$$

We proceed as follows:

$$\frac{3}{8} \cdot \frac{4}{9} = \frac{3 \cdot 4}{8 \cdot 9}$$ We write the products in the numerator and the denominator, but we do not carry them out.

$$= \frac{3 \cdot 2 \cdot 2}{2 \cdot 2 \cdot 2 \cdot 3 \cdot 3}$$ Factoring the numerator and the denominator

$$= \frac{3 \cdot 2 \cdot 2 \cdot 1}{2 \cdot 2 \cdot 2 \cdot 3 \cdot 3}$$ Using the identity property of 1 to insert the number 1 as a factor

$$= \frac{3 \cdot 2 \cdot 2}{3 \cdot 2 \cdot 2} \cdot \frac{1}{2 \cdot 3}$$ Factoring the fraction

$$= 1 \cdot \frac{1}{2 \cdot 3}$$

$$= \frac{1}{2 \cdot 3}$$ Removing a factor of 1

$$= \frac{1}{6}.$$

The procedure could have been shortened had we noticed that 4 is a factor of the 8 in the denominator:

$$\frac{3}{8} \cdot \frac{4}{9} = \frac{3 \cdot 4}{8 \cdot 9} = \frac{3 \cdot 4}{4 \cdot 2 \cdot 3 \cdot 3} = \frac{3 \cdot 4}{3 \cdot 4} \cdot \frac{1}{2 \cdot 3} = 1 \cdot \frac{1}{2 \cdot 3} = \frac{1}{2 \cdot 3} = \frac{1}{6}.$$

STUDY TIPS

ASKING QUESTIONS

Don't be afraid to ask questions in class. Most instructors welcome this and encourage students to ask them. Other students probably have the same questions you do.

To multiply and simplify:

a) Write the products in the numerator and the denominator, but do not carry them out.

b) Factor the numerator and the denominator.

c) Factor the fraction to remove a factor of 1, if possible.

d) Carry out the remaining products.

EXAMPLES   Multiply and simplify.

1. $\dfrac{2}{3} \cdot \dfrac{9}{4} = \dfrac{2 \cdot 9}{3 \cdot 4} = \dfrac{2 \cdot 3 \cdot 3}{3 \cdot 2 \cdot 2} = \dfrac{2 \cdot 3}{2 \cdot 3} \cdot \dfrac{3}{2} = 1 \cdot \dfrac{3}{2} = \dfrac{3}{2}$

2. $\dfrac{6}{7} \cdot \dfrac{5}{3} = \dfrac{6 \cdot 5}{7 \cdot 3} = \dfrac{3 \cdot 2 \cdot 5}{7 \cdot 3} = \dfrac{3}{3} \cdot \dfrac{2 \cdot 5}{7} = 1 \cdot \dfrac{2 \cdot 5}{7} = \dfrac{2 \cdot 5}{7} = \dfrac{10}{7}$

3. $40 \cdot \dfrac{7}{8} = \dfrac{40}{1} \cdot \dfrac{7}{8} = \dfrac{40 \cdot 7}{1 \cdot 8} = \dfrac{8 \cdot 5 \cdot 7}{1 \cdot 8} = \dfrac{8}{8} \cdot \dfrac{5 \cdot 7}{1} = 1 \cdot \dfrac{5 \cdot 7}{1} = \dfrac{5 \cdot 7}{1} = 35$

Answers

Skill to Review:

1. Yes   2. No

----------------------------- *Caution!* -----------------------------

Canceling can be used as follows for these examples.

1. $\dfrac{2}{3} \cdot \dfrac{9}{4} = \dfrac{2 \cdot 9}{3 \cdot 4} = \dfrac{2 \cdot 3 \cdot 3}{3 \cdot 2 \cdot 2} = \dfrac{3}{2}$

Removing a factor of 1:
$\dfrac{2 \cdot 3}{2 \cdot 3} = 1$

2. $\dfrac{6}{7} \cdot \dfrac{5}{3} = \dfrac{6 \cdot 5}{7 \cdot 3} = \dfrac{3 \cdot 2 \cdot 5}{7 \cdot 3} = \dfrac{2 \cdot 5}{7} = \dfrac{10}{7}$

Removing a factor of 1:
$\dfrac{3}{3} = 1$

3. $40 \cdot \dfrac{7}{8} = \dfrac{40 \cdot 7}{1 \cdot 8} = \dfrac{8 \cdot 5 \cdot 7}{1 \cdot 8} = \dfrac{5 \cdot 7}{1} = 35$

Removing a factor of 1:
$\dfrac{8}{8} = 1$

Remember: if you can't factor, you can't cancel!

> Do Exercises 1–4.

Multiply and simplify.

1. $\dfrac{2}{3} \cdot \dfrac{7}{8}$ **2.** $\dfrac{4}{5} \cdot \dfrac{5}{12}$

3. $16 \cdot \dfrac{3}{8}$ **4.** $\dfrac{5}{8} \cdot 4$

b) Applications and Problem Solving

EXAMPLE 4 *Landscaping.* Celina's Landscaping uses $\frac{2}{3}$ lb of peat moss when planting a rosebush. How much will be needed to plant 21 rosebushes?

1. Familiarize.   We first make a drawing or at least visualize the situation. We let $n =$ the number of pounds of peat moss needed.

21 rosebushes

$\frac{2}{3}$ pound of peat moss for each rosebush

2. Translate.   The problem translates to the following equation:

$$n = 21 \cdot \dfrac{2}{3}.$$

3. Solve.   To solve the equation, we carry out the multiplication:

$$n = 21 \cdot \dfrac{2}{3} = \dfrac{21}{1} \cdot \dfrac{2}{3} = \dfrac{21 \cdot 2}{1 \cdot 3} \qquad \text{Multiplying}$$

$$= \dfrac{3 \cdot 7 \cdot 2}{1 \cdot 3} = \dfrac{3}{3} \cdot \dfrac{7 \cdot 2}{1} = 14.$$

4. Check.   We check by repeating the calculation. (This is left to the student.) We can also ask if the answer seems reasonable. We are putting less than a pound of peat moss on each bush, so the answer should be less than 21. Since 14 is less than 21, we have a partial check. The number 14 checks.

5. State.   Celina's Landscaping will need 14 lb of peat moss to plant 21 rosebushes.

> Do Exercise 5.

5. Truffles.   Chocolate Delight sells $\frac{4}{5}$-lb boxes of truffles. How many pounds of truffles will be needed to fill 85 boxes?

Answers

1. $\dfrac{7}{12}$ 2. $\dfrac{1}{3}$ 3. 6 4. $\dfrac{5}{2}$ 5. 68 lb

2.6

Exercise Set

For Extra Help

MyMathLab

Math XL
PRACTICE   WATCH   DOWNLOAD   READ   REVIEW

a Multiply and simplify. | Don't forget to simplify! |

1. $\dfrac{2}{3} \cdot \dfrac{1}{2}$

2. $\dfrac{3}{8} \cdot \dfrac{1}{3}$

3. $\dfrac{7}{8} \cdot \dfrac{1}{7}$

4. $\dfrac{4}{9} \cdot \dfrac{1}{4}$

5. $\dfrac{1}{8} \cdot \dfrac{4}{5}$

6. $\dfrac{2}{5} \cdot \dfrac{1}{6}$

7. $\dfrac{1}{4} \cdot \dfrac{2}{3}$

8. $\dfrac{4}{6} \cdot \dfrac{1}{6}$

9. $\dfrac{12}{5} \cdot \dfrac{9}{8}$

10. $\dfrac{16}{15} \cdot \dfrac{5}{4}$

11. $\dfrac{10}{9} \cdot \dfrac{7}{5}$

12. $\dfrac{25}{12} \cdot \dfrac{4}{3}$

13. $9 \cdot \dfrac{1}{9}$

14. $4 \cdot \dfrac{1}{4}$

15. $\dfrac{1}{3} \cdot 3$

16. $\dfrac{1}{6} \cdot 6$

17. $\dfrac{7}{10} \cdot \dfrac{10}{7}$

18. $\dfrac{8}{9} \cdot \dfrac{9}{8}$

19. $\dfrac{7}{5} \cdot \dfrac{5}{7}$

20. $\dfrac{2}{11} \cdot \dfrac{11}{2}$

21. $\dfrac{1}{4} \cdot 8$

22. $\dfrac{1}{3} \cdot 18$

23. $24 \cdot \dfrac{1}{6}$

24. $16 \cdot \dfrac{1}{2}$

25. $12 \cdot \dfrac{3}{4}$

26. $18 \cdot \dfrac{5}{6}$

27. $\dfrac{3}{8} \cdot 24$

28. $\dfrac{2}{9} \cdot 36$

29. $35 \cdot \dfrac{3}{14}$

30. $15 \cdot \dfrac{1}{6}$

31. $\dfrac{7}{10} \cdot 28$

32. $\dfrac{5}{8} \cdot 34$

33. $\dfrac{1}{6} \cdot 360$

34. $\dfrac{1}{3} \cdot 120$

35. $240 \cdot \dfrac{1}{8}$

36. $150 \cdot \dfrac{1}{5}$

37. $\dfrac{4}{10} \cdot \dfrac{5}{10}$

38. $\dfrac{7}{10} \cdot \dfrac{34}{150}$

39. $\dfrac{8}{10} \cdot \dfrac{45}{100}$

40. $\dfrac{3}{10} \cdot \dfrac{8}{10}$

41. $\dfrac{11}{24} \cdot \dfrac{3}{5}$

42. $\dfrac{15}{22} \cdot \dfrac{4}{7}$

43. $\dfrac{10}{21} \cdot \dfrac{3}{4}$

44. $\dfrac{17}{18} \cdot \dfrac{3}{5}$

 Solve.

The *pitch* of a screw is the distance between its threads. With each complete rotation, the screw goes in or out a distance equal to its pitch. Use this information to do Exercises 45 and 46.

>Pitch = $\frac{1}{16}$ in.

Each rotation moves the screw in or out $\frac{1}{16}$ in.

45. The pitch of a screw is $\frac{1}{16}$ in. How far will it go into a piece of oak when it is turned 10 complete rotations clockwise?

46. The pitch of a screw is $\frac{3}{32}$ in. How far will it come out of a piece of plywood when it is turned 10 complete rotations counterclockwise?

47. *Level of Education and Median Income.* The median yearly income of someone with an associate's degree is approximately $\frac{2}{3}$ of the median income of someone with a bachelor's degree. If the median income for those with bachelor's degrees is $72,420, what is the median income of those with associate's degrees?

Source: U.S. Census Bureau

48. After Jack completes 60 hr of teacher training in college, he can earn $75 for working a full day as a substitute teacher. How much will he receive for working $\frac{3}{5}$ of a day?

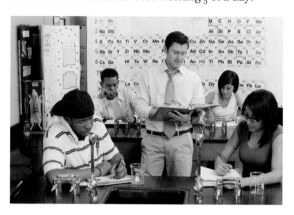

49. *Mailing-List Addresses.* Researchers have determined that $\frac{1}{4}$ of the addresses on a mailing list will change in one year. A business has a mailing list of 2500 people. After one year, how many addresses on that list will be incorrect?

50. *Shy People.* Sociologists have determined that $\frac{2}{5}$ of the people in the world are shy. A sales manager is considering 650 people for an aggressive sales position. How many of these people might be shy?

51. A recipe for piecrust calls for $\frac{2}{3}$ cup of flour. A baker is making $\frac{1}{2}$ of the recipe. How much flour should the baker use?

52. Of the students in the freshman class, $\frac{4}{5}$ have digital cameras; $\frac{1}{4}$ of these students also join the college photography club. What fraction of the students in the freshman class join the photography club?

53. A house worth $154,000 is assessed for $\frac{3}{4}$ of its value. What is the assessed value of the house?

54. Roxanne's tuition was $4600. A loan was obtained for $\frac{3}{4}$ of the tuition. How much was the loan?

55. *Map Scaling.* On a map, 1 in. represents 240 mi. What distance does $\frac{2}{3}$ in. represent?

56. *Map Scaling.* On a map, 1 in. represents 120 mi. What distance does $\frac{3}{4}$ in. represent?

57. *Household Budgets.* A family has an annual income of $42,000. Of this, $\frac{1}{5}$ is spent for food, $\frac{1}{4}$ for housing, $\frac{1}{10}$ for clothing, $\frac{1}{14}$ for savings, $\frac{1}{5}$ for taxes, and the rest for other expenses. How much is spent for each?

58. *Household Budgets.* A family has an annual income of $28,140. Of this, $\frac{1}{5}$ is spent for food, $\frac{1}{4}$ for housing, $\frac{1}{10}$ for clothing, $\frac{1}{14}$ for savings, $\frac{1}{5}$ for taxes, and the rest for other expenses. How much is spent for each?

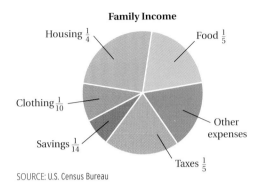

Family Income

Housing $\frac{1}{4}$

Food $\frac{1}{5}$

Clothing $\frac{1}{10}$

Other expenses

Savings $\frac{1}{14}$

Taxes $\frac{1}{5}$

SOURCE: U.S. Census Bureau

Skill Maintenance

Solve. [1.7b]

59. $48 \cdot t = 1680$

60. $74 \cdot x = 6290$

61. $3125 = 25 \cdot t$

62. $2880 = 24 \cdot y$

63. $t + 28 = 5017$

64. $456 + x = 9002$

65. $8797 = y + 2299$

66. $10,000 = 3593 + m$

Subtract. [1.3a]

67. $\begin{array}{r} 9060 \\ -4387 \\ \hline \end{array}$

68. $\begin{array}{r} 7800 \\ -2462 \\ \hline \end{array}$

Synthesis

Multiply and simplify. Use the list of prime numbers on p. 91 or a fraction calculator.

69. ▥ $\dfrac{201}{535} \cdot \dfrac{4601}{6499}$

70. ▥ $\dfrac{5767}{3763} \cdot \dfrac{159}{395}$

71. *College Profile.* Of students entering a college, $\frac{7}{8}$ have completed high school and $\frac{2}{3}$ are older than 20. If $\frac{1}{7}$ of all students are left-handed, what fraction of students entering the college are left-handed high school graduates over the age of 20?

72. *College Profile.* Refer to the information in Exercise 71. If 480 students are entering the college, how many of them are left-handed high school graduates 20 yr old or younger?

73. *College Profile.* Refer to Exercise 71. What fraction of students entering the college did not graduate from high school, are 20 yr old or younger, and are left-handed?

2.7

Division and Applications

OBJECTIVES

a Find the reciprocal of a number.

b Divide and simplify using fraction notation.

c Solve equations of the type $a \cdot x = b$ and $x \cdot a = b$, where a and b may be fractions.

d Solve applied problems involving division of fractions.

SKILL TO REVIEW

Objective 1.7b: Solve equations like $t + 28 = 54$, $28 \cdot x = 168$, and $98 \cdot 2 = y$.

Solve.

1. $88 = 8 \cdot y$

2. $16 \cdot x = 1152$

a Reciprocals

Look at these products:

$$8 \cdot \frac{1}{8} = \frac{8}{1} \cdot \frac{1}{8} = \frac{8 \cdot 1}{1 \cdot 8} = \frac{8}{8} = 1; \qquad \frac{2}{3} \cdot \frac{3}{2} = \frac{2 \cdot 3}{3 \cdot 2} = \frac{6}{6} = 1.$$

RECIPROCALS

If the product of two numbers is 1, we say that they are **reciprocals** of each other. To find the reciprocal of a fraction, interchange the numerator and the denominator.

$$\text{Number: } \frac{3}{4} \longrightarrow \text{Reciprocal: } \frac{4}{3}$$

EXAMPLES Find the reciprocal.

1. The reciprocal of $\frac{4}{5}$ is $\frac{5}{4}$. $\frac{4}{5} \cdot \frac{5}{4} = \frac{20}{20} = 1$

2. The reciprocal of 24 is $\frac{1}{24}$.   Think of 24 as $\frac{24}{1}$: $\frac{24}{1} \cdot \frac{1}{24} = \frac{24}{24} = 1$.

3. The reciprocal of $\frac{1}{3}$ is 3. $\frac{1}{3} \cdot 3 = \frac{1}{3} \cdot \frac{3}{1} = \frac{3}{3} = 1$

Do Margin Exercises 1–4.

Does 0 have a reciprocal? If it did, it would have to be a number x such that $0 \cdot x = 1$. But 0 times any number is 0. Thus we have the following.

0 HAS NO RECIPROCAL

The number 0, or $\frac{0}{n}$, has no reciprocal. $\left(\text{Recall that } \frac{n}{0} \text{ is not defined.}\right)$

Find the reciprocal.

1. $\frac{2}{5}$

2. $\frac{10}{7}$

3. 9

4. $\frac{1}{5}$

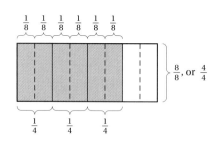

b Division

Consider the division $\frac{3}{4} \div \frac{1}{8}$. We are asking how many $\frac{1}{8}$'s are in $\frac{3}{4}$. We can answer this by looking at the figure at left. We see that there are six $\frac{1}{8}$'s in $\frac{3}{4}$. Thus,

$$\frac{3}{4} \div \frac{1}{8} = 6.$$

We can check this by multiplying:

$$6 \cdot \frac{1}{8} = \frac{6}{1} \cdot \frac{1}{8} = \frac{6}{8} = \frac{2 \cdot 3}{2 \cdot 4} = \frac{2}{2} \cdot \frac{3}{4} = \frac{3}{4}.$$

Answers

Skill to Review:

1. 11   **2.** 72

Margin Exercises:

1. $\frac{5}{2}$ **2.** $\frac{7}{10}$ **3.** $\frac{1}{9}$ **4.** 5

Here is a faster way to do this division:

$$\dfrac{3}{4} \div \dfrac{1}{8} = \dfrac{3}{4} \cdot \dfrac{8}{1} = \dfrac{3 \cdot 8}{4 \cdot 1} = \dfrac{24}{4} = 6.$$
Multiplying by the reciprocal of the divisor

> To divide fractions, multiply the dividend by the reciprocal of the divisor:
>
> $$\dfrac{2}{5} \div \dfrac{3}{4} = \dfrac{2}{5} \cdot \dfrac{4}{3} = \dfrac{2 \cdot 4}{5 \cdot 3} = \dfrac{8}{15}.$$
>
> Multiply by the reciprocal of the divisor.

EXAMPLES Divide and simplify.

4. $\dfrac{5}{6} \div \dfrac{2}{3} = \dfrac{5}{6} \cdot \dfrac{3}{2} = \dfrac{5 \cdot 3}{6 \cdot 2} = \dfrac{5 \cdot 3}{3 \cdot 2 \cdot 2} = \dfrac{3}{3} \cdot \dfrac{5}{2 \cdot 2} = \dfrac{5}{2 \cdot 2} = \dfrac{5}{4}$

5. $\dfrac{7}{8} \div \dfrac{1}{16} = \dfrac{7}{8} \cdot \dfrac{16}{1} = \dfrac{7 \cdot 16}{8 \cdot 1} = \dfrac{7 \cdot 2 \cdot 8}{8 \cdot 1} = \dfrac{8}{8} \cdot \dfrac{7 \cdot 2}{1} = \dfrac{7 \cdot 2}{1} = 14$

6. $\dfrac{2}{5} \div 6 = \dfrac{2}{5} \cdot \dfrac{1}{6} = \dfrac{2 \cdot 1}{5 \cdot 6} = \dfrac{2 \cdot 1}{5 \cdot 2 \cdot 3} = \dfrac{2}{2} \cdot \dfrac{1}{5 \cdot 3} = \dfrac{1}{5 \cdot 3} = \dfrac{1}{15}$

-------------------------------- *Caution!* --------------------------------

Canceling can be used as follows for Examples 4–6.

4. $\dfrac{5}{6} \div \dfrac{2}{3} = \dfrac{5}{6} \cdot \dfrac{3}{2} = \dfrac{5 \cdot 3}{6 \cdot 2} = \dfrac{5 \cdot \cancel{3}}{\cancel{3} \cdot 2 \cdot 2} = \dfrac{5}{2 \cdot 2} = \dfrac{5}{4}$
Removing a factor of 1: $\dfrac{3}{3} = 1$

5. $\dfrac{7}{8} \div \dfrac{1}{16} = \dfrac{7}{8} \cdot \dfrac{16}{1} = \dfrac{7 \cdot 16}{8 \cdot 1} = \dfrac{7 \cdot 2 \cdot \cancel{8}}{\cancel{8} \cdot 1} = \dfrac{7 \cdot 2}{1} = 14$
Removing a factor of 1: $\dfrac{8}{8} = 1$

6. $\dfrac{2}{5} \div 6 = \dfrac{2}{5} \cdot \dfrac{1}{6} = \dfrac{2 \cdot 1}{5 \cdot 6} = \dfrac{\cancel{2} \cdot 1}{5 \cdot \cancel{2} \cdot 3} = \dfrac{1}{5 \cdot 3} = \dfrac{1}{15}$
Removing a factor of 1: $\dfrac{2}{2} = 1$

Remember: if you can't factor, you can't cancel!

⌐ Do Exercises 5–8. ⌐

What is the explanation for multiplying by a reciprocal when dividing? Let's consider $\frac{2}{3} \div \frac{7}{5}$. We multiply by 1. The name for 1 that we will use is $(5/7)/(5/7)$; it comes from the reciprocal of $\frac{7}{5}$.

$$\dfrac{2}{3} \div \dfrac{7}{5} = \dfrac{\frac{2}{3}}{\frac{7}{5}} = \dfrac{\frac{2}{3}}{\frac{7}{5}} \cdot 1 = \dfrac{\frac{2}{3} \cdot \frac{5}{7}}{\frac{7}{5} \cdot \frac{5}{7}} = \dfrac{\frac{2}{3} \cdot \frac{5}{7}}{1} = \dfrac{2}{3} \cdot \dfrac{5}{7} = \dfrac{10}{21}$$

Thus,

$$\dfrac{2}{3} \div \dfrac{7}{5} = \dfrac{2}{3} \cdot \dfrac{5}{7} = \dfrac{10}{21}.$$

⌐ Do Exercise 9. ⌐

Divide and simplify.

5. $\dfrac{6}{7} \div \dfrac{3}{4}$ **6.** $\dfrac{2}{3} \div \dfrac{1}{4}$

7. $\dfrac{4}{5} \div 8$ **8.** $60 \div \dfrac{3}{5}$

9. Divide by multiplying by 1:

$$\dfrac{\frac{4}{5}}{\frac{6}{7}}$$

Answers

5. $\dfrac{8}{7}$ **6.** $\dfrac{8}{3}$ **7.** $\dfrac{1}{10}$ **8.** 100 **9.** $\dfrac{14}{15}$

(c) Solving Equations

Now let's solve equations $a \cdot x = b$ and $x \cdot a = b$, where a and b may be fractions. We proceed as we did with equations involving whole numbers. We divide by a on both sides.

EXAMPLE 7 Solve: $\frac{4}{3} \cdot x = \frac{6}{7}$.

We have

$$\frac{4}{3} \cdot x = \frac{6}{7}$$

$$\frac{\frac{4}{3} \cdot x}{\frac{4}{3}} = \frac{\frac{6}{7}}{\frac{4}{3}} \qquad \text{Dividing by } \tfrac{4}{3} \text{ on both sides}$$

$$x = \frac{6}{7} \cdot \frac{3}{4} \qquad \text{Multiplying by the reciprocal}$$

$$= \frac{6 \cdot 3}{7 \cdot 4} = \frac{2 \cdot 3 \cdot 3}{7 \cdot 2 \cdot 2} = \frac{2}{2} \cdot \frac{3 \cdot 3}{7 \cdot 2} = \frac{3 \cdot 3}{7 \cdot 2} = \frac{9}{14}.$$

The solution is $\frac{9}{14}$.

EXAMPLE 8 Solve: $t \cdot \frac{4}{5} = 80$.

Dividing by $\frac{4}{5}$ on both sides, we get

$$t = 80 \div \frac{4}{5} = 80 \cdot \frac{5}{4} = \frac{80 \cdot 5}{4} = \frac{4 \cdot 20 \cdot 5}{4 \cdot 1} = \frac{4}{4} \cdot \frac{20 \cdot 5}{1} = \frac{20 \cdot 5}{1} = 100.$$

The solution is 100.

Do Exercises 10 and 11.

Solve.

10. $\frac{5}{6} \cdot y = \frac{2}{3}$

11. $\frac{3}{4} \cdot n = 24$

(d) Applications and Problem Solving

EXAMPLE 9 *Doses of an Antibiotic.* How many doses, each containing $\frac{15}{4}$ milliliters (mL), can be obtained from a bottle of a children's antibiotic that contains 60 mL?

1. **Familiarize.** We are asking the question "How many $\frac{15}{4}$'s are in 60?" Repeated addition will apply here. We make a drawing. We let $n =$ the number of doses in all.

$\frac{15}{4}$ milliliter in each dose

n doses in all

2. **Translate.** The equation that corresponds to the situation is

$$n = 60 \div \frac{15}{4}.$$

3. Solve. We solve the equation by carrying out the division:

$$n = 60 \div \frac{15}{4} = 60 \cdot \frac{4}{15} = \frac{60}{1} \cdot \frac{4}{15}$$

$$= \frac{60 \cdot 4}{1 \cdot 15} = \frac{4 \cdot 15 \cdot 4}{1 \cdot 15} = \frac{15}{15} \cdot \frac{4 \cdot 4}{1}$$

$$= 1 \cdot 16 = 16.$$

4. Check. We check by multiplying the number of doses by the size of the dose: $16 \cdot \frac{15}{4} = 60$. The answer checks.

5. State. There are 16 doses in a 60-mL bottle of the antibiotic.

> Do Exercise 12.

EXAMPLE 10 *Brick Walkway.* Logan's Landscape specializes in brick and stone walkways. After they complete 63 ft of a brick walkway between two buildings on campus, $\frac{7}{8}$ of the walkway is complete. What is the total length of this walkway?

1. Familiarize. We ask: "63 ft is $\frac{7}{8}$ of what length?" We make a drawing or at least visualize the problem. We let w = the length of the walkway.

$\frac{7}{8}$ of the walkway
63 ft

2. Translate. We translate to an equation:

$$\underbrace{\frac{7}{8}}_{\substack{\text{Fraction} \\ \text{completed}}} \cdot \underbrace{w}_{\substack{\text{Total} \\ \text{length} \\ \text{of walkway}}} = \underbrace{63.}_{\substack{\text{Amount} \\ \text{already} \\ \text{completed}}}$$

3. Solve. We divide by $\frac{7}{8}$ on both sides and carry out the division:

$$w = 63 \div \frac{7}{8} = \frac{63}{1} \cdot \frac{8}{7} = \frac{63 \cdot 8}{1 \cdot 7} = \frac{7 \cdot 9 \cdot 8}{1 \cdot 7} = \frac{7}{7} \cdot \frac{9 \cdot 8}{1} = 72.$$

4. Check. We determine whether $\frac{7}{8}$ of 72 is 63:

$$\frac{7}{8} \cdot 72 = \frac{7 \cdot 72}{8 \cdot 1} = \frac{7 \cdot 8 \cdot 9}{8 \cdot 1} = \frac{8}{8} \cdot \frac{7 \cdot 9}{1} = 63.$$

The answer, 72, checks.

5. State. The total length of the walkway is 72 ft.

> Do Exercise 13.

12. Each loop in a spring uses $\frac{21}{8}$ in. of wire. How many loops can be made from 210 in. of wire?

13. Sales Trip. John Penna sells soybean seeds to seed companies. After he had driven 210 mi, $\frac{5}{6}$ of his sales trip was completed. How long was the total trip?

$\frac{5}{6}$ of the trip
210 mi

Answers

12. 80 loops   **13.** 252 mi

Translating
for Success

1. *Valentine Boxes.* Jane's Fudge Shop is preparing Valentine boxes. How many pounds of fudge will be needed to fill 80 boxes if each box contains $\frac{5}{16}$ lb?

2. *Gallons of Gasoline.* On the third day of a business trip, a sales representative used $\frac{4}{5}$ of a tank of gasoline. If the tank is a 20-gal tank, how many gallons were used on the third day?

3. *Purchasing a Shirt.* Tom received $36 for his birthday. If he spends $\frac{3}{4}$ of the gift on a new shirt, what is the cost of the shirt?

4. *Checkbook Balance.* The balance in Sam's checking account is $1456. He writes a check for $28 and makes a deposit of $52. What is the new balance?

5. *Valentine Boxes.* Jane's Fudge Shop prepared 80 lb of fudge for Valentine boxes. If each box contains $\frac{5}{16}$ lb, how many boxes can be filled?

The goal of these matching questions is to practice step (2), *Translate*, of the five-step problem-solving process. Translate each problem to an equation and select a correct translation from equations A–O.

A. $x = \dfrac{3}{4} \cdot 36$

B. $28 \cdot x = 52$

C. $x = 80 \cdot \dfrac{5}{16}$

D. $x = 1456 \div 28$

E. $\dfrac{5}{4} \cdot x = 20$

F. $20 = \dfrac{4}{5} \cdot x$

G. $x = 12 \cdot 28$

H. $x = \dfrac{4}{5} \cdot 20$

I. $\dfrac{3}{4} \cdot x = 36$

J. $x = 1456 - 52 - 28$

K. $x \div 28 = 1456$

L. $x = 52 - 28$

M. $x = 52 \cdot 28$

N. $x = 1456 - 28 + 52$

O. $\dfrac{5}{16} \cdot x = 80$

Answers on page A-4

---

6. *Gasoline Tank.* A gasoline tank contains 20 gal when it is $\frac{4}{5}$ full. How many gallons can it hold when full?

7. *Knitting a Scarf.* It takes Rachel 36 hr to knit a scarf. She can knit only $\frac{3}{4}$ hr per day because she is taking 16 hr of college classes. How many days will it take her to knit the scarf?

8. *Bicycle Trip.* On a recent 52-mi bicycle trip, David stopped to make a cell-phone call after completing 28 mi. How many more miles did he bicycle after the call?

9. *Crème de Menthe Thins.* Andes Candies L.P. makes Crème de Menthe Thins. How many 28-piece packages can be filled with 1456 pieces?

10. *Cereal Donations.* The Williams family donates 28 boxes of cereal weekly to the local Family in Crisis Center. How many boxes does this family donate in one year?

a Find the reciprocal of each number.

1. $\dfrac{5}{6}$

2. $\dfrac{7}{8}$

3. 6

4. 4

5. $\dfrac{1}{6}$

6. $\dfrac{1}{4}$

7. $\dfrac{10}{3}$

8. $\dfrac{17}{4}$

b Divide and simplify.   [Don't forget to simplify!]

9. $\dfrac{3}{5} \div \dfrac{3}{4}$

10. $\dfrac{2}{3} \div \dfrac{3}{4}$

11. $\dfrac{3}{5} \div \dfrac{9}{4}$

12. $\dfrac{6}{7} \div \dfrac{3}{5}$

13. $\dfrac{4}{3} \div \dfrac{1}{3}$

14. $\dfrac{10}{9} \div \dfrac{1}{3}$

15. $\dfrac{1}{3} \div \dfrac{1}{6}$

16. $\dfrac{1}{4} \div \dfrac{1}{5}$

17. $\dfrac{3}{8} \div 3$

18. $\dfrac{5}{6} \div 5$

19. $\dfrac{12}{7} \div 4$

20. $\dfrac{18}{5} \div 2$

21. $12 \div \dfrac{3}{2}$

22. $24 \div \dfrac{3}{8}$

23. $28 \div \dfrac{4}{5}$

24. $40 \div \dfrac{2}{3}$

25. $\dfrac{5}{8} \div \dfrac{5}{8}$

26. $\dfrac{2}{5} \div \dfrac{2}{5}$

27. $\dfrac{8}{15} \div \dfrac{4}{5}$

28. $\dfrac{6}{13} \div \dfrac{3}{26}$

29. $\dfrac{9}{5} \div \dfrac{4}{5}$

30. $\dfrac{5}{12} \div \dfrac{25}{36}$

31. $120 \div \dfrac{5}{6}$

32. $360 \div \dfrac{8}{7}$

Solve.

33. $\frac{4}{5} \cdot x = 60$

34. $\frac{3}{2} \cdot t = 90$

35. $\frac{5}{3} \cdot y = \frac{10}{3}$

36. $\frac{4}{9} \cdot m = \frac{8}{3}$

37. $x \cdot \frac{25}{36} = \frac{5}{12}$

38. $p \cdot \frac{4}{5} = \frac{8}{15}$

39. $n \cdot \frac{8}{7} = 360$

40. $y \cdot \frac{5}{6} = 120$

d Solve.

41. *Extension Cords.* An electrical supplier sells rolls of SJO 14-3 cable to a company that makes extension cords. It takes $\frac{7}{3}$ ft of cable to make each cord. How many extension cords can be made with a roll of cable containing 2240 ft of cable?

42. Benny uses $\frac{2}{5}$ gram (g) of toothpaste each time he brushes his teeth. If Benny buys a 30-g tube, how many times will he be able to brush his teeth?

43. A pair of basketball shorts requires $\frac{3}{4}$ yd of nylon. How many pairs of shorts can be made from 24 yd of nylon?

44. A child's baseball shirt requires $\frac{5}{6}$ yd of fabric. How many shirts can be made from 25 yd of fabric?

45. How many $\frac{2}{3}$-cup sugar bowls can be filled from 16 cups of sugar?

46. How many $\frac{2}{3}$-cup cereal bowls can be filled from 10 cups of cornflakes?

47. A bucket had 12 L of water in it when it was $\frac{3}{4}$ full. How much could it hold when full?

48. A tank had 20 L of gasoline in it when it was $\frac{4}{5}$ full. How much could it hold when full?

49. Yoshi Teramoto sells hardware tools. After driving 180 kilometers (km), he has completed $\frac{5}{8}$ of a sales trip. How long is the total trip? How many kilometers are left to drive?

50. A piece of coaxial cable $\frac{4}{5}$ meter (m) long is to be cut into 8 pieces of the same length. What is the length of each piece?

Pitch of a Screw. Refer to Exercises 45 and 46 in Exercise Set 2.6.

51. After a screw has been turned 8 complete rotations, it is extended $\frac{1}{2}$ in. into a piece of wallboard. What is the pitch of the screw?

52. The pitch of a screw is $\frac{3}{32}$ in. How many complete rotations are necessary to drive the screw $\frac{3}{4}$ in. into a piece of pine wood?

Skill Maintenance

In each of Exercises 53–60, fill in the blank with the correct term from the given list. Some of the choices may not be used.

53. The equation $14 + (2 + 30) = (14 + 2) + 30$ illustrates the _____ law of addition. [1.2a]

54. In the product $10 \cdot \frac{3}{4}$, 10 and $\frac{3}{4}$ are called _____. [2.1a], [2.4a]

55. A natural number that has exactly two different factors, only itself and 1, is called a(n) _____ number. [2.1c]

56. In the fraction $\frac{4}{17}$, we call 17 the _____. [2.3a]

57. For any number a, $a + 0 = a$. The number 0 is the _____ identity. [1.2a]

58. The product of 6 and $\frac{1}{6}$ is 1. We say that 6 and $\frac{1}{6}$ are _____. [2.7a]

59. The set of _____ numbers is $0, 1, 2, 3, 4, \ldots$. [1.1b]

60. A sentence with $=$ is called a(n) _____. [1.7a]

associative

commutative

additive

multiplicative

reciprocals

factors

prime

composite

numerator

denominator

equation

variables

whole

natural

Synthesis

Simplify. Use the list of prime numbers on p. 91.

61. ▦ $\dfrac{711}{1957} \div \dfrac{10{,}033}{13{,}081}$

62. ▦ $\dfrac{8633}{7387} \div \dfrac{485}{581}$

63. $\left(\dfrac{9}{10} \div \dfrac{2}{5} \div \dfrac{3}{8} \right)^2$

64. $\dfrac{\left(\dfrac{3}{7}\right)^2 \div \dfrac{12}{5}}{\left(\dfrac{2}{9}\right)\left(\dfrac{9}{2}\right)}$

65. If $\frac{1}{3}$ of a number is $\frac{1}{4}$, what is $\frac{1}{2}$ of the number?

Summary and Review

Key Terms

factor (noun), p. 88
factor (verb), p. 88
factorization, p. 88
multiple, p. 89
divisible, p. 89

prime number, p. 90
composite number, p. 90
prime factorization, p. 91
even number, p. 95
fraction notation, p. 101

numerator, p. 101
denominator, p. 101
equivalent fractions, p. 117
cross products, p. 121

Concept Reinforcement

Determine whether each statement is true or false.

_____ **1.** For any natural number n, $\dfrac{n}{n} > \dfrac{0}{n}$. [2.3b]

_____ **2.** A number is divisible by 10 if its ones digit is 0 or 5. [2.2a]

_____ **3.** If a number is divisible by 9, then it is also divisible by 3. [2.2a]

_____ **4.** The fraction $\dfrac{13}{6}$ is larger than the fraction $\dfrac{11}{6}$. [2.5c]

Important Concepts

Objective 2.1a Find the factors of a number.

Example Find the factors of 84.

We find as many "two-factor" factorizations as we can.

$1 \cdot 84 \qquad 4 \cdot 21$

$2 \cdot 42 \qquad 6 \cdot 14$

$3 \cdot 28 \qquad 7 \cdot 12 \leftarrow$ Since 8, 9, 10, and 11 are not
factors, we are finished.

The factors are 1, 2, 3, 4, 6, 7, 12, 14, 21, 28, 42, and 84.

Practice Exercise

1. Find the factors of 104.

Objective 2.1d Find the prime factorization of a composite number.

Example Find the prime factorization of 84.

To find the prime factorization, we can use either successive divisions or a factor tree.

Thus, $84 = 2 \cdot 2 \cdot 3 \cdot 7$.

Practice Exercise

2. Find the prime factorization of 104.

Objective 2.3b   Simplify fraction notation like n/n to 1, $0/n$ to 0, and $n/1$ to n.

Example   Simplify $\dfrac{6}{6}, \dfrac{0}{6}$, and $\dfrac{6}{1}$.

$$\dfrac{6}{6} = 1, \qquad \dfrac{0}{6} = 0, \qquad \dfrac{6}{1} = 6$$

Practice Exercise

3. Simplify $\dfrac{0}{18}, \dfrac{18}{18}$, and $\dfrac{18}{1}$.

Objective 2.5a   Multiply a number by 1 to find fraction notation with a specified denominator.

Example   Find a name for $\dfrac{2}{7}$ with a denominator of 63.

Since $7 \cdot 9 = 63$, we multiply by $\dfrac{9}{9}$:

$$\dfrac{2}{7} = \dfrac{2}{7} \cdot \dfrac{9}{9} = \dfrac{2 \cdot 9}{7 \cdot 9} = \dfrac{18}{63}.$$

Practice Exercise

4. Find a name for $\dfrac{7}{12}$ with a denominator of 96.

Objective 2.5b   Simplify fraction notation.

Example   Simplify: $\dfrac{315}{1650}$.

Using the test for divisibility by 5, we see that both the numerator and the denominator are divisible by 5:

$$\dfrac{315}{1650} = \dfrac{5 \cdot 63}{5 \cdot 330} = \dfrac{5}{5} \cdot \dfrac{63}{330} = 1 \cdot \dfrac{63}{330}$$

$$= \dfrac{63}{330} = \dfrac{3 \cdot 21}{3 \cdot 110} = \dfrac{3}{3} \cdot \dfrac{21}{110} = 1 \cdot \dfrac{21}{110} = \dfrac{21}{110}.$$

Practice Exercise

5. Simplify: $\dfrac{100}{280}$.

Objective 2.5c   Use the test for equality to determine whether two fractions name the same number.

Example   Use $=$ or $\neq$ for $\square$ to write a true sentence:

$$\dfrac{10}{54} \ \square \ \dfrac{15}{81}.$$

We find the cross products. We multiply 10 and 81: $10 \cdot 81 = 810$. Then we multiply 54 and 15: $54 \cdot 15 = 810$. Because the cross products are the same, we have

$$\dfrac{10}{54} = \dfrac{15}{81}.$$

If the cross products had been different, the fractions would not be equal.

Practice Exercise

6. Use $=$ or $\neq$ for $\square$ to write a true sentence:

$$\dfrac{8}{48} \ \square \ \dfrac{6}{44}.$$

Objective 2.6a   Multiply and simplify using fraction notation.

Example   Multiply and simplify: $\dfrac{7}{16} \cdot \dfrac{40}{49}$.

$$\dfrac{7}{16} \cdot \dfrac{40}{49} = \dfrac{7 \cdot 40}{16 \cdot 49} = \dfrac{7 \cdot 2 \cdot 2 \cdot 2 \cdot 5}{2 \cdot 2 \cdot 2 \cdot 2 \cdot 7 \cdot 7}$$

$$= \dfrac{2 \cdot 2 \cdot 2 \cdot 7}{2 \cdot 2 \cdot 2 \cdot 7} \cdot \dfrac{5}{2 \cdot 7} = 1 \cdot \dfrac{5}{14} = \dfrac{5}{14}$$

Practice Exercise

7. Multiply and simplify: $\dfrac{80}{3} \cdot \dfrac{21}{72}$.

Objective 2.7b   Divide and simplify using fraction notation.

Example   Divide and simplify: $\dfrac{9}{20} \div \dfrac{18}{25}$.

$$\dfrac{9}{20} \div \dfrac{18}{25} = \dfrac{9}{20} \cdot \dfrac{25}{18} = \dfrac{9 \cdot 25}{20 \cdot 18} = \dfrac{3 \cdot 3 \cdot 5 \cdot 5}{2 \cdot 2 \cdot 5 \cdot 2 \cdot 3 \cdot 3}$$

$$= \dfrac{3 \cdot 3 \cdot 5}{3 \cdot 3 \cdot 5} \cdot \dfrac{5}{2 \cdot 2 \cdot 2} = 1 \cdot \dfrac{5}{8} = \dfrac{5}{8}$$

Practice Exercise

8. Divide and simplify: $\dfrac{9}{4} \div \dfrac{45}{14}$.

Objective 2.7d   Solve applied problems involving division of fractions.

Example   A rental car had 18 gal of gasoline when its gas tank was $\frac{6}{7}$ full. How much could the tank hold when full?
The equation that corresponds to the situation is

$$\frac{6}{7} \cdot g = 18.$$

We divide by $\frac{6}{7}$ on both sides and carry out the division:

$$g = 18 \div \frac{6}{7} = \frac{18}{1} \cdot \frac{7}{6} = \frac{18 \cdot 7}{1 \cdot 6}$$

$$= \frac{3 \cdot 6 \cdot 7}{1 \cdot 6} = \frac{6}{6} \cdot \frac{3 \cdot 7}{1} = 1 \cdot \frac{21}{1} = 21.$$

The rental car can hold 21 gal of gasoline.

Practice Exercise

9. A flower vase has $\frac{7}{4}$ cups of water in it when it is $\frac{3}{4}$ full. How much can it hold when full?

Review Exercises

Find all the factors of each number.   [2.1a]

1. 60

2. 176

3. Multiply by 1, 2, 3, and so on, to find ten multiples of 8. [2.1b]

4. Determine whether 924 is divisible by 11.   [2.1b]

5. Determine whether 1800 is divisible by 16.   [2.1b]

Determine whether each number is prime, composite, or neither.   [2.1c]

6. 37

7. 1

8. 91

Find the prime factorization of each number.   [2.1d]

9. 70

10. 30

11. 45

12. 150

13. 648

14. 5250

To do Exercises 15–22, consider the following numbers:

140	716	93	2802
95	2432	330	711
182	4344	255,555	
475	600	780	

Which of the above are divisible by the given number?   [2.2a]

15. 3

16. 2

17. 4

18. 8

19. 5

20. 6

21. 9

22. 10

23. Identify the numerator and the denominator of $\frac{2}{7}$. [2.3a]

What part of each object is shaded?   [2.3a]

24.

25.

26. What part of the set of objects is shaded? [2.3a]

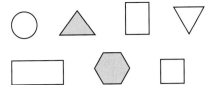

27. For a committee in the United States Senate that consists of 3 Democrats and 5 Republicans, what is the ratio of: [2.3a]

a) Democrats to Republicans?
b) Republicans to Democrats?
c) Democrats to the total number of members of the committee?

Simplify. [2.3b], [2.5b]

28. $\dfrac{12}{30}$ **29.** $\dfrac{7}{28}$ **30.** $\dfrac{23}{23}$

31. $\dfrac{0}{25}$ **32.** $\dfrac{1170}{1200}$ **33.** $\dfrac{18}{1}$

34. $\dfrac{9}{27}$ **35.** $\dfrac{88}{184}$ **36.** $\dfrac{18}{0}$

37. $\dfrac{48}{8}$ **38.** $\dfrac{140}{490}$ **39.** $\dfrac{288}{2025}$

40. Simplify the fractions on this circle graph, if possible. [2.5b]

Museums in the United States

Science $\dfrac{15}{100}$ History $\dfrac{38}{100}$

Art $\dfrac{24}{100}$ Other $\dfrac{23}{100}$

Use = or ≠ for ☐ to write a true sentence. [2.5c]

41. $\dfrac{3}{5}$ ☐ $\dfrac{4}{6}$ **42.** $\dfrac{4}{7}$ ☐ $\dfrac{8}{14}$

43. $\dfrac{4}{5}$ ☐ $\dfrac{5}{6}$ **44.** $\dfrac{4}{3}$ ☐ $\dfrac{28}{21}$

Multiply and simplify. [2.6a]

45. $4 \cdot \dfrac{3}{8}$ **46.** $\dfrac{7}{3} \cdot 24$

47. $9 \cdot \dfrac{5}{18}$ **48.** $\dfrac{6}{5} \cdot 20$

49. $\dfrac{3}{4} \cdot \dfrac{8}{9}$ **50.** $\dfrac{5}{7} \cdot \dfrac{1}{10}$

51. $\dfrac{3}{7} \cdot \dfrac{14}{9}$ **52.** $\dfrac{1}{4} \cdot \dfrac{2}{11}$

53. $\dfrac{4}{25} \cdot \dfrac{15}{16}$ **54.** $\dfrac{11}{3} \cdot \dfrac{30}{77}$

Find the reciprocal. [2.7a]

55. $\dfrac{4}{5}$ **56.** 3

57. $\dfrac{1}{9}$ **58.** $\dfrac{47}{36}$

Divide and simplify. [2.7b]

59. $6 \div \dfrac{4}{3}$ **60.** $\dfrac{5}{9} \div \dfrac{5}{18}$

61. $\dfrac{1}{6} \div \dfrac{1}{11}$ **62.** $\dfrac{3}{14} \div \dfrac{6}{7}$

63. $\dfrac{1}{4} \div \dfrac{1}{9}$ **64.** $180 \div \dfrac{3}{5}$

65. $\dfrac{23}{25} \div \dfrac{23}{25}$ **66.** $\dfrac{2}{3} \div \dfrac{3}{2}$

Solve. [2.7c]

67. $\dfrac{5}{4} \cdot t = \dfrac{3}{8}$ **68.** $x \cdot \dfrac{2}{3} = 160$

Solve. [2.6b], [2.7d]

69. A road crew repaves $\frac{1}{12}$ mi of road each day. How long will it take the crew to repave a $\frac{3}{4}$-mi stretch of road?

70. *Cotton Production.* In 2005–2006, the United States accounted for approximately $\frac{1}{5}$ of the world production of cotton. The total world cotton production was about 114,000,000 metric tons. How much cotton did the United States produce?

Sources: U.S. Department of Agriculture; Foreign Agricultural Service; *World Agricultural Production*, June 2007

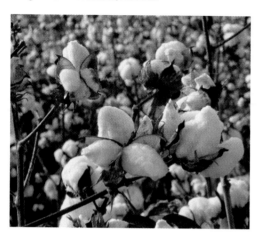

71. After driving 600 km, the Buxton family has completed $\frac{3}{5}$ of their vacation. How long is the total trip?

72. Molly is making a pepper steak recipe that calls for $\frac{2}{3}$ cup of green bell peppers. How much would be needed to make $\frac{1}{2}$ recipe? 3 recipes?

73. Bernardo earns $105 for working a full day. How much does he receive for working $\frac{1}{7}$ of a day?

74. A book bag requires $\frac{4}{5}$ yd of fabric. How many bags can be made from 48 yd?

75. Solve: $\dfrac{2}{13} \cdot x = \dfrac{1}{2}.$ [2.7c]

A. $\dfrac{1}{13}$ **B.** 13

C. $\dfrac{4}{13}$ **D.** $\dfrac{13}{4}$

76. Multiply and simplify: $\dfrac{15}{26} \cdot \dfrac{13}{90}.$ [2.6a]

A. $\dfrac{195}{234}$ **B.** $\dfrac{1}{12}$

C. $\dfrac{3}{36}$ **D.** $\dfrac{13}{156}$

Synthesis

77. 🖩 In the division below, find a and b. [2.7b]

$$\frac{19}{24} \div \frac{a}{b} = \frac{187{,}853}{268{,}224}$$

78. A prime number that remains a prime number when its digits are reversed is called a **palindrome prime**. For example, 17 is a palindrome prime because both 17 and 71 are primes. Which of the following numbers are palindrome primes? [2.1c]

13, 91, 16, 11, 15, 24, 29, 101, 201, 37

Understanding Through Discussion and Writing

1. A student incorrectly insists that $\frac{2}{5} \div \frac{3}{4}$ is $\frac{15}{8}$. What mistake is he probably making? [2.7b]

2. Use the number 9432 to explain why the test for divisibility by 9 works. [2.2a]

3. A student claims that "taking $\frac{1}{2}$ of a number is the same as dividing by $\frac{1}{2}$." Explain the error in this reasoning. [2.7b]

4. On p. 111, we explained, using words and pictures, why $\frac{3}{5} \cdot \frac{3}{4}$ equals $\frac{9}{20}$. Present a similar explanation of why $\frac{2}{3} \cdot \frac{4}{7}$ equals $\frac{8}{21}$. [2.4a]

5. Without performing the division, explain why $5 \div \frac{1}{7}$ is a greater number than $5 \div \frac{2}{3}$. [2.5c], [2.7b]

6. If a fraction's numerator and denominator have no factors (other than 1) in common, can the fraction be simplified? Why or why not? [2.5b]

CHAPTER

2

Test

For Extra Help

CHAPTER
Test Prep
VIDEOS

Step-by-step test solutions are found on the Chapter Test Prep Videos available via the Video Resources on DVD, in **MyMathLab**, and on **YouTube** (search "BittingerBasicMath" and click on "Channels").

1. Find all the factors of 300.

Determine whether each number is prime, composite, or neither.

2. 41

3. 14

Find the prime factorization of the number.

4. 18

5. 60

6. Determine whether 1784 is divisible by 8.

7. Determine whether 784 is divisible by 9.

8. Determine whether 5552 is divisible by 5.

9. Determine whether 2322 is divisible by 6.

10. Identify the numerator and the denominator of $\frac{4}{5}$.

11. What part is shaded?

12. What part of the set is shaded?

13. *Pass Completion Ratio.* In the 2008 regular season, Peyton Manning, of the Indianapolis Colts, completed 371 of 555 passes to earn a third Associated Press NFL Most Valuable Player award.

a) What was the ratio of pass completions to attempts?
b) What was the ratio of incomplete passes to attempts?

Simplify.

14. $\frac{26}{1}$

15. $\frac{12}{12}$

16. $\frac{0}{16}$

17. $\frac{12}{24}$

18. $\frac{42}{7}$

19. $\frac{9}{0}$

20. $\frac{7}{2-2}$

21. $\frac{72}{108}$

Use = or ≠ for □ to write a true sentence.

22. $\frac{3}{4} \square \frac{6}{8}$

23. $\frac{5}{4} \square \frac{9}{7}$

Multiply and simplify.

24. $\frac{4}{3} \cdot 24$

25. $5 \cdot \frac{3}{10}$

26. $\frac{2}{3} \cdot \frac{15}{4}$

27. $\frac{22}{15} \cdot \frac{5}{33}$

Find the reciprocal.

28. $\dfrac{5}{8}$

29. $\dfrac{1}{4}$

30. 18

Divide and simplify.

31. $\dfrac{1}{5} \div \dfrac{1}{8}$

32. $12 \div \dfrac{2}{3}$

33. $\dfrac{24}{5} \div \dfrac{28}{15}$

Solve.

34. $\dfrac{7}{8} \cdot x = 56$

35. $t \cdot \dfrac{2}{5} = \dfrac{7}{10}$

36. There are 7000 students at La Poloma College, and $\frac{5}{8}$ of them live in dorms. How many live in dorms?

37. A strip of taffy $\frac{9}{10}$ m long is cut into 12 equal pieces. What is the length of each piece?

38. A thermos of iced tea held 3 qt of tea when it was $\frac{3}{5}$ full. How much tea could it hold when full?

39. The pitch of a screw is $\frac{1}{8}$ in. How far will it go into a piece of walnut when it is turned 6 complete rotations?

40. In which figure does the shaded part represent $\frac{7}{6}$ of the figure?

A.

B.

C.

D.

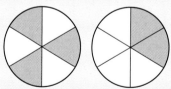

Synthesis

41. Grandma Hammons left $\frac{2}{3}$ of her $\frac{7}{8}$-acre apple farm to Karl. Karl gave $\frac{1}{4}$ of his share to his oldest daughter, Eileen. How much land did Eileen receive?

42. Simplify: $\left(\dfrac{3}{8}\right)^2 \div \dfrac{6}{7} \cdot \dfrac{2}{9} \div 5.$

Fraction Notation and Mixed Numerals

Real-World Application

Black bears typically have two cubs. In January 2007 in northern New Hampshire, a black bear sow gave birth to a litter of 5 cubs. This is so rare that Tom Sears, a wildlife photographer, spent 28 hr per week for six weeks watching for the perfect opportunity to photograph this family of six. At the time of this photo, an observer estimated that the cubs weighed $7\frac{1}{2}$ lb, 8 lb, $9\frac{1}{2}$ lb, $10\frac{5}{8}$ lb, and $11\frac{3}{4}$ lb. What was the average weight of the cubs?

Source: Andrew Timmins, New Hampshire Fish and Game Department, *Northcountry News*, Warren, NH; Tom Sears, photographer

This problem appears as Exercise 25 in Section 3.7.

3.1

Least Common Multiples

OBJECTIVE

a Find the least common multiple, or LCM, of two or more numbers.

SKILL TO REVIEW

Objective 2.1b: Find some multiples of a number.

Multiply by 1, 2, 3, and so on, to find six multiples of each number.

1. 8   **2.** 25

In this chapter, we study addition and subtraction using fraction notation. Suppose we want to add $\frac{2}{3}$ and $\frac{1}{2}$. To do so, we rewrite the fractions with a common denominator. The number we choose for the common denominator is the least common multiple of the denominators, 6; $\frac{2}{3} + \frac{1}{2} = \frac{4}{6} + \frac{3}{6}$. Then we add the numerators and keep the common denominator. In order to do this, we must be able to find the **least common denominator (LCD)**, or **least common multiple (LCM)**, of the denominators.

a Finding Least Common Multiples

> **LEAST COMMON MULTIPLE, LCM**
>
> The **least common multiple**, or LCM, of two natural numbers is the smallest number that is a multiple of both numbers.

EXAMPLE 1   Find the LCM of 20 and 30.

First, we list some multiples of 20 by multiplying 20 by 1, 2, 3, and so on:

$$20, 40, 60, 80, 100, 120, 140, 160, 180, 200, 220, 240, \ldots.$$

Then we list some multiples of 30 by multiplying 30 by 1, 2, 3, and so on:

$$30, 60, 90, 120, 150, 180, 210, 240, \ldots.$$

Now we determine the smallest number *common* to both lists. The LCM of 20 and 30 is 60.

Do Margin Exercise 1.

1. Find the LCM of 9 and 15 by examining lists of multiples.

Next, we develop three methods that are more efficient for finding LCMs. You may choose to learn only one method. (Consult with your instructor.) If you are going to study algebra, you should definitely learn method 2.

Method 1: Finding LCMs Using One List of Multiples

One method for finding LCMs uses *one* list of multiples. Let's consider finding the LCM of 9 and 12. The larger number, 12, is not a multiple of 9, so we check multiples of 12 until we find a number that is also a multiple of 9:

$$1 \cdot 12 = 12, \quad \text{not a multiple of 9;}$$
$$2 \cdot 12 = 24, \quad \text{not a multiple of 9;}$$
$$3 \cdot 12 = 36, \quad \text{a multiple of 9: } 4 \cdot 9 = 36.$$

The LCM of 9 and 12 is 36.

Answers

Skill to Review:
1. 8, 16, 24, 32, 40, 48
2. 25, 50, 75, 100, 125, 150

Margin Exercise:
1. 45

> *Method 1.* To find the LCM of a set of numbers using a list of multiples:
>
> a) Determine whether the largest number is a multiple of the others. If it is, it is the LCM. That is, if the largest number has the others as factors, the LCM is that number.
>
> b) If not, check multiples of the largest number until you get one that is a multiple of each of the others.

EXAMPLE 2 Find the LCM of 12 and 15.

a) 15 is not a multiple of 12.

b) Check multiples of 15:

$1 \cdot 15 = 15,$ Not a multiple of 12. When we divide 15 by 12, we get a nonzero remainder.

$2 \cdot 15 = 30,$ Not a multiple of 12

$3 \cdot 15 = 45,$ Not a multiple of 12

$4 \cdot 15 = 60.$ A multiple of 12: $5 \cdot 12 = 60$

The LCM $= 60$.

Do Exercise 2.

2. Find the LCM of 8 and 10 by examining lists of multiples.

EXAMPLE 3 Find the LCM of 8 and 32.

a) 32 is a multiple of $8 (4 \cdot 8 = 32)$, so the LCM $= 32$.

EXAMPLE 4 Find the LCM of 10, 20, and 50.

a) 50 is a multiple of 10 but not a multiple of 20.

b) Check multiples of 50:

$1 \cdot 50 = 50,$

$2 \cdot 50 = 100.$ A multiple of 10 and of 20: $10 \cdot 10 = 100$ and $5 \cdot 20 = 100$

The LCM $= 100$.

Do Exercises 3–6.

Find the LCM.

3. 10, 15   **4.** 6, 8

5. 5, 10   **6.** 20, 45, 80

Method 2: Finding LCMs Using Prime Factorizations

A second method for finding LCMs uses prime factorizations. Consider again 20 and 30. Their prime factorizations are $20 = 2 \cdot 2 \cdot 5$ and $30 = 2 \cdot 3 \cdot 5$. Let's look at these prime factorizations in order to find the LCM. Any multiple of 20 will have to have *two* 2's as factors and *one* 5 as a factor. Any multiple of 30 will need to have *one* 2, *one* 3, and *one* 5 as factors. The smallest number satisfying these conditions is

Two 2's, one 5; makes 20 a factor

$2 \cdot 2 \cdot 3 \cdot 5.$

One 2, one 3, one 5; makes 30 a factor

Thus the LCM of 20 and 30 is $2 \cdot 2 \cdot 3 \cdot 5$, or 60. It has all the factors of 20 and all the factors of 30, but the factors are not repeated when they are common to both numbers.

Note that the greatest number of times that 2 occurs as a factor of either 20 or 30 is two, and the LCM has 2 as a factor twice. The greatest number of times that 3 occurs as a factor of either 20 or 30 is one, and the LCM has 3 as a factor once. The greatest number of times that 5 occurs as a factor of either 20 or 30 is one, and the LCM has 5 as a factor once.

Method 2. To find the LCM of a set of numbers using prime factorizations:

a) Find the prime factorization of each number.

b) Create a product of factors, using each factor the greatest number of times that it occurs in any one factorization.

EXAMPLE 5 Find the LCM of 6 and 8.

a) Find the prime factorization of each number.

$$6 = 2 \cdot 3, \qquad 8 = 2 \cdot 2 \cdot 2$$

b) Create a product by writing factors that appear in the factorizations of 6 and 8, using each the greatest number of times that it occurs in any one factorization.

Consider the factor 2. The greatest number of times that 2 occurs in any one factorization is three. We write 2 as a factor three times.

$$2 \cdot 2 \cdot 2 \cdot ?$$

Consider the factor 3. The greatest number of times that 3 occurs in any one factorization is one. We write 3 as a factor one time.

$$2 \cdot 2 \cdot 2 \cdot 3 \cdot ?$$

Since there are no other prime factors in either factorization, the

LCM is $2 \cdot 2 \cdot 2 \cdot 3$, or 24.

EXAMPLE 6 Find the LCM of 24 and 36.

a) Find the prime factorization of each number.

$$24 = 2 \cdot 2 \cdot 2 \cdot 3, \qquad 36 = 2 \cdot 2 \cdot 3 \cdot 3$$

b) Create a product by writing factors, using each the greatest number of times that it occurs in any one factorization.

Consider the factor 2. The greatest number of times that 2 occurs in any one factorization is three. We write 2 as a factor three times:

$$2 \cdot 2 \cdot 2 \cdot ?$$

Consider the factor 3. The greatest number of times that 3 occurs in any one factorization is two. We write 3 as a factor two times:

$$2 \cdot 2 \cdot 2 \cdot 3 \cdot 3 \cdot ?$$

Since there are no other prime factors in either factorization, the

LCM is $2 \cdot 2 \cdot 2 \cdot 3 \cdot 3$, or 72.

Use prime factorizations to find each LCM.

7. 8, 10

8. 18, 40

9. 32, 54

Answers

7. 40   8. 360   9. 864

Do Exercises 7–9.

EXAMPLE 7 Find the LCM of 81, 90, and 84.

a) Find the prime factorization of each number.

$$81 = 3 \cdot 3 \cdot 3 \cdot 3, \qquad 90 = 2 \cdot 3 \cdot 3 \cdot 5, \qquad 84 = 2 \cdot 2 \cdot 3 \cdot 7$$

b) Create a product by writing factors, using each the greatest number of times that it occurs in any one factorization.

Consider the factor 2. The greatest number of times that 2 occurs in any one factorization is two. We write 2 as a factor two times:

$$2 \cdot 2 \cdot ?$$

Consider the factor 3. The greatest number of times that 3 occurs in any one factorization is four. We write 3 as a factor four times:

$$2 \cdot 2 \cdot 3 \cdot 3 \cdot 3 \cdot 3 \cdot ?$$

Consider the factor 5. The greatest number of times that 5 occurs in any one factorization is one. We write 5 as a factor one time:

$$2 \cdot 2 \cdot 3 \cdot 3 \cdot 3 \cdot 3 \cdot 5 \cdot ?$$

Consider the factor 7. The greatest number of times that 7 occurs in any one factorization is one. We write 7 as a factor one time:

$$2 \cdot 2 \cdot 3 \cdot 3 \cdot 3 \cdot 3 \cdot 5 \cdot 7 \cdot ?$$

Since there are no other prime factors in any of the factorizations, the

LCM is $2 \cdot 2 \cdot 3 \cdot 3 \cdot 3 \cdot 3 \cdot 5 \cdot 7$,   or   11,340.

Do Exercise 10.

10. Find the LCM of 24, 35, and 45.

Let's reconsider Example 7 using exponents. We want to find the LCM of 81, 90, and 84. We factor and then convert to exponential notation. The largest exponents indicate the greatest number of times that 2, 3, 5, and 7 occur as factors.

$81 = 3 \cdot 3 \cdot 3 \cdot 3 = 3^4;$ 4 is the largest exponent of 3 in any of the factorizations.

$90 = 2 \cdot 3 \cdot 3 \cdot 5 = 2^1 \cdot 3^2 \cdot 5^1;$ 1 is the largest exponent of 5 in any of the factorizations.

$84 = 2 \cdot 2 \cdot 3 \cdot 7 = 2^2 \cdot 3^1 \cdot 7^1.$ 2 is the largest exponent of 2 and 1 is the largest exponent of 7 in any of the factorizations.

Thus the LCM is $2^2 \cdot 3^4 \cdot 5^1 \cdot 7^1$, or 11,340.

Do Exercises 11 and 12.

11. Use exponents to find the LCM of 24, 35, and 45.

12. Redo Margin Exercises 7–9 using exponents.

EXAMPLE 8 Find the LCM of 8 and 9.

We find the prime factorization of each number.

$$8 = 2 \cdot 2 \cdot 2, \qquad 9 = 3 \cdot 3$$

Note that the two numbers, 8 and 9, have no common prime factor. When this is the case, the LCM is just the product of the two numbers. Thus the LCM is $2 \cdot 2 \cdot 2 \cdot 3 \cdot 3$, or $8 \cdot 9$, or 72.

Do Exercises 13 and 14.

Find the LCM.

13. 4, 9

14. 5, 6, 7

Answers

10. 2520   **11.** $2^3 \cdot 3^2 \cdot 5 \cdot 7$, or 2520
12. $2^3 \cdot 5$, or 40; $2^3 \cdot 3^2 \cdot 5$, or 360;
$2^5 \cdot 3^3$, or 864   **13.** 36   **14.** 210

EXAMPLE 9 Find the LCM of 7 and 21.

We find the prime factorization of each number. Because 7 is prime, it has no prime factorization.

$$7 = 7, \qquad 21 = 3 \cdot 7$$

Note that 7 is a factor of 21. We stated earlier that if one number is a factor of another, the LCM is the larger of the numbers. Thus the LCM is $7 \cdot 3$, or 21.

Find the LCM.

15. 3, 18   **16.** 12, 24

Do Exercises 15 and 16.

Method 3: Finding LCMs Using Division by Primes

Here is another method for finding LCMs that is especially useful for three or more numbers. For example, to find the LCM of 48, 72, and 80, we first look for any prime that divides any two of the numbers with no remainder. Then we divide as follows. We repeat the process, bringing down any numbers not divisible by the prime, until we can divide no more—that is, until there are no two numbers divisible by the same prime:

$$
\begin{array}{r|lll}
2 & 48 & 72 & 80 \\
3 & 24 & 36 & 40 \\
2 & 8 & 12 & 40 \\
2 & 4 & 6 & 20 \\
2 & 2 & 3 & 10 \\
\hline
 & 1 & 3 & 5
\end{array}
$$

40 is not divisible by 3.

3 is not divisible by 2.

No two of 1, 3, and 5 are divisible by the same prime. We stop here. The LCM is

$$2 \cdot 3 \cdot 2 \cdot 2 \cdot 2 \cdot 1 \cdot 3 \cdot 5, \quad \text{or} \quad 720.$$

Method 3. To find the LCM using division by primes:

a) First look for any prime that divides at least two of the numbers with no remainder. Then divide, bringing down any numbers not divisible by the prime. If you cannot find a prime that divides at least two of the numbers, then the LCM is the product of the numbers.

b) Repeat the process until you can divide no more—that is, until there are no two numbers divisible by the same prime.

EXAMPLE 10 Find the LCM of 12, 18, 21, and 40.

$$
\begin{array}{r|llll}
3 & 12 & 18 & 21 & 40 \\
2 & 4 & 6 & 7 & 40 \\
2 & 2 & 3 & 7 & 20 \\
\hline
 & 1 & 3 & 7 & 10
\end{array}
$$

No two of 1, 3, 7, and 10 are divisible by the same prime. We stop here.

The LCM is $3 \cdot 2 \cdot 2 \cdot 1 \cdot 3 \cdot 7 \cdot 10$, or 2520.

Find the LCM using division by primes.

17. 81, 90, 84

18. 24, 35, 45

19. 12, 75, 80, 120

Do Exercises 17–19.

Answers

15. 18   **16.** 24   **17.** 11,340   **18.** 2520
19. 1200

a   Find the LCM of each set of numbers.

1. 2, 4

2. 3, 15

3. 10, 25

4. 10, 15

5. 20, 40

6. 8, 12

7. 18, 27

8. 9, 11

9. 30, 50

10. 24, 36

11. 30, 40

12. 21, 27

13. 18, 24

14. 12, 18

15. 60, 70

16. 35, 45

17. 16, 36

18. 18, 20

19. 32, 36

20. 36, 48

21. 2, 3, 5

22. 3, 5, 7

23. 5, 18, 3

24. 6, 12, 18

25. 24, 36, 12

26. 8, 16, 22

27. 5, 12, 15

28. 12, 18, 40

29. 9, 12, 6

30. 8, 16, 12

31. 180, 100, 450, 60

32. 18, 30, 50, 48

33. 8, 48

34. 16, 32

35. 5, 50

36. 12, 72

37. 11, 13

38. 13, 14

39. 12, 35

40. 23, 25

41. 54, 63

42. 56, 72

43. 81, 90

44. 75, 100

45. 36, 54, 80

46. 22, 42, 51

47. 39, 91, 108, 26

48. 625, 75, 500, 25

49. 2000, 3000

50. 300, 4000

Applications of LCMs: Planet Orbits.   Jupiter, Saturn, and Uranus all revolve around the sun. Jupiter takes 12 yr, Saturn 30 yr, and Uranus 84 yr to make a complete revolution. On a certain night, you look at Jupiter, Saturn, and Uranus and wonder how many years it will take before they have the same position again. (*Hint:* To find out, you find the LCM of 12, 30, and 84. It will be that number of years.)

Source: *The Handy Science Answer Book*

51. How often will Jupiter and Saturn appear in the same direction in the night sky as seen from the earth?

52. How often will Jupiter and Uranus appear in the same direction in the night sky as seen from the earth?

53. How often will Saturn and Uranus appear in the same direction in the night sky as seen from the earth?

54. How often will Jupiter, Saturn, and Uranus appear in the same direction in the night sky as seen from the earth?

Skill Maintenance

55. *Tornadoes.*   The United States reports more tornadoes per year than any other country in the world. From January through May 2008, 1007 tornadoes were recorded. This was an increase of 348 tornadoes over the same time period in 2007. How many tornadoes occurred from January through May in 2007?   [1.8a]

Sources: NOAA's National Weather Service; www.tornadoarchive.com/History.aspx

56. *Population of Africa.*   The population of South America is approximately $\frac{2}{5}$ of the population of Africa. In 2007, the population of South America was about 376,000,000. What was the population of Africa in 2007?   [2.7d]

Sources: Population Division/International Programs Center; *The World Almanac* 2008

57. Divide and simplify: $\frac{4}{5} \div \frac{7}{10}$.   [2.7b]

58. Add: $23,456 + 5677 + 4002$.   [1.2a]

59. Multiply: 2118×3001.   [1.4a]

60. Subtract: $80,004 - 2305$.   [1.3a]

Synthesis

61. A pencil company uses two sizes of boxes, 5 in. by 6 in. and 5 in. by 8 in. These boxes are packed in bigger cartons for shipping. Find the width and the length of the smallest carton that will accommodate boxes of either size without any room left over. (Each carton can contain only one type of box and all boxes must point in the same direction.)

62. Consider 8 and 12. Determine whether each of the following is the LCM of 8 and 12. Tell why or why not.
   a) $2 \cdot 2 \cdot 3 \cdot 3$
   b) $2 \cdot 2 \cdot 3$
   c) $2 \cdot 3 \cdot 3$
   d) $2 \cdot 2 \cdot 2 \cdot 3$

3.2 Addition and Applications

a Addition Using Fraction Notation

Like Denominators

Addition using fraction notation corresponds to combining or putting like things together, just as addition with whole numbers does. For example,

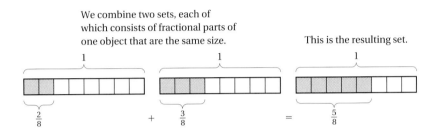

We combine two sets, each of which consists of fractional parts of one object that are the same size.

This is the resulting set.

$$\frac{2}{8} \quad + \quad \frac{3}{8} \quad = \quad \frac{5}{8}$$

2 eighths + 3 eighths = 5 eighths,

or $\quad 2 \cdot \dfrac{1}{8} + 3 \cdot \dfrac{1}{8} = 5 \cdot \dfrac{1}{8},\quad$ or $\quad \dfrac{2}{8} + \dfrac{3}{8} = \dfrac{5}{8}.$

We see that to add when denominators are the same, we add the numerators and keep the denominator.

Do Margin Exercise 1.

To add when denominators are the same,

a) add the numerators,

b) keep the denominator, and

$$\frac{2}{6} + \frac{5}{6} = \frac{2 + 5}{6} = \frac{7}{6}$$

c) simplify, if possible.

EXAMPLES Add and simplify.

1. $\dfrac{2}{4} + \dfrac{1}{4} = \dfrac{2 + 1}{4} = \dfrac{3}{4}\qquad$ No simplifying is possible.

2. $\dfrac{11}{6} + \dfrac{3}{6} = \dfrac{11 + 3}{6} = \dfrac{14}{6} = \dfrac{2 \cdot 7}{2 \cdot 3} = \dfrac{2}{2} \cdot \dfrac{7}{3} = 1 \cdot \dfrac{7}{3} = \dfrac{7}{3}\qquad$ Here we simplified.

3. $\dfrac{3}{12} + \dfrac{5}{12} = \dfrac{3 + 5}{12} = \dfrac{8}{12} = \dfrac{4 \cdot 2}{4 \cdot 3} = \dfrac{4}{4} \cdot \dfrac{2}{3} = 1 \cdot \dfrac{2}{3} = \dfrac{2}{3}$

Do Exercises 2–5.

Different Denominators

What do we do when denominators are different? We can find a common denominator by multiplying by 1. Consider adding $\frac{1}{6}$ and $\frac{3}{4}$. There are many common denominators that can be obtained. Let's look at two possibilities.

SKILL TO REVIEW
Objective 2.1d: Find the prime factorization of a composite number.

Find the prime factorization of each number.

1. 96   **2.** 1400

1. Find $\dfrac{1}{5} + \dfrac{3}{5}$.

Add and simplify.

2. $\dfrac{5}{13} + \dfrac{9}{13}$ **3.** $\dfrac{1}{3} + \dfrac{2}{3}$

4. $\dfrac{5}{12} + \dfrac{1}{12}$ **5.** $\dfrac{9}{16} + \dfrac{3}{16}$

Answers

Skill to Review:
1. $2 \cdot 2 \cdot 2 \cdot 2 \cdot 2 \cdot 3$
2. $2 \cdot 2 \cdot 2 \cdot 5 \cdot 5 \cdot 7$

Margin Exercises:
1. $\dfrac{4}{5}$ 2. $\dfrac{14}{13}$ 3. 1 4. $\dfrac{1}{2}$ 5. $\dfrac{3}{4}$

A. $\dfrac{1}{6} + \dfrac{3}{4} = \dfrac{1}{6} \cdot 1 + \dfrac{3}{4} \cdot 1$

$\qquad\qquad = \dfrac{1}{6} \cdot \dfrac{4}{4} + \dfrac{3}{4} \cdot \dfrac{6}{6}$

$\qquad\qquad = \dfrac{4}{24} + \dfrac{18}{24}$

$\qquad\qquad = \dfrac{22}{24}$

$\qquad\qquad = \dfrac{11}{12} \qquad$ Simplifying

B. $\dfrac{1}{6} + \dfrac{3}{4} = \dfrac{1}{6} \cdot 1 + \dfrac{3}{4} \cdot 1$

$\qquad\qquad = \dfrac{1}{6} \cdot \dfrac{2}{2} + \dfrac{3}{4} \cdot \dfrac{3}{3}$

$\qquad\qquad = \dfrac{2}{12} + \dfrac{9}{12}$

$\qquad\qquad = \dfrac{11}{12}$

We had to simplify in (A) but not in (B). In (B), we used the least common multiple of the denominators, 12, as the common denominator. That number is called the **least common denominator**, or **LCD**. As we will see in Example 6, we may still need to simplify when using the LCD, but it is usually easier than when we use a larger common denominator.

> To add when denominators are different:
>
> **a)** Find the least common multiple of the denominators. That number is the least common denominator, LCD.
>
> **b)** Multiply by 1, using an appropriate notation, n/n, to express each number in terms of the LCD.
>
> **c)** Add the numerators, keeping the same denominator.
>
> **d)** Simplify, if possible.

EXAMPLE 4 Add: $\dfrac{3}{4} + \dfrac{1}{8}$.

The LCD is 8.   4 is a factor of 8, so the LCM of 4 and 8 is 8.

$\dfrac{3}{4} + \dfrac{1}{8} = \dfrac{3}{4} \cdot 1 + \dfrac{1}{8}$ ← This fraction already has the LCD as its denominator.

$\qquad\quad = \dfrac{3}{4} \cdot \dfrac{2}{2} + \dfrac{1}{8}$ *Think*: $4 \times \square = 8$. The answer is 2, so we multiply by 1, using $\frac{2}{2}$.

$\qquad\quad = \dfrac{6}{8} + \dfrac{1}{8} = \dfrac{7}{8}$

EXAMPLE 5 Add: $\dfrac{1}{9} + \dfrac{5}{6}$.

The LCD is 18. $9 = 3 \cdot 3$ and $6 = 2 \cdot 3$, so the LCM of 9 and 6 is $2 \cdot 3 \cdot 3$, or 18.

$\dfrac{1}{9} + \dfrac{5}{6} = \dfrac{1}{9} \cdot 1 + \dfrac{5}{6} \cdot 1 = \dfrac{1}{9} \cdot \dfrac{2}{2} + \dfrac{5}{6} \cdot \dfrac{3}{3}$ *Think*: $6 \times \square = 18$. The answer is 3, so we multiply by 1 using $\frac{3}{3}$.

Think: $9 \times \square = 18$. The answer is 2, so we multiply by 1 using $\frac{2}{2}$.

$\qquad\quad = \dfrac{2}{18} + \dfrac{15}{18} = \dfrac{17}{18}$

Add.

6. $\dfrac{2}{3} + \dfrac{1}{6}$

7. $\dfrac{3}{8} + \dfrac{5}{6}$

Do Exercises 6 and 7.

Answers

6. $\dfrac{5}{6}$ **7.** $\dfrac{29}{24}$

EXAMPLE 6 Add: $\dfrac{5}{9} + \dfrac{11}{18}$.

The LCD is 18.   9 is a factor of 18, so the LCM is 18.

$$\frac{5}{9} + \frac{11}{18} = \frac{5}{9} \cdot \frac{2}{2} + \frac{11}{18} = \frac{10}{18} + \frac{11}{18}$$

$$\left. \begin{aligned} &= \frac{21}{18} = \frac{3 \cdot 7}{3 \cdot 6} = \frac{3}{3} \cdot \frac{7}{6} \\ &= \frac{7}{6} \end{aligned} \right\} \quad \text{Simplifying}$$

Do Exercise 8.

8. Add: $\dfrac{1}{6} + \dfrac{7}{18}$.

EXAMPLE 7 Add: $\dfrac{1}{10} + \dfrac{3}{100} + \dfrac{7}{1000}$.

Since 10 and 100 are factors of 1000, the LCD is 1000. Then

$$\frac{1}{10} + \frac{3}{100} + \frac{7}{1000} = \frac{1}{10} \cdot \frac{100}{100} + \frac{3}{100} \cdot \frac{10}{10} + \frac{7}{1000}$$

$$= \frac{100}{1000} + \frac{30}{1000} + \frac{7}{1000} = \frac{137}{1000}.$$

Do Exercise 9.

9. Add: $\dfrac{4}{10} + \dfrac{1}{100} + \dfrac{3}{1000}$.

When denominators are large, we most often use the prime factorization of each denominator to find the LCD. This is shown in Example 8.

EXAMPLE 8 Add: $\dfrac{13}{70} + \dfrac{11}{21} + \dfrac{8}{15}$.

We have

$$\frac{13}{70} + \frac{11}{21} + \frac{8}{15} = \frac{13}{2 \cdot 5 \cdot 7} + \frac{11}{3 \cdot 7} + \frac{8}{3 \cdot 5}. \quad \text{Factoring denominators}$$

The LCD is $2 \cdot 3 \cdot 5 \cdot 7$, or 210. Then

$$\frac{13}{70} + \frac{11}{21} + \frac{8}{15} = \frac{13}{2 \cdot 5 \cdot 7} \cdot \frac{3}{3} + \frac{11}{3 \cdot 7} \cdot \frac{2 \cdot 5}{2 \cdot 5} + \frac{8}{3 \cdot 5} \cdot \frac{7 \cdot 2}{7 \cdot 2}$$

The LCM of 70, 21, and 15 is $2 \cdot 3 \cdot 5 \cdot 7$. In each case, think of which factors are needed to get the LCD. Then multiply by 1 to obtain the LCD in each denominator.

$$= \frac{13 \cdot 3}{2 \cdot 5 \cdot 7 \cdot 3} + \frac{11 \cdot 2 \cdot 5}{3 \cdot 7 \cdot 2 \cdot 5} + \frac{8 \cdot 7 \cdot 2}{3 \cdot 5 \cdot 7 \cdot 2}$$

$$= \frac{39}{2 \cdot 3 \cdot 5 \cdot 7} + \frac{110}{2 \cdot 3 \cdot 5 \cdot 7} + \frac{112}{2 \cdot 3 \cdot 5 \cdot 7}$$

$$= \frac{261}{2 \cdot 3 \cdot 5 \cdot 7} = \frac{3 \cdot 3 \cdot 29}{2 \cdot 3 \cdot 5 \cdot 7} \quad \text{Factoring the numerator}$$

$$= \frac{3}{3} \cdot \frac{3 \cdot 29}{2 \cdot 5 \cdot 7} = \frac{3 \cdot 29}{2 \cdot 5 \cdot 7} = \frac{87}{70}.$$

Do Exercises 10 and 11.

Add.

10. $\dfrac{7}{10} + \dfrac{2}{21} + \dfrac{6}{7}$

11. $\dfrac{5}{18} + \dfrac{7}{24} + \dfrac{11}{36}$

Answers

8. $\dfrac{5}{9}$ 9. $\dfrac{413}{1000}$ 10. $\dfrac{347}{210}$ 11. $\dfrac{7}{8}$

b Applications and Problem Solving

EXAMPLE 9 *Subflooring.* A contractor requires his subcontractors to use two layers of subflooring under a ceramic tile floor. First, the subcontractors install a $\frac{3}{4}$-in. layer of oriented strand board (OSB). Then a $\frac{1}{2}$-in. sheet of cement board is mortared to the OSB. The mortar is $\frac{1}{8}$-in. thick. What is the total thickness of the two installed subfloors?

1. **Familiarize.** We first make a drawing. We let T = the total thickness of the subfloors.

2. **Translate.** The problem can be translated to an equation as follows.

OSB	plus	Mortar	plus	Cement board	is	Total thickness
$\frac{3}{4}$	$+$	$\frac{1}{8}$	$+$	$\frac{1}{2}$	$=$	T

3. **Solve.** To solve the equation, we carry out the addition. The LCM of the denominators is 8 because 2 and 4 are factors of 8. We multiply by 1 in order to obtain the LCD:

$$\frac{3}{4} + \frac{1}{8} + \frac{1}{2} = T$$

$$\frac{3}{4} \cdot \frac{2}{2} + \frac{1}{8} + \frac{1}{2} \cdot \frac{4}{4} = T$$

$$\frac{6}{8} + \frac{1}{8} + \frac{4}{8} = T$$

$$\frac{11}{8} = T.$$

4. **Check.** We check by repeating the calculation. We also note that the sum should be larger than any of the individual measurements, which it is. This tells us that the answer is reasonable.

5. **State.** The total thickness of the installed subfloors is $\frac{11}{8}$ in.

> Do Exercise 12.

---

12. Fruit Salad. A caterer prepares a mixed berry salad with $\frac{7}{8}$ qt of strawberries, $\frac{3}{4}$ qt of raspberries, and $\frac{5}{16}$ qt of blueberries. What is the total amount of berries in the salad?

Answer

12. $\frac{31}{16}$ qt

a Add and simplify.

1. $\dfrac{7}{8} + \dfrac{1}{8}$

2. $\dfrac{2}{5} + \dfrac{3}{5}$

3. $\dfrac{1}{8} + \dfrac{5}{8}$

4. $\dfrac{3}{10} + \dfrac{3}{10}$

5. $\dfrac{2}{3} + \dfrac{5}{6}$

6. $\dfrac{5}{6} + \dfrac{1}{9}$

7. $\dfrac{1}{8} + \dfrac{1}{6}$

8. $\dfrac{1}{6} + \dfrac{3}{4}$

9. $\dfrac{4}{5} + \dfrac{7}{10}$

10. $\dfrac{3}{4} + \dfrac{1}{12}$

11. $\dfrac{5}{12} + \dfrac{3}{8}$

12. $\dfrac{7}{8} + \dfrac{1}{16}$

13. $\dfrac{3}{20} + \dfrac{3}{4}$

14. $\dfrac{2}{15} + \dfrac{2}{5}$

15. $\dfrac{5}{6} + \dfrac{7}{9}$

16. $\dfrac{5}{8} + \dfrac{5}{6}$

17. $\dfrac{3}{10} + \dfrac{1}{100}$

18. $\dfrac{9}{10} + \dfrac{3}{100}$

19. $\dfrac{5}{12} + \dfrac{4}{15}$

20. $\dfrac{3}{16} + \dfrac{1}{12}$

21. $\dfrac{9}{10} + \dfrac{99}{100}$

22. $\dfrac{3}{10} + \dfrac{27}{100}$

23. $\dfrac{7}{8} + \dfrac{0}{1}$

24. $\dfrac{0}{1} + \dfrac{5}{6}$

25. $\dfrac{3}{8} + \dfrac{1}{6}$

26. $\dfrac{5}{8} + \dfrac{1}{6}$

27. $\dfrac{5}{12} + \dfrac{7}{24}$

28. $\dfrac{1}{18} + \dfrac{7}{12}$

29. $\dfrac{3}{16} + \dfrac{5}{16} + \dfrac{4}{16}$

30. $\dfrac{3}{8} + \dfrac{1}{8} + \dfrac{2}{8}$

31. $\dfrac{8}{10} + \dfrac{7}{100} + \dfrac{4}{1000}$

32. $\dfrac{1}{10} + \dfrac{2}{100} + \dfrac{3}{1000}$

33. $\dfrac{3}{8} + \dfrac{5}{12} + \dfrac{8}{15}$

34. $\dfrac{1}{2} + \dfrac{3}{8} + \dfrac{1}{4}$

35. $\dfrac{15}{24} + \dfrac{7}{36} + \dfrac{91}{48}$

36. $\dfrac{5}{7} + \dfrac{25}{52} + \dfrac{7}{4}$

b Solve.

37. *Riding a Segway®.* Tate rode a Segway® Personal Transporter $\frac{5}{6}$ mi to the library, then $\frac{3}{4}$ mi to class, and then $\frac{3}{2}$ mi to his part-time job. How far did he ride his Segway®?

38. For a community project, an earth science class volunteered one hour per day for three days to join the state highway beautification project. The students collected trash along a $\frac{4}{5}$-mi stretch of highway the first day, $\frac{5}{8}$ mi the second day, and $\frac{1}{2}$ mi the third day. How many miles along the highway did they clean?

39. *Caffeine.* To cut back on their caffeine intake, Michelle and Gerry mix caffeinated and decaffeinated coffee beans before grinding for a customized mix. They mix $\frac{3}{16}$ lb of decaffeinated beans with $\frac{5}{8}$ lb of caffeinated beans. What is the total amount of coffee beans in the mixture?

40. Alyse bought $\frac{1}{3}$ lb of orange pekoe tea and $\frac{1}{2}$ lb of English cinnamon tea. How many pounds of tea did she buy?

41. *Punch Recipe.* A recipe for strawberry punch calls for $\frac{1}{5}$ qt of ginger ale and $\frac{3}{5}$ qt of strawberry soda. How much liquid is needed? If the recipe is doubled, how much liquid is needed? If the recipe is halved, how much liquid is needed?

42. *Concrete Mix.* A cubic meter of concrete mix contains 420 kg of cement, 150 kg of stone, and 120 kg of sand. What is the total weight of a cubic meter of the mix? What part is cement? stone? sand? Add these fractional amounts. What is the result?

43. *Iced Brownies.* The campus culinary arts department is preparing brownies for the international student reception. Students in the catering program iced the $\frac{11}{16}$-in. $\left(\frac{11"}{16}\right)$ butterscotch brownies with a $\frac{5}{32}$-in. $\left(\frac{5"}{32}\right)$ layer of icing. What is the thickness of the iced brownie?

44. *Carpentry.* To cut expenses, a carpenter sometimes glues two kinds of plywood together. He glues a $\frac{1}{4}$-in. $\left(\frac{1"}{4}\right)$ piece of walnut plywood to a $\frac{3}{8}$-in. $\left(\frac{3"}{8}\right)$ piece of less expensive plywood. What is the total thickness of these pieces?

45. A tile $\frac{5}{8}$ in. thick is glued to a board $\frac{7}{8}$ in. thick. The glue is $\frac{3}{32}$ in. thick. How thick is the result?

46. A baker used $\frac{1}{2}$ lb of flour for rolls, $\frac{1}{4}$ lb for donuts, and $\frac{1}{3}$ lb for cookies. How much flour was used?

Skill Maintenance

Multiply. [1.4a]

47. $408 \cdot 516$ **48.** $1125 \cdot 3728$ **49.** $423 \cdot 8009$ **50.** $2025 \cdot 174$

Presidential Elections. The results of the five closest presidential elections are listed in the following table. Use these data for Exercises 51–56. [1.2a], [1.3a]

DATE	PRESIDENT	ELECTORAL VOTES	POPULAR VOTES
1916	Woodrow Wilson (D)	277	9,126,300
	Charles E. Hughes (R)	254	8,546,789
1960	John F. Kennedy (D)	303	34,226,731
	Richard M. Nixon (R)	219	34,108,157
1968	Richard M. Nixon (R)	301	31,785,480
	Hubert H. Humphrey (D)	191	31,275,166
	George C. Wallace (3rd Party)	46	9,906,473
1976	Jimmy Carter (D)	297	40,830,763
	Gerald R. Ford (R)	240	39,147,793
2000	George W. Bush (R)	271	50,459,211
	Albert A. Gore (D)	266	51,003,894

SOURCE: *The World Almanac*, 2008, p. 544

51. How many more popular votes did Albert A. Gore have than George W. Bush in the 2000 presidential election?

52. How many more electoral votes did George W. Bush have than Albert A. Gore in the 2000 presidential election?

53. In the 1960 presidential election, how many more electoral votes did John F. Kennedy have than Richard M. Nixon?

54. In the 1968 presidential election, how many more popular votes did Richard M. Nixon have than Hubert H. Humphrey?

55. How much greater was the total of the popular vote counts in the 2000 presidential election than in the 1976 election?

56. How much greater was the total of the popular vote counts in the 2000 presidential election than in the 1960 election?

57. *Honey Production.* In 2007, 154,907,000 lb of honey were produced in the United States. The two states with the greatest honey production were North Dakota and California. North Dakota produced 25,900,000 lb of honey, and California produced 19,760,000 lb. How many more pounds of honey were produced in North Dakota than in California? [1.8a]

Source: www.cattlenetwork.com

58. The Bingham community garden is to be split into 16 equally sized plots. If the garden occupies $\frac{3}{4}$ acre of land, how large will each plot be? [2.7d]

Synthesis

59. A guitarist's band is booked for Friday and Saturday nights at a local club. The guitarist is part of a trio on Friday and part of a quintet on Saturday. Thus the guitarist is paid one-third of one-half the weekend's pay for Friday and one-fifth of one-half the weekend's pay for Saturday. What fractional part of the band's pay did the guitarist receive for the weekend's work? If the band was paid $1200, how much did the guitarist receive?

3.3 Subtraction, Order, and Applications

OBJECTIVES

a Subtract using fraction notation.

b Use $<$ or $>$ with fraction notation to write a true sentence.

c Solve equations of the type $x + a = b$ and $a + x = b$, where a and b may be fractions.

d Solve applied problems involving subtraction with fraction notation.

SKILL TO REVIEW

Objective 1.6c: Use $<$ or $>$ for $\square$ to write a true sentence in a situation like 6 $\square$ 10.

Use $<$ or $>$ for $\square$ to write a true sentence.

1. 218 $\square$ 128   **2.** 41 $\square$ 95

Subtract and simplify.

1. $\dfrac{7}{8} - \dfrac{3}{8}$

2. $\dfrac{10}{16} - \dfrac{4}{16}$

3. $\dfrac{8}{10} - \dfrac{3}{10}$

a Subtraction Using Fraction Notation

Like Denominators

Let's consider the difference $\frac{4}{8} - \frac{3}{8}$.

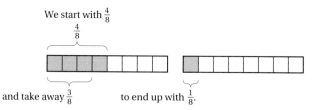

We start with $\frac{4}{8}$

and take away $\frac{3}{8}$ to end up with $\frac{1}{8}$.

We start with 4 eighths and take away 3 eighths:

$$4 \text{ eighths} - 3 \text{ eighths} = 1 \text{ eighth},$$

or $4 \cdot \dfrac{1}{8} - 3 \cdot \dfrac{1}{8} = \dfrac{1}{8},$ or $\dfrac{4}{8} - \dfrac{3}{8} = \dfrac{1}{8}.$

To subtract when denominators are the same,

a) subtract the numerators,

b) keep the denominator, and

c) simplify, if possible.

$$\dfrac{7}{10} - \dfrac{4}{10} = \dfrac{7 - 4}{10} = \dfrac{3}{10}$$

EXAMPLES Subtract and simplify.

1. $\dfrac{32}{12} - \dfrac{25}{12} = \dfrac{32 - 25}{12} = \dfrac{7}{12}$

2. $\dfrac{7}{10} - \dfrac{3}{10} = \dfrac{7 - 3}{10} = \dfrac{4}{10} = \dfrac{2 \cdot 2}{5 \cdot 2} = \dfrac{2}{5} \cdot \dfrac{2}{2} = \dfrac{2}{5} \cdot 1 = \dfrac{2}{5}$

3. $\dfrac{8}{9} - \dfrac{2}{9} = \dfrac{8 - 2}{9} = \dfrac{6}{9} = \dfrac{2 \cdot 3}{3 \cdot 3} = \dfrac{2}{3} \cdot \dfrac{3}{3} = \dfrac{2}{3} \cdot 1 = \dfrac{2}{3}$

Do Margin Exercises 1–3.

Different Denominators

The procedure for subtraction with different denominators is similar to the procedure for addition.

To subtract when denominators are different:

a) Find the least common multiple of the denominators. That number is the least common denominator, LCD.

b) Multiply by 1, using an appropriate notation, n/n, to express each number in terms of the LCD.

c) Subtract the numerators, keeping the same denominator.

d) Simplify, if possible.

Answers

Skill to Review:
1. $>$ 2. $<$

Margin Exercises:
1. $\dfrac{1}{2}$ 2. $\dfrac{3}{8}$ 3. $\dfrac{1}{2}$

EXAMPLE 4 Subtract: $\dfrac{2}{5} - \dfrac{3}{8}$.

The LCM of 5 and 8 is 40, so the LCD is 40.

$$\dfrac{2}{5} - \dfrac{3}{8} = \dfrac{2}{5} \cdot \dfrac{8}{8} - \dfrac{3}{8} \cdot \dfrac{5}{5} \longleftarrow$$
Think: $8 \times \square = 40$. The answer is 5, so we multiply by 1, using $\dfrac{5}{5}$.

Think: $5 \times \square = 40$. The answer is 8, so we multiply by 1, using $\dfrac{8}{8}$.

$$= \dfrac{16}{40} - \dfrac{15}{40} = \dfrac{16 - 15}{40} = \dfrac{1}{40}$$

Do Exercise 4.

4. Subtract: $\dfrac{3}{4} - \dfrac{2}{3}$.

EXAMPLE 5 Subtract: $\dfrac{5}{6} - \dfrac{7}{12}$.

Since 12 is a multiple of 6, the LCM of 6 and 12 is 12. The LCD is 12.

$$\dfrac{5}{6} - \dfrac{7}{12} = \dfrac{5}{6} \cdot \dfrac{2}{2} - \dfrac{7}{12}$$

$$= \dfrac{10}{12} - \dfrac{7}{12} = \dfrac{10 - 7}{12} = \dfrac{3}{12}$$

$$= \dfrac{3 \cdot 1}{3 \cdot 4} = \dfrac{3}{3} \cdot \dfrac{1}{4} = \dfrac{1}{4}$$

Do Exercises 5 and 6.

Subtract.

5. $\dfrac{5}{6} - \dfrac{1}{9}$ **6.** $\dfrac{4}{5} - \dfrac{3}{10}$

EXAMPLE 6 Subtract: $\dfrac{17}{24} - \dfrac{4}{15}$.

We have

$$\dfrac{17}{24} - \dfrac{4}{15} = \dfrac{17}{2 \cdot 2 \cdot 2 \cdot 3} - \dfrac{4}{3 \cdot 5}.$$ Factoring denominators

The LCD is $2 \cdot 2 \cdot 2 \cdot 3 \cdot 5$, or 120. Then

$$\dfrac{17}{24} - \dfrac{4}{15} = \dfrac{17}{2 \cdot 2 \cdot 2 \cdot 3} \cdot \dfrac{5}{5} - \dfrac{4}{3 \cdot 5} \cdot \dfrac{2 \cdot 2 \cdot 2}{2 \cdot 2 \cdot 2}$$

The LCM of 24 and 15 is $2 \cdot 2 \cdot 2 \cdot 3 \cdot 5$. In each case, we multiply by 1 to obtain the LCD.

$$= \dfrac{17 \cdot 5}{2 \cdot 2 \cdot 2 \cdot 3 \cdot 5} - \dfrac{4 \cdot 2 \cdot 2}{3 \cdot 5 \cdot 2 \cdot 2 \cdot 2}$$

$$= \dfrac{85}{120} - \dfrac{32}{120} = \dfrac{53}{120}.$$

Do Exercise 7.

7. Subtract: $\dfrac{11}{28} - \dfrac{5}{16}$.

b Order

We see from this figure that $\dfrac{4}{5}$ is greater than $\dfrac{3}{5}$, and $\dfrac{3}{5}$ is less than $\dfrac{4}{5}$. That is, $\dfrac{4}{5} > \dfrac{3}{5}$, and $\dfrac{3}{5} < \dfrac{4}{5}$.

Answers

4. $\dfrac{1}{12}$ 5. $\dfrac{13}{18}$ 6. $\dfrac{1}{2}$ 7. $\dfrac{9}{112}$

8. Use $<$ or $>$ for $\square$ to write a true sentence:

$$\frac{3}{8} \; \square \; \frac{5}{8}.$$

9. Use $<$ or $>$ for $\square$ to write a true sentence:

$$\frac{7}{10} \; \square \; \frac{6}{10}.$$

To determine which of two numbers is greater when there is a common denominator, compare the numerators:

$$\frac{4}{5}, \frac{3}{5}; \qquad 4 > 3; \qquad \frac{4}{5} > \frac{3}{5}.$$

Do Exercises 8 and 9.

When denominators are different, we cannot compare numerators until we multiply by 1 to make the denominators the same.

EXAMPLE 7 Use $<$ or $>$ for $\square$ to write a true sentence:

$$\frac{2}{5} \; \square \; \frac{3}{4}.$$

The LCD is 20. We have

$$\frac{2}{5} \cdot \frac{4}{4} = \frac{8}{20}; \qquad \text{We multiply by 1 using } \frac{4}{4} \text{ to get the LCD.}$$

$$\frac{3}{4} \cdot \frac{5}{5} = \frac{15}{20}. \qquad \text{We multiply by 1 using } \frac{5}{5} \text{ to get the LCD.}$$

Now that the denominators are the same, 20, we can compare the numerators. Since $8 < 15$, it follows that $\frac{8}{20} < \frac{15}{20}$, so

$$\frac{2}{5} < \frac{3}{4}.$$

EXAMPLE 8 Use $<$ or $>$ for $\square$ to write a true sentence:

$$\frac{9}{10} \; \square \; \frac{89}{100}.$$

The LCD is 100. We write $\frac{9}{10}$ with a denominator of 100 to make the denominators the same.

$$\frac{9}{10} \cdot \frac{10}{10} = \frac{90}{100} \qquad \text{We multiply by } \frac{10}{10} \text{ to get the LCD.}$$

Since $90 > 89$, it follows that $\frac{90}{100} > \frac{89}{100}$, so

$$\frac{9}{10} > \frac{89}{100}.$$

Do Exercises 10–12.

Use $<$ or $>$ for $\square$ to write a true sentence:

10. $\dfrac{2}{3} \; \square \; \dfrac{5}{8}$

11. $\dfrac{3}{4} \; \square \; \dfrac{8}{12}$

12. $\dfrac{5}{6} \; \square \; \dfrac{7}{8}$

(c) Solving Equations

Now let's solve equations of the form $x + a = b$ or $a + x = b$, where a and b may be fractions. Proceeding as we have before, we subtract a on both sides of the equation.

Answers

8. $<$ 9. $>$ 10. $>$ 11. $>$ 12. $<$

EXAMPLE 9 Solve: $x + \frac{1}{4} = \frac{3}{5}$.

$$x + \frac{1}{4} - \frac{1}{4} = \frac{3}{5} - \frac{1}{4}$$ Subtracting $\frac{1}{4}$ on both sides

$$x + 0 = \frac{3}{5} \cdot \frac{4}{4} - \frac{1}{4} \cdot \frac{5}{5}$$ The LCD is 20. We multiply by 1 to get the LCD.

$$x = \frac{12}{20} - \frac{5}{20} = \frac{7}{20}$$

Do Exercises 13 and 14.

Solve.

13. $x + \frac{2}{3} = \frac{5}{6}$

14. $\frac{3}{5} + t = \frac{7}{8}$

(d) Applications and Problem Solving

EXAMPLE 10 *Fraction of the Moon Illuminated.* From anywhere on Earth, the moon appears to be a circular disk. At midnight on July 11, 2008 (Eastern Daylight Time), $\frac{3}{5}$ of the moon appeared illuminated. By July 22, 2008, the illuminated portion had increased to $\frac{17}{20}$. How much more of the moon appeared illuminated on July 22 than on July 11?

Source: Astronomical Applications Department, U.S. Naval Observatory, Washington DC 20392

1. **Familiarize.** We let $m =$ the additional part of the moon that appeared illuminated.

2. **Translate.** We translate to an equation:

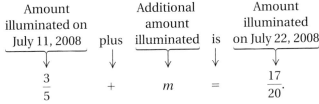

$$\frac{3}{5} + m = \frac{17}{20}.$$

3. **Solve.** To solve the equation, we subtract $\frac{3}{5}$ on both sides:

$$\frac{3}{5} + m = \frac{17}{20}$$

$$\frac{3}{5} + m - \frac{3}{5} = \frac{17}{20} - \frac{3}{5}$$ Subtracting $\frac{3}{5}$ on both sides

$$m + 0 = \frac{17}{20} - \frac{3}{5} \cdot \frac{4}{4}$$ The LCD is 20. We multiply by 1 to obtain the LCD.

$$m = \frac{17}{20} - \frac{12}{20} = \frac{5}{20} = \frac{5 \cdot 1}{5 \cdot 4} = \frac{5}{5} \cdot \frac{1}{4} = \frac{1}{4}.$$

4. **Check.** To check, we add:

$$\frac{3}{5} + \frac{1}{4} = \frac{3}{5} \cdot \frac{4}{4} + \frac{1}{4} \cdot \frac{5}{5} = \frac{12}{20} + \frac{5}{20} = \frac{17}{20}.$$

5. **State.** On July 22, $\frac{1}{4}$ more of the moon was illuminated than on July 11.

Do Exercise 15.

15. Fraction of the Moon Illuminated. At midnight on April 18, 2008 (Eastern Daylight Time), $\frac{19}{20}$ of the moon appeared illuminated. By April 27, 2008, the illuminated part had decreased to $\frac{16}{25}$. How much less of the moon appeared illuminated on April 27 than on April 18?

Source: Astronomical Applications Department, U.S. Naval Observatory, Washington DC 20392

Answers

13. $\frac{1}{6}$ **14.** $\frac{11}{40}$ **15.** $\frac{31}{100}$

Translating
for Success

1. *Bubble Wrap.* One-Stop Postal Center orders bubble wrap in 64-yd rolls. On average, $\frac{3}{4}$ yd is used per small package. How many small packages can be prepared with 2 rolls of bubble wrap?

2. *Distance from College.* The post office is $\frac{7}{9}$ mi from the community college. The medical clinic is $\frac{2}{5}$ as far from the college as the post office is. How far is the clinic from the college?

3. *Swimming.* Andrew swims $\frac{7}{9}$ mi every day. One day he swims $\frac{2}{5}$ mi by 11:00 A.M. How much farther must Andrew swim to reach his daily goal?

4. *Tuition.* The average tuition at Waterside University is $12,000. If a loan is obtained for $\frac{1}{3}$ of the tuition, how much is the loan?

5. *Thermos Bottle Capacity.* A thermos bottle holds $\frac{11}{12}$ gal. How much is in the bottle when it is $\frac{4}{7}$ full?

The goal of these matching questions is to practice step (2), *Translate*, of the five-step problem-solving process. Translate each word problem to an equation and select a correct translation from equations A–O.

A. $\frac{3}{4} \cdot 64 = x$

B. $\frac{1}{3} \cdot 12{,}000 = x$

C. $\frac{1}{3} + \frac{2}{5} = x$

D. $\frac{2}{5} + x = \frac{7}{9}$

E. $\frac{2}{5} \cdot \frac{7}{9} = x$

F. $\frac{3}{4} \cdot x = 64$

G. $\frac{4}{7} = x + \frac{11}{12}$

H. $\frac{2}{5} = x + \frac{7}{9}$

I. $\frac{4}{7} \cdot \frac{11}{12} = x$

J. $\frac{3}{4} \cdot x = 128$

K. $\frac{1}{3} \cdot x = 12{,}000$

L. $\frac{1}{3} + \frac{2}{5} + x = 1$

M. $\frac{4}{3} \cdot 64 = x$

N. $\frac{4}{7} + x = \frac{11}{12}$

O. $\frac{1}{3} + x = \frac{2}{5}$

Answers on page A-5

---

6. *Cutting Rope.* A piece of rope $\frac{11}{12}$ yd long is cut into two pieces. One piece is $\frac{4}{7}$ yd long. How long is the other piece?

7. *Planting Corn.* Each year, Prairie State Farm plants 64 acres of corn. With good weather, $\frac{3}{4}$ of the planting can be completed by April 20. How many acres can be planted with good weather?

8. *Painting Trim.* A painter used $\frac{1}{3}$ gal of white paint for the trim in the library and $\frac{2}{5}$ gal in the family room. How much paint was used for the trim in the two rooms?

9. *Lottery Winnings.* Sally won $12,000 in a state lottery and decided to give the net amount after taxes to three charities. One received $\frac{1}{3}$ of the money, and a second received $\frac{2}{5}$. What fractional part did the third charity receive?

10. *Reading Assignment.* When Lowell had read 64 pages of his political science assignment, he had completed $\frac{3}{4}$ of his required reading. How many total pages were assigned?

3.3 Exercise Set

For Extra Help

MyMathLab

Math XL
PRACTICE

WATCH

DOWNLOAD

READ

REVIEW

a Subtract and simplify.

1. $\dfrac{5}{6} - \dfrac{1}{6}$

2. $\dfrac{5}{8} - \dfrac{3}{8}$

3. $\dfrac{11}{12} - \dfrac{2}{12}$

4. $\dfrac{17}{18} - \dfrac{11}{18}$

5. $\dfrac{3}{4} - \dfrac{1}{8}$

6. $\dfrac{2}{3} - \dfrac{1}{9}$

7. $\dfrac{1}{8} - \dfrac{1}{12}$

8. $\dfrac{1}{6} - \dfrac{1}{8}$

9. $\dfrac{4}{3} - \dfrac{5}{6}$

10. $\dfrac{7}{8} - \dfrac{1}{16}$

11. $\dfrac{3}{4} - \dfrac{3}{28}$

12. $\dfrac{2}{5} - \dfrac{2}{15}$

13. $\dfrac{3}{4} - \dfrac{3}{20}$

14. $\dfrac{5}{6} - \dfrac{1}{2}$

15. $\dfrac{3}{4} - \dfrac{1}{20}$

16. $\dfrac{3}{4} - \dfrac{4}{16}$

17. $\dfrac{5}{12} - \dfrac{2}{15}$

18. $\dfrac{9}{10} - \dfrac{11}{16}$

19. $\dfrac{6}{10} - \dfrac{7}{100}$

20. $\dfrac{9}{10} - \dfrac{3}{100}$

21. $\dfrac{7}{15} - \dfrac{3}{25}$

22. $\dfrac{18}{25} - \dfrac{4}{35}$

23. $\dfrac{99}{100} - \dfrac{9}{10}$

24. $\dfrac{78}{100} - \dfrac{11}{20}$

25. $\dfrac{2}{3} - \dfrac{1}{8}$

26. $\dfrac{3}{4} - \dfrac{1}{2}$

27. $\dfrac{3}{5} - \dfrac{1}{2}$

28. $\dfrac{5}{6} - \dfrac{2}{3}$

29. $\dfrac{5}{12} - \dfrac{3}{8}$

30. $\dfrac{7}{12} - \dfrac{2}{9}$

31. $\dfrac{7}{8} - \dfrac{1}{16}$

32. $\dfrac{5}{12} - \dfrac{5}{16}$

33. $\dfrac{17}{25} - \dfrac{4}{15}$

34. $\dfrac{11}{18} - \dfrac{7}{24}$

35. $\dfrac{23}{25} - \dfrac{112}{150}$

36. $\dfrac{89}{90} - \dfrac{53}{120}$

b Use < or > for ☐ to write a true sentence.

37. $\dfrac{5}{8} \ \square \ \dfrac{6}{8}$

38. $\dfrac{7}{9} \ \square \ \dfrac{5}{9}$

39. $\dfrac{1}{3} \ \square \ \dfrac{1}{4}$

40. $\dfrac{1}{8} \ \square \ \dfrac{1}{6}$

41. $\dfrac{2}{3} \ \square \ \dfrac{5}{7}$

42. $\dfrac{3}{5} \ \square \ \dfrac{4}{7}$

43. $\dfrac{4}{5} \ \square \ \dfrac{5}{6}$

44. $\dfrac{3}{2} \ \square \ \dfrac{7}{5}$

45. $\dfrac{19}{20} \ \square \ \dfrac{4}{5}$

46. $\dfrac{5}{6} \ \square \ \dfrac{13}{16}$

47. $\dfrac{19}{20} \ \square \ \dfrac{9}{10}$

48. $\dfrac{3}{4} \ \square \ \dfrac{11}{15}$

49. $\dfrac{31}{21} \ \square \ \dfrac{41}{13}$

50. $\dfrac{12}{7} \ \square \ \dfrac{132}{49}$

c Solve.

51. $x + \dfrac{1}{30} = \dfrac{1}{10}$

52. $y + \dfrac{9}{12} = \dfrac{11}{12}$

53. $\dfrac{2}{3} + t = \dfrac{4}{5}$

54. $\dfrac{2}{3} + p = \dfrac{7}{8}$

55. $x + \dfrac{1}{3} = \dfrac{5}{6}$

56. $m + \dfrac{5}{6} = \dfrac{9}{10}$

d Solve.

57. For a research paper, Kaitlyn spent $\frac{3}{4}$ hr searching the Internet on google.com and $\frac{1}{3}$ hr on chacha.com. How many more hours did she spend on google.com than on chacha.com?

58. As part of a fitness program, Deb swims $\frac{1}{2}$ mi every day. One day she had already swum $\frac{1}{5}$ mi. How much farther should Deb swim?

59. *Tire Tread.* A long-life tire has a tread depth of $\frac{3}{8}$ in. instead of a more typical $\frac{11}{32}$-in. depth. How much deeper is the long-life tread depth?

Source: *Popular Science*

$\frac{3}{8}$ in.

$\frac{11}{32}$ in.

60. A baker has a dispenser containing $\frac{15}{16}$ cup of icing and puts $\frac{1}{12}$ cup on a cinnamon roll. How much icing remains in the dispenser?

61. At a party, three friends, Ashley, Cole, and Lauren, shared a big tub of popcorn. Within 30 min, the tub was empty. Ashley ate $\frac{7}{12}$ of the tub while Lauren ate only $\frac{1}{6}$ of the tub. How much did Cole eat?

62. An estate was left to four children. One received $\frac{1}{4}$ of the estate, the second $\frac{1}{16}$, and the third $\frac{3}{8}$. How much did the fourth receive?

63. From a $\frac{4}{5}$-lb wheel of cheese, a $\frac{1}{4}$-lb piece was served. How much cheese remained on the wheel?

64. Jovan has an $\frac{11}{10}$-lb mixture of cashews and peanuts that includes $\frac{3}{5}$ lb of cashews. How many pounds of peanuts are in the mixture?

Skill Maintenance

Divide, if possible. If not possible, write "not defined." [1.5a], [2.3b]

65. $\dfrac{38}{38}$

66. $\dfrac{38}{0}$

67. $\dfrac{124}{0}$

68. $\dfrac{124}{31}$

Divide and simplify. [2.7b]

69. $\dfrac{3}{7} \div \dfrac{9}{4}$

70. $\dfrac{9}{10} \div \dfrac{3}{5}$

71. $7 \div \dfrac{1}{3}$

72. $\dfrac{1}{4} \div 8$

73. *Crayons.* The Crayola 64 box of crayons with a built-in sharpener celebrated its 50th birthday in 2008. Since 1958, approximately 200 million Crayola 64 boxes have been sold. How many crayons in the 64 box have been sold altogether? [1.8a]
Source: Crayola LCC

74. A batch of fudge requires $\dfrac{3}{4}$ cup of sugar. How much sugar is needed to make 12 batches? [2.6b]

Synthesis

Solve.

75. ▦ $x + \dfrac{16}{323} = \dfrac{10}{187}$

76. ▦ $x + \dfrac{7}{253} = \dfrac{12}{299}$

77. As part of a rehabilitation program, an athlete must swim and then walk a total of $\dfrac{9}{10}$ km each day. If one lap in the swimming pool is $\dfrac{3}{80}$ km, how far must the athlete walk after swimming 10 laps?

78. *Mountain Climbing.* A mountain climber, beginning at sea level, climbs $\dfrac{3}{5}$ km, descends $\dfrac{1}{4}$ km, climbs $\dfrac{1}{3}$ km, and then descends $\dfrac{1}{7}$ km. At what elevation does the climber finish?

Simplify. Use the rules for order of operations given in Section 1.9.

79. $\dfrac{7}{8} - \dfrac{1}{10} \times \dfrac{5}{6}$

80. $\dfrac{2}{5} + \dfrac{1}{6} \div 3$

81. $\left(\dfrac{2}{3}\right)^2 + \left(\dfrac{3}{4}\right)^2$

82. $5 \times \dfrac{3}{7} - \dfrac{1}{7} \times \dfrac{4}{5}$

Use <, >, or = for ☐ to write a true sentence.

83. ▦ $\dfrac{37}{157} + \dfrac{19}{107}$ ☐ $\dfrac{6941}{16{,}799}$

84. ▦ $\dfrac{12}{97} + \dfrac{67}{139}$ ☐ $\dfrac{8167}{13{,}289}$

85. *Microsoft Interview.* The following is a question taken from an employment interview with Microsoft. Try to answer it. "Given a gold bar that can be cut exactly twice and a contractor who must be paid one-seventh of a gold bar every day for seven days, how should the bar be cut?"
Source: *Fortune Magazine,* January 22, 2001

3.4

OBJECTIVES

a Convert between mixed numerals and fraction notation.

b Divide whole numbers, writing the quotient as a mixed numeral.

SKILL TO REVIEW

Objective 1.5a: Divide whole numbers.

Divide.

1. $735 \div 16$

2. $23\overline{)6023}$

Convert to a mixed numeral.

1. $1 + \dfrac{2}{3} = \square\dfrac{\square}{\square}$

$\underbrace{\frac{1}{3} \quad \frac{1}{3} \quad \frac{1}{3}}_{1} \quad \underbrace{\frac{1}{3} \quad \frac{1}{3}}_{\frac{2}{3}}$

2. $2 + \dfrac{3}{4} = \square\dfrac{\square}{\square}$

3. $12 + \dfrac{2}{7}$

---

a Mixed Numerals

The following figure illustrates the use of a **mixed numeral**. The bolt shown is $2\frac{3}{8}$ in. long. The length is given as a whole-number part, 2, and a fractional part less than 1, $\frac{3}{8}$. We can also represent the measurement of the bolt with fraction notation as $\frac{19}{8}$, but the meaning or interpretation of such a symbol is less understandable or visual than that of mixed numeral notation.

A mixed numeral $2\frac{3}{8}$ represents a sum:

$$2\frac{3}{8} \quad \text{means} \quad 2 + \frac{3}{8}.$$

This is a whole number.   This is a fraction less than 1.

EXAMPLES Convert to a mixed numeral.

1. $7 + \dfrac{2}{5} = 7\dfrac{2}{5}$

2. $4 + \dfrac{3}{10} = 4\dfrac{3}{10}$

> Do Margin Exercises 1–3.

The notation $2\frac{3}{4}$ has a plus sign left out. To aid in understanding, we sometimes write the missing plus sign. This is especially helpful when we convert a mixed numeral to fraction notation.

EXAMPLES Convert to fraction notation.

3. $\quad 2\dfrac{3}{4} = 2 + \dfrac{3}{4}$ Inserting the missing plus sign

$\quad\quad = \dfrac{2}{1} + \dfrac{3}{4}$ $2 = \dfrac{2}{1}$

$\quad\quad = \dfrac{2}{1} \cdot \dfrac{4}{4} + \dfrac{3}{4}$ Finding a common denominator

$\quad\quad = \dfrac{8}{4} + \dfrac{3}{4} = \dfrac{11}{4}$

Answers

Skill to Review:
1. 45 R 15   **2.** 261 R 20

Margin Exercises:
1. $1\frac{2}{3}$ **2.** $2\frac{3}{4}$ **3.** $12\frac{2}{7}$

---

4. $4\dfrac{3}{10} = 4 + \dfrac{3}{10} = \dfrac{4}{1} + \dfrac{3}{10} = \dfrac{4}{1} \cdot \dfrac{10}{10} + \dfrac{3}{10} = \dfrac{40}{10} + \dfrac{3}{10} = \dfrac{43}{10}$

Do Exercises 4 and 5.

Let's now consider a faster method for converting a mixed numeral to fraction notation.

> To convert from a mixed numeral to fraction notation:
>
> (a) Multiply the whole number by the denominator: $4 \cdot 10 = 40$.
> (b) Add the result to the numerator: $40 + 3 = 43$.
> (c) Keep the denominator.
>
> $\overset{\text{(b)}}{\overset{\curvearrowright}{\underset{\curvearrowleft}{4\dfrac{3}{10}}}} = \dfrac{43}{10}$
> (a)   (c)

EXAMPLES Convert to fraction notation.

5. $6\dfrac{2}{3} = \dfrac{20}{3}$ $6 \cdot 3 = 18, 18 + 2 = 20$

6. $8\dfrac{2}{9} = \dfrac{74}{9}$ $8 \cdot 9 = 72, 72 + 2 = 74$

7. $10\dfrac{7}{8} = \dfrac{87}{8}$ $10 \cdot 8 = 80, 80 + 7 = 87$

Do Exercises 6–9.

Writing Mixed Numerals

We can find a mixed numeral for $\frac{5}{3}$ as follows:

$$\dfrac{5}{3} = \dfrac{3}{3} + \dfrac{2}{3} = 1 + \dfrac{2}{3} = 1\dfrac{2}{3}.$$

In terms of objects, we can think of $\frac{5}{3}$ as $\frac{3}{3}$, or 1, plus $\frac{2}{3}$, as shown below.

$\dfrac{5}{3} = \quad \dfrac{3}{3}\text{, or 1} \quad + \quad \dfrac{2}{3}$

Fraction symbols like $\frac{5}{3}$ also indicate division; $\frac{5}{3}$ means $5 \div 3$. Let's divide the numerator by the denominator.

$$\begin{array}{r} 1 \\ 3\overline{)5} \\ \underline{3} \\ 2 \end{array} \leftarrow 2 \div 3 = \tfrac{2}{3}$$

Thus, $\frac{5}{3} = 1\frac{2}{3}$.

Convert to fraction notation.

4. $4\dfrac{2}{5}$ **5.** $6\dfrac{1}{10}$

Convert to fraction notation. Use the faster method.

6. $4\dfrac{5}{6}$ **7.** $9\dfrac{1}{4}$

8. $20\dfrac{2}{3}$ **9.** $1\dfrac{9}{13}$

> **STUDY TIPS**
>
> **WORKED-OUT SOLUTIONS**
>
> The *Student's Solutions Manual* is an excellent resource if you need additional help with an exercise in the exercise sets. It contains step-by-step solutions to the odd-numbered exercises in each exercise set.

Answers

4. $\dfrac{22}{5}$ **5.** $\dfrac{61}{10}$ **6.** $\dfrac{29}{6}$ **7.** $\dfrac{37}{4}$
8. $\dfrac{62}{3}$ **9.** $\dfrac{22}{13}$

To convert from fraction notation to a mixed numeral, divide.

$$\frac{13}{5} \qquad \begin{array}{r} 2 \\ 5\overline{)13} \\ \underline{10} \\ 3 \end{array} \qquad 2\frac{3}{5}$$

The divisor — The quotient — The remainder

EXAMPLES Convert to a mixed numeral.

8. $\dfrac{69}{10}$ $\qquad \begin{array}{r} 6 \\ 10\overline{)69} \\ \underline{60} \\ 9 \end{array} \qquad \dfrac{69}{10} = 6\dfrac{9}{10}$

9. $\dfrac{122}{8}$ $\qquad \begin{array}{r} 15 \\ 8\overline{)122} \\ \underline{8} \\ 42 \\ \underline{40} \\ 2 \end{array} \qquad \dfrac{122}{8} = 15\dfrac{2}{8} = 15\dfrac{1}{4}$

Do Exercises 10–13.

A fraction larger than 1, such as $\frac{27}{8}$, is sometimes referred to as an "improper" fraction. We will not use this terminology because notation such as $\frac{27}{8}$, $\frac{11}{9}$, and $\frac{89}{10}$ is quite "proper" and very common in algebra.

b Writing Mixed Numerals for Quotients

It is quite common when dividing whole numbers to write the quotient using a mixed numeral. The remainder is the numerator of the fraction part of the mixed numeral.

EXAMPLE 10 Divide. Write a mixed numeral for the answer.

$$4\,2\,\overline{)\,8\;9\;1\;5}$$

We first divide as usual.

$$\begin{array}{r} 2\;1\;2 \\ 4\,2\,\overline{)\,8\;9\;1\;5} \\ \underline{8\;4} \\ 5\;1 \\ \underline{4\;2} \\ 9\;5 \\ \underline{8\;4} \\ 1\;1 \end{array} \qquad \dfrac{8915}{42} = 212\dfrac{11}{42}$$

The answer is $212\frac{11}{42}$.

Do Exercises 14 and 15.

Convert to a mixed numeral.

10. $\dfrac{7}{3}$ 　　　**11.** $\dfrac{11}{10}$

12. $\dfrac{110}{6}$ 　　**13.** $\dfrac{229}{18}$

Divide. Write a mixed numeral for the answer.

14. $6\,\overline{)\,4\;8\;4\;6}$

15. $4\,5\,\overline{)\,6\;0\;5\;3}$

3.4

Exercise Set

For Extra Help

MyMathLab

Math XL
PRACTICE

WATCH

DOWNLOAD

READ

REVIEW

a Solve.

1. *Greenhouse Dimensions.* A community college horticulture department builds a greenhouse that measures $32\frac{1}{2}$ ft $\times$ $20\frac{5}{6}$ ft $\times$ $11\frac{3}{4}$ ft. Convert $32\frac{1}{2}$, $20\frac{5}{6}$, and $11\frac{3}{4}$ to fraction notation.

2. *Alarm Clock Dimensions.* A world time clock/radio with NOAA (National Oceanic and Atmospheric Administration) Weather Band measures $14\frac{1}{4}$ in. $\times$ $6\frac{3}{4}$ in. $\times$ $2\frac{1}{4}$ in. Convert $14\frac{1}{4}$, $6\frac{3}{4}$, and $2\frac{1}{4}$ to fraction notation.

3. A quilt design requires three different fabrics. The quilter determines that she needs $\frac{17}{4}$ yd of a dominant fabric, $\frac{10}{3}$ yd of a contrasting fabric, and $\frac{9}{8}$ yd of a border fabric. Convert $\frac{17}{4}$, $\frac{10}{3}$, and $\frac{9}{8}$ to mixed numerals.

4. *Bake Sale.* The Valley Township Fire Department organized a bake sale for a fund-raiser for the local library. Each pie was cut into 8 pieces and each cake into 12 pieces. They sold 73 pieces of pie, or $\frac{73}{8}$ pies, and 55 pieces of cake, or $\frac{55}{12}$ cakes. Convert $\frac{73}{8}$ and $\frac{55}{12}$ to mixed numerals.

Convert to fraction notation.

5. $5\frac{2}{3}$

6. $3\frac{4}{5}$

7. $3\frac{1}{4}$

8. $6\frac{1}{2}$

9. $10\frac{1}{8}$

10. $20\frac{1}{5}$

11. $5\frac{1}{10}$

12. $9\frac{1}{10}$

13. $20\frac{3}{5}$

14. $30\frac{4}{5}$

15. $9\frac{5}{6}$

16. $8\frac{7}{8}$

17. $7\frac{3}{10}$

18. $6\frac{9}{10}$

19. $1\frac{5}{8}$

20. $1\frac{3}{5}$

21. $12\frac{3}{4}$

22. $15\frac{2}{3}$

23. $4\frac{3}{10}$

24. $5\frac{7}{10}$

25. $2\frac{3}{100}$

26. $5\frac{7}{100}$

27. $66\frac{2}{3}$

28. $33\frac{1}{3}$

29. $5\frac{29}{50}$

30. $84\frac{3}{8}$

31. $101\frac{5}{16}$

32. $205\frac{3}{14}$

Convert to a mixed numeral.

33. $\frac{18}{5}$

34. $\frac{17}{4}$

35. $\frac{14}{3}$

36. $\frac{39}{8}$

37. $\frac{27}{6}$

38. $\frac{30}{9}$

39. $\frac{57}{10}$

40. $\frac{89}{10}$

41. $\frac{53}{7}$

42. $\frac{59}{8}$

43. $\frac{45}{6}$

44. $\frac{50}{8}$

45. $\frac{46}{4}$

46. $\frac{39}{9}$

47. $\frac{12}{8}$

48. $\frac{28}{6}$

49. $\frac{757}{100}$

50. $\frac{467}{100}$

51. $\frac{345}{8}$

52. $\frac{223}{4}$

Divide. Write a mixed numeral for the answer.

53. $8\overline{)869}$

54. $3\overline{)2126}$

55. $5\overline{)3091}$

56. $9\overline{)9110}$

57. $21\overline{)852}$

58. $85\overline{)7670}$

59. $102\overline{)5612}$

60. $46\overline{)1081}$

61. $35\overline{)80,243}$

62. $152\overline{)26,107}$

Skill Maintenance

63. Round to the nearest hundred: 45,765.   [1.6a]

64. Round to the nearest ten: 45,765.   [1.6a]

Simplify.   [2.5b]

65. $\dfrac{200}{375}$

66. $\dfrac{63}{75}$

67. $\dfrac{160}{270}$

68. $\dfrac{6996}{8028}$

Multiply and simplify.   [2.4a], [2.6a]

69. $\dfrac{6}{5} \cdot 15$

70. $\dfrac{5}{12} \cdot 6$

71. $\dfrac{7}{10} \cdot \dfrac{5}{14}$

72. $\dfrac{1}{10} \cdot \dfrac{20}{5}$

Divide and simplify.   [2.7b]

73. $\dfrac{2}{3} \div \dfrac{1}{36}$

74. $28 \div \dfrac{4}{7}$

75. $200 \div \dfrac{15}{64}$

76. $\dfrac{3}{4} \div \dfrac{9}{16}$

Synthesis

Write a mixed numeral.

77. 🖩 $\dfrac{128,236}{541}$

78. 🖩 $\dfrac{103,676}{349}$

79. $\dfrac{56}{7} + \dfrac{2}{3}$

80. $\dfrac{72}{12} + \dfrac{5}{6}$

81. There are $\frac{366}{7}$ weeks in a leap year.

82. There are $\frac{365}{7}$ weeks in a year.

Concept Reinforcement

Determine whether each statement is true or false.

_____ **1.** If $\dfrac{a}{b} > \dfrac{c}{b}$, $b \neq 0$, then $a > c$. [3.3b]

_____ **2.** All mixed numerals represent numbers larger than 1. [3.4a]

_____ **3.** The least common multiple of two natural numbers is the smallest number that is a factor of both. [3.1a]

_____ **4.** To add fractions when denominators are the same, we keep the numerator and add the denominators. [3.2a]

Guided Solutions

Fill in each blank with the number that creates a correct solution.

5. Subtract: $\dfrac{11}{42} - \dfrac{3}{35}$. [3.3a]

$$\dfrac{11}{42} - \dfrac{3}{35} = \dfrac{11}{2 \cdot \Box \cdot 7} - \dfrac{3}{\Box \cdot 7} \qquad \text{Factoring the denominators}$$

$$= \dfrac{11}{2 \cdot \Box \cdot 7} \cdot \left(\dfrac{\Box}{\Box}\right) - \dfrac{3}{\Box \cdot 7} \cdot \left(\dfrac{\Box \cdot \Box}{\Box \cdot \Box}\right) \qquad \text{Multiplying by 1 to get the LCD}$$

$$= \dfrac{11 \cdot \Box}{2 \cdot 3 \cdot 7 \cdot \Box} - \dfrac{3 \cdot \Box \cdot \Box}{5 \cdot 7 \cdot \Box \cdot \Box} \qquad \text{Multiplying}$$

$$= \dfrac{\Box}{2 \cdot 3 \cdot 5 \cdot 7} - \dfrac{\Box}{2 \cdot 3 \cdot 5 \cdot 7} \qquad \text{Simplifying}$$

$$= \dfrac{\Box - \Box}{2 \cdot 3 \cdot 5 \cdot 7} = \dfrac{\Box}{\Box} \qquad \text{Subtracting and simplifying}$$

6. Solve: $x + \dfrac{1}{8} = \dfrac{2}{3}$. [3.3c]

$$x + \dfrac{1}{8} = \dfrac{2}{3}$$

$$x + \dfrac{1}{8} - \Box = \dfrac{2}{3} - \Box \qquad \text{Subtracting on both sides}$$

$$x + \Box = \dfrac{2}{3} \cdot \dfrac{\Box}{\Box} - \dfrac{1}{8} \cdot \dfrac{\Box}{\Box} \qquad \text{Multiplying by 1 to get the LCD}$$

$$x = \dfrac{\Box}{\Box} - \dfrac{\Box}{\Box} \qquad \text{Simplifying and multiplying}$$

$$x = \dfrac{\Box}{\Box} \qquad \text{Subtracting}$$

Mixed Review

7. Match each set of numbers in the first column with its least common multiple in the second column by drawing connecting lines.   [3.1a]

45 and 50

50 and 80   120

30 and 24   720

18, 24, and 80   400

30, 45, and 50   450

Calculate and simplify.   [3.2a], [3.3a]

8. $\dfrac{1}{5} + \dfrac{7}{45}$

9. $\dfrac{5}{6} + \dfrac{2}{3} + \dfrac{7}{12}$

10. $\dfrac{2}{9} - \dfrac{1}{6}$

11. $\dfrac{5}{18} - \dfrac{1}{15}$

12. $\dfrac{19}{48} - \dfrac{11}{30}$

13. $\dfrac{3}{7} + \dfrac{15}{17}$

14. $\dfrac{229}{720} - \dfrac{5}{24}$

15. $\dfrac{8}{65} - \dfrac{2}{35}$

Solve.

16. Miguel jogs for $\frac{4}{5}$ mi, rests, and then jogs for another $\frac{2}{3}$ mi. How far does he jog in all?   [3.2b]

17. One weekend, Kirby spent $\frac{39}{5}$ hr playing two iPod games—Brain Challenge: Cerebral Burn and Scrabble: Go for a Triple Word Score. She spent $\frac{11}{4}$ hr playing Scrabble. How many hours did she spend playing Brain Challenge?   [3.3d]

18. Arrange in order from smallest to largest: $\dfrac{4}{9}, \dfrac{3}{10}, \dfrac{2}{7},$ and $\dfrac{1}{5}$.

[3.3b]

19. Solve: $\dfrac{2}{5} + x = \dfrac{9}{16}$.   [3.3c]

20. Divide: $15\overline{)263}$. Write a mixed numeral for the answer. [3.4b]

21. Fraction notation for $9\frac{3}{8}$ is which of the following? [3.4a]

A. $\dfrac{27}{8}$ **B.** $\dfrac{93}{8}$ **C.** $\dfrac{75}{8}$ **D.** $\dfrac{80}{3}$

22. Mixed numeral notation for $\frac{39}{4}$ is which of the following? [3.4a]

A. $35\dfrac{1}{4}$ **B.** $\dfrac{4}{39}$ **C.** $9\dfrac{3}{4}$ **D.** $36\dfrac{3}{4}$

Understanding Through Discussion and Writing

23. Is the LCM of two numbers always larger than either number? Why or why not?   [3.1a]

24. Explain the role of multiplication when adding using fraction notation with different denominators.   [3.2a]

25. A student made the following error:

$$\frac{8}{5} - \frac{8}{2} = \frac{8}{3}.$$

Find at least two ways to convince him of the mistake. [3.3a]

26. Are the numbers $2\frac{1}{3}$ and $2 \cdot \frac{1}{3}$ equal? Why or why not? [3.4a]

3.5 Addition and Subtraction Using Mixed Numerals; Applications

SKILL TO REVIEW

Objective 2.5b: Simplify fraction notation.

Simplify.

1. $\dfrac{18}{32}$

2. $\dfrac{78}{117}$

a Addition Using Mixed Numerals

To find the sum $1\frac{5}{8} + 3\frac{1}{8}$, we first add the fractions. Then we add the whole numbers.

$$
\begin{array}{r}
1\dfrac{5}{8} = \\[2mm]
+\ 3\dfrac{1}{8} = \\[2mm]
\hline
\dfrac{6}{8}
\end{array}
\qquad
\begin{array}{r}
1\dfrac{5}{8} \\[2mm]
+\ 3\dfrac{1}{8} \\[2mm]
\hline
4\dfrac{6}{8} = 4\dfrac{3}{4}
\end{array}
$$

Simplifying: $\dfrac{6}{8} = \dfrac{3}{4}$

Add the fractions. Add the whole numbers.

Sometimes we must write the fractional parts with a common denominator before we can add.

EXAMPLE 1 Add: $5\frac{2}{3} + 3\frac{5}{6}$. Write a mixed numeral for the answer.

The LCD is 6.

$$
\begin{array}{r}
5\dfrac{2}{3} \cdot \dfrac{2}{2} = \quad 5\dfrac{4}{6} \\[2mm]
+\ 3\dfrac{5}{6} \quad = +\ 3\dfrac{5}{6} \\[2mm]
\hline
8\dfrac{9}{6} = 8 + \dfrac{9}{6}
\end{array}
$$

$$= 8 + 1\dfrac{1}{2} \qquad \text{Writing } \tfrac{9}{6} \text{ as a mixed numeral, } 1\tfrac{1}{2}$$

$$= 9\dfrac{1}{2}$$

Do Margin Exercises 1 and 2.

EXAMPLE 2 Add: $10\frac{5}{6} + 7\frac{3}{8}$.

The LCD is 24.

$$
\begin{array}{r}
10\dfrac{5}{6} \cdot \dfrac{4}{4} = \quad 10\dfrac{20}{24} \\[2mm]
+\ 7\dfrac{3}{8} \cdot \dfrac{3}{3} = +\ 7\dfrac{9}{24} \\[2mm]
\hline
17\dfrac{29}{24} = 17 + \dfrac{29}{24}
\end{array}
$$

$$= 17 + 1\dfrac{5}{24} \qquad \text{Writing } \tfrac{29}{24} \text{ as a mixed numeral, } 1\tfrac{5}{24}$$

$$= 18\dfrac{5}{24}$$

Do Exercise 3.

Add.

1.
$$
\begin{array}{r}
2\dfrac{3}{10} \\[2mm]
+\ 5\dfrac{1}{10} \\
\hline
\end{array}
$$

2.
$$
\begin{array}{r}
8\dfrac{2}{5} \\[2mm]
+\ 3\dfrac{7}{10} \\
\hline
\end{array}
$$

3. Add.
$$
\begin{array}{r}
9\dfrac{3}{4} \\[2mm]
+\ 3\dfrac{5}{6} \\
\hline
\end{array}
$$

Answers

Skill to Review:

1. $\dfrac{9}{16}$ 2. $\dfrac{2}{3}$

Margin Exercises:

1. $7\dfrac{2}{5}$ 2. $12\dfrac{1}{10}$ 3. $13\dfrac{7}{12}$

b Subtraction Using Mixed Numerals

Subtraction is a lot like addition; we subtract the fractions and then the whole numbers.

EXAMPLE 3 Subtract: $7\frac{3}{4} - 2\frac{1}{4}$.

$$
\begin{array}{r}
7\,\dfrac{3}{4} \;=\\[4pt]
-\;2\,\dfrac{1}{4} \;=\\[4pt]
\hline
2\\
\overline{4}
\end{array}
\qquad
\begin{array}{r}
7\,\dfrac{3}{4}\\[4pt]
-\;2\,\dfrac{1}{4}\\[4pt]
\hline
5\,\dfrac{2}{4} = 5\dfrac{1}{2}
\end{array}
$$

Simplifying: $\dfrac{2}{4} = \dfrac{1}{2}$

Subtract the fractions.

Subtract the whole numbers.

EXAMPLE 4 Subtract: $9\frac{4}{5} - 3\frac{1}{2}$.

The LCD is 10.

$$
\begin{array}{r}
9\,\dfrac{4}{5}\cdot\dfrac{2}{2} \;=\; 9\,\dfrac{8}{10}\\[8pt]
-\;3\,\dfrac{1}{2}\cdot\dfrac{5}{5} \;=\; -\,3\,\dfrac{5}{10}\\[4pt]
\hline
6\,\dfrac{3}{10}
\end{array}
$$

Do Exercises 4 and 5.

Subtract.

4. $\begin{array}{r} 10\,\dfrac{7}{8}\\[4pt] -\;9\,\dfrac{3}{8}\\ \hline \end{array}$

5. $\begin{array}{r} 8\,\dfrac{2}{3}\\[4pt] -\;5\,\dfrac{1}{2}\\ \hline \end{array}$

EXAMPLE 5 Subtract: $7\frac{1}{6} - 2\frac{1}{4}$.

The LCD is 12.

$$
\begin{array}{r}
7\,\dfrac{1}{6}\cdot\dfrac{2}{2} \;=\; 7\,\dfrac{2}{12}\\[8pt]
-\;2\,\dfrac{1}{4}\cdot\dfrac{3}{3} \;=\; -\,2\,\dfrac{3}{12}\\
\end{array}
$$

We cannot subtract $\frac{3}{12}$ from $\frac{2}{12}$.
$\leftarrow$ We borrow 1, or $\frac{12}{12}$, from 7:
$7\frac{2}{12} = 6 + 1 + \frac{2}{12} = 6 + \frac{12}{12} + \frac{2}{12} = 6\frac{14}{12}$.

We can write this as

$$
\begin{array}{r}
7\,\dfrac{2}{12} \;=\; 6\,\dfrac{14}{12}\\[8pt]
-\;2\,\dfrac{3}{12} \;=\; -\,2\,\dfrac{3}{12}\\[4pt]
\hline
4\,\dfrac{11}{12}.
\end{array}
$$

Do Exercise 6.

6. Subtract.

$$
\begin{array}{r}
8\,\dfrac{1}{9}\\[4pt]
-\;4\,\dfrac{5}{6}\\
\hline
\end{array}
$$

Answers

4. $1\frac{1}{2}$ 5. $3\frac{1}{6}$ 6. $3\frac{5}{18}$

EXAMPLE 6 Subtract: $12 - 9\frac{3}{8}$.

$$
\begin{array}{r}
12 \;=\; 11\frac{8}{8} \\[4pt]
-\;9\frac{3}{8} \;=\; -\;9\frac{3}{8} \\[4pt]
\hline
2\frac{5}{8}
\end{array}
\qquad 12 = 11 + 1 = 11 + \frac{8}{8} = 11\frac{8}{8}
$$

Do Exercise 7.

(c) Applications and Problem Solving

EXAMPLE 7 *Widening a Driveway.* Sherry and Woody are widening their existing $17\frac{1}{4}$-ft driveway by adding $5\frac{9}{10}$ ft on one side. What is the width of the new driveway?

$5\frac{9}{10}$ ft

$17\frac{1}{4}$ ft

1. **Familiarize.** We let $w =$ the width of the new driveway.

2. **Translate.** We translate as follows:

$$
\underbrace{\text{Width of existing driveway}}_{17\frac{1}{4}} \;+\; \underbrace{\text{Width of addition}}_{5\frac{9}{10}} \;=\; \underbrace{\text{Width of new driveway}}_{w}.
$$

3. **Solve.** The translation tells us what to do. We add. The LCD is 20.

$$
\begin{array}{rcccl}
17\frac{1}{4} &=& 17\;\dfrac{1}{4}\cdot\dfrac{5}{5} &=& 17\frac{5}{20} \\[8pt]
+\;5\frac{9}{10} &=& +\;5\;\dfrac{9}{10}\cdot\dfrac{2}{2} &=& +\;5\frac{18}{20} \\[8pt]
\hline
& & & & 22\frac{23}{20} = 23\frac{3}{20}
\end{array}
$$

Thus, $w = 23\frac{3}{20}$.

4. **Check.** We check by repeating the calculation. We also note that the answer is larger than either of the widths, which means that the answer is reasonable.

5. **State.** The width of the new driveway is $23\frac{3}{20}$ ft.

Do Exercise 8.

7. Subtract.

$$
\begin{array}{r}
5 \\
-\;1\frac{1}{3} \\
\hline
\end{array}
$$

8. Travel Distance. On a two-day business trip, Paul drove $213\frac{7}{10}$ mi the first day and $107\frac{5}{8}$ mi the second day. What was the total distance that Paul drove?

Answers

7. $3\frac{2}{3}$ **8.** $321\frac{13}{40}$ mi

EXAMPLE 8 *Communication Cable.* Perfection Cable has two crews who install Internet communication cable. Crew A can install $38\frac{1}{8}$ ft per hour. Crew B can install $31\frac{2}{3}$ ft per hour. How many fewer feet can Crew B install per hour than Crew A?

1. **Familiarize.** The phrase "how many fewer" indicates subtraction. We let c = the difference in the numbers of feet per hour.

2. **Translate.** We translate as follows:

$$
\underbrace{\text{Crew A:} \atop \text{ft/hr}} - \underbrace{\text{Crew B:} \atop \text{ft/hr}} = \underbrace{\text{Difference in} \atop \text{ft/hr}}
$$

$$
38\frac{1}{8} \quad - \quad 31\frac{2}{3} \quad = \quad c.
$$

3. **Solve.** To solve the equation, we carry out the subtraction. The LCD = 24.

$$
\begin{aligned}
38\frac{1}{8} &= 38\frac{1}{8}\cdot\frac{3}{3} = 38\frac{3}{24} = 37\frac{27}{24}\\
-\ 31\frac{2}{3} &= -31\frac{2}{3}\cdot\frac{8}{8} = -31\frac{16}{24} = -31\frac{16}{24}\\
\hline
& \hspace{9.5cm} 6\frac{11}{24}
\end{aligned}
$$

Thus, $c = 6\frac{11}{24}$.

4. **Check.** To check, we add the difference, $6\frac{11}{24}$, to Crew B's rate:

$$
31\frac{2}{3} + 6\frac{11}{24} = 31\frac{16}{24} + 6\frac{11}{24}
$$

$$
= 37\frac{27}{24} = 38\frac{3}{24} = 38\frac{1}{8}.
$$

This checks.

5. **State.** Crew B installs $6\frac{11}{24}$ fewer feet of cable per hour than Crew A.

Do Exercise 9.

9. **Nail Length.** A 30d nail is $4\frac{1}{2}$ in. long. A 5d nail is $1\frac{3}{4}$ in. long. How much longer is the 30d nail than the 5d nail? (The "d" stands for "penny," which was used years ago in England to specify the number of pennies needed to buy 100 nails. Today, "penny" is used only to indicate the length of the nail.)
Source: *Pocket Ref*, 2nd ed., by Thomas J. Glover, p. 280, Sequoia Publishing, Inc., Littleton, CO

$4\frac{1}{2}$ in.

$1\frac{3}{4}$ in.

Multistep Problems

EXAMPLE 9 *Fly Fishing.* Henry is putting together a fly fishing line and uses $58\frac{5}{8}$ ft of slow-sinking fly line and $8\frac{3}{4}$ ft of leader line. The leader line provides a low visibility link between the heavier fly line and the fly. He uses $\frac{3}{8}$ ft of the slow-sinking fly line to connect the two lines. The knot used to connect the fly to the leader line uses $\frac{1}{6}$ ft of the leader line. How long is the finished fly fishing line?

1. **Familiarize.** Let's make a drawing to organize the given information. We let l = the length of the finished fly fishing line, in feet.

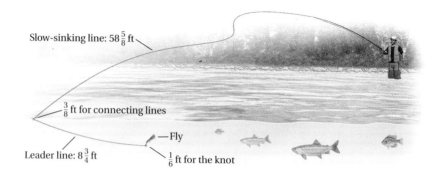

Slow-sinking line: $58\frac{5}{8}$ ft

$\frac{3}{8}$ ft for connecting lines

—Fly

Leader line: $8\frac{3}{4}$ ft

$\frac{1}{6}$ ft for the knot

2. **Translate.** From the drawing, we see that the length l is the sum of the lengths of the two lines, $58\frac{5}{8}$ ft and $8\frac{3}{4}$ ft, minus the sum of the lengths, $\frac{3}{8}$ ft and $\frac{1}{6}$ ft, needed for connecting the lines and attaching the fly with a knot. Thus we have

$$l = \left(58\frac{5}{8} + 8\frac{3}{4}\right) - \left(\frac{3}{8} + \frac{1}{6}\right).$$

3. **Solve.** This is a three-step problem.

   a) We first add the two lengths $58\frac{5}{8}$ and $8\frac{3}{4}$.

$$58\frac{5}{8} = 58\frac{5}{8}$$
$$+\ 8\frac{3}{4} = +\ 8\frac{6}{8}$$
$$\overline{\qquad\qquad 66\frac{11}{8} = 67\frac{3}{8}}$$

   b) Next, we add the two lengths $\frac{3}{8}$ and $\frac{1}{6}$.

$$\frac{3}{8} + \frac{1}{6} = \frac{3}{8}\cdot\frac{3}{3} + \frac{1}{6}\cdot\frac{4}{4} = \frac{9}{24} + \frac{4}{24} = \frac{13}{24}$$

   c) Finally, we subtract $\frac{13}{24}$ from $67\frac{3}{8}$.

$$67\frac{3}{8} = 67\frac{9}{24} = 66\frac{33}{24}$$
$$-\ \frac{13}{24} = -\ \frac{13}{24} = -\ \frac{13}{24}$$
$$\overline{\qquad\qquad\qquad 66\frac{20}{24} = 66\frac{5}{6}}$$

   Thus, $l = 66\frac{5}{6}$.

4. **Check.** We can check by adding the finished length to the lengths needed for the connection and the knot:

$$66\frac{5}{6} + \frac{3}{8} + \frac{1}{6} = \left(66\frac{5}{6} + \frac{1}{6}\right) + \frac{3}{8} = 67 + \frac{3}{8} = 67\frac{3}{8}.$$

   This is the sum of the given lengths.

5. **State.** The finished length of the fly line is $66\frac{5}{6}$ ft.

Do Exercise 10.

10. Liquid Fertilizer. There is $283\frac{5}{8}$ gal of liquid fertilizer in a fertilizer application tank. After applying $178\frac{2}{3}$ gal to a soybean field, the farmer requests that Braden's Farm Supply deliver an additional 250 gal. How many gallons of fertilizer are in the tank after the delivery?

Answer

10. $354\frac{23}{24}$ gal

a Add. Write a mixed numeral for the answer.

1. 20
   $+ \ 8\frac{3}{4}$

2. 37
   $+ \ 18\frac{2}{3}$

3. $129\frac{7}{8}$
   $+ \ 56$

4. $2003\frac{4}{11}$
   $+ \ \ 59$

5. $2\frac{7}{8}$
   $+ \ 3\frac{5}{8}$

6. $4\frac{5}{6}$
   $+ \ 3\frac{5}{6}$

7. $1\frac{1}{4} + 1\frac{2}{3}$

8. $4\frac{1}{3} + 5\frac{2}{9}$

9. $8\frac{3}{4}$
   $+ \ 5\frac{5}{6}$

10. $4\frac{3}{8}$
    $+ \ 6\frac{5}{12}$

11. $3\frac{2}{5}$
    $+ \ 8\frac{7}{10}$

12. $5\frac{1}{2}$
    $+ \ 3\frac{7}{10}$

13. $5\frac{3}{8}$
    $+ \ 10\frac{5}{6}$

14. $\frac{5}{8}$
    $+ \ 1\frac{5}{6}$

15. $12\frac{4}{5}$
    $+ \ 8\frac{7}{10}$

16. $15\frac{5}{8}$
    $+ \ 11\frac{3}{4}$

17. $14\frac{5}{8}$
    $+ \ 13\frac{1}{4}$

18. $16\frac{1}{4}$
    $+ \ 15\frac{7}{8}$

19. $7\frac{1}{8}$
    $9\frac{2}{3}$
    $+ \ 10\frac{3}{4}$

20. $45\frac{2}{3}$
    $31\frac{3}{5}$
    $+ \ 12\frac{1}{4}$

b Subtract. Write a mixed numeral for the answer.

21. $4\frac{1}{5}$
    $- \ 2\frac{3}{5}$

22. $5\frac{1}{8}$
    $- \ 2\frac{3}{8}$

23. $6\frac{3}{5} - 2\frac{1}{2}$

24. $7\frac{2}{3} - 6\frac{1}{2}$

25. $34\frac{1}{3}$
    $- \ 12\frac{5}{8}$

26. $23\frac{5}{16}$
    $- \ 16\frac{3}{4}$

27. 21
    $- \ 8\frac{3}{4}$

28. 42
    $- \ 3\frac{7}{8}$

29.
$$\begin{array}{r} 34 \\ -\ 18\dfrac{5}{8} \\ \hline \end{array}$$

30.
$$\begin{array}{r} 23 \\ -\ 19\dfrac{3}{4} \\ \hline \end{array}$$

31.
$$\begin{array}{r} 21\dfrac{1}{6} \\ -\ 13\dfrac{3}{4} \\ \hline \end{array}$$

32.
$$\begin{array}{r} 42\dfrac{1}{10} \\ -\ 23\dfrac{7}{12} \\ \hline \end{array}$$

33.
$$\begin{array}{r} 14\dfrac{1}{8} \\ -\ \dfrac{3}{4} \\ \hline \end{array}$$

34.
$$\begin{array}{r} 28\dfrac{1}{6} \\ -\ 5 \\ \hline \end{array}$$

35.
$$\begin{array}{r} 25\dfrac{1}{9} \\ -\ 13\dfrac{5}{6} \\ \hline \end{array}$$

36.
$$\begin{array}{r} 23\dfrac{5}{16} \\ -\ 14\dfrac{7}{12} \\ \hline \end{array}$$

 c   Solve.

37. *Planting Flowers.* A landscaper planted $4\frac{1}{2}$ flats of impatiens, $6\frac{2}{3}$ flats of snapdragons, and $3\frac{3}{8}$ flats of phlox. How many flats did she plant altogether?

38. A plumber uses two pipes, each of length $51\frac{5}{16}$ in., and one pipe of length $34\frac{3}{4}$ in. in the installation of a shower. How much pipe was used in all?

39. For a family barbecue, Cayla bought packages of hamburger weighing $1\frac{2}{3}$ lb and $5\frac{3}{4}$ lb. What was the total weight of the meat?

40. Marsha's Butcher Shop sold packages of sliced turkey breast weighing $1\frac{1}{3}$ lb and $4\frac{3}{5}$ lb. What was the total weight of the meat?

41. Tara is 66 in. tall and her son, Tom, is $59\frac{7}{12}$ in. tall. How much taller is Tara?

42. Nicholas is $73\frac{2}{3}$ in. tall and his daughter, Kendra, is $71\frac{5}{16}$ in. tall. How much shorter is Kendra?

43. *Upholstery Fabric.* Executive Car Care sells 45-in. upholstery fabric for car restoration. Art buys $9\frac{1}{4}$ yd and $10\frac{5}{6}$ yd for two car projects. How many yards did Art buy?

44. *Cutco Cutlery.* The Essentials 5-piece set sold by Cutco contains three knives: $7\frac{5}{8}$" Petite Chef, $6\frac{3}{4}$" Petite Carver, and $2\frac{3}{4}$" Paring Knife. How much larger is the blade of the Petite Chef than that of the Petite Carver? than that of the Paring Knife?

Source: Cutco Cutlery Corporation, Fall 2008

45. *Stone Bench.* Baytown Village Stone Creations is making a custom stone bench as shown below. The recommended height for the bench is 18 in. The depth of the stone bench is $3\frac{3}{8}$ in. Each of the two supporting legs is made up of three stacked stones. Two of the stones measure $3\frac{1}{2}$ in. and $5\frac{1}{4}$ in. How much must the third stone measure?

18 in.

$3\frac{3}{8}$ in.

$3\frac{1}{2}$ in.

$5\frac{1}{4}$ in.

?

46. *Winterizing a Swimming Pool.* To winterize their swimming pool, the Jablonskis are draining the water into a nearby field. The distance to the field is $103\frac{1}{2}$ ft. Because their only hose measures $62\frac{3}{4}$ ft, they need to buy an additional hose. How long must the new hose be?

$62\frac{3}{4}$ ft

47. Kim Park is a computer technician. One day, she drove $180\frac{7}{10}$ mi away from Los Angeles for a service call. The next day, she drove $85\frac{1}{2}$ mi back toward Los Angeles for another service call. How far was she then from Los Angeles?

48. Jose is $4\frac{1}{2}$ in. taller than his daughter, Teresa. Teresa is $66\frac{2}{3}$ in. tall. How tall is Jose?

49. Creative Glass sells a framed beveled mirror as shown below. Its dimensions are $30\frac{1}{2}$" wide by $36\frac{5}{8}$" high. What is the perimeter of (total distance around) the framed mirror?

$36\frac{5}{8}$ in.

$30\frac{1}{2}$ in.

50. *Book Size.* One standard book size is $8\frac{1}{2}$ in. by $9\frac{3}{4}$ in. What is the perimeter of (total distance around) the front cover of such a book?

$8\frac{1}{2}$ in.

$9\frac{3}{4}$ in.

51. *Carpentry.* When cutting wood with a saw, a carpenter must take into account the thickness of the saw blade. Suppose that from a piece of wood 36 in. long, a carpenter cuts a $15\frac{3}{4}$-in. length with a saw blade that is $\frac{1}{8}$ in. thick. How long is the piece that remains?

52. *Painting.* A painter used $1\frac{3}{4}$ gal of paint for the Garcias' living room and $1\frac{1}{3}$ gal for their family room. How much paint was used in all?

53. Rene is $5\frac{1}{4}$ in. taller than his son, who is $72\frac{5}{6}$ in. tall. How tall is Rene?

54. A Boeing 767 flew 640 mi on a nonstop flight. On the return flight, it landed after having flown $320\frac{3}{10}$ mi. How far was the plane from its original point of departure?

55. *Interior Design.* Sue worked $10\frac{1}{2}$ hr over a three-day period on an interior design project. If she worked $2\frac{1}{2}$ hr on the first day and $4\frac{1}{5}$ hr on the second, how many hours did Sue work on the third day?

56. *Painting.* Geri had $3\frac{1}{2}$ gal of paint. It took $2\frac{3}{4}$ gal to paint the family room. She estimated that it would take $2\frac{1}{4}$ gal to paint the living room. How much more paint did Geri need?

Find the perimeter of (distance around) each figure.

57.

$5\frac{3}{4}$ yd $5\frac{3}{4}$ yd

$5\frac{3}{4}$ yd $5\frac{3}{4}$ yd

$5\frac{3}{4}$ yd

58.

$3\frac{7}{16}$ ft

$3\frac{7}{16}$ ft

$6\frac{7}{8}$ ft

$6\frac{7}{8}$ ft

Find the length d in each figure.

59.

$2\frac{3}{4}$ ft d $2\frac{3}{4}$ ft

$12\frac{7}{8}$ ft

60.

$2\frac{1}{5}$ in. d $2\frac{1}{5}$ in.

$10\frac{1}{2}$ in.

61. *Sewing from a Pattern.* Suppose you want to make a formal dress with jacket in size 8. Using 45-in. fabric, you need $1\frac{3}{8}$ yd for the dress, $\frac{5}{8}$ yd of contrasting fabric for the band at the bottom, and $3\frac{3}{8}$ yd for the jacket. How many yards of 45-in. fabric are needed to make the outfit?

62. *Sewing from a Pattern.* Suppose you want to make an outfit in size 12. Using 45-in. fabric, you need $2\frac{3}{4}$ yd for the dress and $3\frac{1}{2}$ yd for the jacket. How many yards of 45-in. fabric are needed to make the outfit?

63. *Window Dimensions.* The Sanchez family is replacing a window in their lake house. The original window measures $4\frac{5}{6}$ ft $\times$ $8\frac{1}{4}$ ft. The new window is $2\frac{1}{3}$ ft wider. What are the dimensions of the new window?

64. Find the smallest length of a bolt that will pass through a piece of tubing with an outside diameter of $\frac{1}{2}$ in., a washer $\frac{1}{16}$ in. thick, a piece of tubing with a $\frac{3}{4}$-in. outside diameter, another washer, and a nut $\frac{3}{16}$ in. thick.

Skill Maintenance

Solve.

65. Rick's Market sells Swiss cheese in $\frac{3}{4}$-lb packages. How many packages can be made from a 12-lb slab of cheese? [2.7d]

66. Holstein's Dairy produced 4578 oz of milk one morning. How many 16-oz cartons could be filled? How much milk would be left over? [1.8a]

Determine whether the first number is divisible by the second. [2.2a]

67. 9993 by 3

68. 9993 by 9

69. 2345 by 9

70. 2345 by 5

71. 2335 by 10

72. 7764 by 6

73. 18,888 by 8

74. 18,888 by 4

75. Multiply and simplify: $\dfrac{15}{9} \cdot \dfrac{18}{39}$. [2.6a]

76. Divide and simplify: $\dfrac{12}{25} \div \dfrac{24}{5}$. [2.7b]

Synthesis

Calculate each of the following. Write the result as a mixed numeral.

77. 🖩 $3289\frac{1047}{1189} + 5278\frac{32}{41}$

78. 🖩 $5798\frac{17}{53} - 3909\frac{1957}{2279}$

79. A post for a pier is 29 ft long. Half of the post extends above the water's surface and $8\frac{3}{4}$ ft of the post is buried in mud. How deep is the water at that point?

80. Solve: $47\frac{2}{3} + n = 56\frac{1}{4}$.

3.6

Multiplication and Division Using Mixed Numerals; Applications

OBJECTIVES

a Multiply using mixed numerals.

b Divide using mixed numerals.

c Solve applied problems involving multiplication and division with mixed numerals.

a Multiplication Using Mixed Numerals

Carrying out addition and subtraction with mixed numerals is usually easier if the numbers are left as mixed numerals. With multiplication and division, however, it is easier to convert the numbers to fraction notation first.

> **MULTIPLICATION USING MIXED NUMERALS**
>
> To multiply using mixed numerals, first convert to fraction notation and multiply. Then convert the answer to a mixed numeral, if appropriate.

SKILL TO REVIEW

Objective 2.7b: Divide and simplify using fraction notation.

Divide and simplify.

1. $85 \div \dfrac{17}{5}$ **2.** $\dfrac{7}{65} \div \dfrac{21}{25}$

EXAMPLE 1 Multiply: $6 \cdot 2\dfrac{1}{2}$.

$$6 \cdot 2\frac{1}{2} = \frac{6}{1} \cdot \frac{5}{2} = \frac{6 \cdot 5}{1 \cdot 2} = \frac{2 \cdot 3 \cdot 5}{2 \cdot 1} = \frac{2}{2} \cdot \frac{3 \cdot 5}{1} = 1 \cdot \frac{3 \cdot 5}{1} = 15$$

> Converting the numbers to fraction notation first makes it easier to carry out the multiplication.

EXAMPLE 2 Multiply: $3\dfrac{1}{2} \cdot \dfrac{3}{4}$.

$$3\frac{1}{2} \cdot \frac{3}{4} = \frac{7}{2} \cdot \frac{3}{4} = \frac{21}{8} = 2\frac{5}{8}$$

Here we write fraction notation.

Do Margin Exercises 1 and 2.

Multiply.

1. $6 \cdot 3\dfrac{1}{3}$ **2.** $2\dfrac{1}{2} \cdot \dfrac{3}{4}$

EXAMPLE 3 Multiply: $8 \cdot 4\dfrac{2}{3}$.

$$8 \cdot 4\frac{2}{3} = \frac{8}{1} \cdot \frac{14}{3} = \frac{112}{3} = 37\frac{1}{3}$$

Multiply.

3. $2 \cdot 6\dfrac{2}{5}$ **4.** $3\dfrac{1}{3} \cdot 2\dfrac{1}{2}$

EXAMPLE 4 Multiply: $2\dfrac{1}{4} \cdot 3\dfrac{2}{5}$.

$$2\frac{1}{4} \cdot 3\frac{2}{5} = \frac{9}{4} \cdot \frac{17}{5} = \frac{153}{20} = 7\frac{13}{20}$$

---------- *Caution!* ----------

$2\dfrac{1}{4} \cdot 3\dfrac{2}{5} \neq 6\dfrac{2}{20}$. A common error is to multiply the whole numbers and then the fractions. This does not give the correct answer, $7\dfrac{13}{20}$, which is found by converting to fraction notation first.

Do Exercises 3 and 4.

Answers

Skill to Review:

1. 25   **2.** $\dfrac{5}{39}$

Margin Exercises:

1. 20   **2.** $1\dfrac{7}{8}$ **3.** $12\dfrac{4}{5}$ **4.** $8\dfrac{1}{3}$

b Division Using Mixed Numerals

The division $1\frac{1}{2} \div \frac{1}{6}$ is shown here. *Think:* "How many $\frac{1}{6}$'s are in $1\frac{1}{2}$?" We see that the answer is 9.

When we divide using mixed numerals, we convert to fraction notation first.

$$1\frac{1}{2} \div \frac{1}{6} = \frac{3}{2} \div \frac{1}{6} = \frac{3}{2} \cdot \frac{6}{1}$$

$$= \frac{3 \cdot 6}{2 \cdot 1} = \frac{3 \cdot 3 \cdot 2}{2 \cdot 1} = \frac{3 \cdot 3}{1} \cdot \frac{2}{2} = \frac{3 \cdot 3}{1} \cdot 1 = 9$$

> **DIVISION USING MIXED NUMERALS**
>
> To divide using mixed numerals, first write fraction notation and divide. Then convert the answer to a mixed numeral, if appropriate.

EXAMPLE 5   Divide: $32 \div 3\frac{1}{5}$.

$$32 \div 3\frac{1}{5} = \frac{32}{1} \div \frac{16}{5} \qquad \text{Writing fraction notation}$$

$$= \frac{32}{1} \cdot \frac{5}{16} = \frac{32 \cdot 5}{1 \cdot 16} = \frac{2 \cdot 16 \cdot 5}{1 \cdot 16} = \frac{16}{16} \cdot \frac{2 \cdot 5}{1} = 1 \cdot \frac{2 \cdot 5}{1} = 10$$

Remember to multiply by the reciprocal.

Do Exercise 5.

5. Divide: $84 \div 5\frac{1}{4}$.

EXAMPLE 6   Divide: $35 \div 4\frac{1}{3}$.

$$35 \div 4\frac{1}{3} = \frac{35}{1} \div \frac{13}{3} = \frac{35}{1} \cdot \frac{3}{13} = \frac{105}{13} = 8\frac{1}{13}$$

Do Exercise 6.

6. Divide: $26 \div 3\frac{1}{2}$.

EXAMPLE 7   Divide: $2\frac{1}{3} \div 1\frac{3}{4}$.

$$2\frac{1}{3} \div 1\frac{3}{4} = \frac{7}{3} \div \frac{7}{4} = \frac{7}{3} \cdot \frac{4}{7} = \frac{7 \cdot 4}{3 \cdot 7} = \frac{7}{7} \cdot \frac{4}{3} = 1 \cdot \frac{4}{3} = \frac{4}{3} = 1\frac{1}{3}$$

Divide.

7. $2\frac{1}{4} \div 1\frac{1}{5}$ **8.** $1\frac{3}{4} \div 2\frac{1}{2}$

EXAMPLE 8   Divide: $1\frac{3}{5} \div 3\frac{1}{3}$.

$$1\frac{3}{5} \div 3\frac{1}{3} = \frac{8}{5} \div \frac{10}{3} = \frac{8}{5} \cdot \frac{3}{10} = \frac{8 \cdot 3}{5 \cdot 10} = \frac{2 \cdot 4 \cdot 3}{5 \cdot 2 \cdot 5} = \frac{2}{2} \cdot \frac{4 \cdot 3}{5 \cdot 5} = \frac{12}{25}$$

Do Exercises 7 and 8.

Answers

5. 16   **6.** $7\frac{3}{7}$ **7.** $1\frac{7}{8}$ **8.** $\frac{7}{10}$

(c) Applications and Problem Solving

EXAMPLE 9 *Average Speed in Indianapolis 500.* Arie Luyendyk won the Indianapolis 500 in 1990 with the highest average speed of about 186 mph. This record high through 2008 is about $2\frac{12}{25}$ times the average speed of the first winner, Ray Harroun, in 1911. What was the average speed in the first Indianapolis 500?

Source: Indianapolis Motor Speedway

1. **Familiarize.** We ask the question "186 is $2\frac{12}{25}$ times what number?" We let $s =$ the average speed in 1911. Then the average speed in 1990 was $2\frac{12}{25} \cdot s$.

2. **Translate.** The problem can be translated to an equation as follows:

Average speed in 1990	is	$2\frac{12}{25}$	times	Average speed in 1911
↓	↓	↓	↓	↓
186	=	$2\frac{12}{25}$	·	$s.$

3. **Solve.** To solve the equation, we divide on both sides.

$$186 = \frac{62}{25} \cdot s \qquad \text{Converting } 2\frac{12}{25} \text{ to fraction notation}$$

$$\frac{186}{\frac{62}{25}} = \frac{\frac{62}{25} \cdot s}{\frac{62}{25}} \qquad \text{Dividing by } \frac{62}{25} \text{ on both sides}$$

$$\frac{186}{\frac{62}{25}} = s \qquad \begin{array}{l}\text{Factoring and removing a factor of 1:}\\ (62/25)/(62/25) = 1\end{array}$$

$$186 \cdot \frac{25}{62} = s \qquad \text{Multiplying by the reciprocal}$$

$$75 = s \qquad \text{Simplifying: } 186 \cdot \frac{25}{62} = \frac{3 \cdot 62 \cdot 25}{1 \cdot 62} = 75$$

4. **Check.** If the average speed in 1911 was about 75 mph, we find the average speed in 1990 by multiplying 75 by $2\frac{12}{25}$:

$$2\frac{12}{25} \cdot 75 = \frac{62}{25} \cdot 75 = \frac{62 \cdot 75}{25 \cdot 1} = \frac{62 \cdot 25 \cdot 3}{25 \cdot 1} = 62 \cdot 3 = 186.$$

The answer checks.

5. **State.** The average speed in the first Indianapolis 500 was about 75 mph.

Do Exercises 9 and 10.

Solve.

9. Kyle's pickup truck travels on an interstate highway at 65 mph for $3\frac{1}{2}$ hr. How far does it travel?

10. Holly's minivan travels 302 mi on $15\frac{1}{10}$ gal of gas. How many miles per gallon did it get?

Answers

9. $227\frac{1}{2}$ mi   10. 20 mpg

EXAMPLE 10 *Koi Pond.* Colleen designed a koi fish pond for her backyard. Using the dimensions shown in the diagram below, determine the area of Colleen's backyard before the pond was constructed and the area of the yard remaining after the pond was completed.

Sources: en.wikipedia.org; pondliner.com

1. **Familiarize.** From the diagram, we know that the dimensions of the backyard are $40\frac{2}{3}$ yd by $27\frac{1}{3}$ yd and the dimensions of the pond are $15\frac{3}{4}$ yd by $11\frac{1}{2}$ yd. We let $B =$ the area of the yard before the construction, $P =$ the area of the pond, and $R =$ the area of the yard that remains after the pond is complete.

2. **Translate.** This is a multistep problem. To find R, the area of the yard remaining, we first find B, the area of the yard before construction, and P, the area of the pond. Then R is the area of the yard B minus the area of the pond P:

$$R = B - P.$$

We find each area using the formula for the area of a rectangle, $A = l \times w$.

3. **Solve.** We carry out the calculations as follows:

$$B = \text{length} \times \text{width}$$
$$= 40\frac{2}{3} \cdot 27\frac{1}{3}$$
$$= \frac{122}{3} \cdot \frac{82}{3}$$
$$= \frac{10{,}004}{9} = 1111\frac{5}{9} \text{ sq yd};$$

$$P = \text{length} \times \text{width}$$
$$= 15\frac{3}{4} \cdot 11\frac{1}{2}$$
$$= \frac{63}{4} \cdot \frac{23}{2}$$
$$= \frac{1449}{8} = 181\frac{1}{8} \text{ sq yd.}$$

Then

$$R = B - P$$
$$= 1111\frac{5}{9} - 181\frac{1}{8}$$
$$= 1111\frac{40}{72} - 181\frac{9}{72}$$
$$= 930\frac{31}{72} \text{ sq yd.}$$

The brightly colored koi originated in Japan in the 18th century. The hobby of keeping koi increased greatly after the plastic bag was available for safe shipping.

11. A room measures $22\frac{1}{2}$ ft by $15\frac{1}{2}$ ft. A 9-ft by 12-ft rug is placed in the center of the room. How much floor area is not covered by the rug?

9 ft $15\frac{1}{2}$ ft

12 ft

$22\frac{1}{2}$ ft

4. Check. We can perform a check by repeating the calculations or we can round the measurements to the nearest yard and estimate a solution.

$$B \approx 41 \times 27 \approx 1107;$$
$$P \approx 16 \times 12 \approx 192;$$
$$R \approx 1107 - 192 \approx 915 \text{ sq yd}$$

Since 915 sq yd is close to $930\frac{31}{72}$ sq yd, the answer seems reasonable.

5. State. The area of the yard before construction of the pond was $1111\frac{5}{9}$ sq yd. The area of the yard remaining after the pond was completed was $930\frac{31}{72}$ sq yd.

Do Exercise 11.

Calculator Corner

Operations on Fractions and Mixed Numerals   Fraction calculators can add, subtract, multiply, and divide fractions and mixed numerals. The $\boxed{a^b/c}$ key is used to enter fractions and mixed numerals. To find $\frac{3}{4} + \frac{1}{2}$, for example, we press

$\boxed{3}\ \boxed{a^b/c}\ \boxed{4}\ \boxed{+}\ \boxed{1}\ \boxed{a^b/c}\ \boxed{2}\ \boxed{\div}$. Note that 3/4 and 1/2 appear on the display as $\boxed{\quad 3\ \lrcorner 4\quad}$ and $\boxed{\quad 1\ \lrcorner 2\quad}$, respectively. The result is given as the mixed numeral $1\frac{1}{4}$ and is displayed as $\boxed{\quad 1\ \lrcorner 1\ \lrcorner 4\quad}$. Fraction results that are greater than 1 are always displayed as mixed numerals. To express this result as a fraction, we press $\boxed{\text{SHIFT}}\ \boxed{d/c}$. We get $\boxed{\quad 5\ \lrcorner 4\quad}$, or 5/4.

To find $3\frac{2}{3} \cdot 4\frac{1}{5}$, we press $\boxed{3}\ \boxed{a^b/c}\ \boxed{2}\ \boxed{a^b/c}\ \boxed{3}\ \boxed{\times}\ \boxed{4}\ \boxed{a^b/c}\ \boxed{1}\ \boxed{a^b/c}\ \boxed{5}\ \boxed{=}$. The calculator displays $\boxed{\quad 15\ \lrcorner 2\ \lrcorner 5\quad}$, so the product is $15\frac{2}{5}$.

Some calculators are capable of displaying mixed numerals in the way in which we write them, as shown below.

Exercises: Perform each calculation. Give the answer in fraction notation.

1. $\frac{1}{3} + \frac{1}{4}$

2. $\frac{7}{5} - \frac{3}{10}$

3. $\frac{15}{4} \cdot \frac{7}{12}$

4. $\frac{4}{5} \div \frac{8}{3}$

Perform each calculation. Give the answer as a mixed numeral.

5. $4\frac{1}{3} + 5\frac{4}{5}$

6. $9\frac{2}{7} - 8\frac{1}{4}$

7. $2\frac{1}{3} \cdot 4\frac{3}{5}$

8. $10\frac{7}{10} \div 3\frac{5}{6}$

Answer

11. $240\frac{3}{4} \text{ ft}^2$

Translating for Success

1. *Raffle Tickets.* At the Happy Hollow Camp Fall Festival, Rico and Becca, together, spent $270 on raffle tickets that sell for $\$\frac{9}{20}$ each. How many tickets did they buy?

2. *Irrigation Pipe.* Jed uses two pipes, one of which measures $5\frac{1}{3}$ ft, to repair the irrigation system in the Buxtons' lawn. The total length of the two pipes is $8\frac{7}{12}$ ft. How long is the other pipe?

3. *Vacation Days.* Together, Helmut and Claire have 36 vacation days a year. Helmut has 22 vacation days per year. How many does Claire have?

4. *Enrollment in Japanese Classes.* Last year at the Lakeside Community College, 225 students enrolled in basic mathematics. This number is $4\frac{1}{2}$ times as many as the number who enrolled in Japanese. How many enrolled in Japanese?

5. *Bicycling.* Cole rode his bicycle $5\frac{1}{3}$ mi on Saturday and $8\frac{7}{12}$ mi on Sunday. How far did he ride on the weekend?

The goal of these matching questions is to practice step (2), *Translate*, of the five-step problem-solving process. Translate each word problem to an equation and select a correct translation from equations A–O.

A. $13\frac{11}{12} = x + 5\frac{1}{3}$

B. $\frac{3}{4} \cdot x = 1\frac{2}{3}$

C. $\frac{20}{9} \cdot 270 = x$

D. $225 = 4\frac{1}{2} \cdot x$

E. $98 \div 2\frac{1}{3} = x$

F. $22 + x = 36$

G. $x = 4\frac{1}{2} \cdot 225$

H. $x = 5\frac{1}{3} + 8\frac{7}{12}$

I. $22 \cdot x = 36$

J. $x = \frac{3}{4} \cdot 1\frac{2}{3}$

K. $5\frac{1}{3} + x = 8\frac{7}{12}$

L. $\frac{9}{20} \cdot 270 = x$

M. $1\frac{2}{3} + \frac{3}{4} = x$

N. $98 - 2\frac{1}{3} = x$

O. $\frac{9}{20} \cdot x = 270$

Answers on page A-5

---

6. *Deli Order.* For a promotional open house for contractors last year, the Bayside Builders Association ordered 225 turkey sandwiches. Due to increased registrations this year, $4\frac{1}{2}$ times as many sandwiches are needed. How many sandwiches are ordered?

7. *Dog Ownership.* In Sam's community, $\frac{9}{20}$ of the households own at least one dog. There are 270 households. How many own dogs?

8. *Magic Tricks.* Samantha has 98 ft of rope and needs to cut it into $2\frac{1}{3}$-ft pieces to be used in a magic trick. How many pieces can be cut from the rope?

9. *Painting.* Laura needs $1\frac{2}{3}$ gal of paint to paint the ceiling of the exercise room and $\frac{3}{4}$ gal of the same paint for the bathroom. How much paint does Laura need?

10. *Chocolate Fudge Bars.* A recipe for chocolate fudge bars that serves 16 includes $1\frac{2}{3}$ cups of sugar. How much sugar is needed for $\frac{3}{4}$ of this recipe?

a Multiply. Write a mixed numeral for the answer.

1. $8 \cdot 2\frac{5}{6}$

2. $5 \cdot 3\frac{3}{4}$

3. $3\frac{5}{8} \cdot \frac{2}{3}$

4. $6\frac{2}{3} \cdot \frac{1}{4}$

5. $3\frac{1}{2} \cdot 2\frac{1}{3}$

6. $4\frac{1}{5} \cdot 5\frac{1}{4}$

7. $3\frac{2}{5} \cdot 2\frac{7}{8}$

8. $2\frac{3}{10} \cdot 4\frac{2}{5}$

9. $4\frac{7}{10} \cdot 5\frac{3}{10}$

10. $6\frac{3}{10} \cdot 5\frac{7}{10}$

11. $20\frac{1}{2} \cdot 10\frac{1}{5} \cdot 4\frac{2}{3}$

12. $21\frac{1}{3} \cdot 11\frac{1}{3} \cdot 3\frac{5}{8}$

b Divide. Write a mixed numeral for the answer.

13. $20 \div 3\frac{1}{5}$

14. $18 \div 2\frac{1}{4}$

15. $8\frac{2}{5} \div 7$

16. $3\frac{3}{8} \div 3$

17. $4\frac{3}{4} \div 1\frac{1}{3}$

18. $5\frac{4}{5} \div 2\frac{1}{2}$

19. $1\frac{7}{8} \div 1\frac{2}{3}$

20. $4\frac{3}{8} \div 2\frac{5}{6}$

21. $5\frac{1}{10} \div 4\frac{3}{10}$

22. $4\frac{1}{10} \div 2\frac{1}{10}$

23. $20\frac{1}{4} \div 90$

24. $12\frac{1}{2} \div 50$

c Solve.

25. *Beagles.* There are about 155,000 Labrador retrievers registered with The American Kennel Club. This is $3\frac{4}{9}$ times the number of beagles registered. How many beagles are registered?

Source: The American Kennel Club

26. *Spreading Grass Seed.* Emily seeds lawns for Sam's Superior Lawn Care. When she walks at a rapid pace, the wheel on the broadcast spreader completes $150\frac{2}{3}$ revolutions per minute. How many revolutions does the wheel complete in 15 min?

27. *Population.* The population of Alabama is $6\frac{4}{5}$ times that of Alaska. The population of Alabama is approximately 4,700,000. What is the population of Alaska?

Source: U.S. Census Bureau

28. *Population.* The population of Iowa is $2\frac{1}{3}$ times the population of Hawaii. The population of Hawaii is approximately 1,300,000. What is the population of Iowa?

Source: U.S. Census Bureau

29. *Mural.* Cecilia hired an artist to paint a mural on the wall in her twin sons' bedroom. The dimensions of the mural are $6\frac{2}{3}$ ft by $9\frac{3}{8}$ ft. What is the area of the mural?

30. *Sidewalk.* A sidewalk alongside a garden at the conservatory is to be $14\frac{2}{5}$ yd long. Rectangular stone tiles that are each $1\frac{1}{8}$ yd long are used to form the sidewalk. How many tiles are used?

31. *Sodium Consumption.* The average American woman consumes $1\frac{1}{3}$ tsp of sodium each day. How much sodium do 10 average American women consume in one day?

Source: *Nutrition Action Health Letter,* March 1994

32. *Aeronautics.* Most space shuttles orbit the earth once every $1\frac{1}{2}$ hr. How many orbits are made every 24 hr?

33. *Weight of Water.* The weight of water is $62\frac{1}{2}$ lb per cubic foot. What is the weight of $5\frac{1}{2}$ cubic feet of water?

34. *Weight of Water.* The weight of water is $62\frac{1}{2}$ lb per cubic foot. What is the weight of $2\frac{1}{4}$ cubic feet of water?

35. *Temperature.* Fahrenheit temperature can be obtained from Celsius (Centigrade) temperature by multiplying by $1\frac{4}{5}$ and adding 32°. What Fahrenheit temperature corresponds to a Celsius temperature of 20°?

36. *Temperature.* Fahrenheit temperature can be obtained from Celsius (Centigrade) temperature by multiplying by $1\frac{4}{5}$ and adding 32°. What Fahrenheit temperature corresponds to the Celsius temperature of boiling water, 100°?

37. *Newspaper Circulation.* In 2002, daily circulation of the *Wall Street Journal* (New York) was about $5\frac{1}{4}$ times the daily circulation of the *Star Tribune* (Minneapolis). The circulation of the *Star Tribune* was about 343,000. What was the daily circulation of the *Wall Street Journal*?

Source: *Editor and Publisher International Year Book* 2003

38. *Language Enrollments.* In institutions of higher education in the fall of 2002, the number of students who enrolled in Spanish was $14\frac{3}{10}$ times the number who enrolled in Japanese. About 746,270 students enrolled in Spanish. About how many enrolled in Japanese?

Source: Association of Departments of Foreign Languages at the Modern Language Association

39. *Population.* The population of India is about $3\frac{3}{4}$ times the population of the United States. In 2007, the population of India was approximately 1,130,100,000. What was the population of the United States in 2007?

Source: Population Division/International Programs Center, U.S. Census Bureau, U.S. Dept. of Commerce

40. *Population.* The population of Cleveland is about $1\frac{1}{3}$ times the population of Cincinnati. In 2007, the population of Cincinnati was approximately 332,400. What was the population of Cleveland in 2007?

Source: U.S. Census Bureau

41. *Doubling a Recipe.* The chef of a 5-star hotel is doubling a recipe for chocolate cake. The cake requires $2\frac{3}{4}$ cups of flour and $1\frac{1}{3}$ cups of sugar. How much flour and sugar will she need?

42. *Half of a Recipe.* A caterer is following a salad dressing recipe that calls for $1\frac{7}{8}$ cups of mayonnaise and $1\frac{1}{6}$ cups of sugar. How much mayonnaise and sugar will he need if he prepares $\frac{1}{2}$ of the amount of salad dressing?

43. A car traveled 213 mi on $14\frac{2}{10}$ gal of gas. How many miles per gallon did it get?

44. A car traveled 385 mi on $15\frac{4}{10}$ gal of gas. How many miles per gallon did it get?

45. *Weight of Water.* The weight of water is $62\frac{1}{2}$ lb per cubic foot. How many cubic feet would be occupied by 25,000 lb of water?

46. *Weight of Water.* The weight of water is $8\frac{1}{3}$ lb per gallon. Harry rolls his lawn with an 800-lb capacity roller. Express the water capacity of the roller in gallons.

47. *Servings of Salmon.* A serving of filleted fish is generally considered to be about $\frac{1}{3}$ lb. How many servings can be prepared from $5\frac{1}{2}$ lb of salmon fillet?

48. *Servings of Tuna.* A serving of fish steak (cross section) is generally $\frac{1}{2}$ lb. How many servings can be prepared from a cleaned $18\frac{3}{4}$-lb tuna?

Find the area of each shaded region.

49.

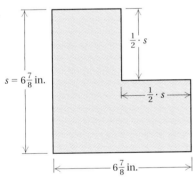

$s = 6\frac{7}{8}$ in.

$\frac{1}{2} \cdot s$

$\frac{1}{2} \cdot s$

$6\frac{7}{8}$ in.

50.

$10\frac{1}{2}$ ft

$8\frac{1}{2}$ ft

4 ft

$10\frac{1}{2}$ ft

51. *Construction.* A rectangular lot has dimensions of $302\frac{1}{2}$ ft by $205\frac{1}{4}$ ft. A building with dimensions of 100 ft by $25\frac{1}{2}$ ft is built on the lot. How much area is left over?

52. *Word Processing.* Kelly wants to create a table using Microsoft® Word. She needs to have two columns each $1\frac{1}{2}$ in. wide and five columns each $\frac{3}{4}$ in. wide. Will this table fit on a piece of standard paper that is $8\frac{1}{2}$ in. wide? If so, how wide will each margin be if the margins on each side are to be of equal width?

Skill Maintenance

In each of Exercises 53–60, fill in the blank with the correct term from the given list. Some of the choices may not be used and some may be used more than once.

53. In the equation $420 \div 60 = 7$, 60 is the _____, 7 the _____ , and 420 the _____. [1.5a]

54. When denominators are the same, we say that fractions have a _____ denominator. [2.5c]

55. The numbers 91, 95, and 111 are examples of _____ numbers. [2.1c]

56. The number 22,223,133 is _____ by 9 because the sum of its digits is _____ by 9. [2.2a]

57. When simplifying $24 \div 4 + 4 \times 12 - 6 \div 2$, do all _____ and _____ in order from left to right before doing all _____ and _____ in order from left to right. [1.9c]

58. In the equation $2 + 3 = 5$, 2 and 3 are called _____. [1.2a]

59. In the expression $\dfrac{c}{d}$, we call c the _____. [2.3a]

60. The number 0 has no _____. [2.7a]

identity
reciprocal
divisor
dividend
quotient
additions
subtractions
multiplications
divisions
numerator
denominator
equal
common
prime
composite
product
divisible
digits
factors
addends

Synthesis

Multiply. Write the answer as a mixed numeral whenever possible.

61. 🔲 $15\dfrac{2}{11} \cdot 23\dfrac{31}{43}$

62. 🔲 $17\dfrac{23}{31} \cdot 19\dfrac{13}{15}$

Simplify.

63. $8 \div \dfrac{1}{2} + \dfrac{3}{4} + \left(5 - \dfrac{5}{8}\right)^2$

64. $\left(\dfrac{5}{9} - \dfrac{1}{4}\right) \times 12 + \left(4 - \dfrac{3}{4}\right)^2$

65. $\dfrac{1}{3} \div \left(\dfrac{1}{2} - \dfrac{1}{5}\right) \times \dfrac{1}{4} + \dfrac{1}{6}$

66. $\dfrac{7}{8} - 1\dfrac{1}{8} \times \dfrac{2}{3} + \dfrac{9}{10} \div \dfrac{3}{5}$

67. $4\dfrac{1}{2} \div 2\dfrac{1}{2} + 8 - 4 \div \dfrac{1}{2}$

68. $6 - 2\dfrac{1}{3} \times \dfrac{3}{4} + \dfrac{5}{8} \div \dfrac{2}{3}$

3.7 Order of Operations; Estimation

a. Order of Operations; Fraction Notation and Mixed Numerals

The rules for order of operations that we use with whole numbers (see Section 1.9) apply when we are simplifying expressions involving fraction notation and mixed numerals. For review, these rules are listed below.

> **RULES FOR ORDER OF OPERATIONS**
> 1. Do all calculations within parentheses before operations outside.
> 2. Evaluate all exponential expressions.
> 3. Do all multiplications and divisions in order from left to right.
> 4. Do all additions and subtractions in order from left to right.

EXAMPLE 1 Simplify: $\dfrac{1}{6} + \dfrac{2}{3} \div \dfrac{1}{2} \cdot \dfrac{5}{8}$.

$$\dfrac{1}{6} + \dfrac{2}{3} \div \dfrac{1}{2} \cdot \dfrac{5}{8} = \dfrac{1}{6} + \dfrac{2}{3} \cdot \dfrac{2}{1} \cdot \dfrac{5}{8}$$

Doing the division first by multiplying by the reciprocal of $\dfrac{1}{2}$

$$= \dfrac{1}{6} + \dfrac{2 \cdot 2 \cdot 5}{3 \cdot 1 \cdot 8}$$

Doing the multiplications in order from left to right

$$= \dfrac{1}{6} + \dfrac{2 \cdot 2 \cdot 5}{3 \cdot 1 \cdot 2 \cdot 2 \cdot 2}$$

Factoring in order to simplify

$$= \dfrac{1}{6} + \dfrac{5}{6}$$

Removing a factor of 1: $\dfrac{2 \cdot 2}{2 \cdot 2} = 1$; simplifying

$$= \dfrac{6}{6}, \text{ or } 1$$

Doing the addition

Do Margin Exercises 1 and 2.

EXAMPLE 2 Simplify: $\dfrac{2}{3} \cdot 24 - 11\dfrac{1}{2}$.

$$\dfrac{2}{3} \cdot 24 - 11\dfrac{1}{2} = \dfrac{2 \cdot 24}{3 \cdot 1} - 11\dfrac{1}{2}$$

Doing the multiplication first

$$= \dfrac{2 \cdot 3 \cdot 8}{3 \cdot 1} - 11\dfrac{1}{2}$$

Factoring the fraction

$$= 2 \cdot 8 - 11\dfrac{1}{2}$$

Removing a factor of 1: $\dfrac{3}{3} = 1$

$$= 16 - 11\dfrac{1}{2}$$

Completing the multiplication

$$= 4\dfrac{1}{2}, \text{ or } \dfrac{9}{2}$$

Doing the subtraction

Do Exercise 3.

OBJECTIVES

a. Simplify expressions using the rules for order of operations.

b. Estimate with fraction notation and mixed numerals.

SKILL TO REVIEW
Objective 1.9c: Simplify expressions using the rules for order of operations.

Simplify.
1. $22 - 3 \cdot 4$
2. $6(4 + 1)^2 - 3^4 \div 3$

Simplify.
1. $\dfrac{2}{5} \cdot \dfrac{5}{8} + \dfrac{1}{4}$
2. $\dfrac{1}{3} \cdot \dfrac{3}{4} \div \dfrac{5}{8} - \dfrac{1}{10}$

3. Simplify: $\dfrac{3}{4} \cdot 16 + 8\dfrac{2}{3}$.

Answers

Skill to Review:
1. 10   2. 123

Margin Exercises:
1. $\dfrac{1}{2}$ 2. $\dfrac{3}{10}$ 3. $20\dfrac{2}{3}$, or $\dfrac{62}{3}$

3.7 Order of Operations; Estimation   **199**

EXAMPLE 3 *Harvesting Walnut Trees.* A woodland owner decided to harvest five walnut trees in order to improve the growing conditions of the remaining trees. The logs she sold measured $7\frac{5}{8}$ ft, $8\frac{1}{4}$ ft, $8\frac{3}{4}$ ft, $9\frac{1}{8}$ ft, and $10\frac{1}{2}$ ft. What is the average length of the logs?

Recall that to compute an average, we add the numbers and then divide the sum by the number of addends. (See Section 1.9.) We have

$$\frac{7\frac{5}{8} + 8\frac{1}{4} + 8\frac{3}{4} + 9\frac{1}{8} + 10\frac{1}{2}}{5} = \frac{7\frac{5}{8} + 8\frac{2}{8} + 8\frac{6}{8} + 9\frac{1}{8} + 10\frac{4}{8}}{5}$$

$$= \frac{42\frac{18}{8}}{5} = \frac{42\frac{9}{4}}{5} = \frac{\frac{177}{4}}{5} \qquad \begin{array}{l}\text{Adding,}\\ \text{simplifying, and}\\ \text{converting}\\ \text{to fraction notation}\end{array}$$

$$= \frac{177}{4} \cdot \frac{1}{5} \qquad \begin{array}{l}\text{Multiplying}\\ \text{by the reciprocal}\end{array}$$

$$= \frac{177}{20} = 8\frac{17}{20}. \qquad \begin{array}{l}\text{Converting}\\ \text{to a mixed numeral}\end{array}$$

The average length of the logs is $8\frac{17}{20}$ ft.

> Do Exercises 4–6.

4. Rachel has triplets. Their birth weights are $3\frac{1}{2}$ lb, $2\frac{3}{4}$ lb, and $3\frac{1}{8}$ lb. What is the average weight of her babies?

5. Find the average of
$$\frac{1}{2}, \frac{1}{3}, \text{ and } \frac{5}{6}.$$

6. Find the average of $\frac{3}{4}$ and $\frac{4}{5}$.

EXAMPLE 4 Simplify: $\left(\frac{7}{8} - \frac{1}{3}\right) \times 48 + \left(13 + \frac{4}{5}\right)^2.$

$$\left(\frac{7}{8} - \frac{1}{3}\right) \times 48 + \left(13 + \frac{4}{5}\right)^2$$

$$= \left(\frac{7}{8} \cdot \frac{3}{3} - \frac{1}{3} \cdot \frac{8}{8}\right) \times 48 + \left(13 \cdot \frac{5}{5} + \frac{4}{5}\right)^2 \qquad \begin{array}{l}\text{Carrying out operations}\\ \text{inside parentheses first.}\\ \text{To do so, we first multiply}\\ \text{by 1 to obtain the LCD.}\end{array}$$

$$= \left(\frac{21}{24} - \frac{8}{24}\right) \times 48 + \left(\frac{65}{5} + \frac{4}{5}\right)^2$$

$$= \frac{13}{24} \times 48 + \left(\frac{69}{5}\right)^2 \qquad \text{Completing the operations within parentheses}$$

$$= \frac{13}{24} \times 48 + \frac{4761}{25} \qquad \text{Evaluating the exponential expression next}$$

$$= 26 + \frac{4761}{25} \qquad \text{Doing the multiplication}$$

$$= 26 + 190\frac{11}{25} \qquad \text{Converting to a mixed numeral}$$

$$= 216\frac{11}{25}, \quad \text{or} \quad \frac{5411}{25} \qquad \text{Adding}$$

Answers can be given using either fraction notation or mixed numerals.

> Do Exercise 7.

7. Simplify:
$$\left(\frac{2}{3} + \frac{3}{4}\right) \div 2\frac{1}{3} - \left(\frac{1}{2}\right)^3.$$

Answers

4. $3\frac{1}{8}$ lb   **5.** $\frac{5}{9}$ **6.** $\frac{31}{40}$ **7.** $\frac{27}{56}$

Estimation with Fraction Notation and Mixed Numerals

We now estimate with fraction notation and mixed numerals.

EXAMPLES Estimate each of the following as 0, $\frac{1}{2}$, or 1.

5. $\dfrac{2}{17}$

A fraction is close to 0 when the numerator is small in comparison to the denominator. Thus, 0 is an estimate for $\frac{2}{17}$ because 2 is small in comparison to 17. Thus, $\frac{2}{17} \approx 0$.

6. $\dfrac{11}{23}$

A fraction is close to $\frac{1}{2}$ when the denominator is about twice the numerator. Thus, $\frac{1}{2}$ is an estimate for $\frac{11}{23}$ because $2 \cdot 11 = 22$ and 22 is close to 23. Thus, $\frac{11}{23} \approx \frac{1}{2}$.

7. $\dfrac{37}{38}$

A fraction is close to 1 when the numerator is nearly equal to the denominator. Thus, 1 is an estimate for $\frac{37}{38}$ because 37 is nearly equal to 38. Thus, $\frac{37}{38} \approx 1$.

8. $\dfrac{43}{41}$

As in the preceding example, the numerator 43 is nearly equal to the denominator 41. Thus, $\frac{43}{41} \approx 1$.

> Do Exercises 8–11.

Estimate each of the following as 0, $\frac{1}{2}$, or 1.

8. $\dfrac{3}{59}$ **9.** $\dfrac{61}{59}$

10. $\dfrac{29}{59}$ **11.** $\dfrac{57}{59}$

EXAMPLE 9 Estimate $16\frac{8}{9} + 11\frac{2}{13} - 4\frac{22}{43}$ by estimating each mixed numeral as a whole number or as a mixed numeral where the fractional part is $\frac{1}{2}$.

We estimate each fraction as 0, $\frac{1}{2}$, or 1. Then we calculate:

$$16\frac{8}{9} + 11\frac{2}{13} - 4\frac{22}{43} \approx 17 + 11 - 4\frac{1}{2}$$

$$= 28 - 4\frac{1}{2}$$

$$= 23\frac{1}{2}.$$

> Do Exercises 12–14.

Estimate each of the following by estimating each mixed numeral as a whole number or as a mixed numeral where the fractional part is $\frac{1}{2}$ and each fraction as 0, $\frac{1}{2}$, or 1.

12. $5\dfrac{9}{10} + 26\dfrac{1}{2} - 10\dfrac{3}{29}$

13. $10\dfrac{7}{8} \cdot \left(25\dfrac{11}{13} - 14\dfrac{1}{9} \right)$

14. $\left(10\dfrac{4}{5} + 7\dfrac{5}{9} \right) \div \dfrac{17}{30}$

Answers

8. 0   **9.** 1   **10.** $\dfrac{1}{2}$ **11.** 1

12. $22\dfrac{1}{2}$ **13.** 132   **14.** 37

3.7 **Exercise Set**

For Extra Help

MyMathLab

Math XL
PRACTICE

WATCH

DOWNLOAD

READ

REVIEW

a Simplify.

1. $\dfrac{5}{8} \div \dfrac{1}{4} - \dfrac{2}{3} \cdot \dfrac{4}{5}$

2. $\dfrac{4}{7} \cdot \dfrac{7}{15} + \dfrac{2}{3} \div 8$

3. $\dfrac{3}{4} - \dfrac{2}{3} \cdot \left(\dfrac{1}{2} + \dfrac{2}{5}\right)$

4. $\dfrac{3}{4} \div \dfrac{1}{2} \cdot \left(\dfrac{8}{9} - \dfrac{2}{3}\right)$

5. $28\dfrac{1}{8} - 5\dfrac{1}{4} + 3\dfrac{1}{2}$

6. $10\dfrac{3}{5} - 4\dfrac{1}{10} - 1\dfrac{1}{2}$

7. $\dfrac{7}{8} \div \dfrac{1}{2} \cdot \dfrac{1}{4}$

8. $\dfrac{7}{10} \cdot \dfrac{4}{5} \div \dfrac{2}{3}$

9. $\left(\dfrac{2}{3}\right)^2 - \dfrac{1}{3} \cdot 1\dfrac{1}{4}$

10. $\left(\dfrac{3}{4}\right)^2 + 3\dfrac{1}{2} \div 1\dfrac{1}{4}$

11. $\dfrac{1}{2} - \left(\dfrac{1}{2}\right)^2 + \left(\dfrac{1}{2}\right)^3$

12. $1 + \dfrac{1}{4} + \left(\dfrac{1}{4}\right)^2 - \left(\dfrac{1}{4}\right)^3$

13. $\left(\dfrac{2}{3} + \dfrac{3}{4}\right) \div \left(\dfrac{5}{6} - \dfrac{1}{3}\right)$

14. $\left(\dfrac{3}{5} - \dfrac{1}{2}\right) \div \left(\dfrac{3}{4} - \dfrac{3}{10}\right)$

15. $\left(\dfrac{1}{2} + \dfrac{1}{3}\right)^2 \cdot 144 - \dfrac{5}{8} \div 10\dfrac{1}{2}$

16. $\left(3\dfrac{1}{2} - 2\dfrac{1}{3}\right)^2 + 6 \cdot 2\dfrac{1}{2} \div 32$

17. Find the average of $\dfrac{2}{3}$ and $\dfrac{7}{8}$.

18. Find the average of $\dfrac{1}{4}$ and $\dfrac{1}{5}$.

19. Find the average of $\dfrac{1}{6}, \dfrac{1}{8}$, and $\dfrac{3}{4}$.

20. Find the average of $\dfrac{4}{5}, \dfrac{1}{2}$, and $\dfrac{1}{10}$.

21. Find the average of $3\dfrac{1}{2}$ and $9\dfrac{3}{8}$.

22. Find the average of $10\dfrac{2}{3}$ and $24\dfrac{5}{6}$.

23. *Hiking the Appalachian Trail.* Ellen camped and hiked for three consecutive days along a section of the Appalachian Trail. The distances she hiked on the three days were $15\frac{5}{32}$ mi, $20\frac{3}{16}$ mi, and $12\frac{7}{8}$ mi. Find the average of these distances.

24. *Vertical Leaps.* Eight-year-old Zachary registered vertical leaps of $12\frac{3}{4}$ in., $13\frac{3}{4}$ in., $13\frac{1}{2}$ in., and 14 in. Find his average vertical leap.

25. *Black Bear Cubs.* Black bears typically have two cubs. In January 2007 in northern New Hampshire, a black bear sow gave birth to a litter of 5 cubs. This is so rare that Tom Sears, a wildlife photographer, spent 28 hr per week for six weeks watching for the perfect opportunity to photograph this family of six. At the time of this photo, an observer estimated that the cubs weighed $7\frac{1}{2}$ lb, 8 lb, $9\frac{1}{2}$ lb, $10\frac{5}{8}$ lb, and $11\frac{3}{4}$ lb. What was the average weight of the cubs?

Source: Andrew Timmins, New Hampshire Fish and Game Department, *Northcountry News*, Warren, NH; Tom Sears, photographer

26. *Acceleration.* The results of a *Road & Track* road acceleration test for five cars are given in the graph below. The test measures the time in seconds required to go from 0 mph to 60 mph. What was the average time?

Acceleration: 0 mph to 60 mph

SOURCE: *Road & Track*, October 2008, pp.156–157

b Estimate each of the following as 0, $\frac{1}{2}$, or 1.

27. $\dfrac{2}{47}$

28. $\dfrac{5}{12}$

29. $\dfrac{7}{8}$

30. $\dfrac{1}{13}$

31. $\dfrac{6}{11}$

32. $\dfrac{11}{13}$

33. $\dfrac{7}{15}$

34. $\dfrac{1}{16}$

35. $\dfrac{7}{100}$

36. $\dfrac{5}{9}$

37. $\dfrac{19}{20}$

38. $\dfrac{4}{5}$

Estimate each of the following by estimating each mixed numeral as a whole number or as a mixed numeral where the fractional part is $\frac{1}{2}$ and by estimating each fraction as 0, $\frac{1}{2}$, or 1.

39. $2\dfrac{7}{8}$

40. $1\dfrac{1}{3}$

41. $12\dfrac{5}{6}$

42. $26\dfrac{6}{13}$

43. $\dfrac{4}{5} + \dfrac{7}{8}$

44. $\dfrac{1}{12} \cdot \dfrac{7}{15}$

45. $\dfrac{3}{8} + \dfrac{7}{13} + \dfrac{5}{9}$

46. $\dfrac{8}{9} + \dfrac{4}{5} + \dfrac{11}{12}$

47. $\dfrac{43}{100} + \dfrac{1}{10} - \dfrac{11}{1000}$

48. $\dfrac{23}{24} + \dfrac{37}{39} + \dfrac{51}{50}$

49. $7\dfrac{29}{60} + 10\dfrac{12}{13} \cdot 24\dfrac{2}{17}$

50. $5\dfrac{13}{14} - 1\dfrac{5}{8} + 1\dfrac{23}{28} \cdot 6\dfrac{35}{74}$

51. $24 \div 7\dfrac{8}{9}$

52. $43\dfrac{16}{17} \div 11\dfrac{2}{13}$

53. $76\dfrac{3}{14} + 23\dfrac{19}{20}$

54. $76\dfrac{13}{14} \cdot 23\dfrac{17}{20}$

55. $16\dfrac{1}{5} \div 2\dfrac{1}{11} + 25\dfrac{9}{10} - 4\dfrac{11}{23}$

56. $96\dfrac{2}{13} \div 5\dfrac{19}{20} + 3\dfrac{1}{7} \cdot 5\dfrac{18}{21}$

Skill Maintenance

Divide and simplify. [2.7b]

57. $\dfrac{4}{5} \div \dfrac{3}{10}$

58. $\dfrac{3}{10} \div \dfrac{4}{5}$

59. Classify the given numbers as prime, composite, or neither.

$1, 5, 7, 9, 14, 23, 43$ [2.1c]

60. Divide: $7865 \div 132$. [1.5a]

Solve.

61. *Luncheon Servings.* Ian purchased 6 lb of cold cuts for a luncheon. If he has allowed $\frac{3}{8}$ lb per person, how many people did he invite to the luncheon? [2.7d]

62. *Cholesterol.* A 3-oz serving of crabmeat contains 85 milligrams (mg) of cholesterol. A 3-oz serving of shrimp contains 128 mg of cholesterol. How much more cholesterol is in the shrimp? [1.8a]

Synthesis

63. ▦ In the sum below, a and b are digits. Find a and b.

$$\dfrac{a}{17} + \dfrac{1b}{23} = \dfrac{35a}{391}$$

64. ▦ Consider only the numbers 3, 4, 5, and 6. Assume each can be placed in a blank in the following.

$$\square + \dfrac{\square}{\square} \cdot \square = ?$$

What placement of the numbers in the blanks yields the largest sum?

65. ▦ Consider only the numbers 2, 3, 4, and 5. Assume each is placed in a blank in the following.

$$\dfrac{\square}{\square} + \dfrac{\square}{\square} = ?$$

What placement of the numbers in the blanks yields the largest sum?

66. ▦ Use a calculator to arrange the following in order from smallest to largest.

$$\dfrac{3}{4}, \dfrac{17}{21}, \dfrac{13}{15}, \dfrac{7}{9}, \dfrac{15}{17}, \dfrac{13}{12}, \dfrac{19}{22}$$

Summary and Review

Key Terms

least common multiple, p. 148
least common denominator, p. 148
mixed numeral, p. 170

Concept Reinforcement

Determine whether each statement is true or false.

_____ **1.** The mixed numeral $5\frac{2}{3}$ can be represented by the sum $5 \cdot \frac{3}{3} + \frac{2}{3}$. [3.4a]

_____ **2.** The least common multiple of two numbers is always larger than or equal to the larger number. [3.1a]

_____ **3.** If $\frac{a}{b} < 1$, then $a > b$. [3.7b]

_____ **4.** The product of any two mixed numerals is greater than 1. [3.6a]

Important Concepts

Objective 3.1a Find the least common multiple, or LCM, of two or more numbers.

Example Find the LCM of 105 and 90.

$$105 = 3 \cdot 5 \cdot 7,$$
$$90 = 2 \cdot 3 \cdot 3 \cdot 5;$$
$$\text{LCM} = 2 \cdot 3 \cdot 3 \cdot 5 \cdot 7 = 630$$

Practice Exercise

1. Find the LCM of 52 and 78.

Objective 3.2a Add using fraction notation.

Example Add: $\dfrac{5}{24} + \dfrac{7}{45}$.

$$\frac{5}{24} + \frac{7}{45} = \frac{5}{2 \cdot 2 \cdot 2 \cdot 3} + \frac{7}{3 \cdot 3 \cdot 5}$$

$$= \frac{5}{2 \cdot 2 \cdot 2 \cdot 3} \cdot \frac{3 \cdot 5}{3 \cdot 5} + \frac{7}{3 \cdot 3 \cdot 5} \cdot \frac{2 \cdot 2 \cdot 2}{2 \cdot 2 \cdot 2}$$

$$= \frac{75}{360} + \frac{56}{360} = \frac{131}{360}$$

Practice Exercise

2. Add: $\dfrac{19}{60} + \dfrac{11}{36}$.

Objective 3.3a Subtract using fraction notation.

Example Subtract: $\dfrac{7}{12} - \dfrac{11}{60}$.

$$\frac{7}{12} - \frac{11}{60} = \frac{7}{12} \cdot \frac{5}{5} - \frac{11}{60} = \frac{35}{60} - \frac{11}{60}$$

$$= \frac{35 - 11}{60} = \frac{24}{60} = \frac{2 \cdot \cancel{12}}{5 \cdot \cancel{12}} = \frac{2}{5}$$

Practice Exercise

3. Subtract: $\dfrac{29}{35} - \dfrac{5}{7}$.

Objective 3.3b Use < or > with fraction notation to write a true sentence.

Example Use < or > for ☐ to write a true sentence:

$$\frac{5}{12} \ \square \ \frac{9}{16}.$$

The LCD = 48. Thus we have

$$\frac{5}{12} \cdot \frac{4}{4} \ \square \ \frac{9}{16} \cdot \frac{3}{3}$$

$$\frac{20}{48} \ \square \ \frac{27}{48}.$$

Since $20 < 27$, $\dfrac{20}{48} < \dfrac{27}{48}$ and thus $\dfrac{5}{12} < \dfrac{9}{16}$.

Practice Exercise

4. Use < or > for ☐ to write a true sentence:

$$\frac{3}{13} \ \square \ \frac{5}{12}.$$

Objective 3.3c Solve equations of the type $x + a = b$ and $a + x = b$, where a and b may be fractions.

Example Solve: $x + \dfrac{1}{6} = \dfrac{5}{8}$.

$$x + \frac{1}{6} = \frac{5}{8}$$

$$x + \frac{1}{6} - \frac{1}{6} = \frac{5}{8} - \frac{1}{6}$$

$$x = \frac{5}{8} \cdot \frac{3}{3} - \frac{1}{6} \cdot \frac{4}{4}$$

$$x = \frac{15}{24} - \frac{4}{24} = \frac{11}{24}$$

Practice Exercise

5. Solve: $\dfrac{2}{9} + x = \dfrac{9}{11}$.

Objective 3.4a Convert between mixed numerals and fraction notation.

Example Convert $2\dfrac{5}{13}$ to fraction notation: $2\dfrac{5}{13} = \dfrac{31}{13}$.

Example Convert $\dfrac{40}{9}$ to a mixed numeral: $\dfrac{40}{9} = 4\dfrac{4}{9}$.

Practice Exercises

6. Convert $8\dfrac{2}{3}$ to fraction notation.

7. Convert $\dfrac{47}{6}$ to a mixed numeral.

Objective 3.5b Subtract using mixed numerals.

Example Subtract: $3\dfrac{3}{8} - 1\dfrac{4}{5}$.

$$3\frac{3}{8} = \ \ 3\frac{15}{40} \ = \ \ 2\frac{55}{40}$$

$$-1\frac{4}{5} = -1\frac{32}{40} \ = -1\frac{32}{40}$$

$$\overline{\phantom{-1\frac{32}{40} = } 1\frac{23}{40}}$$

Practice Exercise

8. Subtract: $10\dfrac{5}{7} - 2\dfrac{3}{4}$.

Objective 3.6a Multiply using mixed numerals.

Example Multiply: $7\frac{1}{4} \cdot 5\frac{3}{10}$. Write a mixed numeral for the answer.

$$7\frac{1}{4} \cdot 5\frac{3}{10} = \frac{29}{4} \cdot \frac{53}{10}$$

$$= \frac{1537}{40} = 38\frac{17}{40}$$

Practice Exercise

9. Multiply: $4\frac{1}{5} \cdot 3\frac{7}{15}$.

Objective 3.6c Solve applied problems involving multiplication and division with mixed numerals.

Example The population of New York is $3\frac{1}{3}$ times that of Missouri. The population of New York is approximately 19,000,000. What is the population of Missouri?

Translate:

$$\underbrace{\text{Population of New York}}_{\big\downarrow} \quad \underset{\big\downarrow}{\text{is}} \quad \underset{\big\downarrow}{3\frac{1}{3}} \quad \underset{\big\downarrow}{\text{times}} \quad \underbrace{\text{Population of Missouri}}_{\big\downarrow}$$

$$19{,}000{,}000 \quad = \quad 3\frac{1}{3} \quad \cdot \quad x.$$

Solve:

$$19{,}000{,}000 = \frac{10}{3} \cdot x$$

$$\frac{19{,}000{,}000}{\frac{10}{3}} = \frac{\frac{10}{3} \cdot x}{\frac{10}{3}}$$

$$19{,}000{,}000 \cdot \frac{3}{10} = x$$

$$5{,}700{,}000 = x.$$

The population of Missouri is about 5,700,000.

Practice Exercise

10. The population of Louisiana is $2\frac{1}{2}$ times the population of West Virginia. The population of West Virginia is approximately 1,800,000. What is the population of Louisiana?

Objective 3.7a Simplify expressions using the rules for order of operations.

Example Simplify: $\left(\frac{4}{5}\right)^2 - \frac{1}{5} \cdot 2\frac{1}{8}$.

$$\left(\frac{4}{5}\right)^2 - \frac{1}{5} \cdot 2\frac{1}{8} = \frac{16}{25} - \frac{1}{5} \cdot \frac{17}{8}$$

$$= \frac{16}{25} - \frac{17}{40} = \frac{16}{25} \cdot \frac{8}{8} - \frac{17}{40} \cdot \frac{5}{5}$$

$$= \frac{128}{200} - \frac{85}{200} = \frac{43}{200}$$

Practice Exercise

11. Simplify: $\frac{3}{2} \cdot 1\frac{1}{3} \div \left(\frac{2}{3}\right)^2$.

Objective 3.7b Estimate with fraction notation and mixed numerals.

Example Estimate

$$3\frac{2}{5} + \frac{7}{10} + 6\frac{5}{9}$$

by estimating each mixed numeral as a whole number or as a mixed numeral where the fractional part is $\frac{1}{2}$ and by estimating each fraction as 0, $\frac{1}{2}$, or 1.

$$3\frac{2}{5} + \frac{7}{10} + 6\frac{5}{9} \approx 3\frac{1}{2} + \frac{1}{2} + 6\frac{1}{2} = 10\frac{1}{2}$$

Practice Exercise

12. Estimate

$$1\frac{19}{20} + 3\frac{1}{8} - \frac{8}{17}$$

by estimating each mixed numeral as a whole number or as a mixed numeral where the fractional part is $\frac{1}{2}$ and by estimating each fraction as 0, $\frac{1}{2}$, or 1.

Review Exercises

Find the LCM. [3.1a]

1. 12 and 18

2. 18 and 45

3. 3, 6, and 30

4. 26, 36, and 54

Add and simplify. [3.2a]

5. $\dfrac{6}{5} + \dfrac{3}{8}$

6. $\dfrac{5}{16} + \dfrac{1}{12}$

7. $\dfrac{6}{5} + \dfrac{11}{15} + \dfrac{3}{20}$

8. $\dfrac{1}{1000} + \dfrac{19}{100} + \dfrac{7}{10}$

Subtract and simplify. [3.3a]

9. $\dfrac{5}{9} - \dfrac{2}{9}$

10. $\dfrac{7}{8} - \dfrac{3}{4}$

11. $\dfrac{11}{27} - \dfrac{2}{9}$

12. $\dfrac{5}{6} - \dfrac{2}{9}$

Use $<$ or $>$ for $\square$ to write a true sentence. [3.3b]

13. $\dfrac{4}{7} \,\square\, \dfrac{5}{9}$

14. $\dfrac{8}{9} \,\square\, \dfrac{11}{13}$

Solve. [3.3c]

15. $x + \dfrac{2}{5} = \dfrac{7}{8}$

16. $\dfrac{1}{2} + y = \dfrac{9}{10}$

Convert to fraction notation. [3.4a]

17. $7\dfrac{1}{2}$

18. $8\dfrac{3}{8}$

19. $4\dfrac{1}{3}$

20. $10\dfrac{5}{7}$

Convert to a mixed numeral. [3.4a]

21. $\dfrac{7}{3}$

22. $\dfrac{27}{4}$

23. $\dfrac{63}{5}$

24. $\dfrac{7}{2}$

Divide. Write a mixed numeral for the answer. [3.4b]

25. $9 \,\overline{)\, 7\ 8\ 9\ 6}$

26. $2\ 3 \,\overline{)\, 1\ 0,4\ 9\ 3}$

Add. Write a mixed numeral for the answer where appropriate. [3.5a]

27. $\begin{array}{r} 5\frac{3}{5} \\ + 4\frac{4}{5} \\ \hline \end{array}$

28. $\begin{array}{r} 8\frac{1}{3} \\ + 3\frac{2}{5} \\ \hline \end{array}$

29. $\begin{array}{r} 5\frac{5}{6} \\ + 4\frac{5}{6} \\ \hline \end{array}$

30. $\begin{array}{r} 2\frac{3}{4} \\ + 5\frac{1}{2} \\ \hline \end{array}$

Subtract. Write a mixed numeral for the answer where appropriate. [3.5b]

31. $\begin{array}{r} 12 \\ - 4\frac{2}{9} \\ \hline \end{array}$

32. $\begin{array}{r} 9\frac{3}{5} \\ - 4\frac{13}{15} \\ \hline \end{array}$

33. $\begin{array}{r} 10\frac{1}{4} \\ - 6\frac{1}{10} \\ \hline \end{array}$

34. $\begin{array}{r} 24 \\ - 10\frac{5}{8} \\ \hline \end{array}$

Multiply. Write a mixed numeral for the answer where appropriate. [3.6a]

35. $6 \cdot 2\dfrac{2}{3}$

36. $5\dfrac{1}{4} \cdot \dfrac{2}{3}$

37. $2\dfrac{1}{5} \cdot 1\dfrac{1}{10}$

38. $2\dfrac{2}{5} \cdot 2\dfrac{1}{2}$

Divide. Write a mixed numeral for the answer where appropriate. [3.6b]

39. $27 \div 2\frac{1}{4}$

40. $2\frac{2}{5} \div 1\frac{7}{10}$

41. $3\frac{1}{4} \div 26$

42. $4\frac{1}{5} \div 4\frac{2}{3}$

Solve. [3.2b], [3.5c], [3.6c]

43. *Sewing.* Gloria wants to make a dress and a jacket. She needs $1\frac{5}{8}$ yd of 60-in. fabric for the dress and $2\frac{5}{8}$ yd for the jacket. How many yards in all does Gloria need to make the outfit?

44. What is the sum of the areas in the figure below?

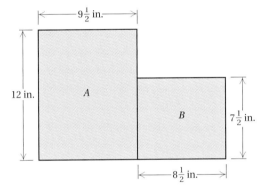

45. In the figure above, how much larger is the area of rectangle *A* than the area of rectangle *B*?

46. *Carpentry.* A board $\frac{9}{10}$ in. thick is glued to a board $\frac{4}{5}$ in. thick. The glue is $\frac{3}{100}$ in. thick. How thick is the result?

47. *Turkey Servings.* Turkey contains $1\frac{1}{3}$ servings per pound. How many pounds are needed for 32 servings?

48. *Painting a Border.* Katie hired an artist to paint a decorative border around the top of her son's bedroom. The artist charges $20 per foot. The room measures $11\frac{3}{4}$ ft $\times$ $9\frac{1}{2}$ ft. What is Katie's cost for the project?

$9\frac{1}{2}$ ft

$11\frac{3}{4}$ ft

49. *Cake Recipe.* A wedding-cake recipe requires 12 cups of shortening. Being calorie-conscious, the wedding couple decides to reduce the shortening by $3\frac{5}{8}$ cups and replace it with prune purée. How many cups of shortening are used in their new recipe?

50. *Building a Ziggurat.* The dimensions of all of the square bricks that King Nebuchadnezzar used over 2500 yr ago to build ziggurats were $13\frac{1}{4}$ in. $\times$ $13\frac{1}{4}$ in. $\times$ $3\frac{1}{4}$ in. What is the perimeter and the area of the $13\frac{1}{4}$ in. $\times$ $13\frac{1}{4}$ in. side? of the $13\frac{1}{4}$ in. $\times$ $3\frac{1}{4}$ in. side?
Source: www.eartharchitecture.org

51. *Firefighters' Pie Sale.* Green River's Humane Society recently hosted its annual ice cream social. Each of the 83 pies donated was cut into 6 pieces. At the end of the evening, 382 pieces of pie had been sold. How many pies were sold? How many were left over? Express your answers in mixed numerals.

Simplify each expression using the rules for order of operations. [3.7a]

52. $\dfrac{1}{8} \div \dfrac{1}{4} + \dfrac{1}{2}$

53. $\dfrac{4}{5} - \dfrac{1}{2} \cdot \left(1 + \dfrac{1}{4}\right)$

54. $20\dfrac{3}{4} - 1\dfrac{1}{2} \times 12 + \left(\dfrac{1}{2}\right)^2$

55. Find the average of $\dfrac{1}{2}, \dfrac{1}{4}, \dfrac{1}{3}$, and $\dfrac{1}{5}$. [3.7a]

Estimate each of the following as $0, \frac{1}{2}$, or 1. [3.7b]

56. $\dfrac{29}{59}$ **57.** $\dfrac{2}{59}$ **58.** $\dfrac{61}{59}$

Estimate by estimating each mixed numeral as a whole number or as a mixed numeral where the fractional part is $\frac{1}{2}$ and by estimating each fraction as $0, \frac{1}{2}$, or 1. [3.7b]

59. $6\dfrac{7}{8}$ **60.** $10\dfrac{2}{17}$

61. $\dfrac{11}{12} \cdot 5\dfrac{6}{13}$ **62.** $\dfrac{1}{15} \cdot \dfrac{2}{3}$

63. $\dfrac{6}{11} + \dfrac{5}{6} + \dfrac{31}{29}$

64. $32\dfrac{14}{15} + 27\dfrac{6}{7} - 4\dfrac{25}{28} \cdot 6\dfrac{37}{76}$

65. Simplify: $\dfrac{1}{4} + \dfrac{2}{5} \div 5^2$. [3.7a]

    A. $\dfrac{133}{500}$ **B.** $\dfrac{3}{500}$

    C. $\dfrac{117}{500}$ **D.** $\dfrac{5}{2}$

66. Solve: $x + \dfrac{2}{3} = 5$. [3.3c]

    A. $\dfrac{15}{2}$ **B.** $5\dfrac{2}{3}$

    C. $\dfrac{10}{3}$ **D.** $4\dfrac{1}{3}$

Synthesis

67. *Orangutan Circus Act.* Yuri and Olga are orangutans that perform in a circus by riding bicycles around a circular track. It takes Yuri 6 min and Olga 4 min to make one trip around the track. Suppose they start at the same point and then complete their act when they again reach the same point. How long is their act? [3.1a]

68. Place the numbers 3, 4, 5, and 6 in the boxes in order to make a true equation: [3.5a]

$$\dfrac{\Box}{\Box} + \dfrac{\Box}{\Box} = 3\dfrac{1}{4}.$$

Understanding Through Discussion and Writing

1. Is the sum of two mixed numerals always a mixed numeral? Why or why not? [3.5a]

2. Write a problem for a classmate to solve. Design the problem so that its solution is found by performing the multiplication $4\frac{1}{2} \cdot 33\frac{1}{3}$. [3.6c]

3. A student insists that $3\frac{2}{5} \cdot 1\frac{3}{7} = 3\frac{6}{35}$. What mistake is he making and how should he have proceeded? [3.6a]

4. Discuss the role of least common multiples in adding and subtracting with fraction notation. [3.2a], [3.3a]

5. Find a real-world situation that fits this equation:

$$2 \cdot 15\dfrac{3}{4} + 2 \cdot 28\dfrac{5}{8} = 88\dfrac{3}{4}. \quad [3.5c], [3.6c]$$

6. A student insists that $5 \cdot 3\frac{2}{7} = (5 \cdot 3) \cdot \left(5 \cdot \frac{2}{7}\right)$. What mistake is she making and how should she have proceeded? [3.7a]

Cumulative Review

Solve.

1. *Cross-Country Skiing.* During a three-day holiday weekend trip, David and Sally Jean cross-country skied $3\frac{2}{3}$ mi on Friday, $6\frac{1}{8}$ mi on Saturday, and $4\frac{3}{4}$ mi on Sunday.
   a) Find the total number of miles that they skied.
   b) Find the average number of miles that they skied per day. Express your answer as a mixed numeral.

2. How many people can receive equal $16 shares from a total of $496?

3. A recipe calls for $\frac{4}{5}$ tsp of salt. How much salt should be used for $\frac{1}{2}$ recipe? for 5 recipes?

4. How many pieces, each $2\frac{3}{8}$ ft long, can be cut from a piece of wire 38 ft long?

5. An emergency food pantry fund contains $423. From this fund, $148 and $167 are withdrawn for expenses. How much is left in the fund?

6. In a walkathon, Jermaine walked $\frac{9}{10}$ mi and Oleta walked $\frac{3}{4}$ mi. What was the total distance that they walked?

What part is shaded?

7.

8.

Calculate and simplify.

9. $\begin{array}{r} 3\ 7\ 0\ 4 \\ +\ 5\ 2\ 7\ 8 \end{array}$

10. $\begin{array}{r} 7\ 6\ 0\ 5 \\ -\ 3\ 0\ 8\ 7 \end{array}$

11. $\begin{array}{r} 2\ 7\ 8 \\ \times\ \ 1\ 8 \end{array}$

12. $\begin{array}{r} 8\ 9\ 4 \\ \times\ 3\ 2\ 8 \end{array}$

13. $\frac{3}{8} + \frac{1}{24}$

14. $\begin{array}{r} 2\frac{3}{4} \\ +\ 5\frac{1}{2} \\ \hline \end{array}$

15. $\frac{3}{4} - \frac{1}{3}$

16. $\begin{array}{r} 2\frac{1}{3} \\ -\ 1\frac{1}{6} \\ \hline \end{array}$

17. $\frac{9}{10} \cdot \frac{5}{3}$

18. $18 \cdot \frac{5}{6}$

19. $\frac{2}{5} \div \frac{7}{10}$

20. $2\frac{1}{5} \div \frac{3}{10}$

Divide. Write the answer in the form 34 R 7.

21. $6 \overline{)\ 4\ 2\ 9\ 0}$ **22.** $4\ 5\ \overline{)\ 2\ 5\ 3\ 1}$

23. Write a mixed numeral for the answer in Exercise 22.

24. In the number 2753, what digit names tens?

25. *Room Carpeting.* The Chandlers are carpeting an L-shaped family room consisting of one rectangle that is $8\frac{1}{2}$ ft by 11 ft and another rectangle that is $6\frac{1}{2}$ ft by $7\frac{1}{2}$ ft.

  a) Find the area of the carpet.
  b) Find the perimeter of the carpet.

26. Round 38,478 to the nearest hundred.

27. Find the LCM of 18 and 24.

28. Simplify:
$$\left(\frac{1}{2} + \frac{2}{5}\right)^2 \div 3 + 6 \times \left(2 + \frac{1}{4}\right).$$

Use $<$, $>$, or $=$ for ☐ to write a true sentence.

29. $\frac{4}{5} \ \square \ \frac{4}{6}$ **30.** $\frac{3}{13} \ \square \ \frac{9}{39}$ **31.** $\frac{5}{12} \ \square \ \frac{3}{7}$

Estimate each of the following as $0, \frac{1}{2}$, or 1.

32. $\frac{29}{30}$ **33.** $\frac{15}{29}$ **34.** $\frac{2}{43}$

Simplify.

35. $\frac{36}{45}$ **36.** $\frac{0}{27}$ **37.** $\frac{320}{10}$

38. Convert to fraction notation: $4\frac{5}{8}$.

39. Convert to a mixed numeral: $\frac{17}{3}$.

Solve.

40. $x + 24 = 117$ **41.** $x + \frac{7}{9} = \frac{4}{3}$

42. $\frac{7}{9} \cdot t = \frac{4}{3}$ **43.** $y = 32,580 \div 36$

44. *Matching.* Match each item in the first column with the appropriate item in the second column by drawing connecting lines. There can be more than one correct correspondence for an item.

Factors of 68	12, 54, 72, 300
Factorization of 68	2, 3, 17, 19, 23, 31, 47, 101
Prime factorization of 68	$2 \cdot 2 \cdot 17$
Numbers divisible by 6	$2 \cdot 34$
Numbers divisible by 8	8, 16, 24, 32, 40, 48, 64, 864
Numbers divisible by 5	1, 2, 4, 17, 34, 68
Prime numbers	70, 95, 215

Synthesis

45. Find the smallest prime number that is larger than 2000.

Decimal Notation

Real-World Application

Camping in national parks has declined steadily in recent years. The numbers of overnight camping stays in Park Service campgrounds in the National Park System from 2002 to 2006 were 5.8 million, 5.7 million, 5.4 million, 5.2 million, and 5.0 million, respectively. Find the average number of stays per year during this period.

Source: U.S. National Park Service

This problem appears as Exercise 77 in Section 4.4.

4.1

Decimal Notation, Order, and Rounding

OBJECTIVES

a Given decimal notation, write a word name.

b Convert between decimal notation and fraction notation.

c Given a pair of numbers in decimal notation, tell which is larger.

d Round decimal notation to the nearest thousandth, hundredth, tenth, one, ten, hundred, or thousand.

SKILL TO REVIEW
Objective 1.6a: Round to the nearest ten, hundred, or thousand.

Round 4735 to the nearest:

1. Ten.   **2.** Hundred.

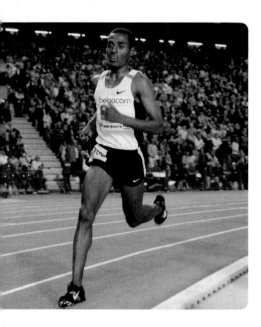

The set of **arithmetic numbers**, or **nonnegative rational numbers**, consists of the whole numbers 0, 1, 2, 3, 4, 5, 6, 7, 8, 9, 10, and so on, and fractions like $\frac{1}{2}, \frac{2}{3}, \frac{7}{8}, \frac{17}{10}$, and so on. Note that we can write the whole numbers using fraction notation. For example, 3 can be written as $\frac{3}{1}$. We studied the use of fraction notation for arithmetic numbers in Chapters 2 and 3.

In Chapter 4, we will study the use of *decimal notation*. The word *decimal* comes from the Latin word *decima,* meaning a tenth part. Although we are using different notation, we are still considering the nonnegative rational numbers. Using decimal notation, we can write 0.875 for $\frac{7}{8}$, for example, or 48.97 for $48\frac{97}{100}$.

a   Decimal Notation and Word Names

One model of the Magellan GPS navigation system sells for $249.98. The dot in $249.98 is called a **decimal point**. Since $0.98, or 98¢, is $\frac{98}{100}$ of a dollar, it follows that

$$\$249.98 = 249 + \frac{98}{100} \text{ dollars.}$$

Also, since $0.98, or 98¢, has the same value as

9 dimes + 8 cents

and 1 dime is $\frac{1}{10}$ of a dollar and 1 cent is $\frac{1}{100}$ of a dollar, we can write

$$249.98 = 2 \cdot 100 + 4 \cdot 10 + 9 \cdot 1 + 9 \cdot \frac{1}{10} + 8 \cdot \frac{1}{100}.$$

This is an extension of the expanded notation for whole numbers that we used in Chapter 1. The place values are 100, 10, 1, $\frac{1}{10}$, $\frac{1}{100}$, and so on. We can see this on a **place-value chart**. The value of each place is $\frac{1}{10}$ as large as that of the one to its left.

Let's consider decimal notation using a place-value chart to represent 26.2922 min, the men's 10,000-meter run record held by Kenenisa Bekele from Ethiopia.

PLACE-VALUE CHART							
Hundreds	Tens	Ones	Tenths	Hundredths	Thousandths	Ten-Thousandths	Hundred-Thousandths
100	10	1	$\frac{1}{10}$	$\frac{1}{100}$	$\frac{1}{1000}$	$\frac{1}{10,000}$	$\frac{1}{100,000}$
	2	6 .	2	9	2	2	

The decimal notation 26.2922 means

$$20 + 6 + \frac{2}{10} + \frac{9}{100} + \frac{2}{1000} + \frac{2}{10,000}, \quad \text{or} \quad 26\frac{2922}{10,000}.$$

We read both 26.2922 and $26\frac{2922}{10,000}$ as

"Twenty-six *and* two thousand, nine hundred twenty-two ten-thousandths."

We can also read 26.2922 as

"Two six *point* two nine two two," or "twenty-six *point* two nine two two."

To write a word name from decimal notation,

a) write a word name for the whole number (the number named to the left of the decimal point),

397.685 → Three hundred ninety-seven

b) write the word "and" for the decimal point, and

397.685 Three hundred ninety-seven and

c) write a word name for the number named to the right of the decimal point, followed by the place value of the last digit.

397.685 Three hundred ninety-seven and six hundred eighty-five *thousandths*

EXAMPLE 1 *Birth Rate.* There were 14.3 births per 1000 Americans in a recent year. Write a word name for 14.3.
Source: U.S. Centers for Disease Control and Prevention

Fourteen and three tenths

EXAMPLE 2 Write a word name for 410.87.

Four hundred ten and eighty-seven hundredths

1. Life Expectancy. The life expectancy at birth in the United States is projected to be 79.2 yr in 2015. Write a word name for 79.2.
Source: U.S. Census Bureau

2. Olympic Swimming. American swimmer Michael Phelps won eight gold medals in the 2008 Summer Olympics, including one for the 200-m freestyle, which he won in the world record time of 1.716 min. Write a word name for 1.716.
Source: beijing2008.cn

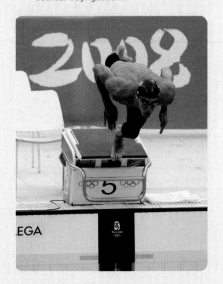

Write a word name for each number.

3. 245.89

4. 34.0064

5. 31,079.756

EXAMPLE 3 *Indianapolis 500.* In the 2009 Indianapolis 500-mile race, winner Helio Castroneves had a 1.9819-second margin of victory over second-place finisher Dan Wheldon. Write a word name for 1.9819.
Source: Indianapolis Motor Speedway, historian Donald Davidson

One and nine thousand, eight hundred nineteen ten-thousandths

EXAMPLE 4 Write a word name for 1788.045.

One thousand, seven hundred eighty-eight and forty-five thousandths

Do Exercises 1–5.

b Converting Between Decimal Notation and Fraction Notation

Given decimal notation, we can convert to fraction notation as follows:

$$9.875 = 9 + \frac{8}{10} + \frac{7}{100} + \frac{5}{1000}$$

$$= 9 \cdot \frac{1000}{1000} + \frac{8}{10} \cdot \frac{100}{100} + \frac{7}{100} \cdot \frac{10}{10} + \frac{5}{1000}$$

$$= \frac{9000}{1000} + \frac{800}{1000} + \frac{70}{1000} + \frac{5}{1000} = \frac{9875}{1000}.$$

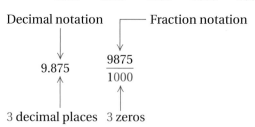

Decimal notation ⟶ Fraction notation

9.875 $\dfrac{9875}{1000}$

3 decimal places   3 zeros

To convert from decimal to fraction notation,

a) count the number of decimal places,

4.98 — 2 places

b) move the decimal point that many places to the right, and

4.98. Move 2 places.

c) write the answer over a denominator with a 1 followed by that number of zeros.

$\dfrac{498}{100}$ 2 zeros

EXAMPLE 5 Write fraction notation for 0.876. Do not simplify.

0.876   0.876. $0.876 = \dfrac{876}{1000}$

3 places   3 zeros

For a number like 0.876, we generally write a 0 before the decimal point to draw attention to the presence of the decimal point.

EXAMPLE 6 Write fraction notation for 56.23. Do not simplify.

56.23   56.23. $56.23 = \dfrac{5623}{100}$

2 places   2 zeros

EXAMPLE 7 Write fraction notation for 1.5018. Do not simplify.

1.5018   1.5018. $1.5018 = \dfrac{15{,}018}{10{,}000}$

4 places   4 zeros

Do Exercises 6–9.

> Write fraction notation.
>
> **6.** 0.896   **7.** 23.78
>
> **8.** 5.6789   **9.** 1.9

If fraction notation has a denominator that is a power of ten, such as 10, 100, 1000, and so on, we convert to decimal notation by reversing the procedure that we used before.

To convert from fraction notation to decimal notation when the denominator is 10, 100, 1000, and so on,

a) count the number of zeros, and

$\dfrac{8679}{1000}$ — 3 zeros

b) move the decimal point that number of places to the left. Leave off the denominator.

8.679.   Move 3 places.

$\dfrac{8679}{1000} = 8.679$

Answers

6. $\dfrac{896}{1000}$ **7.** $\dfrac{2378}{100}$ **8.** $\dfrac{56{,}789}{10{,}000}$ **9.** $\dfrac{19}{10}$

EXAMPLE 8 Write decimal notation for $\frac{47}{10}$.

$$\frac{47}{10} \qquad 4.7. \qquad \frac{47}{10} = 4.7$$

↑ 1 zero   1 place

EXAMPLE 9 Write decimal notation for $\frac{123,067}{10,000}$.

$$\frac{123,067}{10,000} \qquad 12.3067. \qquad \frac{123,067}{10,000} = 12.3067$$

↑ 4 zeros   4 places

EXAMPLE 10 Write decimal notation for $\frac{13}{1000}$.

$$\frac{13}{1000} \qquad 0.013. \qquad \frac{13}{1000} = 0.013$$

↑ 3 zeros   3 places

EXAMPLE 11 Write decimal notation for $\frac{570}{100,000}$.

$$\frac{570}{100,000} \qquad 0.00570. \qquad \frac{570}{100,000} = 0.0057$$

↑ 5 zeros   5 places

Note that when we write decimal notation, it is not necessary to include the 0 that follows the 7.

Do Exercises 10–15.

When denominators are numbers other than 10, 100, and so on, we will use another method for conversion. It will be considered in Section 4.5.

If a mixed numeral has a fractional part with a denominator that is a power of ten, such as 10, 100, or 1000, and so on, we first write the mixed numeral as a sum of a whole number and a fraction. Then we convert to decimal notation.

EXAMPLE 12 Write decimal notation for $23\frac{59}{100}$.

$$23\frac{59}{100} = 23 + \frac{59}{100} = 23 \text{ and } \frac{59}{100} = 23.59$$

EXAMPLE 13 Write decimal notation for $772\frac{129}{10,000}$.

$$772\frac{129}{10,000} = 772 + \frac{129}{10,000} = 772 \text{ and } \frac{129}{10,000} = 772.0129$$

Do Exercises 16–18.

Write decimal notation.

10. $\frac{743}{100}$ **11.** $\frac{406}{1000}$

12. $\frac{67,089}{10,000}$ **13.** $\frac{9}{10}$

14. $\frac{57}{1000}$ **15.** $\frac{830}{10,000}$

Write decimal notation.

16. $4\frac{3}{10}$

17. $283\frac{71}{100}$

18. $456\frac{13}{1000}$

Answers

10. 7.43 **11.** 0.406 **12.** 6.7089
13. 0.9 **14.** 0.057 **15.** 0.083 **16.** 4.3
17. 283.71 **18.** 456.013

(c) Order

To understand how to compare numbers in decimal notation, consider 0.85 and 0.9. First note that $0.9 = 0.90$ because $\frac{9}{10} = \frac{90}{100}$. Then $0.85 = \frac{85}{100}$ and $0.90 = \frac{90}{100}$. Since $\frac{85}{100} < \frac{90}{100}$, it follows that $0.85 < 0.90$. This leads us to a quick way to compare two numbers in decimal notation.

> **COMPARING NUMBERS IN DECIMAL NOTATION**
>
> To compare two numbers in decimal notation, start at the left and compare corresponding digits moving from left to right. If two digits differ, the number with the larger digit is the larger of the two numbers. To ease the comparison, extra zeros can be written to the right of the last decimal place.

EXAMPLE 14   Which of 2.109 and 2.1 is larger?

Think.

2.109   2.109
2.1   →   2.100

Same   └─Different; $9 > 0$

Thus, 2.109 is larger than 2.1. That is, $2.109 > 2.1$.

EXAMPLE 15   Which of 0.09 and 0.108 is larger?

Think.

0.09   0.090
0.108   →   0.108

Same   └─ Different; $1 > 0$

Thus, 0.108 is larger than 0.09. That is, $0.108 > 0.09$.

Do Exercises 19–24.

Which number is larger?

19. 2.04, 2.039

20. 0.06, 0.008

21. 0.5, 0.58

22. 1, 0.9999

23. 0.8989, 0.09898

24. 21.006, 21.05

(d) Rounding

Rounding is done as for whole numbers. To understand, we first consider an example using the number line. It might help to review Section 1.6.

EXAMPLE 16   Round 0.37 to the nearest tenth.

Here is part of a number line.

We see that 0.37 is closer to 0.40 than to 0.30. Thus, 0.37 rounded to the nearest tenth is 0.4.

ROUNDING DECIMAL NOTATION

To round to a certain place:

a) Locate the digit in that place.

b) Consider the next digit to the right.

c) If the digit to the right is 5 or higher, round up. If the digit to the right is 4 or lower, round down.

EXAMPLE 17 Round 3872.2459 to the nearest tenth.

a) Locate the digit in the tenths place, 2.

3 8 7 2.2 4 5 9
        ↑

b) Consider the next digit to the right, 4.

3 8 7 2.2 4 5 9
          ↑

c) Since that digit, 4, is 4 or lower, round down.

3 8 7 2.2 ← This is the answer.

-------- Caution! --------

3872.3 is not a correct answer to Example 17. It is *incorrect* to round from the ten-thousandths digit over to the tenths digit, as follows:

3872.246 → 3872.25 → 3872.3.

EXAMPLE 18 Round 3872.2459 to the nearest thousandth, hundredth, tenth, one, ten, hundred, and thousand.

Thousandth:	3872.246	Ten:	3870
Hundredth:	3872.25	Hundred:	3900
Tenth:	3872.2	Thousand:	4000
One:	3872		

EXAMPLE 19 Round 14.8973 to the nearest hundredth.

a) Locate the digit in the hundredths place, 9.　　　1 4.8 9 7 3
　　　　　　　　　　　　　　　　　　　　　　　　　　　　↑

b) Consider the next digit to the right, 7.　　　　　1 4.8 9 7 3
　　　　　　　　　　　　　　　　　　　　　　　　　　　　　↑

c) Since that digit, 7, is 5 or higher, round up. When we make the hundredths digit a 10, we carry 1 to the tenths place.

The answer is 14.90. Note that the 0 in 14.90 indicates that the answer is correct to the nearest hundredth.

EXAMPLE 20 Round 0.008 to the nearest tenth.

a) Locate the digit in the tenths place, 0.　　　　　0.0 0 8
　　　　　　　　　　　　　　　　　　　　　　　　　　　↑

b) Consider the next digit to the right, 0.　　　　　0.0 0 8
　　　　　　　　　　　　　　　　　　　　　　　　　　　　↑

c) Since that digit, 0, is less than 5, round down.

The answer is 0.0.

Do Exercises 25–43.

Round to the nearest tenth.

25. 2.76

26. 13.85

27. 234.448

28. 7.009

Round to the nearest hundredth.

29. 0.636

30. 7.834

31. 34.675

32. 0.025

Round to the nearest thousandth.

33. 0.9434

34. 8.0038

35. 43.1119

36. 37.4005

Round 7459.3548 to the nearest:

37. Thousandth.

38. Hundredth.

39. Tenth.　　　**40.** One.

41. Ten. (*Caution:* "Tens" are not "tenths.")

42. Hundred.　　**43.** Thousand.

Answers

25. 2.8　**26.** 13.9　**27.** 234.4　**28.** 7.0
29. 0.64　**30.** 7.83　**31.** 34.68　**32.** 0.03
33. 0.943　**34.** 8.004　**35.** 43.112
36. 37.401　**37.** 7459.355　**38.** 7459.35
39. 7459.4　**40.** 7459　**41.** 7460
42. 7500　**43.** 7000

a Write a word name for the number in each sentence.

1. *Stock Price.* Google's stock was priced at $486.34 per share.
Source: New York Stock Exchange

2. *Soft-Drink Consumption.* The average American consumed 51.5 gal of carbonated soft drinks in a recent year.
Source: U.S. Department of Agriculture

3. *Currency Conversion.* One Chinese yuan was worth about $0.146 in U.S. currency recently.
Source: Interactive Data

4. *Currency Conversion.* One U.S. dollar was worth 0.6714 euros recently.
Source: Interactive Data

5. *Game System.* The Nintendo Wii game system sold for $249.89 recently.
Source: Best Buy

6. *Cell-Phone Plan.* The Sprint Talk/Message cell-phone plan with unlimited anytime minutes was priced at $89.99 per month recently.
Source: Sprint

7. One gallon of paint is equal to 3.785 L of paint.

8. *Water Weight.* One gallon of water weighs 8.35 lb.

Write a word name.

9. 27.1245

10. 34.891

b Write fraction notation. Do not simplify.

11. 8.3

12. 203.6

13. 3.56

14. 0.17

15. 20.003

16. 1.509

17. 1.0008

18. 2.0114

19. 37.2

20. 4567.2

21. 0.00013

22. 6.14057

Write decimal notation.

23. $\dfrac{8}{10}$

24. $\dfrac{51}{10}$

25. $\dfrac{3798}{1000}$

26. $\dfrac{780}{1000}$

27. $\dfrac{889}{100}$

28. $\dfrac{92}{100}$

29. $\dfrac{19}{100,000}$

30. $\dfrac{56,788}{100,000}$

31. $\dfrac{78}{10,000}$

32. $\dfrac{13,463}{10,000}$

33. $\dfrac{376,193}{1,000,000}$

34. $\dfrac{8,953,074}{1,000,000}$

35. $2\dfrac{9}{10}$

36. $9243\dfrac{1}{10}$

37. $3\dfrac{98}{1000}$

38. $4\dfrac{67}{1000}$

39. $99\dfrac{44}{100}$

40. $67\dfrac{83}{100}$

41. $2\dfrac{1739}{10,000}$

42. $2256\dfrac{3059}{10,000}$

c Which number is larger?

43. 0.06, 0.58

44. 0.008, 0.8

45. 0.905, 0.91

46. 42.06, 42.1

47. 0.0009, 0.001

48. 7.067, 7.054

49. 234.07, 235.07

50. 0.99999, 1

51. 0.004, $\dfrac{4}{100}$

52. $\dfrac{73}{10}$, 0.73

53. 0.432, 0.4325

54. 0.8437, 0.84384

d Round to the nearest tenth.

55. 0.11

56. 0.85

57. 0.49

58. 0.5394

59. 2.7449

60. 4.78

61. 123.65

62. 36.049

Round to the nearest hundredth.

63. 0.893

64. 0.675

65. 0.6666

66. 6.524

67. 0.995

68. 207.9976

69. 0.094

70. 11.4246

Round to the nearest thousandth.

71. 0.3246

72. 0.6666

73. 17.0014

74. 123.4562

75. 10.1011

76. 0.1161

77. 9.9989

78. 67.100602

Round 809.5732 to the nearest:

79. Hundred.

80. Tenth.

81. Thousandth.

82. Hundredth.

83. One.

84. Ten.

Round 34.54389 to the nearest:

85. Ten-thousandth.

86. Thousandth.

87. Hundredth.

88. Tenth.

89. One.

90. Ten.

Skill Maintenance

Round 6172 to the nearest: [1.6a]

91. Ten.

92. Hundred.

93. Thousand.

Find the prime factorization.   [2.1d]

94. 2000

95. 1530

96. 2002

97. 4312

Synthesis

98. Arrange the following numbers in order from smallest to largest.

0.99, 0.099, 1, 0.9999, 0.89999, 1.00009, 0.909, 0.9889

99. Arrange the following numbers in order from smallest to largest.

2.1, 2.109, 2.108, 2.018, 2.0119, 2.0302, 2.000001

Truncating.   There are other methods of rounding decimal notation. To round using **truncating**, we drop off all decimal places past the rounding place, which is the same as changing all digits to the right to zeros. For example, rounding 6.78093456285102 to the ninth decimal place, using truncating, gives us 6.780934562. Use truncating to round each of the following to the fifth decimal place—that is, the hundred-thousandth place.

100. 6.78346623

101. 6.783465902

102. 99.999999999

103. 0.030303030303

4.2

Addition and Subtraction

OBJECTIVES

a Add using decimal notation.

b Subtract using decimal notation.

c Solve equations of the type $x + a = b$ and $a + x = b$, where a and b may be in decimal notation.

d Balance a checkbook.

SKILLS TO REVIEW
Objective 1.2a: Add whole numbers.
Objective 1.3a: Subtract whole numbers.

Add or subtract.

1.
```
   3 8 7
 + 2 5 4
```

2.
```
   4 0 2 3
 - 1 6 6 7
```

Add.

1.
```
     0.8 4 7
 + 1 0.0 7
```

2.
```
     2.1
     0.7 3
 + 3 1.3 6 8
```

Add.

3. $0.02 + 4.3 + 0.649$

4. $0.12 + 3.006 + 0.4357$

5. $0.4591 + 0.2374 + 8.70894$

a Addition

Adding with decimal notation is similar to adding whole numbers. First, we line up the decimal points so that we can add corresponding place-value digits. Then we add digits from the right. For example, we add the thousandths, then the hundredths, and so on, carrying if necessary. If desired, we can write extra zeros to the right of the decimal point so that the number of places is the same in all of the addends.

EXAMPLE 1 Add: $56.314 + 17.78$.

```
   5 6 . 3 1 4      Lining up the decimal points in order to add
 + 1 7 . 7 8 0      Writing an extra zero to the right
                    of the decimal point
```

```
   5 6 . 3 1 4
 + 1 7 . 7 8 0
               4    Adding thousandths
```

```
   5 6 . 3 1 4
 + 1 7 . 7 8 0
             9 4    Adding hundredths
```

```
       1
   5 6 . 3 1 4      Adding tenths
 + 1 7 . 7 8 0      We get 10 tenths = 1 one + 0 tenths,
       . 0 9 4      so we carry the 1 to the ones column.
                    Writing a decimal point in the answer.
```

```
   1 1
   5 6 . 3 1 4      Adding ones
 + 1 7 . 7 8 0      We get 14 ones = 1 ten + 4 ones,
     4 . 0 9 4      so we carry the 1 to the tens column.
```

```
   1 1
   5 6 . 3 1 4
 + 1 7 . 7 8 0
   7 4 . 0 9 4      Adding tens
```

Do Margin Exercises 1 and 2.

EXAMPLE 2 Add: $3.42 + 0.237 + 14.1$.

```
     3.4 2 0     Lining up the decimal points
     0.2 3 7     and writing extra zeros
 + 1 4.1 0 0
   1 7.7 5 7     Adding
```

Do Exercises 3–5.

Answers

Skills to Review:
1. 641   2. 2356

Margin Exercises:
1. 10.917   2. 34.198   3. 4.969
4. 3.5617   5. 9.40544

Now we consider the addition 3456 + 19.347. Keep in mind that any whole number has an "unwritten" decimal point at the right that can be followed by zeros. For example, 3456 can also be written 3456.000. When adding, we can always write in the decimal point and extra zeros if desired.

EXAMPLE 3 Add: 3456 + 19.347.

$$
\begin{array}{r}
3\ 4\ 5\overset{1}{6}.0\ 0\ 0 \\
+\qquad 1\ 9.3\ 4\ 7 \\
\hline
3\ 4\ 7\ 5.3\ 4\ 7
\end{array}
$$

   Writing in the decimal point and extra zeros
   Lining up the decimal points
   Adding

Do Exercises 6 and 7.

Add.

6. 789 + 123.67

7. 45.78 + 2467 + 1.993

(b) Subtraction

Subtracting with decimal notation is similar to subtracting whole numbers. First, we line up the decimal points so that we can subtract corresponding place-value digits. Then we subtract digits from the right. For example, we subtract the thousandths, then the hundredths, the tenths, and so on, borrowing if necessary.

EXAMPLE 4 Subtract: 56.314 − 17.78.

$$
\begin{array}{r}
5\ 6.3\ 1\ 4 \\
-\ 1\ 7.7\ 8\ 0 \\
\end{array}
$$
   Lining up the decimal points in order to subtract
   Writing an extra 0

$$
\begin{array}{r}
5\ 6.3\ 1\ 4 \\
-\ 1\ 7.7\ 8\ 0 \\
\hline
4
\end{array}
$$
   Subtracting thousandths

$$
\begin{array}{r}
5\ 6.3\ \overset{2\ 11}{\cancel{1}}\ 4 \\
-\ 1\ 7.7\ 8\ 0 \\
\hline
3\ 4
\end{array}
$$
   Borrowing tenths to subtract hundredths
   Subtracting hundredths

$$
\begin{array}{r}
5\ \overset{12}{6}.\overset{2}{\cancel{3}}\ \overset{11}{\cancel{1}}\ 4 \\
-\ 1\ 7.7\ 8\ 0 \\
\hline
.5\ 3\ 4
\end{array}
$$
   Borrowing ones to subtract tenths
   Subtracting tenths; writing a decimal point

$$
\begin{array}{r}
\overset{15}{\cancel{5}}\ \overset{12}{\cancel{6}}.\overset{2}{\cancel{3}}\ \overset{11}{\cancel{1}}\ 4 \\
-\ 1\ 7.7\ 8\ 0 \\
\hline
8.5\ 3\ 4
\end{array}
$$
   Borrowing tens to subtract ones
   Subtracting ones

$$
\begin{array}{r}
\overset{15}{\cancel{5}}\ \overset{12}{\cancel{6}}.\overset{2}{\cancel{3}}\ \overset{11}{\cancel{1}}\ 4 \\
-\ 1\ 7.7\ 8\ 0 \\
\hline
3\ 8.5\ 3\ 4
\end{array}
$$
   Subtracting tens

Check by adding:
$$
\begin{array}{r}
3\ \overset{1}{8}.\overset{1}{5}\ \overset{1}{3}\ 4 \\
+\ 1\ 7.7\ 8\ 0 \\
\hline
5\ 6.3\ 1\ 4
\end{array}
$$
The answer checks because this is the top number in the subtraction.

Subtract.

8. 37.428 − 26.674

9.
$$
\begin{array}{r}
0.3\ 4\ 7 \\
-\ 0.0\ 0\ 8
\end{array}
$$

Do Exercises 8 and 9.

Answers
6. 912.67   7. 2514.773   8. 10.754
9. 0.339

4.2 Addition and Subtraction   **227**

EXAMPLE 5 Subtract: 13.07 − 9.205.

$$
\begin{array}{r}
\overset{12}{\cancel{2}} \;{}^{10}\,{}^{6}\,{}^{10} \\
\cancel{1}\,3.0\,7\,\cancel{0} \\
-\quad 9.2\,0\,5 \\
\hline
3.8\,6\,5
\end{array}
$$

Writing an extra zero

Subtracting

EXAMPLE 6 Subtract: 23.08 − 5.0053.

$$
\begin{array}{r}
{}^{1}\;{}^{13}\quad\;{}^{7}\,{}^{9}\,{}^{10} \\
2\,\cancel{3}.0\,\cancel{8}\,\cancel{0}\,\cancel{0} \\
-\quad 5.0\,0\,5\,3 \\
\hline
1\,8.0\,7\,4\,7
\end{array}
$$

Writing two extra zeros

Subtracting

Check by adding:

$$
\begin{array}{r}
{}^{1}\qquad\;{}^{1}\,{}^{1} \\
1\,8.0\,7\,4\,7 \\
+\quad 5.0\,0\,5\,3 \\
\hline
2\,3.0\,8\,0\,0
\end{array}
$$

Do Exercises 10–12.

When subtraction involves a whole number, again keep in mind that there is an "unwritten" decimal point that can be written in if desired. Extra zeros can also be written in to the right of the decimal point.

EXAMPLE 7 Subtract: 456 − 2.467.

$$
\begin{array}{r}
\quad\;{}^{5}\,{}^{9}\,{}^{9}\,{}^{10} \\
4\,5\,\cancel{6}.\cancel{0}\,\cancel{0}\,\cancel{0} \\
-\quad 2.4\,6\,7 \\
\hline
4\,5\,3.5\,3\,3
\end{array}
$$

Writing in the decimal point and extra zeros

Subtracting

Do Exercises 13 and 14.

Subtract.

10. 1.2345 − 0.7

11. 0.9564 − 0.4392

12. 7.37 − 0.00008

Subtract.

13. 1277 − 82.78

14. 5 − 0.0089

Calculator Corner

Addition and Subtraction with Decimal Notation To use a calculator to add and subtract with decimal notation, we use the $\boxed{\cdot}$, $\boxed{+}$, $\boxed{-}$, and $\boxed{=}$ keys. To find 47.046 − 28.193, for example, we press $\boxed{4}$ $\boxed{7}$ $\boxed{\cdot}$ $\boxed{0}$ $\boxed{4}$ $\boxed{6}$ $\boxed{-}$ $\boxed{2}$ $\boxed{8}$ $\boxed{\cdot}$ $\boxed{1}$ $\boxed{9}$ $\boxed{3}$ $\boxed{=}$. The display reads $\boxed{18.853}$, so 47.046 − 28.193 = 18.853.

Exercises:

Use a calculator to add.

1. 274.159
    + 43.486

2.   19.805
    + 486.748

3. 1.7 + 14.56 + 0.89

4. 3.4 + 45 + 0.68

Use a calculator to subtract.

5.   9.2
    − 4.8

6. 52.34
    − 18.51

7. 489 − 34.26

8. 6.09 − 5.1

Answers

10. 0.5345   **11.** 0.5172   **12.** 7.36992
13. 1194.22   **14.** 4.9911

(c) Solving Equations

Now let's solve equations $x + a = b$ and $a + x = b$, where a and b may be in decimal notation. Proceeding as we have before, we subtract a on both sides.

EXAMPLE 8 Solve: $x + 28.89 = 74.567$.

We have

$$x + 28.89 - 28.89 = 74.567 - 28.89 \qquad \text{Subtracting 28.89 on both sides}$$
$$x = 45.677.$$

$$
\begin{array}{r}
\overset{\overset{13\ 14}{6\ \cancel{3}\ \cancel{4}\ \overset{16}{\ }}}{7\ 4.\cancel{5}\ \cancel{6}\ 7} \\
-\ 2\ 8.8\ 9\ 0 \\
\hline
4\ 5.6\ 7\ 7
\end{array}
$$

The solution is 45.677.

EXAMPLE 9 Solve: $0.8879 + y = 9.0026$.

We have

$$0.8879 + y - 0.8879 = 9.0026 - 0.8879 \qquad \text{Subtracting 0.8879 on both sides}$$
$$y = 8.1147.$$

$$
\begin{array}{r}
\overset{\overset{11}{8\ 9\ 9\ \cancel{1}\ \overset{16}{\ }}}{9.0\ 0\ 2\ \cancel{6}} \\
-\ 0.8\ 8\ 7\ 9 \\
\hline
8.1\ 1\ 4\ 7
\end{array}
$$

The solution is 8.1147.

Do Exercises 15 and 16.

EXAMPLE 10 Solve: $120 + x = 4380.6$.

We have

$$120 + x - 120 = 4380.6 - 120 \qquad \text{Subtracting 120 on both sides}$$
$$x = 4260.6.$$

$$
\begin{array}{r}
4\ 3\ 8\ 0.6 \\
-\ 1\ 2\ 0.0 \\
\hline
4\ 2\ 6\ 0.6
\end{array}
$$

The solution is 4260.6.

Do Exercise 17.

Solve.

15. $x + 17.78 = 56.314$

16. $8.906 + t = 23.07$

17. Solve: $241 + y = 2374.5$.

(d) Balancing a Checkbook

Let's use addition and subtraction with decimals to balance a checkbook.

EXAMPLE 11 Find the errors, if any, in the balances in this checkbook.

DATE	CHECK NUMBER	TRANSACTION DESCRIPTION	√ T	(−) PAYMENT/ DEBIT	(+ OR −) OTHER	(+) DEPOSIT/ CREDIT	BALANCE FORWARD
							8767 73
8/16	432	Burch Laundry		23 56			8744 16
8/19	433	Rogers Bakery		20 49			8764 65
8/20		Deposit				85 00	8848 65
8/21	434	Galaxy Video		48 60			8801 05
8/22	435	Benson's Garage		267 95			8533 09

20___   RECORD ALL CHARGES OR CREDITS THAT AFFECT YOUR ACCOUNT

Answers
15. 38.534   **16.** 14.164   **17.** 2133.5

There are two ways to determine whether there are errors. We assume that the amount $8767.73 in the "Balance forward" column is correct. If we can determine that the ending balance is correct, we have some assurance that the checkbook is correct. But two errors could offset each other to give us that balance.

METHOD 1

a) We add the debits: $23.56 + 20.49 + 48.60 + 267.95 = 360.60$.

b) We add the deposits/credits. In this case, there is only one deposit, 85.00.

c) We add the total of the deposits to the balance brought forward:

$$8767.73 + 85.00 = 8852.73.$$

d) We subtract the total of the debits: $8852.73 - 360.60 = 8492.13$.

The result should be the ending balance, 8533.09. We see that $8492.13 \neq 8533.09$. Since the numbers are not equal, we proceed to method 2.

METHOD 2   We successively add or subtract deposit/credits and debits, and check the result in the "Balance forward" column.

$$8767.73 - 23.56 = 8744.17.$$

We have found our first error. The subtraction was incorrect. We correct it and continue, using 8744.17 as the corrected balance forward:

$$8744.17 - 20.49 = 8723.68.$$

It looks as though 20.49 was added instead of subtracted. Actually, we would have to correct this line even if it had been subtracted, because the error of 1¢ in the first step has been carried through successive calculations. We correct that balance line and continue, using 8723.68 as the balance and adding the deposit 85.00:

$$8723.68 + 85.00 = 8808.68.$$

We make the correction and continue subtracting the last two debits:

$$8808.68 - 48.60 = 8760.08.$$

Then

$$8760.08 - 267.95 = 8492.13.$$

The corrected checkbook is below.

---

18. Find the errors, if any, in this checkbook.

20 ___

DATE	CHECK NUMBER	TRANSACTION DESCRIPTION	√T	PAYMENT/DEBIT	(+ OR −) OTHER	(+) DEPOSIT/CREDIT	BALANCE FORWARD
							3078 92
12/1	888	H.H. Gregg Appliances		340 69			2738 23
12/3	889	Marie Callendar's Pies		78 56			2659 66
12/5	890	Deposit <Paycheck>				230 80	2890 46
12/6	891	Chili's Restaurant		13 17			2877 32
12/8	891	Stonycreek Golf Course		48 00			2829 32
12/8		Deposit <Molly>				39 58	2848 90
12/10	892	Frontier Trading Post		102 87			2766 83
12/14	893	Goody's Music		48 59			2697 45
12/15	894	Salvation Army		100 00			2497 45

---

20 ___   RECORD ALL CHARGES OR CREDITS THAT AFFECT YOUR ACCOUNT

DATE	CHECK NUMBER	TRANSACTION DESCRIPTION	√T	(−) PAYMENT/DEBIT	(+ OR −) OTHER	(+) DEPOSIT/CREDIT	BALANCE FORWARD	
							8767 73	
8/16	432	Burch Laundry		23 56			8744 16	→ 8744 17
8/19	433	Rogers Bakery		20 49			8764 65	→ 8723 68
8/20		Deposit				85 00	8848 65	→ 8808 68
8/21	434	Galaxy Video		48 60			8801 05	→ 8760 08
8/22	435	Benson's Garage		267 95			8533 09	→ 8492 13

Do Exercise 18.

There are other ways in which errors can be made in checkbooks, such as forgetting to record a transaction or writing the amounts incorrectly, but we will not consider those here.

Answer

18. The "Balance forward" column should read:

$3078.92
2738.23
2659.67
2890.47
2877.33
2829.33
2868.91
2766.04
2697.45
2597.45

4.2

Exercise Set

For Extra Help

MyMathLab

Math XL
PRACTICE

WATCH

DOWNLOAD

READ

REVIEW

a Add.

1. 3 1 6.2 5
    + 1 8.1 2

2. 6 4 1.8 0 3
    + 1 4.9 3 5

3. 6 5 9.4 0 3
    + 9 1 6.8 1 2

4. 4 2 0 3.2 8
    +   3.3 9

5.   9.1 0 4
    + 1 2 3.4 5 6

6. 8 1.0 0 8
    +   3.4 0 9

7. 20.0124 + 30.0124

8. 0.687 + 0.9

9. 39 + 1.007

10. 2.3 + 0.729 + 23

11.   4 7.8
     2 1 9.8 5 2
        4 3.5 9
   + 6 6 6.7 1 3

12.   1 3.7 2
          9.1 1 2
    6 5 4 2.7 9 0 8
   +   2 3.9 0 1

13. 0.34 + 3.5 + 0.127 + 768

14. 17 + 3.24 + 0.256 + 0.3689

15. 99.6001 + 7285.18 + 500.042 + 870

16. 65.987 + 9.4703 + 6744.02 + 1.0003 + 200.895

b Subtract.

17. 5 1.3 1
    −   2.2 9

18. 4 4.3 4 5
    −   3.1 0 5

19. 9 2.3 4 1
    −   6.4 2

20. 9 7.0 1
    −   3.1 5

21. 2.5
    − 0.0 0 2 5

22. 3 9.0
    −   0.2 8

23. 3.4
    − 0.0 0 3

24. 2.8
    − 2.0 8

25. 28.2 − 19.35

26. 100.16 − 0.118

27. 34.07 − 30.7

28. 36.2 − 16.28

29. $8.45 - 7.405$

30. $3.801 - 2.81$

31. $6.003 - 2.3$

32. $9.087 - 8.807$

33. $1 - 0.0098$

34. $2 - 1.0908$

35. $100 - 0.34$

36. $624 - 18.79$

37. $7.48 - 2.6$

38. $18.4 - 5.92$

39. $3 - 2.006$

40. $263.7 - 102.08$

41. $19 - 1.198$

42. $2548.98 - 2.007$

43. $65 - 13.87$

44. $45 - 0.999$

45.
$$\begin{array}{r} 3\,2.7\,9\,7\,8 \\ -\ \ \ 0.0\,5\,9\,2 \\ \hline \end{array}$$

46.
$$\begin{array}{r} 0.4\,9\,6\,3\,4 \\ -\ 0.1\,2\,6\,7\,8 \\ \hline \end{array}$$

47.
$$\begin{array}{r} 6.0\,7 \\ -\ 2.0\,0\,7\,8 \\ \hline \end{array}$$

48.
$$\begin{array}{r} 1.0 \\ -\ 0.9\,9\,9\,9 \\ \hline \end{array}$$

c Solve.

49. $x + 17.5 = 29.15$

50. $t + 50.7 = 54.07$

51. $17.95 + p = 402.63$

52. $w + 1.3004 = 47.8$

53. $13{,}083.3 = x + 12{,}500.33$

54. $100.23 = 67.8 + z$

55. $x + 2349 = 17{,}684.3$

56. $1830.4 + t = 23{,}067$

(d) Find the errors, if any, in each checkbook.

57.

DATE	CHECK NUMBER	TRANSACTION DESCRIPTION	√T	(-) PAYMENT/DEBIT	(+ OR -) OTHER	(+) DEPOSIT/CREDIT	BALANCE FORWARD
20		RECORD ALL CHARGES OR CREDITS THAT AFFECT YOUR ACCOUNT					9704 56
8/8	342	Bill Rydman		27 44			9677 12
8/9		Deposit				1000 00	10,677 12
8/12	343	Chuck Taylor		123 95			10,553 17
8/14	344	Jennifer Crum		124 02			10,677 19
8/22	345	Neon Johnny's Pizza		12 43			10,664 76
8/24		Deposit				2500 00	13,164 76
8/29	346	Border's Bookstore		137 78			13,302 54
9/2		Deposit				18 88	13,283 66
9/3	347	Fireman's Fund		2800 00			10,483 66

58.

DATE	CHECK NUMBER	TRANSACTION DESCRIPTION	√T	(-) PAYMENT/DEBIT	(+ OR -) OTHER	(+) DEPOSIT/CREDIT	BALANCE FORWARD
20		RECORD ALL CHARGES OR CREDITS THAT AFFECT YOUR ACCOUNT					1876 43
4/1	500	Bart Kaufman		500 12			1376 31
4/3	501	Jim Lawler		28 56			1347 75
4/3		Deposit				10,000 00	11,347 75
4/3	502	Victoria Montoya		464 00			10,883 75
4/3		Deposit				2500 00	8383 75
4/4	503	Art's Garage		1600 00			6783 75
4/8	504	Golf Galaxy		1349 98			5433 77
4/12	505	Don Mitchell Pro Shops		658 97			4774 80
4/13		Deposit				900 00	5674 80
4/15	506	Kroger		185 58			5479 22

Skill Maintenance

59. Round 34,567 to the nearest thousand.   [1.6a]

60. Round 34,496 to the nearest thousand.   [1.6a]

Subtract.

61. $\frac{13}{24} - \frac{3}{8}$ [3.3a]

62. $\frac{8}{9} - \frac{2}{15}$ [3.3a]

63. $8805 - 2639$ [1.3a]

64. $8005 - 2639$ [1.3a]

Solve.

65. A serving of filleted fish is generally considered to be about $\frac{1}{3}$ lb. How many servings can be prepared from $5\frac{1}{2}$ lb of flounder fillet?   [3.6c]

66. A photocopier technician drove $125\frac{7}{10}$ mi away from Scottsdale for a repair call. The next day he drove $65\frac{1}{2}$ mi back toward Scottsdale for another service call. How far was the technician from Scottsdale?   [3.5c]

Synthesis

67. A student presses the wrong button when using a calculator and adds 235.7 instead of subtracting it. The incorrect answer is 817.2. What is the correct answer?

Multiplication

OBJECTIVES

a Multiply using decimal notation.

b Convert from notation like 45.7 million to standard notation, and convert between dollars and cents.

SKILL TO REVIEW
Objective 1.4a: Multiply whole numbers.

Multiply.

1.   4 2
     × 6 3

2.   7 1 6
     ×   5 8

STUDY TIPS

HAVE A POSITIVE ATTITUDE

You can choose to improve your attitude and raise the academic goals that you have set for yourself. Projecting a positive attitude toward your study of mathematics and expecting a positive outcome can make it easier for you to learn and to perform well in this course.

a Multiplication

Let's find the product

$$2.3 \times 1.12.$$

To understand how we find such a product, we first convert each factor to fraction notation. Next, we multiply the numerators and then divide by the product of the denominators.

$$2.3 \times 1.12 = \frac{23}{10} \times \frac{112}{100} = \frac{23 \times 112}{10 \times 100} = \frac{2576}{1000} = 2.576$$

Note the number of decimal places.

$$
\begin{array}{rl}
1.1\ 2 & \text{(2 decimal places)} \\
\times\quad 2.3 & \text{(1 decimal place)} \\
\hline
2.5\ 7\ 6 & \text{(3 decimal places)}
\end{array}
$$

Now consider

$$0.011 \times 15.0002 = \frac{11}{1000} \times \frac{150{,}002}{10{,}000} = \frac{1{,}650{,}022}{10{,}000{,}000} = 0.1650022.$$

Note the number of decimal places.

$$
\begin{array}{rl}
1\ 5.0\ 0\ 0\ 2 & \text{(4 decimal places)} \\
\times\qquad 0.0\ 1\ 1 & \text{(3 decimal places)} \\
\hline
0.1\ 6\ 5\ 0\ 0\ 2\ 2 & \text{(7 decimal places)}
\end{array}
$$

To multiply using decimals:

a) Ignore the decimal points and multiply as though both factors were whole numbers.

b) Then place the decimal point in the result. The number of decimal places in the product is the sum of the numbers of places in the factors. (Count places from the right.)

0.8×0.43

$$
\begin{array}{rl}
\overset{2}{0.4}\ 3 & \\
\times\quad 0.8 & \text{Ignore the decimal} \\
\hline
3\ 4\ 4 & \text{points for now.}
\end{array}
$$

$$
\begin{array}{rl}
0.4\ 3 & \text{(2 decimal places)} \\
\times\quad 0.8 & \text{(1 decimal place)} \\
\hline
0.3\ 4\ 4 & \text{(3 decimal places)}
\end{array}
$$

EXAMPLE 1 Multiply: 8.3×74.6.

a) We ignore the decimal points and multiply as though factors were whole numbers.

$$
\begin{array}{r}
7\ 4.6 \\
\times\quad 8.3 \\
\hline
2\ 2\ 3\ 8 \\
5\ 9\ 6\ 8\ 0 \\
\hline
6\ 1\ 9\ 1\ 8
\end{array}
$$

b) We place the decimal point in the result. The number of decimal places in the product is the sum of the numbers of places in the factors, $1 + 1$, or 2.

$$
\begin{array}{r}
7\ 4.6 \quad \text{(1 decimal place)} \\
\times \qquad 8.3 \quad \text{(1 decimal place)} \\
\hline
2\ 2\ 3\ 8 \\
5\ 9\ 6\ 8\ 0 \\
\hline
6\ 1\ 9.1\ 8 \quad \text{(2 decimal places)}
\end{array}
$$

Do Exercise 1.

EXAMPLE 2 Multiply: 0.0032×2148.

$$
\begin{array}{r}
2\ 1\ 4\ 8 \quad \text{(0 decimal places)} \\
\times\ 0.0\ 0\ 3\ 2 \quad \text{(4 decimal places)} \\
\hline
4\ 2\ 9\ 6 \\
6\ 4\ 4\ 4\ 0 \\
\hline
6.8\ 7\ 3\ 6 \quad \text{(4 decimal places)}
\end{array}
$$

EXAMPLE 3 Multiply: 0.14×0.867.

$$
\begin{array}{r}
0.8\ 6\ 7 \quad \text{(3 decimal places)} \\
\times \qquad 0.1\ 4 \quad \text{(2 decimal places)} \\
\hline
3\ 4\ 6\ 8 \\
8\ 6\ 7\ 0 \\
\hline
0.1\ 2\ 1\ 3\ 8 \quad \text{(5 decimal places)}
\end{array}
$$

Do Exercises 2 and 3.

1. Multiply.
$$
\begin{array}{r}
8\ 5.4 \\
\times \quad 6.2 \\
\hline
\end{array}
$$

Multiply.

2.
$$
\begin{array}{r}
1\ 2\ 3\ 4 \\
\times\ 0.0\ 0\ 4\ 1 \\
\hline
\end{array}
$$

3.
$$
\begin{array}{r}
4\ 2.6\ 5 \\
\times\ 0.8\ 0\ 4 \\
\hline
\end{array}
$$

Multiplying by 0.1, 0.01, 0.001, and So On

Now let's consider some special kinds of products. The first involves multiplying by a tenth, hundredth, thousandth, ten-thousandth, and so on. Let's look at those products.

$$
0.1 \times 38 = \frac{1}{10} \times 38 = \frac{38}{10} = 3.8
$$

$$
0.01 \times 38 = \frac{1}{100} \times 38 = \frac{38}{100} = 0.38
$$

$$
0.001 \times 38 = \frac{1}{1000} \times 38 = \frac{38}{1000} = 0.038
$$

$$
0.0001 \times 38 = \frac{1}{10,000} \times 38 = \frac{38}{10,000} = 0.0038
$$

Note in each case that the product is *smaller* than 38. That is, the decimal point in each product is farther to the left than the unwritten decimal point in 38.

Answers

1. 529.48   2. 5.0594   3. 34.2906

To multiply any number by 0.1, 0.01, 0.001, and so on,

a) count the number of decimal places in the tenth, hundredth, or thousandth, and so on, and

$$0.001 \times 34.45678$$
$$\longrightarrow 3 \text{ places}$$

b) move the decimal point in the other number that many places to the left.

$$0.001 \times 34.45678 = 0.034.45678$$
Move 3 places to the left.

$$0.001 \times 34.45678 = 0.03445678$$

EXAMPLES Multiply.

4. $0.1 \times 14.605 = 1.4605$ $\quad 1.4.605$

5. $0.01 \times 14.605 = 0.14605$

6. $0.001 \times 14.605 = 0.014605$ We write an extra zero.

7. $0.0001 \times 14.605 = 0.0014605$ We write two extra zeros.

Do Exercises 4–7.

Multiply.

4. 0.1×3.48

5. 0.01×3.48

6. 0.001×3.48

7. 0.0001×3.48

Multiplying by 10, 100, 1000, and So On

Next, let's consider multiplying by 10, 100, 1000, and so on. Let's look at those products.

$$10 \times 97.34 = 973.4$$
$$100 \times 97.34 = 9734$$
$$1000 \times 97.34 = 97{,}340$$

Note in each case that the product is *larger* than 97.34. That is, the decimal point in each product is farther to the right than the decimal point in 97.34.

To multiply any number by 10, 100, 1000, and so on,

a) count the number of zeros, and

$$1000 \times 34.45678$$
$$\longrightarrow 3 \text{ zeros}$$

b) move the decimal point in the other number that many places to the right.

$$1000 \times 34.45678 = 34.456.78$$
Move 3 places to the right.

$$1000 \times 34.45678 = 34{,}456.78$$

Multiply.

8. 10×3.48

9. 100×3.48

10. 1000×3.48

11. $10{,}000 \times 3.48$

EXAMPLES Multiply.

8. $10 \times 14.605 = 146.05$ $\quad 14.6.05$

9. $100 \times 14.605 = 1460.5$

10. $1000 \times 14.605 = 14{,}605$

11. $10{,}000 \times 14.605 = 146{,}050$ $\quad 14.6050.$ We write an extra zero.

Do Exercises 8–11.

Answers

4. 0.348 **5.** 0.0348 **6.** 0.00348
7. 0.000348 **8.** 34.8 **9.** 348
10. 3480 **11.** 34,800

b Naming Large Numbers; Money Conversion

Naming Large Numbers

We often see notation like the following in newspapers and magazines and on television and the Internet.

- U.S. venture-capital firms invested $2.6 billion in alternative-energy startup enterprises during the first three quarters of 2007.
  Source: National Venture Capital Association

- Each day about 39.7 billion person-to-person e-mails, 17.1 billion automated e-mails, and 40.5 billion pieces of spam are sent worldwide.
  Source: IDC

- The largest building in the world is the Pentagon, which has 3.7 million square feet of floor space.

To understand such notation, consider the information in the following table.

NAMING LARGE NUMBERS

1 hundred = 100 = $10 \cdot 10 = 10^2$ → 2 zeros

1 thousand = 1000 = $10 \cdot 10 \cdot 10 = 10^3$ → 3 zeros

1 million = 1,000,000 = $10 \cdot 10 \cdot 10 \cdot 10 \cdot 10 \cdot 10 = 10^6$ → 6 zeros

1 billion = 1,000,000,000 = 10^9 → 9 zeros

1 trillion = 1,000,000,000,000 = 10^{12} → 12 zeros

To convert a large number to standard notation, we proceed as follows.

EXAMPLE 12 Losses due to identity fraud totaled $49.3 billion in 2007. Convert 49.3 billion to standard notation.

Source: Javelin Strategy & Research

$$49.3 \text{ billion} = 49.3 \times 1 \text{ billion}$$
$$= 49.3 \times 1,000,000,000$$

→ 9 zeros

$$= 49,300,000,000 \quad \text{Moving the decimal point 9 places to the right}$$

Do Exercises 12 and 13.

Convert the number in each sentence to standard notation.

12. The largest building in the world is the Pentagon, which has 3.7 million square feet of floor space.

13. Each day about 39.7 billion personal e-mails are sent.

Money Conversion

Converting from dollars to cents is like multiplying by 100. To see why, consider $19.43.

$$\$19.43 = 19.43 \times \$1 \quad \text{We think of } \$19.43 \text{ as } 19.43 \times 1 \text{ dollar, or } 19.43 \times \$1.$$
$$= 19.43 \times 100¢ \quad \text{Substituting } 100¢ \text{ for } \$1: \ \$1 = 100¢$$
$$= 1943¢ \quad \text{Multiplying}$$

> **DOLLARS TO CENTS**
>
> To convert from dollars to cents, move the decimal point two places to the right and change the $ sign in front to a ¢ sign at the end.

EXAMPLES Convert from dollars to cents.

13. $189.64 = 18,964¢

14. $0.75 = 75¢

Do Exercises 14 and 15.

Convert from dollars to cents.
14. $15.69   **15.** $0.17

Converting from cents to dollars is like multiplying by 0.01. To see why, consider 65¢.

$$65¢ = 65 \times 1¢ \quad \text{We think of } 65¢ \text{ as } 65 \times 1 \text{ cent, or } 65 \times 1¢.$$
$$= 65 \times \$0.01 \quad \text{Substituting } \$0.01 \text{ for } 1¢: \ 1¢ = \$0.01$$
$$= \$0.65 \quad \text{Multiplying}$$

> **CENTS TO DOLLARS**
>
> To convert from cents to dollars, move the decimal point two places to the left and change the ¢ sign at the end to a $ sign in front.

Convert from cents to dollars.
16. 35¢   **17.** 577¢

EXAMPLES Convert from cents to dollars.

15. 395¢ = $3.95

16. 8503¢ = $85.03

Do Exercises 16 and 17.

Answers

12. 3,700,000   **13.** 39,700,000,000
14. 1569¢   **15.** 17¢   **16.** $0.35   **17.** $5.77

a Multiply.

1. 8.6
   $\times$ 7

2. 5.7
   $\times$ 0.8

3. 0.8 4
   $\times$ 8

4. 9.4
   $\times$ 0.6

5.   6.3
   $\times$ 0.0 4

6.   9.8
   $\times$ 0.0 8

7.   8 7
   $\times$ 0.0 0 6

8.   1 8.4
   $\times$ 0.0 7

9. 10×23.76

10. 100×3.8798

11. 1000×583.686852

12. 0.34×1000

13. 7.8×100

14. 0.00238×10

15. 0.1×89.23

16. 0.01×789.235

17. 0.001×97.68

18. 8976.23×0.001

19. 78.2×0.01

20. 0.0235×0.1

21. 3 2.6
   $\times$ 1 6

22. 9.2 8
   $\times$ 8.6

23. 0.9 8 4
   $\times$ 3.3

24. 8.4 8 9
   $\times$ 7.4

25. 3 7 4
   $\times$ 2.4

26. 8 6 5
   $\times$ 1.0 8

27. 7 4 9
   $\times$ 0.4 3

28. 9 7 8
   $\times$ 2 0.5

29. 0.8 7
   $\times$ 6 4

30. 7.2 5
   $\times$ 6 0

31. 4 6.5 0
   $\times$ 7 5

32. 8.2 4
   $\times$ 7 0 3

33.   8 1.7
      × 0.6 1 2

34.   3 1.8 2
      ×   7.1 5

35.   1 0.1 0 5
      × 1 1.3 2 4

36.   1 5 1.2
      × 4.5 5 5

37.   1 2.3
      × 1.0 8

38.   7.8 2
      × 0.0 2 4

39.   3 2.4
      ×   2.8

40.   8.0 9
      × 0.0 0 7 5

41.   0.0 0 3 4 2
      ×   0.8 4

42.   2.0 0 5 6
      ×   3.8

43.   0.3 4 7
      ×   2.0 9

44.   2.5 3 2
      × 1.0 6 7

45.   3.0 0 5
      × 0.6 2 3

46.   1 6.3 4
      × 0.0 0 0 5 1 2

47. 1000×45.678

48. 0.001×45.678

b)   Convert from dollars to cents.

49. $28.88

50. $67.43

51. $0.66

52. $1.78

Convert from cents to dollars.

53. 34¢

54. 95¢

55. 3445¢

56. 933¢

57. *Back-to-School Spending.* It is estimated that the amount spent on back-to-school supplies by college students and their families was $47.3 billion in 2007. Convert 47.3 billion to standard notation.

Source: National Retail Federation

58. *Cell-Phone Ads.* Global spending on advertisements appearing on cell phones is projected to be $16.2 billion in 2011. Convert 16.2 billion to standard notation.

Source: eMarketer

59. *Identity Fraud.* About 9.3 million adults were victims of identity fraud in the United States in 2007. Convert 9.3 million to standard notation.

Source: Javelin Strategy & Research

60. *Low-Cost Laptops.* It is estimated that 9.2 million low-cost laptop computers will be shipped worldwide in 2012. Convert 9.2 million to standard notation.

Source: IDC

61. *DVD Sales.* Spending on DVDs totaled $23.4 billion in 2007. Convert 23.4 billion to standard notation.

Source: The Digital Entertainment Group

62. *Text Messages.* In June 2007, 28.9 billion text messages were sent in the United States. Convert 28.9 billion to standard notation.

Source: CTIA

Skill Maintenance

Calculate.

63. $2\frac{1}{3} \cdot 4\frac{4}{5}$ [3.6a]

64. $2\frac{1}{3} \div 4\frac{4}{5}$ [3.6b]

65. $4\frac{4}{5} - 2\frac{1}{3}$ [3.5b]

66. $4\frac{4}{5} + 2\frac{1}{3}$ [3.5a]

Divide. [1.5a]

67. $2\ 4\ )\ \overline{8\ 2\ 0\ 8}$

68. $4\ )\ \overline{3\ 4\ 8}$

69. $7\ )\ \overline{3\ 1,9\ 6\ 2}$

70. $1\ 8\ )\ \overline{2\ 2,6\ 2\ 6}$

71. $4\ 0\ )\ \overline{3\ 4\ 8\ 0}$

72. $1\ 7\ )\ \overline{2\ 0,0\ 0\ 6}$

Synthesis

Consider the following names for large numbers in addition to those already discussed in this section:

1 quadrillion = 1,000,000,000,000,000 = 10^{15};

1 quintillion = 1,000,000,000,000,000,000 = 10^{18};

1 sextillion = 1,000,000,000,000,000,000,000 = 10^{21};

1 septillion = 1,000,000,000,000,000,000,000,000 = 10^{24}.

Find each of the following. Express the answer with a name that is a power of 10.

73. (1 trillion) $\cdot$ (1 billion)

74. (1 million) $\cdot$ (1 billion)

75. (1 trillion) $\cdot$ (1 trillion)

76. Is a billion millions the same as a million billions? Explain.

4.4

Division

OBJECTIVES

a Divide using decimal notation.

b Solve equations of the type $a \cdot x = b$, where a and b may be in decimal notation.

c Simplify expressions using the rules for order of operations.

a Division

Whole-Number Divisors

We use the following method when we divide a decimal quantity by a whole number.

To divide by a whole number,

a) place the decimal point directly above the decimal point in the dividend, and

b) divide as though dividing whole numbers.

$$
\begin{array}{r}
0.8\ 4 \leftarrow \text{Quotient} \\
\text{Divisor} \rightarrow 7\,\overline{)\,5.8\ 8} \leftarrow \text{Dividend} \\
5\ 6 \\
\hline
2\ 8 \\
2\ 8 \\
\hline
0 \leftarrow \text{Remainder}
\end{array}
$$

SKILL TO REVIEW

Objective 1.5a: Divide whole numbers.

Divide.

1. $5\,\overline{)\,2\ 4\ 5}$

2. $2\ 3\,\overline{)\,1\ 9\ 7\ 8}$

EXAMPLE 1 Divide: $379.2 \div 8$.

Place the decimal point.

Divide as though dividing whole numbers.

EXAMPLE 2 Divide: $82.08 \div 24$.

Place the decimal point.

$$
\begin{array}{r}
3.4\ 2 \\
2\ 4\,\overline{)\,8\ 2.0\ 8} \\
7\ 2 \\
\hline
1\ 0\ 0 \\
9\ 6 \\
\hline
4\ 8 \\
4\ 8 \\
\hline
0
\end{array}
$$

Divide as though dividing whole numbers.

Do Margin Exercises 1–3.

Divide.

1. $9\,\overline{)\,5.4}$

2. $1\ 5\,\overline{)\,2\ 2.5}$

3. $8\ 2\,\overline{)\,3\ 8.5\ 4}$

Answers

Skill to Review:
1. 49   **2.** 86

Margin Exercises:
1. 0.6   **2.** 1.5   **3.** 0.47

We can think of a whole-number dividend as having a decimal point at the end with as many 0's as we wish after the decimal point. For example, $12 = 12. = 12.0 = 12.00 = 12.000$, and so on. We can also add 0's after the last digit in the decimal portion of a number: $3.6 = 3.60 = 3.600$, and so on.

EXAMPLE 3 Divide: $30 \div 8$.

$$
\begin{array}{r}
3. \\
8 \overline{)\ 3\ 0.} \\
2\ 4 \\
\hline
6
\end{array}
$$
Place the decimal point and divide to find how many ones.

$$
\begin{array}{r}
3. \\
8 \overline{)\ 3\ 0.0} \\
2\ 4 \downarrow \\
\hline
6\ 0
\end{array}
$$
Write an extra zero.

$$
\begin{array}{r}
3.7 \\
8 \overline{)\ 3\ 0.0} \\
2\ 4 \\
\hline
6\ 0 \\
5\ 6 \\
\hline
4
\end{array}
$$
Divide to find how many tenths.

$$
\begin{array}{r}
3.7 \\
8 \overline{)\ 3\ 0.0\ 0} \\
2\ 4 \\
\hline
6\ 0 \\
5\ 6 \downarrow \\
\hline
4\ 0
\end{array}
$$
Write another zero.

$$
\begin{array}{r}
3.7\ 5 \\
8 \overline{)\ 3\ 0.0\ 0} \\
2\ 4 \\
\hline
6\ 0 \\
5\ 6 \\
\hline
4\ 0 \\
4\ 0 \\
\hline
0
\end{array}
$$
Divide to find how many hundredths.

Check:
$$
\begin{array}{r}
{}^{6}\ {}^{4} \\
3.7\ 5 \\
\times \quad\quad 8 \\
\hline
3\ 0.0\ 0
\end{array}
$$

EXAMPLE 4 Divide: $4.5 \div 25$.

$$
\begin{array}{r}
0.1\ 8 \\
2\ 5 \overline{)\ 4.5\ 0} \\
2\ 5 \\
\hline
2\ 0\ 0 \\
2\ 0\ 0 \\
\hline
0
\end{array}
$$

Check:
$$
\begin{array}{r}
0.1\ 8 \\
\times \quad\ 2\ 5 \\
\hline
9\ 0 \\
3\ 6\ 0 \\
\hline
4.5\ 0
\end{array}
$$

Do Exercises 4–6.

Divide.

4. $2\ 5 \overline{)\ 8}$

5. $4 \overline{)\ 1\ 5}$

6. $8\ 6 \overline{)\ 2\ 1.5}$

Answers
4. 0.32   **5.** 3.75   **6.** 0.25

Divisors That Are Not Whole Numbers

Consider the division

$$0.2\,4\,\overline{)\,8.2\,0\,8}$$

We write the division as $\dfrac{8.208}{0.24}$. Then we multiply by 1 to change to a whole-number divisor:

The division $0.24\overline{)8.208}$ is the same as $24\overline{)820.8}$.

$$\frac{8.208}{0.24} = \frac{8.208}{0.24} \times \frac{100}{100} = \frac{820.8}{24}.$$

The divisor is now a whole number.

To divide when the divisor is not a whole number,

a) move the decimal point (multiply by 10, 100, and so on) to make the divisor a whole number;

$$0.2\,4\,\overline{)\,8.2\,0\,8}$$

Move 2 places to the right.

b) move the decimal point in the dividend the same number of places (multiply the same way); and

$$0.2\,4\,\overline{)\,8.2\,0\,8}$$

Move 2 places to the right.

c) place the decimal point directly above the new decimal point in the dividend and divide as though dividing whole numbers.

$$
\begin{array}{r}
3\,4.2 \\
0.2\,4\,\overline{)\,8.2\,0_\wedge 8} \\
7\,2 \\
\overline{1\,0\,0} \\
9\,6 \\
\overline{4\,8} \\
4\,8 \\
\overline{0}
\end{array}
$$

(The new decimal point in the dividend is indicated by a caret.)

EXAMPLE 5 Divide: $5.848 \div 8.6$.

$$8.6\,\overline{)\,5.8\,4\,8}$$

Multiply the divisor by 10. (Move the decimal point 1 place.) Multiply the same way in the dividend. (Move 1 place.)

$$
\begin{array}{r}
0.6\,8 \\
8.6\,\overline{)\,5.8_\wedge 4\,8} \\
5\,1\,6 \\
\overline{6\,8\,8} \\
6\,8\,8 \\
\overline{0}
\end{array}
$$

Place a decimal point above the new decimal point and then divide.

Note: $\dfrac{5.848}{8.6} = \dfrac{5.848}{8.6} \cdot \dfrac{10}{10} = \dfrac{58.48}{86}.$

Do Exercises 7–9.

7. a) Complete.

$$\frac{3.75}{0.25} = \frac{3.75}{0.25} \times \frac{100}{100}$$

$$= \frac{(\quad)}{25}$$

b) Divide.

$$0.2\,5\,\overline{)\,3.7\,5}$$

Divide.

8. $0.8\,3\,\overline{)\,4.0\,6\,7}$

9. $3.5\,\overline{)\,4\,4.8}$

Answers

7. (a) 375; **(b)** 15   **8.** 4.9   **9.** 12.8

EXAMPLE 6   Divide: $12 \div 0.64$.

$$0.6\,4 \,\overline{)\, 1\,2.}$$

Place a decimal point at the end of the whole number.

$$0.6\,4 \,\overline{)\, 1\,2.0\,0}$$

Multiply the divisor by 100. (Move the decimal point 2 places.) Multiply the same way in the dividend. (Move 2 places after adding extra zeros.)

$$
\begin{array}{r}
1\,8.7\,5 \\
0.6\,4 \,\overline{)\, 1\,2.0\,0\,{}_\wedge 0\,0} \\
6\,4 \\ \hline
5\,6\,0 \\
5\,1\,2 \\ \hline
4\,8\,0 \\
4\,4\,8 \\ \hline
3\,2\,0 \\
3\,2\,0 \\ \hline
0
\end{array}
$$

Place a decimal point above the new decimal point and then divide.

Do Exercise 10.

10. Divide.

$$1.6 \,\overline{)\, 2\,5}$$

Dividing by 10, 100, 1000, and So On

It is often helpful to be able to divide quickly by a ten, hundred, or thousand, or by a tenth, hundredth, or thousandth. Each procedure we use is based on multiplying by 1. Consider the following example:

$$\frac{23.789}{1000} = \frac{23.789}{1000} \cdot \frac{1000}{1000} = \frac{23,789}{1,000,000} = 0.023789.$$

We are dividing by a number greater than 1: The result is *smaller* than 23.789.

To divide by 10, 100, 1000, and so on,

a) count the number of zeros in the divisor, and

$$\frac{713.49}{100}$$

→ 2 zeros

b) write the quotient by moving the decimal point in the dividend that number of places to the left.

$$\frac{713.49}{100}, \quad 7.13.49, \quad \frac{713.49}{100} = 7.1349$$

2 places to the left

EXAMPLE 7   Divide: $\dfrac{0.0104}{10}$.

$$\frac{0.0104}{10}, \quad 0.0.0104, \quad \frac{0.0104}{10} = 0.00104$$

1 zero   1 place to the left

Dividing by 0.1, 0.01, 0.001, and So On

Now consider the following example:

$$\frac{23.789}{0.01} = \frac{23.789}{0.01} \cdot \frac{100}{100} = \frac{2378.9}{1} = 2378.9.$$

We are dividing by a number less than 1: The result is *larger* than 23.789.
We use the following procedure.

> To divide by 0.1, 0.01, 0.001, and so on,
>
> **a)** count the number of decimal places in the divisor, and
>
> $$\frac{713.49}{0.001}$$
>
> $\longrightarrow$ 3 places
>
> **b)** write the quotient by moving the decimal point in the dividend that number of places to the right.
>
> $$\frac{713.49}{0.001}, \qquad 713.490. \qquad \frac{713.49}{0.001} = 713,490$$
>
> 3 places to the right to change 0.001 to 1

Divide.

11. $\dfrac{0.1278}{0.01}$ **12.** $\dfrac{0.1278}{100}$

13. $\dfrac{98.47}{1000}$ **14.** $\dfrac{6.7832}{0.1}$

EXAMPLE 8 Divide: $\dfrac{23.738}{0.001}$.

$$\frac{23.738}{0.001}, \qquad 23.738. \qquad \frac{23.738}{0.001} = 23,738$$

3 places   3 places to the right

Do Exercises 11–14.

b Solving Equations

Now let's solve equations of the type $a \cdot x = b$, where a and b may be in decimal notation. Proceeding as before, we divide by a on both sides.

EXAMPLE 9 Solve: $8 \cdot x = 27.2$.

We have

$$\frac{8 \cdot x}{8} = \frac{27.2}{8} \qquad \text{Dividing by 8 on both sides}$$

$$x = 3.4.$$

$$\begin{array}{r} 3.4 \\ 8\overline{)27.2} \\ 2\ 4 \\ \hline 3\ 2 \\ 3\ 2 \\ \hline 0 \end{array}$$

The solution is 3.4.

EXAMPLE 10 Solve: $2.9 \cdot t = 0.14616$.

We have

$$\frac{2.9 \cdot t}{2.9} = \frac{0.14616}{2.9} \qquad \text{Dividing by 2.9 on both sides}$$

$$t = 0.0504.$$

```
        0.0 5 0 4
2.9 ) 0.1ᴧ4 6 1 6
        1 4 5
          1 1 6
          1 1 6
              0
```

The solution is 0.0504.

Do Exercises 15 and 16.

Solve.

15. $100 \cdot x = 78.314$

16. $0.25 \cdot y = 276.4$

c Order of Operations: Decimal Notation

The same rules for order of operations used with whole numbers and fraction notation apply when simplifying expressions with decimal notation.

> **RULES FOR ORDER OF OPERATIONS**
>
> 1. Do all calculations within grouping symbols before operations outside.
> 2. Evaluate all exponential expressions.
> 3. Do all multiplications and divisions in order from left to right.
> 4. Do all additions and subtractions in order from left to right.

EXAMPLE 11 Simplify: $2.56 \times 25.6 \div 25{,}600 \times 256$.

There are no exponents or parentheses, so we multiply and divide from left to right:

$$
\begin{aligned}
2.56 \times 25.6 \div 25{,}600 \times 256 &= 65.536 \div 25{,}600 \times 256 \\
&= 0.00256 \times 256 \\
&= 0.65536.
\end{aligned}
$$

Doing all multiplications and divisions in order from left to right

EXAMPLE 12 Simplify: $(5 - 0.06) \div 2 + 3.42 \times 0.1$.

$$
\begin{aligned}
(5 - 0.06) \div 2 + 3.42 \times 0.1 &= 4.94 \div 2 + 3.42 \times 0.1 \\
&= 2.47 + 0.342 \\
&= 2.812
\end{aligned}
$$

Carrying out the operation inside parentheses

Doing all multiplications and divisions in order from left to right

Answers

15. 0.78314 **16.** 1105.6

EXAMPLE 13 Simplify: $10^2 \times \{[(3 - 0.24) \div 2.4] - (0.21 - 0.092)\}$.

$10^2 \times \{[(3 - 0.24) \div 2.4] - (0.21 - 0.092)\}$

$= 10^2 \times \{[2.76 \div 2.4] - 0.118\}$ Doing the calculations in the innermost parentheses first

$= 10^2 \times \{1.15 - 0.118\}$ Again, doing the calculations in the innermost grouping symbols

$= 10^2 \times 1.032$ Subtracting inside the grouping symbols

$= 100 \times 1.032$ Evaluating the exponential expression

$= 103.2$

Do Exercises 17–19.

Simplify.

17. $625 \div 62.5 \times 25 \div 6250$

18. $0.25 \cdot (1 + 0.08) - 0.0274$

19. $20^2 - 3.4^2 +$
$\{2.5[20(9.2 - 5.6)] + 5(10 - 5)\}$

STUDY TIPS

HOMEWORK TIPS

Prepare for your homework assignment by reading the explanations of concepts and by following the step-by-step solutions of examples in the text. The time that you spend preparing will save valuable time when you do your assignment.

EXAMPLE 14 *Mountains in Peru.* The figure below shows a range of very high mountains in Peru, together with their altitudes, given both in feet and in meters. Find the average height of these mountains, in feet.
Source: *National Geographic*, July 1968, p. 130

Nev. Sara Sara, 18,060 ft 5,505 m
Nev. Coropuna, 21,079 ft 6,425 m
Nevado Ampato, 20,700 ft 6,309 m
Nev. Chachani, 19,931 ft 6,075 m
Volcan Misti, 19,101 ft 5,822 m
Nev. Pichu Pichu, 18,600 ft 5,669 m
Arequipa
Pacific Ocean
Peru
South America
Area enlarged
Scale varies in this perspective.
SOURCE: WOOD RONASVILLE HARLIN INC/NGS Image Collection

The **average** of a set of numbers is the sum of the numbers divided by the number of addends. (See Section 1.9.) We find the sum of the heights divided by the number of addends, 6:

$$\frac{18{,}060 + 21{,}079 + 20{,}700 + 19{,}931 + 19{,}101 + 18{,}600}{6} = \frac{117{,}471}{6} = 19{,}578.5.$$

Thus the average height of these mountains is 19,578.5 ft.

Do Exercise 20.

20. Mountains in Peru. Refer to the figure in Example 14. Find the average height of the mountains, in meters.

Answers

17. 0.04   **18.** 0.2426   **19.** 593.44
20. 5967.5 m

a Divide.

1. $2 \overline{)\, 5.9\, 8}$

2. $5 \overline{)\, 1\, 8}$

3. $4 \overline{)\, 9\, 5.1\, 2}$

4. $8 \overline{)\, 2\, 5.9\, 2}$

5. $1\, 2 \overline{)\, 8\, 9.7\, 6}$

6. $2\, 3 \overline{)\, 2\, 5.0\, 7}$

7. $3\, 3 \overline{)\, 2\, 3\, 7.6}$

8. $5\, 4 \overline{)\, 4\, 4\, 8.2}$

9. $9.144 \div 8$

10. $4.5 \div 9$

11. $12.123 \div 3$

12. $12.4 \div 4$

13. $0.0\, 6 \overline{)\, 3.3\, 6}$

14. $0.0\, 4 \overline{)\, 1.6\, 8}$

15. $0.1\, 2 \overline{)\, 8.4}$

16. $0.3\, 6 \overline{)\, 2.8\, 8}$

17. $3.4 \overline{)\, 6\, 8}$

18. $0.2\, 5 \overline{)\, 5}$

19. $1\, 5 \overline{)\, 6}$

20. $1\, 2 \overline{)\, 1.8}$

21. $3\, 6 \overline{)\, 1\, 4.7\, 6}$

22. $5\, 2 \overline{)\, 1\, 1\, 9.6}$

23. $3.2 \overline{)\, 2\, 7.2}$

24. $8.5 \overline{)\, 2\, 7.2}$

25. $4.2 \overline{)\, 3\, 9.0\, 6}$

26. $4.8 \overline{)\, 0.1\, 1\, 0\, 4}$

27. $8 \overline{)\, 5}$

28. $8 \overline{)\, 3}$

29. $0.4\, 7 \overline{)\, 0.1\, 2\, 2\, 2}$

30. $1.0\, 8 \overline{)\, 0.5\, 4}$

31. $4.8 \overline{)\, 7\, 5}$

32. $0.2\,8\,\overline{)\,6\,3}$

33. $0.0\,3\,2\,\overline{)\,0.0\,7\,4\,8\,8}$

34. $0.0\,1\,7\,\overline{)\,1.5\,8\,1}$

35. $8\,2\,\overline{)\,3\,8.5\,4}$

36. $3\,4\,\overline{)\,0.1\,4\,6\,2}$

37. $\dfrac{213.4567}{1000}$

38. $\dfrac{769.3265}{1000}$

39. $\dfrac{23.59}{10}$

40. $\dfrac{83.57}{10}$

41. $\dfrac{426.487}{100}$

42. $\dfrac{591.348}{100}$

43. $\dfrac{16.94}{0.1}$

44. $\dfrac{100.7604}{0.1}$

45. $\dfrac{1.0237}{0.001}$

46. $\dfrac{3.4029}{0.001}$

47. $\dfrac{42.561}{0.01}$

48. $\dfrac{98.473}{0.01}$

b Solve.

49. $4.2 \cdot x = 39.06$

50. $36 \cdot y = 14.76$

51. $1000 \cdot y = 9.0678$

52. $789.23 = 0.25 \cdot q$

53. $1048.8 = 23 \cdot t$

54. $28.2 \cdot x = 423$

c Simplify.

55. $14 \times (82.6 + 67.9)$

56. $(26.2 - 14.8) \times 12$

57. $0.003 + 3.03 \div 0.01$

58. $9.94 + 4.26 \div (6.02 - 4.6) - 0.9$

59. $42 \times (10.6 + 0.024)$

60. $(18.6 - 4.9) \times 13$

61. $4.2 \times 5.7 + 0.7 \div 3.5$

62. $123.3 - 4.24 \times 1.01$

63. $9.0072 + 0.04 \div 0.1^2$

64. $12 \div 0.03 - 12 \times 0.03^2$

65. $(8 - 0.04)^2 \div 4 + 8.7 \times 0.4$

66. $(5 - 2.5)^2 \div 100 + 0.1 \times 6.5$

67. $86.7 + 4.22 \times (9.6 - 0.03)^2$

68. $2.48 \div (1 - 0.504) + 24.3 - 11 \times 2$

69. $4 \div 0.4 + 0.1 \times 5 - 0.1^2$

70. $6 \times 0.9 + 0.1 \div 4 - 0.2^3$

71. $5.5^2 \times [(6 - 4.2) \div 0.06 + 0.12]$

72. $12^2 \div (12 + 2.4) - [(2 - 1.6) \div 0.8]$

73. $200 \times \{[(4 - 0.25) \div 2.5] - (4.5 - 4.025)\}$

74. $0.03 \times \{1 \times 50.2 - [(8 - 7.5) \div 0.05]\}$

75. Find the average of $1276.59, $1350.49, $1123.78, and $1402.58.

76. Find the average weight of two wrestlers who weigh 308 lb and 296.4 lb.

77. *Camping in National Parks.* Camping in national parks has declined steadily in recent years. The graph below shows the numbers of overnight camping stays in Park Service campgrounds in the National Park System from 2002 to 2006. Find the average number of stays per year during this period.

Camping in Park Service Campgrounds

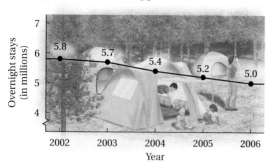

SOURCE: U.S. National Park Service

78. *Unclaimed Prizes.* Each year millions of lottery prize dollars go unclaimed. The table below shows the five states with the most unclaimed prize money for a recent year. Find the average amount that went unclaimed in these states.

STATE	UNCLAIMED PRIZES (in millions)
New York	$73.6
Texas	58.9
Florida	48.4
Ohio	33.4
California	29.4

SOURCE: Individual state lotteries

Skill Maintenance

Simplify. [2.5b]

79. $\dfrac{36}{42}$

80. $\dfrac{56}{64}$

81. $\dfrac{38}{146}$

82. $\dfrac{92}{124}$

Find the prime factorization. [2.1d]

83. 684

84. 162

85. 2007

86. 2005

87. Add: $10\frac{1}{2} + 4\frac{5}{8}$. [3.5a]

88. Subtract: $10\frac{1}{2} - 4\frac{5}{8}$. [3.5b]

Synthesis

Simplify.

89. ▦ $9.0534 - 2.041^2 \times 0.731 \div 1.043^2$

90. ▦ $23.042(7 - 4.037 \times 1.46 - 0.932^2)$

In Exercises 91–94, find the missing value.

91. $439.57 \times 0.01 \div 1000 \times \square = 4.3957$

92. $5.2738 \div 0.01 \times 1000 \div \square = 52.738$

93. $0.0329 \div 0.001 \times 10^4 \div \square = 3290$

94. $0.0047 \times 0.01 \div 10^4 \times \square = 4.7$

Mid-Chapter Review

Concept Reinforcement

Determine whether each statement is true or false.

_____ **1.** In the number 308.00567, the digit 6 names the tens place.   [4.1a]

_____ **2.** When writing a word name for decimal notation, we write the word "and" for the decimal point. [4.1a]

_____ **3.** To multiply any number by 10, 100, 1000, and so on, count the number of zeros and move the decimal point that many places to the right. [4.3a]

Guided Solutions

Fill in each blank with the number that creates a correct statement or solution.

4. Solve: $y + 12.8 = 23.35$.   [4.2c]

$$y + 12.8 = 23.35$$
$$y + 12.8 - \square = 23.35 - \square \quad \text{Subtracting 12.8 on both sides}$$
$$y + \square = \square \quad\quad\quad\quad \text{Carrying out the subtraction}$$
$$y = \square$$

5. Simplify: $5.6 + 4.3 \times (6.5 - 0.25)^2$.   [4.4c]

$$5.6 + 4.3 \times (6.5 - 0.25)^2 = 5.6 + 4.3 \times (\square)^2 \quad \text{Carrying out the operation inside parentheses}$$
$$= 5.6 + 4.3 \times \square \quad\quad \text{Evaluating the exponential expression}$$
$$= 5.6 + \square \quad\quad\quad\quad \text{Multiplying}$$
$$= \square \quad\quad\quad\quad\quad\quad \text{Adding}$$

Mixed Review

6. Usain Bolt of Jamaica set a world record of 9.69 sec in the men's 100-m race at the 2008 Summer Olympics. Write a word name for 9.69.   [4.1a]
Source: beijing2008.cn

7. After removing coins from their pockets at security check-points in U.S. airports, travelers forgot to reclaim coins totaling about $1.05 million from 2005 to 2007. Convert 1.05 million to standard notation.   [4.3b]
Source: Transportation Security Administration

Write fraction notation.   [4.1b]

8. 4.53

9. 0.287

Which number is larger?   [4.1c]

10. 0.07, 0.13

11. 5.2, 5.09

Write decimal notation.   [4.1b]

12. $\dfrac{7}{10}$

13. $\dfrac{639}{100}$

14. $35\dfrac{67}{100}$

15. $8\dfrac{2}{1000}$

Round 28.4615 to the nearest: [4.1d]

16. Thousandth.

17. Hundredth.

18. Tenth.

19. One

Add. [4.2a]

20.
$$\begin{array}{r} 4\ 7.6\ 3\ 8 \\ +\ \ \ 2.4\ 5\ 7 \\ \hline \end{array}$$

21.
$$\begin{array}{r} 1\ 5.6 \\ 2\ 3\ 4.7\ 2\ 9 \\ 3.0\ 8 \\ +\ 9\ 6\ 1.4\ 5\ 3 \\ \hline \end{array}$$

22. $4.5 + 0.728$

23. $16 + 0.34 + 1.9$

Subtract. [4.2b]

24.
$$\begin{array}{r} 3\ 2\ 1.5\ 7 \\ -\ \ \ 4\ 9.3\ 8 \\ \hline \end{array}$$

25.
$$\begin{array}{r} 5.6 \\ -\ 0.0\ 0\ 7 \\ \hline \end{array}$$

26. $34.3 - 18.75$

27. $49.07 - 9.7$

Multiply. [4.3a]

28.
$$\begin{array}{r} 4.6 \\ \times\ 0.9 \\ \hline \end{array}$$

29.
$$\begin{array}{r} 1\ 5.3 \\ \times\ \ \ 6.0\ 7 \\ \hline \end{array}$$

30. 100×81.236

31. 0.1×29.37

Divide. [4.4a]

32. $20.24 \div 4$

33. $21.76 \div 6.8$

34. $76.34 \div 0.1$

35. $914.036 \div 1000$

36. Convert \$20.45 to cents. [4.3b]

37. Convert 147¢ to dollars. [4.3b]

38. Solve: $46.3 + x = 59$. [4.2c]

39. Solve: $42.84 = 5.1 \cdot y$. [4.4b]

Simplify. [4.4c]

40. $6.594 + 0.5318 \div 0.01$

41. $7.3 \times 4.6 - 0.8 \div 3.2$

Understanding Through Discussion and Writing

42. A fellow student rounds 236.448 to the nearest one and gets 237. Explain the possible error. [4.1d]

43. Explain the error in the following: [4.2b]
Subtract.
$$73.089 - 5.0061 = 2.3028$$

44. Explain why $10 \div 0.2 = 100 \div 2$. [4.4a]

45. Kayla made these two computational mistakes:
$$0.247 \div 0.1 = 0.0247; \quad 0.247 \div 10 = 2.47.$$
In each case, how could you convince her that a mistake has been made? [4.4a]

4.5 Converting from Fraction Notation to Decimal Notation

a Fraction Notation to Decimal Notation

When a denominator has no prime factors other than 2's and 5's, we can find decimal notation by multiplying by 1. We multiply to get a denominator that is a power of ten, like 10, 100, or 1000.

EXAMPLE 1 Find decimal notation for $\frac{3}{5}$.

$$\frac{3}{5} = \frac{3}{5} \cdot \frac{2}{2} = \frac{6}{10} = 0.6$$ $5 \cdot 2 = 10$, so we use $\frac{2}{2}$ for 1 to get a denominator of 10.

EXAMPLE 2 Find decimal notation for $\frac{7}{20}$.

$$\frac{7}{20} = \frac{7}{20} \cdot \frac{5}{5} = \frac{35}{100} = 0.35$$ $20 \cdot 5 = 100$, so we use $\frac{5}{5}$ for 1 to get a denominator of 100.

EXAMPLE 3 Find decimal notation for $\frac{87}{25}$.

$$\frac{87}{25} = \frac{87}{25} \cdot \frac{4}{4} = \frac{348}{100} = 3.48$$ $25 \cdot 4 = 100$, so we use $\frac{4}{4}$ for 1 to get a denominator of 100.

EXAMPLE 4 Find decimal notation for $\frac{9}{40}$.

$$\frac{9}{40} = \frac{9}{40} \cdot \frac{25}{25} = \frac{225}{1000} = 0.225$$ $40 \cdot 25 = 1000$, so we use $\frac{25}{25}$ for 1 to get a denominator of 1000.

Do Margin Exercises 1–4.

We can always divide to find decimal notation.

EXAMPLE 5 Find decimal notation for $\frac{3}{5}$.

$$\frac{3}{5} = 3 \div 5$$
$$\begin{array}{r} 0.6 \\ 5 \overline{)\ 3.0} \\ 3\ 0 \\ \hline 0 \end{array}$$ $\frac{3}{5} = 0.6$

EXAMPLE 6 Find decimal notation for $\frac{7}{8}$.

$$\frac{7}{8} = 7 \div 8$$
$$\begin{array}{r} 0.8\ 7\ 5 \\ 8 \overline{)\ 7.0\ 0\ 0} \\ 6\ 4 \\ \hline 6\ 0 \\ 5\ 6 \\ \hline 4\ 0 \\ 4\ 0 \\ \hline 0 \end{array}$$ $\frac{7}{8} = 0.875$

Do Exercises 5 and 6.

OBJECTIVES

a Convert from fraction notation to decimal notation.

b Round numbers named by repeating decimals in problem solving.

c Calculate using fraction notation and decimal notation together.

SKILL TO REVIEW
Objective 4.4a: Divide using decimal notation.

Divide.
  1. $3 \div 4$ **2.** $25 \div 8$

Find decimal notation. Use multiplying by 1.

1. $\frac{4}{5}$ **2.** $\frac{9}{20}$

3. $\frac{11}{40}$ **4.** $\frac{33}{25}$

Find decimal notation.

5. $\frac{2}{5}$ **6.** $\frac{3}{8}$

Answers

Skill to Review:
1. 0.75   **2.** 3.125

Margin Exercises:
1. 0.8   **2.** 0.45   **3.** 0.275
4. 1.32   **5.** 0.4   **6.** 0.375

In Examples 5 and 6, the division *terminated,* meaning that eventually we got a remainder of 0. A **terminating decimal** occurs when the denominator has only 2's or 5's, or both, as factors, as in $\frac{17}{25}$, $\frac{5}{8}$, or $\frac{83}{100}$. This assumes that the fraction notation has been simplified.

Consider a different situation:

$$\frac{5}{6}, \quad \text{or} \quad \frac{5}{2 \cdot 3}.$$

Since 6 has a 3 as a factor, the division will not terminate. Although we can still use division to get decimal notation, the answer will be a **repeating decimal**, as follows.

EXAMPLE 7 Find decimal notation for $\frac{5}{6}$.

$$\frac{5}{6} = 5 \div 6$$

$$
\begin{array}{r}
0.8\ 3\ 3 \\
6\ )\overline{5.0\ 0\ 0} \\
4\ 8 \\
\hline
2\ 0 \\
1\ 8 \\
\hline
2\ 0 \\
1\ 8 \\
\hline
2
\end{array}
$$

Since 2 keeps reappearing as a remainder, the digits repeat and will continue to do so; therefore,

$$\frac{5}{6} = 0.83333\ldots.$$

The red dots indicate an endless sequence of digits in the quotient. When there is a repeating pattern, the dots are often replaced by a bar to indicate the repeating part—in this case, only the 3:

$$\frac{5}{6} = 0.8\overline{3}.$$

Do Exercises 7 and 8.

EXAMPLE 8 Find decimal notation for $\frac{4}{11}$.

$$\frac{4}{11} = 4 \div 11$$

$$
\begin{array}{r}
0.3\ 6\ 3\ 6 \\
1\ 1\ )\overline{4.0\ 0\ 0\ 0} \\
3\ 3 \\
\hline
7\ 0 \\
6\ 6 \\
\hline
4\ 0 \\
3\ 3 \\
\hline
7\ 0 \\
6\ 6 \\
\hline
4
\end{array}
$$

Since 7 and 4 keep repeating as remainders, the sequence of digits "36" repeats in the quotient, and

$$\frac{4}{11} = 0.363636\ldots, \quad \text{or} \quad 0.\overline{36}.$$

Do Exercises 9 and 10.

Find decimal notation.

7. $\frac{1}{6}$ **8.** $\frac{2}{3}$

Find decimal notation.

9. $\frac{5}{11}$ **10.** $\frac{12}{11}$

Answers

7. $0.1\overline{6}$ **8.** $0.\overline{6}$ **9.** $0.\overline{45}$ **10.** $1.\overline{09}$

EXAMPLE 9 Find decimal notation for $\frac{5}{7}$.

$$
\begin{array}{r}
0.7\ 1\ 4\ 2\ 8\ 5 \\
7\overline{)5.0\ 0\ 0\ 0\ 0\ 0} \\
\underline{4\ 9} \\
1\ 0 \\
\underline{7} \\
3\ 0 \\
\underline{2\ 8} \\
2\ 0 \\
\underline{1\ 4} \\
6\ 0 \\
\underline{5\ 6} \\
4\ 0 \\
\underline{3\ 5} \\
5
\end{array}
$$

Since 5 appears again as a remainder, the sequence of digits "714285" repeats in the quotient, and

$$\frac{5}{7} = 0.714285714285\ldots, \quad \text{or} \quad 0.\overline{714285}.$$

The length of a repeating part can be very long—too long to find on a calculator. An example is $\frac{5}{97}$, which has a repeating part of 96 digits.

Do Exercise 11.

11. Find decimal notation for $\frac{3}{7}$.

b Rounding in Problem Solving

In applied problems, repeating decimals are rounded to get approximate answers. To round a repeating decimal, we can extend the decimal notation at least one place past the rounding digit, and then round as before.

EXAMPLES Round each of the following to the nearest tenth, hundredth, and thousandth.

		Nearest tenth	Nearest hundredth	Nearest thousandth
10.	$0.8\overline{3} = 0.83333\ldots$	0.8	0.83	0.833
11.	$0.0\overline{9} = 0.090909\ldots$	0.1	0.09	0.091
12.	$0.\overline{714285} = 0.714285714285\ldots$	0.7	0.71	0.714

Do Exercises 12–14.

Round each to the nearest tenth, hundredth, and thousandth.

12. $0.\overline{6}$

13. $0.8\overline{0}$

14. $6.2\overline{45}$

Converting Ratios to Decimal Notation

When solving applied problems, we often convert ratios to decimal notation.

EXAMPLE 13 *Gas Mileage.* A car travels 457 mi on 16.4 gal of gasoline. The ratio of number of miles driven to amount of gasoline used is *gas mileage.* Find the gas mileage and convert the ratio to decimal notation rounded to the nearest tenth.

$$\frac{\text{Miles driven}}{\text{Gasoline used}} = \frac{457}{16.4} \approx 27.86 \quad \text{Dividing to 2 decimal places}$$

$$\approx 27.9 \quad \text{Rounding to 1 decimal place}$$

The gas mileage is 27.9 miles to the gallon.

Answers

11. $0.\overline{428571}$ **12.** $0.7; 0.67; 0.667$
13. $0.8; 0.81; 0.808$ **14.** $6.2; 6.25; 6.245$

15. Coin Tossing. A coin is tossed 51 times. It lands heads 26 times. Find the ratio of heads to tosses and convert it to decimal notation rounded to the nearest thousandth. (This is also the experimental probability of getting heads.)

16. Gas Mileage. A car travels 380 mi on 15.7 gal of gasoline. Find the gasoline mileage and convert the ratio to decimal notation rounded to the nearest tenth.

EXAMPLE 14 *Wildfires.*
A severe thunderstorm system that moved through northern and central California on June 20, 2008, produced over 6000 lightning strikes. This sparked 2096 wildfires that burned about 1,200,000 acres. Find the ratio of the number of acres burned to the number of fires and convert it to decimal notation rounded to the nearest thousandth.
Source: California Department of Forestry and Fire Prevention

We have

$$\frac{\text{Acres burned}}{\text{Number of fires}} = \frac{1{,}200{,}000 \text{ acres}}{2096 \text{ fires}}$$
$$\approx 572.51908.$$

About 572.519 acres burned per fire.

Do Exercises 15 and 16.

Averages

When finding an average, we may at times need to round an answer.

EXAMPLE 15 *Peer-to-Peer Loans.* Peer-to-peer lending, in which individuals make loans to other individuals via the Internet, is beginning to soar. The graph below shows the amounts of peer-to-peer loans made from 2005 to 2010. Find the average amount loaned per year during this period. Round to the nearest thousandth.

Peer-to-Peer Loans

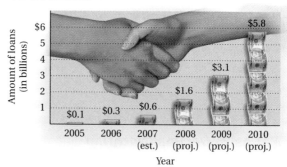

SOURCE: Celent

17. Peer-to-Peer Loans. Refer to the data on peer-to-peer loans in Example 15. Find the average amount loaned per year from 2007 to 2009. Round to the nearest thousandth.

We add the amounts of the loans and divide by the number of addends, 6. The average is

$$\frac{0.1 + 0.3 + 0.6 + 1.6 + 3.1 + 5.8}{6} = \frac{11.5}{6} = 1.91\overline{6} \approx 1.917.$$

Peer-to-peer loans made from 2005 to 2010 averaged a total of about $1.917 billion per year.

Do Exercise 17.

Answers

15. 0.510   **16.** 24.2 miles per gallon
17. About $1.767 billion

(c) Calculations with Fraction Notation and Decimal Notation Together

In certain kinds of calculations, fraction notation and decimal notation might occur together. In such cases, there are at least three ways in which we might proceed.

EXAMPLE 16 Calculate: $\frac{2}{3} \times 0.576$.

METHOD 1: One way to do this calculation is to convert the fraction notation to decimal notation so that both numbers are in decimal notation. Since $\frac{2}{3}$ converts to repeating decimal notation, it is first rounded to some chosen decimal place. We choose three decimal places because 0.576 has three decimal places. Then, using decimal notation, we multiply.

$$\frac{2}{3} \times 0.576 = 0.\overline{6} \times 0.576 \approx 0.667 \times 0.576 = 0.384192$$

METHOD 2: A second way to do this calculation is to convert the decimal notation to fraction notation so that both numbers are in fraction notation. The answer can be left in fraction notation and simplified, or we can convert to decimal notation and round, if appropriate.

$$\frac{2}{3} \times 0.576 = \frac{2}{3} \cdot \frac{576}{1000} = \frac{2 \cdot 576}{3 \cdot 1000}$$

$$= \frac{2 \cdot 2 \cdot 2 \cdot 2 \cdot 2 \cdot 2 \cdot 2 \cdot 3 \cdot 3}{2 \cdot 2 \cdot 2 \cdot 3 \cdot 5 \cdot 5 \cdot 5}$$

$$= \frac{2 \cdot 2 \cdot 2 \cdot 3}{2 \cdot 2 \cdot 2 \cdot 3} \cdot \frac{2 \cdot 2 \cdot 2 \cdot 2 \cdot 3}{5 \cdot 5 \cdot 5}$$

$$= 1 \cdot \frac{2 \cdot 2 \cdot 2 \cdot 2 \cdot 3}{5 \cdot 5 \cdot 5}$$

$$= \frac{2 \cdot 2 \cdot 2 \cdot 2 \cdot 3}{5 \cdot 5 \cdot 5} = \frac{48}{125}, \text{ or } 0.384$$

METHOD 3: A third way to do this calculation is to treat 0.576 as $\frac{0.576}{1}$. Then we multiply 0.576 by 2, and divide the result by 3.

$$\frac{2}{3} \times 0.576 = \frac{2}{3} \times \frac{0.576}{1} = \frac{2 \times 0.576}{3} = \frac{1.152}{3} = 0.384$$

Note that we get an exact answer with methods 2 and 3, but method 1 gives an approximation since we rounded decimal notation for $\frac{2}{3}$.

Do Exercise 18.

18. Calculate: $\frac{5}{6} \times 0.864$.

EXAMPLE 17 Calculate: $\frac{2}{3} \times 0.576 + 3.287 \div \frac{4}{5}$.

We use the rules for order of operations, doing first the multiplication and then the division. Then we add.

$$\frac{2}{3} \times 0.576 + 3.287 \div \frac{4}{5} = 0.384 + 3.287 \cdot \frac{5}{4}$$

$$= 0.384 + 4.10875$$

$$= 4.49275$$

Method 3:
$\frac{2}{3} \times \frac{0.576}{1} = 0.384$;
$\frac{3.287}{1} \times \frac{5}{4} = 4.10875$

Do Exercises 19 and 20.

Calculate.

19. $\frac{1}{3} \times 0.384 + \frac{5}{8} \times 0.6784$

20. $\frac{5}{6} \times 0.864 + 14.3 \div \frac{8}{5}$

Answers
18. 0.72   **19.** 0.552   **20.** 9.6575

4.5

Exercise Set

For Extra Help

MyMathLab

Math XL
PRACTICE   WATCH   DOWNLOAD   READ   REVIEW

a Find decimal notation.

1. $\dfrac{23}{100}$

2. $\dfrac{9}{100}$

3. $\dfrac{3}{5}$

4. $\dfrac{19}{20}$

5. $\dfrac{13}{40}$

6. $\dfrac{3}{16}$

7. $\dfrac{1}{5}$

8. $\dfrac{4}{5}$

9. $\dfrac{17}{20}$

10. $\dfrac{11}{20}$

11. $\dfrac{3}{8}$

12. $\dfrac{7}{8}$

13. $\dfrac{39}{40}$

14. $\dfrac{31}{40}$

15. $\dfrac{13}{25}$

16. $\dfrac{61}{125}$

17. $\dfrac{2502}{125}$

18. $\dfrac{181}{200}$

19. $\dfrac{1}{4}$

20. $\dfrac{1}{2}$

21. $\dfrac{29}{25}$

22. $\dfrac{37}{25}$

23. $\dfrac{19}{16}$

24. $\dfrac{5}{8}$

25. $\dfrac{4}{15}$

26. $\dfrac{7}{9}$

27. $\dfrac{1}{3}$

28. $\dfrac{1}{9}$

29. $\dfrac{4}{3}$

30. $\dfrac{8}{9}$

31. $\dfrac{7}{6}$

32. $\dfrac{7}{11}$

33. $\dfrac{4}{7}$

34. $\dfrac{14}{11}$

35. $\dfrac{11}{12}$

36. $\dfrac{5}{12}$

b

37.–47. *Odds.* Round each answer of the odd-numbered Exercises 25–35 to the nearest tenth, hundredth, and thousandth.

38.–48. *Evens.* Round each answer of the even-numbered Exercises 26–36 to the nearest tenth, hundredth, and thousandth.

Round each to the nearest tenth, hundredth, and thousandth.

49. $0.1\overline{8}$

50. $0.8\overline{3}$

51. $0.2\overline{7}$

52. $3.5\overline{4}$

53. For this set of people, what is the ratio, in decimal notation rounded to the nearest thousandth, where appropriate, of:

a) women to the total number of people?
b) women to men?
c) men to the total number of people?
d) men to women?

54. For this set of pennies and quarters, what is the ratio, in decimal notation rounded to the nearest thousandth, where appropriate, of:

a) pennies to quarters?
b) quarters to pennies?
c) pennies to total number of coins?
d) total number of coins to pennies?

Gas Mileage. In each of Exercises 55–58, find the gas mileage rounded to the nearest tenth.

55. 285 mi; 18 gal

56. 396 mi; 17 gal

57. 324.8 mi; 18.2 gal

58. 264.8 mi; 12.7 gal

59. *Windy Cities.* Although nicknamed the Windy City, Chicago is not the windiest city in the United States. Listed in the table below are the six windiest cities and their average wind speeds. Find the average of these wind speeds and round your answer to the nearest tenth.

Source: *The Handy Geography Answer Book*

CITY	AVERAGE WIND SPEED (in miles per hour)
Mt. Washington, NH	35.3
Boston, MA	12.5
Honolulu, HI	11.3
Dallas, TX	10.7
Kansas City, MO	10.7
Chicago, IL	10.4

60. *Areas of the New England States.* The table below lists the areas of the New England states. Find the average area and round your answer to the nearest tenth.

Source: *The New York Times Almanac*

STATE	TOTAL AREA (in square miles)
Maine	33,265
New Hampshire	9,279
Vermont	9,614
Massachusetts	8,284
Connecticut	5,018
Rhode Island	1,211

Stock Prices. At one time, stock prices were given using mixed numerals involving halfs, fourths, eighths, and so on. The Securities and Exchange Commission now mandates the use of decimal notation. Thus a price of $23\frac{13}{16}$ is expressed in decimal notation, rounded to the nearest hundredth, as $23.81. Complete the following table.

	STOCK	PRICE PER SHARE	DECIMAL NOTATION	ROUNDED TO NEAREST HUNDREDTH
61.	General Electric	$29\frac{9}{16}$		
62.	Intel	$24\frac{7}{16}$		
63.	Microsoft	$27\frac{7}{8}$		
64.	Home Depot	$27\frac{1}{8}$		
65.	Verisign	$32\frac{31}{64}$		
66.	Hewlett Packard	$45\frac{53}{64}$		

SOURCE: New York Stock Exchange

c Calculate.

67. $\frac{7}{8} \times 12.64$

68. $\frac{4}{5} \times 384.8$

69. $2\frac{3}{4} + 5.65$

70. $4\frac{4}{5} + 3.25$

71. $\frac{47}{9} \times 79.95$

72. $\frac{7}{11} \times 2.7873$

73. $\frac{1}{2} - 0.5$

74. $3\frac{1}{8} - 2.75$

75. $4.875 - 2\frac{1}{16}$

76. $55\frac{3}{5} - 12.22$

77. $\frac{5}{6} \times 0.0765 + \frac{5}{4} \times 0.1124$

78. $\frac{3}{5} \times 6384.1 - \frac{3}{8} \times 156.56$

79. $\frac{4}{5} \times 384.8 + 24.8 \div \frac{8}{3}$

80. $102.4 \div \frac{2}{5} - 12 \times \frac{5}{6}$

81. $\dfrac{7}{8} \times 0.86 - 0.76 \times \dfrac{3}{4}$

82. $17.95 \div \dfrac{5}{8} + \dfrac{3}{4} \times 16.2$

83. $3.375 \times 5\dfrac{1}{3}$

84. $2.5 \times 3\dfrac{5}{8}$

85. $6.84 \div 2\dfrac{1}{2}$

86. $8\dfrac{1}{2} \div 2.125$

Skill Maintenance

Multiply. [3.6a]

87. $9 \cdot 2\dfrac{1}{3}$

88. $10\dfrac{1}{2} \cdot 22\dfrac{3}{4}$

Divide. [3.6b]

89. $84 \div 8\dfrac{2}{5}$

90. $8\dfrac{3}{5} \div 10\dfrac{2}{5}$

Add. [3.5a]

91. $17\dfrac{5}{6} + 32\dfrac{3}{8}$

92. $14\dfrac{3}{5} + 16\dfrac{1}{10}$

Subtract. [3.5b]

93. $16\dfrac{1}{10} - 14\dfrac{3}{5}$

94. $32\dfrac{3}{8} - 17\dfrac{5}{6}$

Solve. [3.5c]

95. A recipe for bread calls for $\dfrac{2}{3}$ cup of water, $\dfrac{1}{4}$ cup of milk, and $\dfrac{1}{8}$ cup of oil. How many cups of liquid ingredients does the recipe call for?

96. A board $\dfrac{7}{10}$ in. thick is glued to a board $\dfrac{3}{5}$ in. thick. The glue is $\dfrac{3}{100}$ in. thick. How thick is the result?

Synthesis

▦ Find decimal notation.

97. $\dfrac{1}{7}$

98. $\dfrac{2}{7}$

99. $\dfrac{3}{7}$

100. $\dfrac{4}{7}$

101. $\dfrac{5}{7}$

102. ▦ From the pattern of Exercises 97–101, guess the decimal notation for $\dfrac{6}{7}$. Check on your calculator.

▦ Find decimal notation.

103. $\dfrac{1}{9}$

104. $\dfrac{1}{99}$

105. $\dfrac{1}{999}$

106. ▦ From the pattern of Exercises 103–105, guess the decimal notation for $\dfrac{1}{9999}$. Check on your calculator.

4.6

Estimating

OBJECTIVE

a Estimate sums, differences, products, and quotients.

SKILL TO REVIEW

Objective 1.6b: Estimate sums, differences, products, and quotients by rounding.

Estimate by first rounding to the nearest ten.

1. 4 6 7
    − 2 8 4

2.   5 4
    × 2 9

---

1. Estimate by rounding to the nearest ten the total cost of one grill and one camera. Which of the following is an appropriate estimate?
   a) $43   b) $400
   c) $410   d) $430

2. About how much more does the TV cost than the grill? Estimate by rounding to the nearest ten. Which of the following is an appropriate estimate?
   a) $100   b) $150
   c) $160   d) $300

---

a Estimating Sums, Differences, Products, and Quotients

Estimating has many uses. It can be done before a problem is even attempted in order to get an idea of the answer. It can be done afterward as a check, even when we are using a calculator. In many situations, an estimate is all we need. We usually estimate by rounding the numbers so that there are one or two nonzero digits, depending on how accurate we want our estimate. Consider the following prices for Examples 1–3.

EXAMPLE 1 Estimate by rounding to the nearest ten the total cost of one grill and one TV.

We are estimating the sum

$289.95 + $449.99 = Total cost.

The estimate found by rounding the addends to the nearest ten is

$290 + $450 = $740.   (Estimated total cost)

Do Margin Exercise 1.

EXAMPLE 2 About how much more does the TV cost than the camera? Estimate by rounding to the nearest ten.

We are estimating the difference

$449.99 − $139.97 = Price difference.

The estimate, found by rounding the minuend and the subtrahend to the nearest ten, is

$450 − $140 = $310.   (Estimated price difference)

Do Exercise 2.

Answers

Skill to Review:
1. 190   **2.** 1500

Margin Exercises:
1. (d)   **2.** (c)

EXAMPLE 3 Estimate the total cost of 4 cameras.

We are estimating the product

$4 \times \$139.97 = $ Total cost.

The estimate is found by rounding 139.97 to the nearest ten:

$4 \times \$140 = \$560.$

Do Exercise 3.

EXAMPLE 4 About how many Blu-ray discs at $29.99 each can be purchased for $154?

We estimate the quotient

$\$154 \div \$29.99.$

Since we want a whole-number estimate, we choose our rounding appropriately. Rounding $29.99 to the nearest one, we get $30. Since $154 is close to $150, which is a multiple of 30, we estimate

$\$150 \div \$30,$

so the answer is 5.

Do Exercise 4.

EXAMPLE 5 Estimate: 4.8×52. Do not find the actual product. Which of the following is an appropriate estimate?

a) 25   **b)** 250   **c)** 2500   **d)** 360

We round 4.8 to the nearest one and 52 to the nearest ten:

$5 \times 50 = 250.$ (Estimated product)

Thus an approximate estimate is (b).

Other estimates that we might have used in Example 5 are

$5 \times 52 = 260$ or $4.8 \times 50 = 240.$

The estimate in Example 5, $5 \times 50 = 250$, is the easiest to do because the factors have the fewest nonzero digits. You could probably do it mentally. In general, we try to round so that a computation has as few nonzero digits as possible while still keeping the estimated value close to the original value.

Do Exercises 5–10.

3. Estimate the total cost of 6 TVs. Which of the following is an appropriate estimate?
a) $450   b) $2700
c) $4500   d) $27,000

4. About how many Blu-ray discs can be purchased for $485? Choose an appropriate estimate from the following.
a) 16   b) 21
c) 25   d) 30

Estimate each product. Do not find the actual product. Which of the following is an appropriate estimate?

5. 2.4×8
a) 16   b) 34
c) 125   d) 5

6. 24×0.6
a) 200   b) 5
c) 110   d) 20

7. 0.86×0.432
a) 0.04   b) 0.4
c) 1.1   d) 4

8. 0.82×0.1
a) 800   b) 8
c) 0.08   d) 80

9. 0.12×18.248
a) 180   b) 1.8
c) 0.018   d) 18

10. 24.234×5.2
a) 200   b) 120
c) 12.5   d) 234

Answers
3. (b)   **4.** (a)   **5.** (a)   **6.** (d)   **7.** (b)
8. (c)   **9.** (b)   **10.** (b)

EXAMPLE 6 Estimate: $82.08 \div 24$. Which of the following is an appropriate estimate?

a) 400   **b)** 16   **c)** 40   **d)** 4

This is about $80 \div 20$, so the answer is about 4. Thus an appropriate estimate is (d).

EXAMPLE 7 Estimate: $94.18 \div 3.2$. Which of the following is an appropriate estimate?

a) 30   **b)** 300   **c)** 3   **d)** 60

This is about $90 \div 3$, so the answer is about 30. Thus an appropriate estimate is (a).

EXAMPLE 8 Estimate: $0.0156 \div 1.3$. Which of the following is an appropriate estimate?

a) 0.2   **b)** 0.002   **c)** 0.02   **d)** 20

This is about $0.02 \div 1$, so the answer is about 0.02. Thus an appropriate estimate is (c).

> Do Exercises 11-13.

In some cases, it is easier to estimate a quotient directly rather than by rounding the divisor and the dividend.

EXAMPLE 9 Estimate: $0.0074 \div 0.23$. Which of the following is an appropriate estimate?

a) 0.3   **b)** 0.03   **c)** 300   **d)** 3

We estimate 3 for a quotient. We check by multiplying.

$$0.23 \times 3 = 0.69$$

We make the estimate smaller. We estimate 0.3 and check by multiplying.

$$0.23 \times 0.3 = 0.069$$

We make the estimate smaller. We estimate 0.03 and check by multiplying.

$$0.23 \times 0.03 = 0.0069$$

This is about 0.0074, so the quotient is about 0.03. Thus an appropriate estimate is (b).

> Do Exercise 14.

Estimate each quotient. Which of the following is an appropriate estimate?

11. $59.78 \div 29.1$

  a) 200   **b)** 20

  c) 2   **d)** 0.2

12. $82.08 \div 2.4$

  a) 40   **b)** 4.0

  c) 400   **d)** 0.4

13. $0.1768 \div 0.08$

  a) 8   **b)** 10

  c) 2   **d)** 20

14. Estimate: $0.0069 \div 0.15$. Which of the following is an appropriate estimate?

  a) 0.5   **b)** 50

  c) 0.05   **d)** 0.004

Answers

11. (c)   **12.** (a)   **13.** (c)   **14.** (c)

4.6 **Exercise Set**

For Extra Help

MyMathLab

Math XL
PRACTICE

WATCH

DOWNLOAD

READ

REVIEW

a Consider the following prices for Exercises 1–8. Estimate the sums, differences, products, or quotients involved in these problems. Indicate which of the choices is an appropriate estimate.

2 Quart Ice Cream Maker $79.⁹⁹

Satellite Radio Receiver $149.⁹⁹

Black & White Multifunction Printer/Copier/Scanner $279.⁸⁹

1. Estimate the total cost of one printer and one satellite radio.

   a) $43   **b)** $4300   **c)** $360   **d)** $430

2. Estimate the total cost of one satellite radio and one ice cream maker.

   a) $230   **b)** $23   **c)** $2300   **d)** $400

3. About how much more does the printer cost than the satellite radio?

   a) $1300   **b)** $200   **c)** $130   **d)** $13

4. About how much more does the satellite radio cost than the ice cream maker?

   a) $7000   **b)** $70   **c)** $130   **d)** $700

5. Estimate the total cost of 6 ice cream makers.

   a) $480   **b)** $48   **c)** $240   **d)** $4800

6. Estimate the total cost of 4 printers.

   a) $1200   **b)** $1120   **c)** $11,200   **d)** $600

7. About how many ice cream makers can be purchased for $830?

   a) 120   **b)** 100   **c)** 10   **d)** 1000

8. About how many printers can be purchased for $5627?

   a) 200   **b)** 20   **c)** 1800   **d)** 2000

Estimate by rounding as directed.

9. 0.02 + 1.31 + 0.34;
   nearest tenth

10. 0.88 + 2.07 + 1.54;
   nearest one

11. 6.03 + 0.007 + 0.214;
   nearest one

12. 1.11 + 8.888 + 99.94;
   nearest one

13. 52.367 + 1.307 + 7.324;
   nearest one

14. 12.9882 + 1.2115;
   nearest tenth

15. $2.678 - 0.445$;
nearest tenth

16. $12.9882 - 1.0115$;
nearest one

17. $198.67432 - 24.5007$;
nearest ten

Estimate. Choose a rounding digit that gives one or two nonzero digits. Indicate which of the choices is an appropriate estimate.

18. $234.12321 - 200.3223$

   a) 600   **b)** 60
   c) 300   **d)** 30

19. 49×7.89

   a) 400   **b)** 40
   c) 4   **d)** 0.4

20. 7.4×8.9

   a) 95   **b)** 63
   c) 124   **d)** 6

21. 98.4×0.083

   a) 80   **b)** 12
   c) 8   **d)** 0.8

22. 78×5.3

   a) 400   **b)** 800
   c) 40   **d)** 8

23. $3.6 \div 4$

   a) 10   **b)** 1
   c) 0.1   **d)** 0.01

24. $0.0713 \div 1.94$

   a) 3.5   **b)** 0.35
   c) 0.035   **d)** 35

25. $74.68 \div 24.7$

   a) 9   **b)** 3
   c) 12   **d)** 120

26. $914 \div 0.921$

   a) 10   **b)** 100
   c) 1000   **d)** 1

27. *Fence Posts.* A zoo plans to construct a fence around its proposed animals of the Great Plains exhibit. The perimeter of the area to be fenced is 1760 ft. Estimate the number of wooden fence posts needed if the posts are placed 8.625 ft apart.

28. *Ticketmaster.* Recently, Ticketmaster stock sold for $22.25 per share. Estimate how many shares can be purchased for $4400.

29. *Day-Care Supplies.* Helen wants to buy 12 boxes of crayons at $1.89 per box for the day care center that she runs. Estimate the total cost of the crayons.

30. *Batteries.* Oscar buys 6 packages of AAA batteries at $5.29 per package. Estimate the total cost of the purchase.

Skill Maintenance

In each of Exercises 31–38, fill in the blank with the correct term from the given list. Some of the choices may not be used and some may be used more than once.

31. The decimal $0.57\overline{3}$ is an example of a(n) _____ decimal. [4.5a]

32. The least common _____ of two natural numbers is the smallest number that is a multiple of both. [3.1a]

33. The sentence $5(3 + 8) = 5 \cdot 3 + 5 \cdot 8$ illustrates the _____ law. [1.4a]

34. A _____ of an equation is a replacement for the variable that makes the equation true. [1.7a]

35. The number 1 is the _____ identity. [1.4a]

36. The sentence $13 + 7 = 7 + 13$ illustrates the _____ law of addition. [1.2a]

37. The least common _____ of two or more fractions is the least common _____ of their denominators. [3.2a]

38. The number 3728 is _____ by 4 if the number named by the last two digits is _____ by 4. [2.2a]

additive

multiplicative

numerator

denominator

commutative

associative

distributive

solution

divisible

terminating

repeating

multiple

factor

Synthesis

The following were done on a calculator. Estimate to determine whether the decimal point was placed correctly.

39. $178.9462 \times 61.78 = 11,055.29624$

40. $14,973.35 \div 298.75 = 501.2$

41. $19.7236 - 1.4738 \times 4.1097 = 1.366672414$

42. $28.46901 \div 4.9187 - 2.5081 = 3.279813473$

43. ▦ Use one of $+, -, \times,$ and $\div$ in each blank to make a true sentence.

   a) $(0.37 \,\square\, 18.78) \,\square\, 2^{13} = 156{,}876.8$

   b) $2.56 \,\square\, 6.4 \,\square\, 51.2 \,\square\, 17.4 = 312.84$

4.7

Applications and Problem Solving

OBJECTIVE

a Solve applied problems involving decimals.

a Solving Applied Problems

Solving applied problems with decimals is like solving applied problems with whole numbers. We translate first to an equation that corresponds to the situation. Then we solve the equation.

EXAMPLE 1 *Canals.* The Panama Canal in Panama is 50.7 mi long. The Suez Canal in Egypt is 119.9 mi long. How much longer is the Suez Canal?

Panama Canal   Suez Canal

1. **Debit-Card Transactions.**
Debit-card transactions in 2000 totaled 9.8 billion. The number of transactions projected to be made in 2010 is 42.5 billion. How many more debit-card transactions are expected in 2010 than in 2000?

Source: *The Nilson Report*

1. **Familiarize.** We let l = the distance in miles that the length of the longer canal differs from the length of the shorter canal.

2. **Translate.** We translate as follows, using the given information:

Length of Panama Canal, the shorter canal	plus	Additional length	is	Length of Suez Canal, the longer canal
50.7 mi	+	l	=	119.9 mi.

3. **Solve.** We solve the equation by subtracting 50.7 mi on both sides:

$$50.7 + l = 119.9$$
$$50.7 + l - 50.7 = 119.9 - 50.7$$
$$l = 69.2.$$

4. **Check.** We can check by adding.

$$\begin{array}{r} 5\,0.7 \\ +\ 6\,9.2 \\ \hline 1\,1\,9.9 \end{array}$$

The answer checks.

5. **State.** The Suez Canal is 69.2 mi longer than the Panama Canal.

Do Exercise 1.

Answer

1. 32.7 billion transactions

EXAMPLE 2 *Meat and Seafood Consumption.* In a recent year, the average American consumed about 62.8 lb of beef, 47.3 lb of pork and lamb, 73.6 lb of poultry, and 16.1 lb of seafood. Find the total amount of meat and seafood consumed by the average American per year.

Meat and Seafood Consumption

SOURCE: U.S. Department of Agriculture

1. **Familiarize.** The figure above helps us visualize the situation. We let t = the total amount of meat and seafood consumed by the average American per year.

2. **Translate.** We are combining amounts. We translate to an equation that does this.

Beef	plus	Pork and lamb	plus	Poultry	plus	Seafood	is	Total
62.8	+	47.3	+	73.6	+	16.1	=	t

3. **Solve.** To solve, we carry out the addition.

$$
\begin{array}{r}
{\scriptstyle 1\ 1} \\
6\ 2.8 \\
4\ 7.3 \\
7\ 3.6 \\
+\ \ 1\ 6.1 \\
\hline
1\ 9\ 9.8
\end{array}
$$

Thus, $t = 199.8$.

4. **Check.** We can check by repeating our addition. We can also see whether our answer is reasonable by first noting that it is indeed larger than any of the numbers being added. We can also do a check by rounding and estimating:

$$62.8 + 47.3 + 73.6 + 16.1 \approx 60 + 50 + 70 + 20$$
$$\approx 200 \approx 199.8.$$

If we had gotten an answer like 19.98 or 1998, then the estimate, 200, would have told us that we had made a mistake, such as not lining up the decimal points.

5. **State.** The average American consumes about 199.8 lb of meat and seafood per year.

Do Exercise 2.

2. **Amount of Medication.** Over a 24-hr period, a patient received injections of 2.8 mL, 1.35 mL, 2.0 mL, and 1.88 mL of a medication. What was the total amount of medication received?

2.8 mL   1.35 mL   2.0 mL   1.88 mL

Answer

2. 8.03 mL

EXAMPLE 3 *IRS Driving Allowance.* The Internal Revenue Service allowed a tax deduction of 58.5¢ per mile driven for business purposes in the last six months of 2008. What deduction, in dollars, would be allowed for driving 9143 mi during this time period?

Source: Internal Revenue Service

1. **Familiarize.** We first make a drawing or at least visualize the situation. Repeated addition fits this situation. We let d = the deduction, in dollars, allowed for driving 9143 mi.

9143 mi

2. **Translate.** We translate as follows, writing 58.5¢ as $0.585:

Deduction for each mile	times	Number of miles driven	is	Total deduction
$0.585	×	9143	=	d.

3. **Solve.** To solve the equation, we carry out the multiplication.

$$
\begin{array}{r}
9\,1\,4\,3 \\
\times\ 0.5\,8\,5 \\
\hline
4\,5\,7\,1\,5 \\
7\,3\,1\,4\,4\,0 \\
4\,5\,7\,1\,5\,0\,0 \\
\hline
5\,3\,4\,8.6\,5\,5
\end{array}
$$

Thus, $d = 5348.655 \approx 5348.66$.

4. **Check.** We can obtain a partial check by rounding and estimating:

$$9143 \times 0.585 \approx 9000 \times 0.6 \approx 5400 \approx 5348.66.$$

5. **State.** The total allowable deduction would be $5348.66.

Do Exercise 3.

3. Printing Costs. At Kwik Copy, the cost of copying is 11 cents per page. How much, in dollars, would it cost to make 466 copies?

EXAMPLE 4 *Loan Payments.* A car loan of $7382.52 is to be paid off in 36 equal monthly payments. How much is each payment?

1. **Familiarize.** We first make a drawing. We let n = the amount of each payment.

There may be some fractional part of $1.

$7382.52

36 payments. How much in each payment?

n = the amount of payment in each row

Answer

3. $51.26

2. Translate.   The problem can be translated to the following equation, thinking that

(Total loan) ÷ (Number of payments) = Amount of each payment

$$\$7382.52 \div 36 = n.$$

3. Solve.   To solve the equation, we carry out the division.

```
        2 0 5.0 7
3 6 ) 7 3 8 2.5 2
      7 2
        1 8 2
        1 8 0
            2 5 2
            2 5 2
                0
```

Thus, $n = 205.07$.

4. Check.   A partial check can be obtained by estimating the quotient: $\$7382.56 \div 36 \approx 8000 \div 40 \approx 200 \approx 205.07$. The estimate checks.

5. State.   Each payment is $205.07.

Do Exercise 4.

4. Loan Payments.   A loan of $4425 is to be paid off in 12 equal monthly payments. How much is each payment?

EXAMPLE 5 *Jackie Robinson Poster.*   A limited-edition poster by sports artist Leroy Neiman commemorates Jackie Robinson's entrance into Major League Baseball in 1947. Robinson was the first African-American player in the major leagues. The dimensions of the poster are 19.3 in. by 27.4 in. Find the area.

Source: Barton L. Kaufman, private collection

1. Familiarize.   We first make a drawing. We let $A =$ the area.

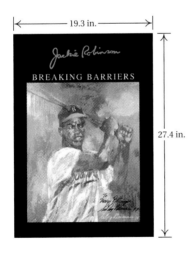

2. Translate.   We use the formula $A = l \cdot w$ and substitute:

$$A = l \cdot w$$
$$A = 27.4 \times 19.3.$$

Answer

4. $368.75

3. Solve. We solve by carrying out the multiplication.

$$
\begin{array}{r}
2\ 7.4 \\
\times\ 1\ 9.3 \\
\hline
8\ 2\ 2 \\
2\ 4\ 6\ 6\ 0 \\
2\ 7\ 4\ 0\ 0 \\
\hline
5\ 2\ 8.8\ 2
\end{array}
$$

Thus, $A = 528.82$.

4. Check. We obtain a partial check by estimating the product:

$$A = 27.4 \times 19.3 \approx 25 \times 20 \approx 500.$$

This estimate is close to 528.82, so the answer is probably correct.

5. State. The area of the Jackie Robinson poster is 528.82 in^2.

Do Exercise 5.

EXAMPLE 6 *Digital Camera Purchase.* Dawson Real Estate spent $10,399.74 on 26 Nikon Coolpix 10.1 megapixel digital cameras, so that its agents could place photos of properties on the firm's Web site. How much did each camera cost?

1. Familiarize. We let $c =$ the cost of each camera.

2. Translate. We translate as follows:

Cost of each camera	is	Total cost of purchase	divided by	Number of cameras purchased
↓	↓	↓	↓	↓
c	$=$	$10,399.74	÷	26.

3. Solve. To solve, we carry out the division.

$$
\begin{array}{r}
3\ 9\ 9.9\ 9 \\
2\ 6\ )\overline{1\ 0,3\ 9\ 9.7\ 4} \\
\underline{7\ 8} \\
2\ 5\ 9 \\
\underline{2\ 3\ 4} \\
2\ 5\ 9 \\
\underline{2\ 3\ 4} \\
2\ 5\ 7 \\
\underline{2\ 3\ 4} \\
2\ 3\ 4 \\
\underline{2\ 3\ 4} \\
0
\end{array}
$$

4. Check. We check by estimating

$$10,399.74 \div 26 \approx 10,000 \div 25 = 400.$$

Since 400 is close to 399.99, the answer is probably correct.

5. State. The cost of each camera was $399.99.

Do Exercise 6.

---

5. Index Cards. A standard-size index card measures 12.7 cm by 7.6 cm. What is its area?

7.6 cm

12.7 cm

6. One pound of lean boneless ham contains 4.5 servings. It costs $7.99 per pound. What is the cost per serving? Round to the nearest cent.

Answers

5. 96.52 cm^2 **6.** $1.78

Multistep Problems

EXAMPLE 7 *Gas Mileage.* Ava filled her gas tank and noted that the odometer read 67,507.8. After the next filling, the odometer read 68,006.1. It took 16.5 gal to fill the tank. How many miles per gallon did Ava get?

1. **Familiarize.** We first make a drawing.

n miles, 16.5 gallons

This is a two-step problem. First, we find the number of miles that have been driven between fillups. We let n = the number of miles driven.

2., 3. **Translate** and **Solve.** We translate and solve as follows:

First odometer reading	plus	Number of miles driven	is	Second odometer reading
67,507.8	+	n	=	68,006.1.

To solve the equation, we subtract 67,507.8 on both sides:

$$n = 68,006.1 - 67,507.8$$
$$= 498.3.$$

$$\begin{array}{r} 6\,8,0\,0\,6.1 \\ -\ 6\,7,5\,0\,7.8 \\ \hline 4\,9\,8.3 \end{array}$$

Next, we divide the total number of miles driven by the number of gallons. This gives us m = the number of miles per gallon—that is, the mileage. The division that corresponds to the situation is

$$498.3 \div 16.5 = m.$$

To find the number m, we divide.

$$\begin{array}{r} 3\,0.2 \\ 1\,6.5\,)\overline{4\,9\,8.3{\scriptstyle\wedge}0} \\ 4\,9\,5 \\ \hline 3\,3\,0 \\ 3\,3\,0 \\ \hline 0 \end{array}$$

Thus, $m = 30.2$.

4. **Check.** To check, we first multiply the number of miles per gallon times the number of gallons to find the number of miles driven:

$$16.5 \times 30.2 = 498.3.$$

Then we add 498.3 to 67,507.8 to find the new odometer reading:

$$67,507.8 + 498.3 = 68,006.1.$$

The mileage 30.2 checks.

5. **State.** Ava got 30.2 miles per gallon.

Do Exercise 7.

7. Gas Mileage. John filled his gas tank and noted that the odometer read 38,320.8. After the next filling, the odometer read 38,735.5. It took 14.5 gal to fill the tank. How many miles per gallon did John get?

Answer

7. 28.6 mpg

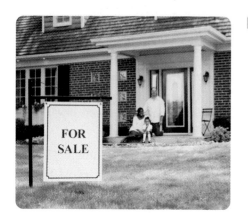

EXAMPLE 8 *Home-Cost Comparison.* Suppose you own a home like the one shown here and it is valued at $225,000 in Tulsa, Oklahoma. What would it cost to buy a similar home in Boston, Massachusetts? To find out, we can use an index table prepared by Coldwell Banker Real Estate Corporation. (For a complete index table, contact your local representative.) We use the following formula:

$$\left(\begin{array}{c}\text{Cost of your}\\\text{home in new city}\end{array}\right) = \dfrac{\left(\begin{array}{c}\text{Value of}\\\text{your home}\end{array}\right) \times \left(\begin{array}{c}\text{Index of}\\\text{new city}\end{array}\right)}{\left(\begin{array}{c}\text{Index of}\\\text{your city}\end{array}\right)}.$$

Find the cost of your Tulsa home in Boston.

STATE	CITY	INDEX	STATE	CITY	INDEX
Massachusetts	Boston	327	Texas	Dallas	72
Illinois	Chicago	173	Oklahoma	Tulsa	36
California	Palo Alto	397	Florida	Miami	151
	Sacramento	90		Orlando	96
	San Diego	147		Key West	220
Alaska	Juneau	110	Colorado	Boulder	146
Kentucky	Louisville	56	North Carolina	Charlotte	61

SOURCE: Coldwell Banker Real Estate Corporation

1. **Familiarize.** We let C = the cost of the home in Boston. We use the table and look up the indexes of Tulsa and the new city, Boston. We see that Tulsa's index is 36 and Boston's index is 327.

2. **Translate.** Using the formula, we translate to the following equation:

$$C = \frac{\$225,000 \times 327}{36}.$$

3. **Solve.** To solve, we carry out the computations using the rules for order of operations. (See Section 4.4.)

$$C = \frac{\$225,000 \times 327}{36}$$

$$= \frac{\$73,575,000}{36} \qquad \text{Carrying out the multiplication first}$$

$$= \$2,043,750 \qquad \text{Carrying out the division}$$

On a calculator, the computation could be done in one step.

4. **Check.** We can repeat our computations.

5. **State.** A home that sells for $225,000 in Tulsa would cost about $2,043,750 in Boston.

Do Exercises 8 and 9.

Refer to the table in Example 8 to answer Exercises 8 and 9. Round to the nearest dollar.

8. Home-Cost Comparison. Suppose your home in Charlotte is valued at $180,000. What would it cost to buy a similar home in Chicago?

9. Home-Cost Comparison. Suppose your home in Boulder is valued at $315,000. What would it cost to buy a similar home in Louisville?

Answers

8. $510,492   **9.** $120,822

Translating
for Success

1. *Gas Mileage.* Art filled his SUV's gas tank and noted that the odometer read 38,271.8. At the next filling, the odometer read 38,677.92. It took 28.4 gal to fill the tank. How many miles per gallon did the SUV get?

2. *Dimensions of a Parking Lot.* Seals' parking lot is a rectangle that measures 85.2 ft by 52.3 ft. What is the area of the parking lot?

3. *Game Snacks.* Three students pay $18.40 for snacks at a football game. What is each person's share?

4. *Electrical Wiring.* An electrician needs 1314 ft of wiring cut into $2\frac{1}{2}$-ft pieces. How many pieces will she have?

5. *College Tuition.* Wayne needs $4638 for the fall semester's tuition. On the day of registration, he has only $3092. How much does he need to borrow?

The goal of these matching questions is to practice step (2), *Translate,* of the five-step problem-solving process. Translate each word problem to an equation and select a correct translation from equations A–O.

A. $2\frac{1}{2} \cdot n = 1314$

B. $18.4 \times 3.87 = n$

C. $n = 85.2 \times 52.3$

D. $1314.28 - 437 = n$

E. $3 \times 18.40 = n$

F. $2\frac{1}{2} \cdot 1314 = n$

G. $3092 + n = 4638$

H. $18.4 \cdot n = 3.87$

I. $\dfrac{406.12}{28.4} = n$

J. $52.3 \cdot n = 85.2$

K. $n = 1314.28 + 437$

L. $52.3 + n = 85.2$

M. $3092 + 4638 = n$

N. $3 \cdot n = 18.40$

O. $85.2 + 52.3 = n$

Answers on page A-7

6. *Cost of Gasoline.* What is the cost, in dollars, of 18.4 gal of gasoline at $3.87 per gallon?

7. *Savings Account Balance.* Margaret has $1314.28 in her savings account. Before using her debit card to buy an office chair, she transferred $437 to her checking account. How much was left in her savings account?

8. *Acres Planted.* This season Sam planted 85.2 acres of corn and 52.3 acres of soybeans. Find the total number of acres that he planted.

9. *Amount Inherited.* Tara inherited $2\frac{1}{2}$ times as much as her cousin. Her cousin received $1314. How much did Tara receive?

10. *Travel Funds.* The athletic department needs travel funds of $4638 for the tennis team and $3092 for the golf team. What is the total amount needed for travel?

a Solve.

Hurricane Damage. The amount of damage caused by the five most costly Atlantic hurricanes in the United States is shown in the table below. Use this table to do Exercises 1 and 2.

Most Costly Hurricanes

RANK	HURRICANE	YEAR	COST IN 2007 DOLLARS (in billions)
1	Katrina	2005	$81.2
2	Andrew	1992	38.1
3	Wilma	2005	30.4
4	Ivan	2004	18.1
5	Charley	2004	16.2

SOURCE: National Hurricane Center

1. How much more costly was Hurricane Katrina than Hurricane Andrew?

2. What was the total cost of the two hurricanes that occurred in 2005?

Top American Sky Routes. The figure below shows the numbers of airline passengers carried on the top five sky routes in the United States from December 2006 to November 2007. Note that these routes go in both directions. Use this figure to do Exercises 3 and 4.

Top American Sky Routes

New York City–Chicago 3.47
Washington–Chicago 2.82
Atlanta–Orlando 2.77
Los Angeles–Chicago 2.71
Atlanta–New York City 2.69

Number of passengers (in millions)

SOURCE: U.S. Bureau of Transportation Statistics

3. What is the total number of passengers carried on the top two routes?

4. How many more passengers were carried on the New York City–Chicago route than on the Los Angeles–Chicago route?

5. *Record Movie Openings.* The movie *The Dark Knight* took in $155.34 million on its first weekend. This topped the previous high opening weekend revenue set by *Spider-Man 3* by $4.24 million. How much did *Spider-Man 3* take in on its opening weekend?

Source: Associated Press

6. *Counterfeit Money.* The amount of counterfeit money passed in the United States is on the rise. About $62.0 million was passed in 2006. This amount was $22.8 million more than entered circulation in 1999. How much counterfeit money was passed in 1999?

Source: U.S. Secret Service

7. *Cost of Bottled Water.*   The cost of a year's supply of a popular brand of bottled water, based on the recommended consumption of 64 oz per day, at the supermarket price of $3.99 for a six-pack of half-liter bottles is $918.82. This is $918.31 more than the cost of drinking the same amount of tap water for a year. What is the cost of drinking tap water for a year?

Source: American Water Works Association

8. *Bottled Water Consumption.*   The annual consumption of bottled water in the United States was 27.6 gal per person in 2006. This was an increase of 26 gal per person over the amount consumed in 1976. What was the annual consumption of bottled water in 1976?

Source: Beverage Marketing Corporation

9. *Body Temperature.*   Normal body temperature is 98.6°F. During an illness, a patient's temperature rose 4.2°. What was the new temperature?

10. *Gasoline Cost.*   What is the cost, in dollars, of 12.6 gal of gasoline at $3.79 per gallon? Round the answer to the nearest cent.

11. *Lottery Winnings.*   The largest lotto jackpot ever won in California totaled $193,000,000 and was shared equally by 3 winners. How much was each winner's share? Round to the nearest cent.

Source: California State Lottery

12. *Lunch Costs.*   A group of 4 students pays $47.84 for lunch and splits the cost equally. What is each person's share?

13. *Stamp.*   Find the area and the perimeter of the stamp shown here.

2.5 cm

3.25 cm

14. *Pole Vault Pit.*   Find the area and the perimeter of the landing area of the pole vault pit shown here.

16.4 ft

16.4 ft

Landing area

15. *Odometer Reading.*   The Binford family's odometer reads 22,456.8 at the beginning of a trip. The family's online driving directions tell them that they will be driving 234.7 mi. What will the odometer read at the end of the trip?

16. *Miles Driven.*   Petra bought gasoline when the odometer read 14,296.3. At the next gasoline purchase, the odometer read 14,515.8. How many miles had been driven?

17. *Gas Mileage.*   Peggy filled her van's gas tank and noted that the odometer read 26,342.8. After the next filling, the odometer read 26,736.7. It took 19.5 gal to fill the tank. How many miles per gallon did the van get?

18. *Gas Mileage.*   Henry filled his Honda's gas tank and noted that the odometer read 18,943.2. After the next filling, the odometer read 19,306.2. It took 13.2 gal to fill the tank. How many miles per gallon did the car get?

19. *Chemistry.* The water in a filled tank weighs 748.45 lb. One cubic foot of water weighs 62.5 lb. How many cubic feet of water does the tank hold?

20. *Highway Routes.* You can drive from home to work using either of two routes:

> *Route A*: Via interstate highway, 7.6 mi, with a speed limit of 65 mph.

> *Route B*: Via a country road, 5.6 mi, with a speed limit of 50 mph.

Assuming you drive at the posted speed limit, which route takes less time? (Use the formula *Distance = Speed × Time*.)

Find the perimeter of each figure.

21.

8.9 cm   23.8 cm   4.7 cm   18.6 cm   22.1 cm

22.

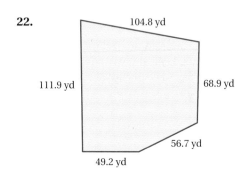

104.8 yd   111.9 yd   68.9 yd   56.7 yd   49.2 yd

23.

2.5 cm

2.25 cm

24.

2.5 cm   4.0 cm

25. Andrew bought a DVD of the movie *Horton Hears a Who* for his nephew for $23.99 plus $1.68 sales tax. He paid for it with a $50 bill. How much change did he receive?

26. Claire bought a copy of the book *Make Way for Ducklings* for her daughter for $16.95 plus $0.85 sales tax. She paid for it with a $20 bill. How much change did she receive?

Find the length *d* in each figure.

27.

0.8 cm — — 0.8 cm
d
← 3.91 cm →

28.

0.9 cm — — 0.9 cm
d
← 4.52 cm →

29. *Calories Burned Mowing.* A person weighing 150 lb burns 7.3 calories per minute while mowing a lawn with a power lawnmower. How many calories would be burned in 2 hr of mowing?
Source: *The Handy Science Answer Book*

30. Lot A measures 250.1 ft by 302.7 ft. Lot B measures 389.4 ft by 566.2 ft. What is the total area of the two lots?

31. Holly had $1123.56 in her checking account. She used her debit card to make purchases of $23.82, $507.88, and $98.32. She then deposited a bonus check of $678.20. How much was in her account after these changes?

32. Natalie had $185.00 to spend for fall clothes: $44.95 was spent on shoes, $71.95 for a jacket, and $55.35 for pants. How much was left?

33. A rectangular yard is 20 ft by 15 ft. The yard is covered with grass except for an 8.5-ft-square flower garden. How much grass is in the yard?

34. Rita earns a gross paycheck (before deductions) of $495.72. Her deductions are $59.60 for federal income tax, $29.00 for FICA, and $29.00 for medical insurance. What is her take-home paycheck?

35. *Batting Average.* Chipper Jones of the Atlanta Braves won the 2008 National League batting title with 160 hits in 439 times at bat. What part of his at-bats were hits? Give decimal notation rounded to the nearest thousandth. (This is a *batting average*.)
Source: Major League Baseball

36. *Batting Average.* Joe Mauer of the Minnesota Twins won the 2008 American League batting title with 176 hits in 536 times at bat. What part of his at-bats were hits? Give decimal notation rounded to the nearest thousandth.
Source: Major League Baseball

37. *Field Dimensions.* The dimensions of a World Cup soccer field are 114.9 yd by 74.4 yd. The dimensions of a standard football field are 120 yd by 53.3 yd. How much greater is the area of a World Cup soccer field?

38. *Loan Payment.* In order to make money on loans, financial institutions are paid back more money than they loan. Suppose you borrow $120,000 to buy a house and agree to make monthly payments of $880.52 for 30 yr. How much do you pay back altogether? How much more do you pay back than the amount of the loan?

World Cup Soccer Field

Football Field

114.9 yd

74.4 yd

120 yd

53.3 yd

39. *Egg Costs.*   A restaurant owner bought 20 dozen eggs for $25.80. Find the cost of each egg to the nearest tenth of a cent (thousandth of a dollar).

40. *Weight Loss.*   A person weighing 170 lb burns 8.6 calories per minute while mowing a lawn. One must burn about 3500 calories in order to lose 1 lb. How many pounds would be lost by mowing for 2 hr? Round to the nearest tenth.

41. *Construction Pay.*   A construction worker is paid $18.50 per hour for the first 40 hr of work, and time and a half, or $27.75 per hour, for any overtime exceeding 40 hr per week. One week she works 46 hr. How much is her pay?

42. *Summer Work.*   Zachary worked 53 hr during a week one summer. He earned $7.50 per hour for the first 40 hr and $11.25 per hour for overtime (hours exceeding 40). How much did Zachary earn during the week?

43. *Projected World Population.*   Using the information in the bar graph below, determine the average population of the world for the years 1950 through 2050. Round to the nearest thousandth of a billion.

World Population (historical and projected)

World population (in billions)

2.556 3.042 3.712 4.453 5.282 6.085 6.867 7.659 8.373 9.003 9.539

1950 1960 1970 1980 1990 2000 2010* 2020* 2030* 2040* 2050*

Year

*Projected

SOURCE: U.S. Census Bureau

44. *Number of Volunteers.*   The table at right lists the number of Americans in various age groups who do volunteer work. What is the average number of volunteers in all the age groups shown?

AGE	NUMBER OF VOLUNTEERS (in millions)
16–24yr	8.044
25–34 yr	9.096
35–44 yr	13.308
45–54 yr	13.415
55–64 yr	8.819
65 yr and older	8.518

SOURCE: U.S. Census Bureau

45. *Body Temperature.*   Normal body temperature is 98.6°F. A baby's bath water should be 100°F. How many degrees above normal body temperature is this?

46. *Body Temperature.*   Normal body temperature is 98.6°F. The lowest temperature at which a patient has survived is 69°F. How many degrees below normal is this?

Home-Cost Comparison.   Use the table and the formula in Example 8 to do Exercises 47–54. In each of the following cases, find the value of the house in the new location. Round to the nearest dollar.

Source: Coldwell Banker Real Estate Corporation

	VALUE	PRESENT LOCATION	NEW LOCATION	NEW VALUE
47.	$125,000	Dallas	Miami	
48.	$180,000	San Diego	Key West	
49.	$96,000	Juneau	Orlando	
50.	$300,000	Palo Alto	Dallas	
51.	$240,000	Louisville	Boston	
52.	$160,000	Charlotte	Sacramento	
53.	$140,000	Chicago	Tulsa	
54.	$99,000	Boulder	Orlando	

55. *Cell-Phone Plan.*   In 2008, at&t offered an individual cell-phone plan with 450 anytime minutes for a monthly access fee of $39.99. Minutes in excess of 450 were charged at the rate of $0.45 per minute. In June, Leila used her cell phone for 479 min. What was she charged?

Source: at&t

56. *Cell-Phone Plan.*   In 2008, at&t offered an individual cell-phone plan with 900 anytime minutes per month for a monthly access fee of $59.99. Minutes in excess of 900 were charged at the rate of $0.40 per minute. One month Jeff used his cell phone for 946 min. What was the charge?

Source: at&t

IRS Driving Allowance.   The Internal Revenue Service allowed a tax deduction of 50.5¢ per mile driven for business purposes from January 1, 2008, through June 30, 2008. In response to increasing gas prices, this deduction was increased to 58.5¢ per mile driven from July 1, 2008, through December 31, 2008. Use this information for Exercises 57 and 58.

Source: Internal Revenue Service

57. Sheila drove 3156 mi for business purposes between January 1, 2008, and June 30, 2008. She drove an additional 3678 mi between July 1, 2008, and December 31, 2008. What deduction, in dollars, was she allowed to take?

58. Manolo drove 1250 mi for business purposes between January 1, 2008, and June 30, 2008. He drove an additional 1872 mi between July 1, 2008, and December 31, 2008. What deduction, in dollars, was he allowed to take?

59. *Property Taxes.*   The Brunners own a house with an assessed value of $165,500. For every $1000 of assessed value, they pay $8.50 in property taxes each year. How much do they pay in property taxes each year?

60. *Property Taxes.*   The Colavitos own a house with an assessed value of $184,500. For every $1000 of assessed value, they pay $7.68 in property taxes each year. How much do they pay in property taxes each year?

Skill Maintenance

Add.

61. $4569 + 1766$ [1.2a]

62. $\dfrac{2}{3} + \dfrac{5}{8}$ [3.2a]

63. $4\dfrac{1}{3} + 2\dfrac{1}{2}$ [3.5a]

64. $\dfrac{5}{6} + \dfrac{7}{10}$ [3.2a]

Subtract.

65. $\dfrac{2}{3} - \dfrac{5}{8}$ [3.3a]

66. $4569 - 1766$ [1.3a]

67. $\dfrac{5}{6} - \dfrac{7}{10}$ [3.3a]

68. $4\dfrac{1}{3} - 2\dfrac{1}{2}$ [3.5b]

Simplify.   [2.5b]

69. $\dfrac{3225}{6275}$

70. $\dfrac{125}{400}$

71. $\dfrac{325}{625}$

72. $\dfrac{625}{475}$

Solve.

73. If a water wheel made 469 revolutions at a rate of $16\dfrac{3}{4}$ revolutions per minute, for how long did it rotate? [3.6c]

74. If a bicycle wheel made 480 revolutions at a rate of $66\dfrac{2}{3}$ revolutions per minute, for how long did it rotate?

[3.6c]

75. *Calories in Pie.* A piece of pecan pie $\left(\dfrac{1}{8} \text{ of a 9-in. pie}\right)$ has 502 calories. A piece of pumpkin pie has 316 calories. How many more calories does a piece of pecan pie have than a piece of pumpkin pie?   [1.8a]

76. *Calories in Turkey.* Dark meat turkey contains 187 calories per 3.5-oz serving. White meat turkey contains 157 calories per 3.5-oz serving. How many more calories per 3.5-oz serving does the dark turkey meat contain than the white turkey meat?   [1.8a]

Synthesis

77. Suppose you buy a half-dozen packs of basketball cards with a dozen cards in each pack. The cost is twelve dozen cents for each half-dozen cards. How much do you pay for the cards?

Key Terms

arithmetic numbers, p. 216   terminating decimal, p. 256   repeating decimal, p. 256

Concept Reinforcement

Determine whether each statement is true or false.

_____ **1.** One thousand billion is one trillion.   [4.3b]

_____ **2.** The number of decimal places in the product of two numbers is the product of the numbers of places in the factors.   [4.3a]

_____ **3.** When we divide a number by 0.1, 0.01, 0.001, and so on, the quotient is larger than the divisor.   [4.4a]

_____ **4.** For a fraction with a factor other than 2 or 5 in its denominator, decimal notation terminates.   [4.5a]

_____ **5.** An estimate found by rounding to the nearest ten is usually more accurate than one found by rounding to the nearest hundred.   [4.6a]

Important Concepts

Objective 4.1b   Convert between decimal notation and fraction notation.

Example   Write fraction notation for 5.347.

$$5.347 \qquad 5.347. \qquad \frac{5347}{1000}$$

3 decimal places   Move 3 places   3 zeros
                        to the right.

$$5.347 = \frac{5347}{1000}$$

Practice Exercise

1. Write fraction notation for 50.93.

Example   Write decimal notation for $\frac{29}{1000}$.

$$\frac{29}{1000} \qquad 0.029.$$

3 zeros   Move 3 places to the left.

$$\frac{29}{1000} = 0.029$$

Practice Exercise

2. Write decimal notation for $\frac{817}{10}$.

Example   Write decimal notation for $4\frac{63}{100}$.

$$4\frac{63}{100} = 4 + \frac{63}{100} = 4 \text{ and } \frac{63}{100} = 4.63$$

Practice Exercise

3. Write decimal notation for $42\frac{159}{1000}$.

Objective 4.1d Round decimal notation to the nearest thousandth, hundredth, tenth, one, ten, hundred, or thousand.

Example Round 19.7625 to the nearest hundredth.

Locate the digit in the hundredths place, 6. Consider the next digit to the right, 2. Since that digit, 2, is 4 or lower, round down.

19.7625
↓
19.76

Practice Exercise

4. Round 153.346 to the nearest hundredth.

Objective 4.2a Add using decimal notation.

Example Add: 14.26 + 63.589.

$$
\begin{array}{r}
1 \\
1\ 4.2\ 6\ 0 \\
+\ 6\ 3.5\ 8\ 9 \\
\hline
7\ 7.8\ 4\ 9
\end{array}
$$
Writing an extra zero

Practice Exercise

5. Add: 5.54 + 33.071.

Objective 4.2b Subtract using decimal notation.

Example Subtract: 67.345 − 24.28.

$$
\begin{array}{r}
{}^{2}\ {}^{14} \\
6\ 7.\cancel{3}\ \cancel{4}\ 5 \\
-\ 2\ 4.2\ 8\ 0 \\
\hline
4\ 3.0\ 6\ 5
\end{array}
$$
Writing an extra zero

Practice Exercise

6. Subtract: 221.04 − 13.192.

Objective 4.3a Multiply using decimal notation.

Example Multiply: 1.8 × 0.04.

$$
\begin{array}{r}
1.8 \\
\times\ 0.0\ 4 \\
\hline
0.0\ 7\ 2
\end{array}
$$
(1 decimal place)
(2 decimal places)
(3 decimal places)

Example Multiply: 0.001 × 87.1.

0.001 × 87.1   0.087.1
↑   ↰

3 decimal places   Move 3 places to the left.
                      We write an extra zero.

0.001 × 87.1 = 0.0871

Example Multiply: 63.4 × 100.

63.4 × 100   63.40.
↑   ↰

2 zeros   Move 2 places to the right.
                  We write an extra zero.

63.4 × 100 = 6340

Practice Exercise

7. Multiply: 5.46 × 3.5.

Practice Exercise

8. Multiply: 17.6 × 0.01.

Practice Exercise

9. Multiply: 1000 × 60.437.

Objective 4.4a Divide using decimal notation.

Example Divide: $21.35 \div 6.1$.

$$
\begin{array}{r}
3.5 \\
6.1\overline{\smash{)}2\ 1.3{\scriptstyle\wedge}5} \\
\underline{1\ 8\ 3} \\
3\ 0\ 5 \\
\underline{3\ 0\ 5} \\
0
\end{array}
$$

Example Divide: $\dfrac{16.7}{1000}$.

$\dfrac{16.7}{1000}$ 0.016.7

3 zeros   Move 3 places to the left.

$\dfrac{16.7}{1000} = 0.0167$

Example Divide: $\dfrac{42.93}{0.001}$.

$\dfrac{42.93}{0.001}$ 42.930.

3 decimal places   Move 3 places to the right.

$\dfrac{42.93}{0.001} = 42{,}930$

Practice Exercise

10. Divide: $26.64 \div 3.6$.

Practice Exercise

11. Divide: $\dfrac{4.7}{100}$.

Practice Exercise

12. Divide: $\dfrac{156.9}{0.01}$.

Review Exercises

Convert the number in each sentence to standard notation. [4.3b]

1. Russia has the largest total area of any country in the world, at 6.59 million square miles.

2. Americans eat more than 3.1 billion lb of chocolate each year.
Source: Chocolate Manufacturers' Association

Write a word name. [4.1a]

3. 3.47

4. 0.031

5. 27.0001

6. 0.9

Write fraction notation. [4.1b]

7. 0.09

8. 4.561

9. 0.089

10. 3.0227

Write decimal notation. [4.1b]

11. $\dfrac{34}{1000}$

12. $\dfrac{42{,}603}{10{,}000}$

13. $27\dfrac{91}{100}$

14. $867\dfrac{6}{1000}$

Which number is larger? [4.1c]

15. 0.034, 0.0185

16. 0.91, 0.19

17. 0.741, 0.6943

18. 1.038, 1.041

Round 17.4287 to the nearest: [4.1d]

19. Tenth.

20. Hundredth.

21. Thousandth.

22. One.

Add.   [4.2a]

23.
```
      2.0 4 8
     6 5.3 7 1
   + 5 0 7.1
```

24.
```
    0.6
    0.0 0 4
    0.0 7
  + 0.0 0 9 8
```

25. 219.3 + 2.8 + 7

26. 0.41 + 4.1 + 41 + 0.041

Subtract.   [4.2b]

27.
```
    3 0.0
  -  0.7 9 0 8
```

28.
```
    8 4 5.0 8
  -   5 4.7 9
```

29. 37.645 − 8.497

30. 70.8 − 0.0109

Multiply.   [4.3a]

31.
```
      4 8
  × 0.2 7
```

32.
```
    0.1 7 4
  ×   0.8 3
```

33. 100 × 0.043

34. 0.001 × 24.68

Divide.   [4.4a]

35. 8 $)\overline{6\ 0}$

36. 5 2 $)\overline{2\ 3.4}$

37. 2.6 $)\overline{1\ 1\ 7.5\ 2}$

38. 2.1 4 $)\overline{2.1\ 8\ 7\ 0\ 8}$

39. $\dfrac{276.3}{1000}$

40. $\dfrac{13.892}{0.01}$

Solve.   [4.2c], [4.4b]

41. $x + 51.748 = 548.0275$

42. $3 \cdot x = 20.85$

43. $10 \cdot y = 425.4$

44. $0.0089 + y = 5$

Solve.   [4.7a]

45. Stacia earned $620.80 working as a coronary intensive-care nurse during a 40-hr week. What is her hourly wage?

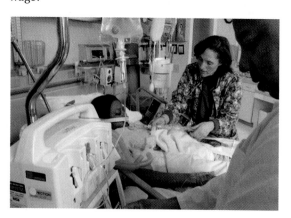

46. *Nutrition.*   The average person eats 688.6 lb of fruits and vegetables in a year. What is the average consumption in one day? (Use 1 year = 365 days.) Round to the nearest tenth of a pound.
Source: U.S. Department of Agriculture

47. Derek had $1034.46 in his checking account. He used his debit card to buy a Wii game system for $249.99. How much was left in his account?

48. *Cell-Phone Plan.*   In 2008, at&t offered its FamilyTalk cell-phone plan with 2 lines and 700 anytime minutes for a monthly access fee of $69.99. Minutes in excess of 700 were charged at the rate of $0.45 per minute. One month, Mr. and Mrs. Wu used their cell phones for 925 min. What was the charge?
Source: at&t

49. *Gas Mileage.*   Ellie wants to estimate the gas mileage of her car. At 36,057.1 mi, she fills the tank with 10.7 gal. At 36,217.6 mi, she fills the tank with 11.1 gal. Find the mileage per gallon. Round to the nearest tenth.

50. *Seafood Consumption.*   The graph below shows the annual consumption, in pounds, of seafood per person in the United States in recent years.

a) Find the total consumption per person for the seven years.

b) Find the average consumption per person for the years shown, rounded to the nearest tenth.

Seafood Consumption

SOURCE: U.S. Department of Agriculture

Estimate each of the following.   [4.6a]

51. The product 7.82×34.487 by rounding to the nearest one

52. The difference $219.875 - 4.478$ by rounding to the nearest one

53. The sum $\$45.78 + \$78.99$ by rounding to the nearest one

Find decimal notation. Use multiplying by 1.   [4.5a]

54. $\dfrac{13}{25}$ **55.** $\dfrac{9}{20}$ **56.** $\dfrac{11}{4}$

Find decimal notation. Use division.   [4.5a]

57. $\dfrac{13}{4}$ **58.** $\dfrac{7}{6}$ **59.** $\dfrac{17}{11}$

Round the answer to Exercise 59 to the nearest: [4.5b]

60. Tenth.   **61.** Hundredth.   **62.** Thousandth.

Convert from cents to dollars.   [4.3b]

63. 8273¢   **64.** 487¢

Convert from dollars to cents.   [4.3b]

65. $24.93   **66.** $9.86

Calculate.   [4.4c], [4.5c]

67. $(8 - 1.23) \div 4 + 5.6 \times 0.02$

68. $(1 + 0.07)^2 + 10^3 \div 10^2$
$+ [4(10.1 - 5.6) + 8(11.3 - 7.8)]$

69. $\dfrac{3}{4} \times 20.85$

70. Divide: $\dfrac{346.295}{0.001}$.   [4.4a]

    A. 0.346295   **B.** 3.46295
    C. 34,629.5   **D.** 346,295

71. Estimate the quotient $82.304 \div 17.287$ by rounding to the nearest ten.   [4.6a]

    A. 0.4   **B.** 4
    C. 40   **D.** 400

Synthesis

72. ▦ In each of the following, use one of $+$, $-$, $\times$, and $\div$ in each blank to make a true sentence.   [4.4c]

    a) $2.56 \ \square \ 6.4 \ \square \ 51.2 \ \square \ 17.4 \ \square \ 89.7 = 72.62$
    b) $(11.12 \ \square \ 0.29) \ \square \ 3^4 = 877.23$

73. Use the fact that $\frac{1}{3} = 0.\overline{3}$ to find repeating decimal notation for 1. Explain how you got your answer. [4.5a]

Understanding Through Discussion and Writing

1. Describe in your own words a procedure for converting from decimal notation to fraction notation.   [4.1b]

2. A student insists that $346.708 \times 0.1 = 3467.08$. How could you convince him that a mistake had been made without checking on a calculator?   [4.3a]

3. When is long division *not* the fastest way to convert from fraction notation to decimal notation?   [4.5a]

4. Consider finding decimal notation for $\frac{44}{125}$. Discuss as many ways as you can for finding such notation and give the answer.   [4.5a]

Test

For Extra Help

CHAPTER
Test Prep
VIDEOS

Step-by-step test solutions are found on the Chapter Test Prep Videos available via the Video Resources on DVD, in *MyMathLab* , and on You Tube (search "BittingerBasicMath" and click on "Channels").

Convert the number in each sentence to standard notation.

1. The United States issued 18.4 million passports in 2007, an all-time high, due in part to new rules requiring a passport for travel to Canada and Mexico.
Source: Associated Press

2. An ethanol boom helped to deliver a corn crop of a record 13.1 billion bushels in 2007.
Source: U.S. Department of Agriculture

Write a word name.

3. 2.34

4. 105.0005

Write fraction notation.

5. 0.91

6. 2.769

Write decimal notation.

7. $\dfrac{74}{1000}$

8. $\dfrac{37,047}{10,000}$

9. $756\dfrac{9}{100}$

10. $91\dfrac{703}{1000}$

Which number is larger?

11. 0.07, 0.162

12. 0.078, 0.06

13. 0.09, 0.9

Round 5.6783 to the nearest:

14. One.

15. Hundredth.

16. Thousandth.

17. Tenth.

Calculate.

18.
```
   0.7
   0.0 8
   0.0 0 9
 + 0.0 0 1 2
```

19. $102.4 + 6.1 + 78$

20. $0.93 + 9.3 + 93 + 930$

21.
```
  5 2.6 7 8
 −   4.3 2 1
```

22.
```
  2 0.0
 −  0.9 0 9 9
```

23. $234.6788 - 81.7854$

24.
```
   0.1 2 5
 ×   0.2 4
```

25. 0.001×213.45

26. 1000×73.962

27. $4\overline{)19}$

28. $3.3 \overline{)\ 1\ 0\ 0.3\ 2}$

29. $8\ 2 \overline{)\ 1\ 5.5\ 8}$

30. $\dfrac{346.89}{1000}$

31. $\dfrac{346.89}{0.01}$

Solve.

32. $4.8 \cdot y = 404.448$

33. $x + 0.018 = 9$

34. *Cell-Phone Plan.*   In 2008, at&t offered its FamilyTalk cell-phone plan with 2 lines and 1400 anytime minutes for a monthly access fee of $89.99. Minutes in excess of 1400 were charged at the rate of $0.40 per minute. One month, Mr. and Mrs. Tews used their cell phones for 1510 min. What was the charge?
Source: at&t

35. *Gas Mileage.*   Tina wants to estimate the gas mileage in her economy car. At 76,843 mi, she fills the tank with 14.3 gal of gasoline. At 77,310 mi, she fills the tank with 16.5 gal of gasoline. Find the mileage per gallon. Round to the nearest tenth.

36. *Checking Account Balance.*   Nicholas has a balance of $820 in his checking account before making purchases of $123.89, $56.68, and $46.98 with his debit card. What was the balance after the purchases had been made?

37. The office manager for the Drake, Smith, and Hartner law firm buys 7 cases of copy paper at $41.99 per case. What is the total cost?

38. *Busiest Airports.*   The graph below shows the numbers of passengers in 2007 who traveled through the country's busiest airports. Find the average total number of passengers through these airports.

Busiest Airports in the United States

Atlanta (Hartsfield) — 89.4
Chicago (O'Hare) — 76.2
Los Angeles — 61.9
Dallas/Fort Worth — 59.8
Denver — 49.9

Passengers (in millions)

SOURCE: Airports Council International

Estimate each of the following.

39. The product 8.91 × 22.457 by rounding to the nearest one

40. The quotient 78.2209 ÷ 16.09 by rounding to the nearest ten

Find decimal notation. Use multiplying by 1.

41. $\dfrac{7}{20}$

42. $\dfrac{22}{25}$

43. $\dfrac{21}{4}$

Find decimal notation. Use division.

44. $\dfrac{3}{4}$

45. $\dfrac{11}{9}$

46. $\dfrac{15}{7}$

Round the answer to Exercise 46 to the nearest:

47. Tenth.

48. Hundredth.

49. Thousandth.

Calculate.

50. $256 \div 3.2 \div 2 - 1.56 + 78.325 \times 0.02$

51. $(1 - 0.08)^2 + 6[5(12.1 - 8.7) + 10(14.3 - 9.6)]$

52. $\dfrac{7}{8} \times 345.6$

53. Convert from cents to dollars: 949¢.

   A. 0.949¢   **B.** $9.49   **C.** $94.90   **D.** $949

Synthesis

54. The Silver's Health Club charges a $79 membership fee and $42.50 a month. Allise has a coupon that will allow her to join the club for $299 for six months. How much will Allise save if she uses the coupon?

55. ▦ Arrange from smallest to largest.

$$\frac{2}{3}, \frac{15}{19}, \frac{11}{13}, \frac{5}{7}, \frac{13}{15}, \frac{17}{20}$$

Cumulative Review

Convert to fraction notation.

1. $2\dfrac{2}{9}$ **2.** 3.051

Find decimal notation.

3. $\dfrac{7}{5}$ **4.** $\dfrac{6}{11}$

5. Determine whether 43 is prime, composite, or neither.

6. Determine whether 2,053,752 is divisible by 4.

Calculate.

7. $48 + 12 \div 4 - 10 \times 2 + 6892 \div 4$

8. $4.7 - \{0.1[1.2(3.95 - 1.65) + 1.5 \div 2.5]\}$

Round to the nearest hundredth.

9. 584.973   **10.** $218.\overline{5}$

11. Estimate the product 16.392×9.715 by rounding to the nearest one.

12. Estimate by rounding to the nearest tenth:
$2.714 + 4.562 - 3.31 - 0.0023$.

13. Estimate the product 6418×1984 by rounding to the nearest hundred.

14. Estimate the quotient $717.832 \div 124.998$ by rounding to the nearest ten.

Add and simplify.

15.
$$\begin{array}{r} 2\dfrac{1}{4} \\[4pt] + 3\dfrac{4}{5} \\ \hline \end{array}$$

16.
$$\begin{array}{r} 34{,}921 \\ 93{,}092 \\ + 11{,}103 \\ \hline \end{array}$$

17. $\dfrac{1}{6} + \dfrac{2}{3} + \dfrac{8}{9}$ **18.** $143.9 + 2.053$

Subtract and simplify.

19. $723{,}041 - 12{,}904$ **20.** $19 - 5.903$

21. $5\dfrac{1}{7} - 4\dfrac{3}{7}$ **22.** $\dfrac{10}{11} - \dfrac{9}{10}$

Multiply and simplify.

23. $\dfrac{3}{8} \cdot \dfrac{4}{9}$

24.
$$\begin{array}{r} 2532 \\ \times\ 2100 \\ \hline \end{array}$$

25.
$$\begin{array}{r} 23.9 \\ \times\ \ 0.2 \\ \hline \end{array}$$

26.
$$\begin{array}{r} 27.9431 \\ \times\ \ \ 0.001 \\ \hline \end{array}$$

Divide and simplify.

27. $16.5\overline{)35.013}$ **28.** $26\overline{)47{,}918}$

29. $13.8621 \div 0.001$

30. $\dfrac{4}{9} \div \dfrac{8}{15}$

Solve.

31. $8.32 + x = 9.1$

32. $75 \cdot x = 2100$

33. $y \cdot 9.47 = 81.6314$

34. $1062 + y = 368,313$

35. $t + \dfrac{5}{6} = \dfrac{8}{9}$

36. $\dfrac{7}{8} \cdot t = \dfrac{7}{16}$

37. *Organ Transplants.* There were 16,646 kidney transplants, 6136 liver transplants, and 2147 heart transplants in the United States in a recent year. How many transplants of these three organs were performed that year?

Source: U.S. Department of Health and Human Services

38. *Hospital Revenue.* Hospitals received $265 billion in revenue from private health insurance payments in a recent year. This was $88 billion more than was received from Medicare payments. How much revenue was received from Medicare payments?

Source: U.S. Census Bureau

39. After Lauren made a $450 down payment on a sofa, $\dfrac{3}{10}$ of the total cost was paid. How much did the sofa cost?

40. Joshua's tuition was $3600. He obtained a loan for $\dfrac{2}{3}$ of the tuition. How much was the loan?

41. The balance in Elliott's checking account is $314.79. After a check is written for $56.02, what is the balance in the account?

42. A clerk in Leah's Delicatessen sold $1\dfrac{1}{2}$ lb of ham, $2\dfrac{3}{4}$ lb of turkey, and $2\dfrac{1}{4}$ lb of roast beef. How many pounds of meat were sold altogether?

43. A baker used $\dfrac{1}{2}$ lb of sugar for cookies, $\dfrac{2}{3}$ lb of sugar for pie, and $\dfrac{5}{6}$ lb of sugar for cake. How much sugar was used in all?

44. The Currys' rectangular family room measures 19.8 ft by 23.6 ft. What is its area?

Simplify.

45. $\left(\dfrac{3}{4}\right)^2 - \dfrac{1}{8} \cdot \left(3 - 1\dfrac{1}{2}\right)^2$

46. $1.2 \times 12.2 \div 0.1 \times 3.6$

Synthesis

47. Using a manufacturer's coupon, Lucy bought 2 cartons of orange juice and received a third carton free. The price of each carton was $3.59. What was the cost per carton with the coupon? Round to the nearest cent.

48. A carton of gelatin mix packages weighs $15\dfrac{3}{4}$ lb. Each package weighs $1\dfrac{3}{4}$ oz. How many packages are in the carton? ($1\,\text{lb} = 16\,\text{oz}$)

Ratio and Proportion

Real-World Application

Of the 71 shark attacks recorded worldwide in 2007, 50 occurred in U.S. waters. What is the ratio of the number of shark attacks in U.S. waters to the number of shark attacks worldwide?

Source: University of Florida

This problem appears as Example 7 in Section 5.1.

5.1

Introduction to Ratios

OBJECTIVES

a Find fraction notation for ratios.

b Simplify ratios.

SKILL TO REVIEW
Objective 2.5b: Simplify fraction notation.

Simplify.

1. $\dfrac{16}{64}$ 2. $\dfrac{40}{24}$

a Ratios

> **RATIO**
>
> A **ratio** is the quotient of two quantities.

In the 2007–2008 regular basketball season, the Boston Celtics scored a total of 8245 points and allowed their opponents a total of 7404 points. The *ratio* of points scored to points allowed is given by the fraction notation

$$\dfrac{8245}{7404} \begin{array}{l} \leftarrow \text{Points scored} \\ \leftarrow \text{Points allowed} \end{array}$$

or by the colon notation

$$8245 : 7404.$$

Points scored ⟶   ⟵ Points allowed

We read both forms of notation as "the ratio of 8245 to 7404."

> **RATIO NOTATION**
>
> The **ratio** of a to b is given by the fraction notation $\frac{a}{b}$, where a is the numerator and b is the denominator, or by the colon notation $a : b$.

EXAMPLE 1 Find the ratio of 7 to 8.

The ratio is $\dfrac{7}{8}$, or 7 : 8.

EXAMPLE 2 Find the ratio of 31.4 to 100.

The ratio is $\dfrac{31.4}{100}$, or 31.4 : 100.

EXAMPLE 3 Find the ratio of $4\frac{2}{3}$ to $5\frac{7}{8}$. You need not simplify.

The ratio is $\dfrac{4\frac{2}{3}}{5\frac{7}{8}}$, or $4\frac{2}{3} : 5\frac{7}{8}$.

1. Find the ratio of 5 to 11.

2. Find the ratio of 57.3 to 86.1.

3. Find the ratio of $6\frac{3}{4}$ to $7\frac{2}{5}$.

Do Margin Exercises 1–3.

In most of our work, we will use fraction notation for ratios.

Answers

Skill to Review:

1. $\dfrac{1}{4}$ 2. $\dfrac{5}{3}$

Margin Exercises:

1. $\dfrac{5}{11}$, or 5 : 11   2. $\dfrac{57.3}{86.1}$, or 57.3 : 86.1

3. $\dfrac{6\frac{3}{4}}{7\frac{2}{5}}$, or $6\frac{3}{4} : 7\frac{2}{5}$

EXAMPLE 4 *Wind Speeds.* The average wind speed in Chicago is 10.4 mph. The average wind speed in Boston is 12.5 mph. Find the ratio of the wind speed in Chicago to the wind speed in Boston.

Source: *The Handy Geography Answer Book*

The ratio is $\dfrac{10.4}{12.5}$.

EXAMPLE 5 *Record Rainfall.* The greatest rainfall ever recorded in the United States during a 12-month period was 739 in. in Kukui, Maui, Hawaii, from December 1981 to December 1982. What is the ratio of amount of rainfall, in inches, to time, in months? of time, in months, to amount of rainfall, in inches?

Source: *Time Almanac*

The ratio of amount of rainfall, in inches, to time, in months, is

$$\dfrac{739}{12}. \quad \begin{array}{l}\leftarrow \text{Rainfall} \\ \leftarrow \text{Time}\end{array}$$

The ratio of time, in months, to amount of rainfall, in inches, is

$$\dfrac{12}{739}. \quad \begin{array}{l}\leftarrow \text{Time} \\ \leftarrow \text{Rainfall}\end{array}$$

EXAMPLE 6 Refer to the triangle below.

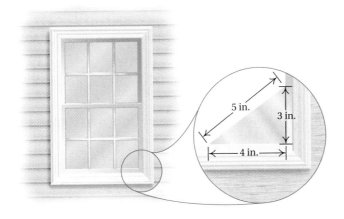

a) What is the ratio of the length of the longest side to the length of the shortest side?

$$\dfrac{5}{3} \quad \begin{array}{l}\leftarrow \text{Longest side} \\ \leftarrow \text{Shortest side}\end{array}$$

b) What is the ratio of the length of the shortest side to the length of the longest side?

$$\dfrac{3}{5} \quad \begin{array}{l}\leftarrow \text{Shortest side} \\ \leftarrow \text{Longest side}\end{array}$$

> Do Exercises 4–6; Exercise 6 is on the next page.

4. Record Snowfall. The greatest snowfall recorded in North America during a 24-hr period was 76 in. in Silver Lake, Colorado, on April 14–15, 1921. What is the ratio of amount of snowfall, in inches, to time, in hours?

Source: U.S. Army Corps of Engineers

5. Frozen Fruit Drinks. A Berries & Kreme Chiller from Krispy Kreme contains 960 calories, while Smoothie King's MangoFest drink contains 258 calories. What is the ratio of the number of calories in the Krispy Kreme drink to the number of calories in the Smoothie King drink? of the number of calories in the Smoothie King drink to the number of calories in the Krispy Kreme drink?

Source: Physicians Committee for Responsible Medicine

Answers

4. $\dfrac{76}{24}$ 5. $\dfrac{960}{258}; \dfrac{258}{960}$

6. In the triangle below, what is the ratio of the length of the shortest side to the length of the longest side?

55.5 yd

38.2 yd

40.3 yd

EXAMPLE 7 *Shark Attacks.* Of the 71 shark attacks recorded worldwide in 2007, 50 occurred in U.S. waters. The bar graph below shows the breakdown by state.

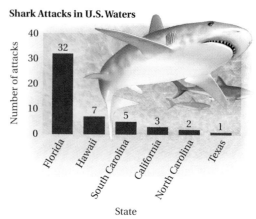

Shark Attacks in U.S. Waters

Number of attacks

Florida 32
Hawaii 7
South Carolina 5
California 3
North Carolina 2
Texas 1

State

SOURCE: University of Florida

a) What is the ratio of the number of shark attacks in U.S. waters to the number of shark attacks worldwide?

b) What is the ratio of the number of shark attacks in North Carolina to the number of shark attacks in South Carolina?

c) What is the ratio of the number of shark attacks in Florida to the total number of shark attacks in the other five states?

a) The ratio of the number of shark attacks in U.S. waters to the number of shark attacks worldwide is

$$\frac{50}{71}. \quad \begin{array}{l} \leftarrow \text{Attacks in U.S. waters} \\ \leftarrow \text{Attacks worldwide} \end{array}$$

b) The ratio of the number of shark attacks in North Carolina to the number of shark attacks in South Carolina is

$$\frac{2}{5}. \quad \begin{array}{l} \leftarrow \text{Attacks in North Carolina} \\ \leftarrow \text{Attacks in South Carolina} \end{array}$$

c) Of the 50 shark attacks in U.S. waters, 32 occurred in Florida. We subtract to determine how many shark attacks took place in the other five states. We have

$$50 - 32 = 18.$$

Thus the ratio of the number of shark attacks in Florida to the number of shark attacks in the other five states is

$$\frac{32}{18}. \quad \begin{array}{l} \leftarrow \text{Attacks in Florida} \\ \leftarrow \text{Attacks in other five states} \end{array}$$

Do Exercise 7.

7. Soap Box Derby. Of the 538 participants in the 2008 All-American Soap Box Derby, 296 were boys and 242 were girls. What was the ratio of girls to boys? of boys to girls? of boys to total number of participants?
Source: All-American Soap Box Derby

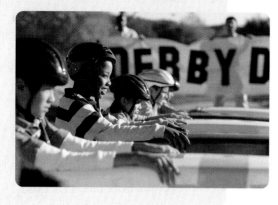

b Simplifying Notation for Ratios

Sometimes a ratio can be simplified. This provides a means of finding other numbers with the same ratio.

EXAMPLE 8 Find the ratio of 6 to 8. Then simplify to find two other numbers in the same ratio.

Answers

6. $\frac{38.2}{55.5}$ **7.** $\frac{242}{296}; \frac{296}{242}; \frac{296}{538}$

We write the ratio in fraction notation and then simplify:

$$\frac{6}{8} = \frac{2 \cdot 3}{2 \cdot 4} = \frac{2}{2} \cdot \frac{3}{4} = 1 \cdot \frac{3}{4} = \frac{3}{4}.$$

Thus, 3 and 4 have the same ratio as 6 and 8. We can express this by saying "6 is to 8 as 3 is to 4."

Do Exercise 8.

EXAMPLE 9 Find the ratio of 2.4 to 10. Then simplify to find two other numbers in the same ratio.

We first write the ratio in fraction notation. Next, we multiply by 1 to clear the decimal from the numerator. Then we simplify.

$$\frac{2.4}{10} = \frac{2.4}{10} \cdot \frac{10}{10} = \frac{24}{100} = \frac{4 \cdot 6}{4 \cdot 25} = \frac{4}{4} \cdot \frac{6}{25} = \frac{6}{25}$$

Thus, 2.4 is to 10 as 6 is to 25.

Do Exercises 9 and 10.

EXAMPLE 10 An HDTV screen that measures 46 in. diagonally has a width of 40 in. and a height of $22\frac{1}{2}$ in. Find the ratio of width to height and simplify.

The ratio is
$$\frac{40}{22\frac{1}{2}} = \frac{40}{22.5} = \frac{40}{22.5} \cdot \frac{10}{10} = \frac{400}{225}$$
$$= \frac{25 \cdot 16}{25 \cdot 9} = \frac{25}{25} \cdot \frac{16}{9}$$
$$= \frac{16}{9}.$$

Thus we can say that the ratio of width to height is 16 to 9, which can also be expressed as 16:9.

Do Exercise 11.

8. Find the ratio of 18 to 27. Then simplify to find two other numbers in the same ratio.

9. Find the ratio of 3.6 to 12. Then simplify to find two other numbers in the same ratio.

10. Find the ratio of 1.2 to 1.5. Then simplify to find two other numbers in the same ratio.

11. An HDTV screen that measures 44 in. diagonally has a width of 38.4 in. and a height of 21.6 in. Find the ratio of height to width and simplify.

Answers

8. 18 is to 27 as 2 is to 3.   **9.** 3.6 is to 12 as 3 is to 10.   **10.** 1.2 is to 1.5 as 4 is to 5.   **11.** $\frac{9}{16}$

a   Find fraction notation for each ratio. You need not simplify.

1. 4 to 5

2. 3 to 2

3. 178 to 572

4. 329 to 967

5. 0.4 to 12

6. 2.3 to 22

7. 3.8 to 7.4

8. 0.6 to 0.7

9. 56.78 to 98.35

10. 456.2 to 333.1

11. $8\frac{3}{4}$ to $9\frac{5}{6}$

12. $10\frac{1}{2}$ to $43\frac{1}{4}$

13. *Space Plane.* It is estimated that it will take 4 hr to fly from London to Sydney on the beyond-the-atmosphere space plane being developed in Europe. This 10,600-mi trip currently takes about 21 hr. What is the ratio of the time of the current trip to the time of the trip on the space plane? of the time of the trip on the space plane to the time of the current trip?
Source: EADS Atrium

14. *Silicon in the Earth's Crust.* Every 100 tons of the earth's crust contains about 28 tons of silicon. What is the ratio of the weight of silicon to the weight of crust? of the weight of crust to the weight of silicon?
Source: *The Handy Science Answer Book*

15. *Health Insurance Coverage.* Of every 100 people in the United States, 84.7 are covered by health insurance. What is the ratio of all people to those covered by health insurance? of those covered by health insurance to all people?
Source: U.S. Census Bureau

16. *Price of a Book.* The list price of the book *The Last Lecture* by Randy Pausch is $21.95, but the book recently sold for $12.07 on amazon.com. What is the ratio of the list price to the Amazon price? of the Amazon price to the list price?
Source: amazon.com

17. *Tax Freedom Day.* Of the 366 days in 2008 (a leap year), the average American worked 113 days to pay his or her federal, state, and local taxes. Find the ratio of the number of days worked to pay taxes in 2008 to the number of days in the year.
Source: Tax Foundation

18. *Spending Categories.* The average American worked 60 days in 2008 to pay housing and household operating expenses and 35 days to pay for food. Find the ratio of the number of days worked to pay for food to the number of days worked to pay housing and household operating expenses.
Source: Tax Foundation

19. *Population Estimates.* It is estimated that, of every 1000 people in the United States in 2050, 126 will be from 20 to 29 years old. What is the ratio of all people to those in the 20- to 29-year age group? of those in the 20- to 29-year age group to all people?

Source: U.S. Census Bureau

20. *Population Estimates.* It is estimated that, of every 1000 people in the United States in 2050, 118 will be age 75 and older. What is the ratio of all people to those age 75 and older? of those age 75 and older to all people?

Source: U.S. Census Bureau

21. *Field Hockey.* A diagram of the playing area for field hockey is shown below. What is the ratio of width to length? of length to width?

Source: *Sports: The Complete Visual Reference*

Shooting circle
Sideline
Officials' table
Penalty spot
100 yd
60 yd

22. *The Leaning Tower of Pisa.* The Leaning Tower of Pisa was reopened to the public in 2001 following a 10-yr stabilization project. The 184.5-ft tower now leans about 13 ft from its base. What is the ratio of the distance that it leans to its height? of its height to the distance that it leans?

Source: CNN

184.5 ft

13 ft

b Find the ratio of the first number to the second and simplify.

23. 4 to 6

24. 6 to 10

25. 18 to 24

26. 28 to 36

27. 4.8 to 10

28. 5.6 to 10

29. 2.8 to 3.6

30. 4.8 to 6.4

31. 20 to 30

32. 40 to 60

33. 56 to 100

34. 42 to 100

35. 128 to 256

36. 232 to 116

37. 0.48 to 0.64

38. 0.32 to 0.96

39. In this rectangle, find the ratios of length to width and of width to length.

478 ft

213 ft

40. In this right triangle, find the ratios of shortest length to longest length and of longest length to shortest length.

107.3 m

47.5 m

96.2 m

Skill Maintenance

Use = or ≠ for ☐ to write a true sentence.   [2.5c]

41. $\dfrac{12}{8} \;\square\; \dfrac{6}{4}$

42. $\dfrac{4}{7} \;\square\; \dfrac{5}{9}$

43. $\dfrac{7}{2} \;\square\; \dfrac{31}{9}$

44. $\dfrac{17}{25} \;\square\; \dfrac{68}{100}$

Divide. Write decimal notation for the answer.   [4.4a]

45. 200 ÷ 4

46. 95 ÷ 10

47. 232 ÷ 16

48. 342 ÷ 2.25

Solve.   [3.5c]

49. Rocky is $187\frac{1}{10}$ cm tall and his daughter is $180\frac{3}{4}$ cm tall. How much taller is Rocky?

50. Aunt Louise is $168\frac{1}{4}$ cm tall and her son is $150\frac{7}{10}$ cm tall. How much taller is Aunt Louise?

Synthesis

51. Find the ratio of $3\frac{3}{4}$ to $5\frac{7}{8}$ and simplify.

Fertilizer.   Exercises 52 and 53 refer to a common lawn fertilizer known as "5, 10, 15." This mixture contains 5 parts of potassium for every 10 parts of phosphorus and 15 parts of nitrogen. (This is often denoted 5 : 10 : 15.)

52. Find and simplify the ratio of potassium to nitrogen and of nitrogen to phosphorus.

53. Simplify the ratio 5 : 10 : 15.

5.2

Rates and Unit Prices

a Rates

A 2008 Kia Sportage EX can travel 414 mi on 18 gal of gasoline. Let's consider the ratio of miles to gallons:

Source: Kia Motors America, Inc.

$$\frac{414 \text{ mi}}{18 \text{ gal}} = \frac{414}{18} \frac{\text{miles}}{\text{gallon}} = \frac{23}{1} \frac{\text{miles}}{\text{gallon}}$$
$$= 23 \text{ miles per gallon} = 23 \text{ mpg.}$$

> "per" means "division," or "for each."

The ratio

$$\frac{414 \text{ mi}}{18 \text{ gal}}, \quad \text{or} \quad \frac{414}{18} \frac{\text{mi}}{\text{gal}}, \quad \text{or } 23 \text{ mpg,}$$

is called a **rate**.

OBJECTIVES

a Give the ratio of two different measures as a rate.

b Find unit prices and use them to compare purchases.

> **RATE**
>
> When a ratio is used to compare two different kinds of measure, we call it a **rate**.

Suppose David says his car travels 392.4 mi on 16.8 gal of gasoline. Is the mpg (mileage) of his car better than that of the Kia Sportage above? To determine this, it helps to convert the ratio to decimal notation and perhaps round. Then we have

$$\frac{392.4 \text{ miles}}{16.8 \text{ gallons}} = \frac{392.4}{16.8} \text{ mpg} \approx 23.357 \text{ mpg.}$$

Since $23.357 > 23$, David's car gets better mileage than the Kia Sportage does.

EXAMPLE 1 It takes 60 oz of grass seed to seed 3000 sq ft of lawn. What is the rate in ounces per square foot?

$$\frac{60 \text{ oz}}{3000 \text{ sq ft}} = \frac{1}{50} \frac{\text{oz}}{\text{sq ft}}, \quad \text{or} \quad 0.02 \frac{\text{oz}}{\text{sq ft}}$$

EXAMPLE 2 Martina bought 5 lb of organic russet potatoes for $4.99. What was the rate in cents per pound?

$$\frac{\$4.99}{5 \text{ lb}} = \frac{499 \text{ cents}}{5 \text{ lb}} = 99.8¢/\text{lb}$$

STUDY TIPS

RELYING ON THE ANSWER SECTION

Don't begin solving a homework problem by working backward from the answer in the answer section at the back of the text. If you are having trouble getting the correct answer to an exercise, you might need to reread the section preceding the exercise set, paying particular attention to the example that corresponds to the type of exercise you are doing, and/or work more slowly and carefully. Keep in mind that when you take quizzes and tests you have no answer section to rely on.

EXAMPLE 3 A pharmacy student employed as a pharmacist's assistant earned $3930 for working 3 months one summer. What was the rate of pay per month?

The rate of pay is the ratio of money earned to length of time worked, or

$$\frac{\$3930}{3 \text{ mo}} = 1310 \frac{\text{dollars}}{\text{month}}, \quad \text{or} \quad \$1310 \text{ per month.}$$

EXAMPLE 4 *Ratio of Strikeouts to Home Runs.* In the 2008 baseball season, Dustin Pedroia of the Boston Red Sox had 52 strikeouts and 17 home runs. What was his strikeout to home-run rate?

Source: Major League Baseball

$$\frac{52 \text{ strikeouts}}{17 \text{ home runs}} = \frac{52}{17} \frac{\text{strikeouts}}{\text{home runs}} = \frac{52}{17} \text{ strikeouts per home run}$$

$$\approx 3.06 \text{ strikeouts per home run}$$

Do Exercises 1–5.

b Unit Pricing

UNIT PRICE

A **unit price**, or **unit rate**, is the ratio of price to the number of units.

EXAMPLE 5 *Unit Price of Pears.* Ruth bought a 15.2-oz can of pears for $1.30. What is the unit price in cents per ounce?

$$\text{Unit price} = \frac{\text{Price}}{\text{Number of units}}$$

$$= \frac{\$1.30}{15.2 \text{ oz}} = \frac{130 \text{ cents}}{15.2 \text{ oz}} = \frac{130}{15.2} \frac{\text{cents}}{\text{oz}}$$

$$\approx 8.553 \text{ cents per ounce.}$$

Do Exercise 6.

A ratio of distance traveled to time is called *speed*. What is the rate, or speed, in miles per hour?

1. 45 mi, 9 hr

2. 120 mi, 10 hr

What is the rate, or speed, in feet per second?

3. 2200 ft, 2 sec

4. 52 ft, 13 sec

5. Babe Ruth. In his baseball career, Babe Ruth had 1330 strikeouts and 714 home runs. What was his home-run to strikeout rate?

Source: Major League Baseball

6. Unit Price of Pasta Sauce. Gregory bought a 26-oz jar of pasta sauce for $2.79. What is the unit price in cents per ounce?

Answers

1. 5 mi/hr, or 5 mph   **2.** 12 mi/hr, or 12 mph
3. 1100 ft/sec   **4.** 4 ft/sec   **5.** $\frac{714}{1330}$ home runs per strikeout ≈ 0.537 home run per strikeout   **6.** 10.731¢/oz

Unit prices enable us to do comparison shopping and determine the best buy of a product on the basis of price. It is often helpful to change all prices to cents so that we can compare unit prices more easily.

EXAMPLE 6 *Unit Price of Salad Dressing.* At the request of his customers, Angelo started bottling and selling the popular basil and lemon salad dressing that he serves in his neighborhood café. The dressing is sold in four sizes of containers as listed in the table below. Compute the unit price of each size of container and determine which size is the best buy on the basis of unit price alone.

SIZE	PRICE	UNIT PRICE
10 oz	$2.49	24.900¢/oz
16 oz	$3.59	22.438¢/oz
20 oz	$4.09	20.450¢/oz Lowest unit price
32 oz	$6.79	21.219¢/oz

We compute the unit price of each size and fill in the chart:

10 oz: $\dfrac{\$2.49}{10 \text{ oz}} = \dfrac{249 \text{ cents}}{10 \text{ oz}} = \dfrac{249}{10}\dfrac{\text{cents}}{\text{oz}} = 24.900\text{¢/oz};$

16 oz: $\dfrac{\$3.59}{16 \text{ oz}} = \dfrac{359 \text{ cents}}{16 \text{ oz}} = \dfrac{359}{16}\dfrac{\text{cents}}{\text{oz}} \approx 22.438\text{¢/oz};$

20 oz: $\dfrac{\$4.09}{20 \text{ oz}} = \dfrac{409 \text{ cents}}{20 \text{ oz}} = \dfrac{409}{20}\dfrac{\text{cents}}{\text{oz}} = 20.450\text{¢/oz};$

32 oz: $\dfrac{\$6.79}{32 \text{ oz}} = \dfrac{679 \text{ cents}}{32 \text{ oz}} = \dfrac{679}{32}\dfrac{\text{cents}}{\text{oz}} \approx 21.219\text{¢/oz}.$

On the basis of unit price alone, we see that the 20-oz container is the best buy.

Do Exercise 7.

Although we often think that "bigger is cheaper," this is not always the case, as we see in Example 6. In addition, even when a larger package has a lower unit price than a smaller package, it still might not be the best buy for you. For example, some of the food in a large package could spoil before it is used, or you might not have room to store a large package.

7. Cost of Mayonnaise. Complete the following table for Hellmann's mayonnaise sold on an online shopping site. Which size has the lowest unit price?

Source: Peapod

SIZE	PRICE	UNIT PRICE
8 oz	$2.79	
10 oz	$3.69	
30 oz	$5.39	

a In Exercises 1–4, find each rate, or speed, as a ratio of distance to time. Round to the nearest hundredth where appropriate.

1. 120 km, 3 hr

2. 18 mi, 9 hr

3. 217 mi, 29 sec

4. 443 m, 48 sec

5. *Chevrolet Cobalt LS—City Driving.* A 2008 Chevrolet Cobalt LS will travel 312.5 mi on 12.5 gal of gasoline in city driving. What is the rate in miles per gallon?
Source: Chevrolet

6. *Mazda A3—City Driving.* A 2008 Mazda A3 will travel 348 mi on 14.5 gal of gasoline in city driving. What is the rate in miles per gallon?
Source: Ford Motor Company

7. *Mazda A3—Highway Driving.* A 2008 Mazda A3 will travel 624 mi on 19.5 gal of gasoline in highway driving. What is the rate in miles per gallon?
Source: Ford Motor Company

8. *Chevrolet Cobalt LS—Highway Driving.* A 2008 Chevrolet Cobalt LS will travel 486 mi on 13.5 gal of gasoline in highway driving. What is the rate in miles per gallon?
Source: Chevrolet

9. *Population Density of Monaco.* Monaco is a tiny country on the Mediterranean coast of France. It has an area of 0.75 square mile and a population of 32,796 people. What is the rate of number of people per square mile? The rate per square mile is called the *population density.* Monaco has the highest population density of any country in the world.
Source: The World Factbook

10. *Population Density of Australia.* The continent of Australia, with the island state of Tasmania, has an area of 2,967,893 sq mi and a population of 21,007,310 people. What is the rate of number of people per square mile? The rate per square mile is called the *population density.* Australia has one of the lowest population densities in the world.
Source: The World Factbook

11. A car is driven 500 mi in 20 hr. What is the rate in miles per hour? in hours per mile?

12. A student eats 3 hamburgers in 15 min. What is the rate in hamburgers per minute? in minutes per hamburger?

13. *Points per Game.* Dwayne Wade of the Miami Heat scored 1254 points in 51 games during the 2007–2008 basketball season. What was the rate in points per game?

Source: National Basketball Association

14. *Rebounds per Game.* Dwight Howard of the Orlando Magic got 1161 rebounds in 82 games during the 2007–2008 basketball season. What was the rate in rebounds per game?

Source: National Basketball Association

15. *Lawn Watering.* Watering a lawn adequately requires 623 gal of water for every 1000 ft^2. What is the rate in gallons per square foot?

16. A car is driven 200 km on 40 L of gasoline. What is the rate in kilometers per liter?

17. *Speed of Light.* Light travels 186,000 mi in 1 sec. What is its rate, or speed, in miles per second?

Source: *The Handy Science Answer Book*

18. *Speed of Sound.* Sound travels 1100 ft in 1 sec. What is its rate, or speed, in feet per second?

Source: *The Handy Science Answer Book*

19. Impulses in nerve fibers travel 310 km in 2.5 hr. What is the rate, or speed, in kilometers per hour?

20. A black racer snake can travel 4.6 km in 2 hr. What is its rate, or speed, in kilometers per hour?

21. *Elephant Heart Rate.* The heart of an elephant, at rest, will beat an average of 1500 beats in 60 min. What is the rate in beats per minute?

Source: *The Handy Science Answer Book*

22. *Human Heart Rate.* The heart of a human, at rest, will beat an average of 4200 beats in 60 min. What is the rate in beats per minute?

Source: *The Handy Science Answer Book*

b Find each unit price in Exercises 23–30. Then, in each exercise, determine which size is the better buy based on unit price alone.

23. *Hidden Valley Ranch Dressing.*

SIZE	PRICE	UNIT PRICE
16 oz	$4.19	
20 oz	$5.29	

24. *Miracle Whip.*

SIZE	PRICE	UNIT PRICE
32 oz	$5.29	
48 oz	$8.29	

25. *Cascade Powder Detergent.*

SIZE	PRICE	UNIT PRICE
45 oz	$3.99	
75 oz	$5.49	

26. *Bush's Homestyle Baked Beans.*

SIZE	PRICE	UNIT PRICE
16 oz	$1.59	
28 oz	$2.09	

27. *Jif Creamy Peanut Butter.*

SIZE	PRICE	UNIT PRICE
18 oz	$2.50	
28 oz	$4.89	

28. *Hills Brothers Coffee.*

SIZE	PRICE	UNIT PRICE
26 oz	$8.99	
39 oz	$12.99	

29. *Campbell's Condensed Tomato Soup.*

SIZE	PRICE	UNIT PRICE
10.7 oz	$1.39	
26 oz	$2.69	

30. *Nabisco Saltines.*

SIZE	PRICE	UNIT PRICE
16 oz	$3.59	
32 oz	$5.19	

Find the unit price of each brand in Exercises 31–34. Then, in each exercise, determine which brand is the best buy based on unit price alone.

31. *Vanilla Ice Cream.*

BRAND	SIZE	PRICE
B	32 oz	$5.99
E	48 oz	$6.99

32. *Orange Juice.*

BRAND	SIZE	PRICE
M	54 oz	$4.79
T	59 oz	$5.99

33. *Tomato Ketchup.*

BRAND	SIZE	PRICE
A	24 oz	$2.49
B	36 oz	$3.29
H	46 oz	$3.69

34. *Yellow Mustard.*

BRAND	SIZE	PRICE
F	14 oz	$1.29
G	19 oz	$1.99
P	20 oz	$2.49

Skill Maintenance

Solve.

35. *Drive-in Movie Theaters.* The number of outdoor movie theaters has declined steadily since the 1950s. In 2007, there were 405 drive-in theaters. This is 3658 fewer than in 1958. How many drive-in movie theaters were there in 1958? [1.8a]

Source: Drive-Ins.com

36. A serving of fish steak (cross section) is generally $\frac{1}{2}$ lb. How many servings can be prepared from a cleaned $18\frac{3}{4}$-lb tuna? [3.6c]

37. *Gift-Card Sales.* Gift-card sales rose from $17.2 billion in 2003 to $26.3 billion in 2007. By how much did gift-card sales rise from 2003 to 2007? [4.7a]

Source: National Retail Federation

38. *Shifting Music Sales.* Sales of individual digital music tracks outpaced sales of albums by 176.9 million in the first quarter of 2008. During that quarter, 104.5 million albums were sold. How many individual tracks were sold during the same time period? [4.7a]

Source: Nielsen SoundScan

Multiply. [4.3a]

39.
$$\begin{array}{r} 4\ 5.6\ 7 \\ \times\ \ \ \ \ 2.4 \\ \hline \end{array}$$

40.
$$\begin{array}{r} 6\ 7\ 8.1\ 9 \\ \times\ \ \ \ \ 1\ 0\ 0 \\ \hline \end{array}$$

41. 84.3×69.2

42. 1002.56×465

Synthesis

43. Recently, some manufacturers have been changing the size of their containers in such a way that the consumer thinks the price of a product has been lowered when, in reality, a higher unit price is being charged.

Some aluminum juice cans are now concave (curved in) on the bottom. Suppose the volume of the can in the figure has been reduced from a fluid capacity of 6 oz to 5.5 oz, and the price of each can has been reduced from 65¢ to 60¢. Find the unit price of each container in cents per ounce.

5.3

Proportions

OBJECTIVES

a Determine whether two pairs of numbers are proportional.

b Solve proportions.

SKILL TO REVIEW

Objective 2.5c: Use the test for equality to determine whether two fractions name the same number.

Use = or ≠ for ☐ to write a true sentence.

1. $\frac{18}{12} \ \Box \ \frac{9}{6}$ **2.** $\frac{4}{7} \ \Box \ \frac{5}{8}$

a Proportions

When two pairs of numbers, such as 3, 2 and 6, 4, have the same ratio, we say that they are **proportional**. The equation

$$\frac{3}{2} = \frac{6}{4}$$

states that the pairs 3, 2 and 6, 4 are proportional. Such an equation is called a **proportion**. We sometimes read $\frac{3}{2} = \frac{6}{4}$ as "3 is to 2 as 6 is to 4."

Since ratios can be written using fraction notation, we can use the test for equality of fractions discussed in Section 2.5 to determine whether two ratios are the same.

EXAMPLE 1   Determine whether 1, 2 and 3, 6 are proportional.

We can use cross products:

$$1 \cdot 6 = 6 \qquad \frac{1}{2} \overset{?}{=} \frac{3}{6} \qquad 2 \cdot 3 = 6.$$

Since the cross products are the same, $6 = 6$, we know that $\frac{1}{2} = \frac{3}{6}$, so the numbers are proportional.

EXAMPLE 2   Determine whether 2, 5 and 4, 7 are proportional.

We can use cross products:

$$2 \cdot 7 = 14 \qquad \frac{2}{5} \overset{?}{=} \frac{4}{7} \qquad 5 \cdot 4 = 20.$$

Since the cross products are not the same, $14 \neq 20$, we know that $\frac{2}{5} \neq \frac{4}{7}$, so the numbers are not proportional.

Do Margin Exercises 1–3.

Determine whether the two pairs of numbers are proportional.

1. 3, 4 and 6, 8

2. 1, 4 and 10, 39

3. 1, 2 and 20, 39

EXAMPLE 3   Determine whether 3.2, 4.8 and 0.16, 0.24 are proportional.

We can use cross products:

$$3.2 \times 0.24 = 0.768 \qquad \frac{3.2}{4.8} \overset{?}{=} \frac{0.16}{0.24} \qquad 4.8 \times 0.16 = 0.768.$$

Since the cross products are the same, $0.768 = 0.768$, we know that $\frac{3.2}{4.8} = \frac{0.16}{0.24}$, so the numbers are proportional.

Determine whether the two pairs of numbers are proportional.

4. 6.4, 12.8 and 5.3, 10.6

5. 6.8, 7.4 and 3.4, 4.2

Do Exercises 4 and 5.

Answers

Skill to Review:
1. =   **2.** ≠

Margin Exercises:
1. Yes   **2.** No   **3.** No   **4.** Yes   **5.** No

EXAMPLE 4 Determine whether $4\frac{2}{3}$, $5\frac{1}{2}$ and $8\frac{7}{8}$, $16\frac{1}{3}$ are proportional.

We can use cross products:

$$4\frac{2}{3} \cdot 16\frac{1}{3} = \frac{14}{3} \cdot \frac{49}{3} \qquad \frac{4\frac{2}{3}}{5\frac{1}{2}} \stackrel{?}{=} \frac{8\frac{7}{8}}{16\frac{1}{3}} \qquad 5\frac{1}{2} \cdot 8\frac{7}{8} = \frac{11}{2} \cdot \frac{71}{8}$$

$$= \frac{686}{9} \qquad\qquad\qquad\qquad = \frac{781}{16}$$

$$= 76\frac{2}{9}; \qquad\qquad\qquad\qquad = 48\frac{13}{16}.$$

Since the cross products are not the same, $76\frac{2}{9} \neq 48\frac{13}{16}$, we know that the numbers are not proportional.

> Do Exercise 6.

6. Determine whether $4\frac{2}{3}$, $5\frac{1}{2}$ and 14, $16\frac{1}{2}$ are proportional.

b Solving Proportions

Let's now look at solving proportions. Consider the proportion

$$\frac{x}{3} = \frac{4}{6}.$$

One way to solve a proportion is to use cross products. Then we can divide on both sides to get the variable alone:

$$\frac{x}{3} = \frac{4}{6}$$

$$x \cdot 6 = 3 \cdot 4 \qquad \text{Equating cross products (finding cross products and setting them equal)}$$

$$\frac{x \cdot 6}{6} = \frac{3 \cdot 4}{6} \qquad \text{Dividing by 6 on both sides}$$

$$x = \frac{3 \cdot 4}{6} = \frac{12}{6} = 2.$$

We can check that 2 is the solution by replacing x with 2 and finding cross products:

$$2 \cdot 6 = 12 \qquad \frac{2}{3} \stackrel{?}{=} \frac{4}{6} \qquad 3 \cdot 4 = 12.$$

Since the cross products are the same, it follows that $\frac{2}{3} = \frac{4}{6}$. Thus the pairs of numbers 2, 3 and 4, 6 are proportional, and 2 is the solution of the equation.

> **SOLVING PROPORTIONS**
>
> To solve $\frac{x}{a} = \frac{c}{d}$, equate *cross products* and divide on both sides to get x alone.

> Do Exercise 7.

7. Solve: $\frac{x}{63} = \frac{2}{9}$.

Answers

6. Yes   **7.** 14

EXAMPLE 5 Solve: $\frac{x}{7} = \frac{5}{3}$. Write a mixed numeral for the answer.

We have

$$\frac{x}{7} = \frac{5}{3}$$

$$x \cdot 3 = 7 \cdot 5 \qquad \text{Equating cross products}$$

$$\frac{x \cdot 3}{3} = \frac{7 \cdot 5}{3} \qquad \text{Dividing by 3}$$

$$x = \frac{7 \cdot 5}{3} = \frac{35}{3}, \text{ or } 11\frac{2}{3}.$$

The solution is $11\frac{2}{3}$.

Do Exercise 8.

8. Solve: $\frac{x}{9} = \frac{5}{4}$. Write a mixed numeral for the answer.

EXAMPLE 6 Solve: $\frac{7.7}{15.4} = \frac{y}{2.2}$.

We have

$$\frac{7.7}{15.4} = \frac{y}{2.2}$$

$$7.7 \times 2.2 = 15.4 \times y \qquad \text{Equating cross products}$$

$$\frac{7.7 \times 2.2}{15.4} = \frac{15.4 \times y}{15.4} \qquad \text{Dividing by 15.4}$$

$$\frac{7.7 \times 2.2}{15.4} = y$$

$$\frac{16.94}{15.4} = y \qquad \text{Multiplying}$$

$$1.1 = y. \qquad \text{Dividing:}$$

$$\begin{array}{r} 1.1 \\ 1\ 5.4\ )\overline{1\ 6.9_\wedge 4} \\ 1\ 5\ 4 \\ \hline 1\ 5\ 4 \\ 1\ 5\ 4 \\ \hline 0 \end{array}$$

The solution is 1.1.

EXAMPLE 7 Solve: $\frac{8}{x} = \frac{5}{3}$. Write decimal notation for the answer.

We have

$$\frac{8}{x} = \frac{5}{3}$$

$$8 \cdot 3 = x \cdot 5 \qquad \text{Equating cross products}$$

$$\frac{8 \cdot 3}{5} = x \qquad \text{Dividing by 5}$$

$$\frac{24}{5} = x \qquad \text{Multiplying}$$

$$4.8 = x. \qquad \text{Simplifying}$$

The solution is 4.8.

9. Solve: $\frac{21}{5} = \frac{n}{2.5}$.

10. Solve: $\frac{6}{x} = \frac{25}{11}$. Write decimal notation for the answer.

Answers

8. $11\frac{1}{4}$ **9.** 10.5   **10.** 2.64

Do Exercises 9 and 10.

EXAMPLE 8 Solve: $\dfrac{3.4}{4.93} = \dfrac{10}{n}$.

We have

$$\frac{3.4}{4.93} = \frac{10}{n}$$

$3.4 \times n = 4.93 \times 10$ \qquad Equating cross products

$\dfrac{3.4 \times n}{3.4} = \dfrac{4.93 \times 10}{3.4}$ \qquad Dividing by 3.4

$n = \dfrac{4.93 \times 10}{3.4}$

$n = \dfrac{49.3}{3.4}$ \qquad Multiplying

$n = 14.5.$ \qquad Dividing

The solution is 14.5.

> Do Exercise 11.

11. Solve: $\dfrac{0.4}{0.9} = \dfrac{4.8}{t}$.

EXAMPLE 9 Solve: $\dfrac{4\frac{2}{3}}{5\frac{1}{2}} = \dfrac{14}{x}$.

We have

$$\frac{4\frac{2}{3}}{5\frac{1}{2}} = \frac{14}{x}$$

$4\dfrac{2}{3} \cdot x = 5\dfrac{1}{2} \cdot 14$ \qquad Equating cross products

$\dfrac{14}{3} \cdot x = \dfrac{11}{2} \cdot 14$ \qquad Converting to fraction notation

$\dfrac{\dfrac{14}{3} \cdot x}{\dfrac{14}{3}} = \dfrac{\dfrac{11}{2} \cdot 14}{\dfrac{14}{3}}$ \qquad Dividing by $\dfrac{14}{3}$

$x = \dfrac{11}{2} \cdot 14 \div \dfrac{14}{3}$

$x = \dfrac{11}{2} \cdot 14 \cdot \dfrac{3}{14}$ \qquad Multiplying by the reciprocal of the divisor

$x = \dfrac{11 \cdot 3}{2}$ \qquad Simplifying by removing a factor of 1: $\dfrac{14}{14} = 1$

$x = \dfrac{33}{2}$, or $16\dfrac{1}{2}$.

The solution is $\frac{33}{2}$, or $16\frac{1}{2}$.

> Do Exercise 12.

12. Solve:

$$\frac{8\frac{1}{3}}{x} = \frac{10\frac{1}{2}}{3\frac{3}{4}}.$$

a Determine whether the two pairs of numbers are proportional.

1. 5, 6 and 7, 9

2. 7, 5 and 6, 4

3. 1, 2 and 10, 20

4. 7, 3 and 21, 9

5. 2.4, 3.6 and 1.8, 2.7

6. 4.5, 3.8 and 6.7, 5.2

7. $5\frac{1}{3}, 8\frac{1}{4}$ and $2\frac{1}{5}, 9\frac{1}{2}$

8. $2\frac{1}{3}, 3\frac{1}{2}$ and 14, 21

b Solve.

9. $\dfrac{18}{4} = \dfrac{x}{10}$

10. $\dfrac{x}{45} = \dfrac{20}{25}$

11. $\dfrac{x}{8} = \dfrac{9}{6}$

12. $\dfrac{8}{10} = \dfrac{n}{5}$

13. $\dfrac{t}{12} = \dfrac{5}{6}$

14. $\dfrac{12}{4} = \dfrac{x}{3}$

15. $\dfrac{2}{5} = \dfrac{8}{n}$

16. $\dfrac{10}{6} = \dfrac{5}{x}$

17. $\dfrac{n}{15} = \dfrac{10}{30}$

18. $\dfrac{2}{24} = \dfrac{x}{36}$

19. $\dfrac{16}{12} = \dfrac{24}{x}$

20. $\dfrac{8}{12} = \dfrac{20}{x}$

21. $\dfrac{6}{11} = \dfrac{12}{x}$

22. $\dfrac{8}{9} = \dfrac{32}{n}$

23. $\dfrac{20}{7} = \dfrac{80}{x}$

24. $\dfrac{36}{x} = \dfrac{9}{5}$

25. $\dfrac{12}{9} = \dfrac{x}{7}$

26. $\dfrac{x}{20} = \dfrac{16}{15}$

27. $\dfrac{x}{13} = \dfrac{2}{9}$

28. $\dfrac{8}{11} = \dfrac{x}{5}$

29. $\dfrac{100}{25} = \dfrac{20}{n}$

30. $\dfrac{35}{125} = \dfrac{7}{m}$

31. $\dfrac{6}{y} = \dfrac{18}{15}$

32. $\dfrac{15}{y} = \dfrac{3}{4}$

33. $\dfrac{x}{3} = \dfrac{0}{9}$

34. $\dfrac{x}{6} = \dfrac{1}{6}$

35. $\dfrac{1}{2} = \dfrac{7}{x}$

36. $\dfrac{2}{5} = \dfrac{12}{x}$

37. $\dfrac{1.2}{4} = \dfrac{x}{9}$

38. $\dfrac{x}{11} = \dfrac{7.1}{2}$

39. $\dfrac{8}{2.4} = \dfrac{6}{y}$

40. $\dfrac{3}{y} = \dfrac{5}{4.5}$

41. $\dfrac{t}{0.16} = \dfrac{0.15}{0.40}$

42. $\dfrac{0.12}{0.04} = \dfrac{t}{0.32}$

43. $\dfrac{0.5}{n} = \dfrac{2.5}{3.5}$

44. $\dfrac{6.3}{0.9} = \dfrac{0.7}{n}$

45. $\dfrac{1.28}{3.76} = \dfrac{4.28}{y}$

46. $\dfrac{10.4}{12.4} = \dfrac{6.76}{t}$

47. $\dfrac{7}{\frac{1}{4}} = \dfrac{28}{x}$

48. $\dfrac{5}{\frac{1}{3}} = \dfrac{3}{x}$

49. $\dfrac{\frac{1}{5}}{\frac{1}{10}} = \dfrac{\frac{1}{10}}{x}$

50. $\dfrac{\frac{1}{4}}{\frac{1}{2}} = \dfrac{\frac{1}{2}}{x}$

51. $\dfrac{y}{\frac{3}{5}} = \dfrac{\frac{7}{12}}{\frac{14}{15}}$

52. $\dfrac{\frac{5}{8}}{\frac{5}{4}} = \dfrac{y}{\frac{3}{2}}$

53. $\dfrac{x}{1\frac{3}{5}} = \dfrac{2}{15}$

54. $\dfrac{1}{7} = \dfrac{x}{4\frac{1}{2}}$

55. $\dfrac{2\frac{1}{2}}{3\frac{1}{3}} = \dfrac{x}{4\frac{1}{4}}$

56. $\dfrac{3\frac{1}{2}}{y} = \dfrac{6\frac{1}{2}}{4\frac{2}{3}}$

57. $\dfrac{5\frac{1}{5}}{6\frac{1}{6}} = \dfrac{y}{3\frac{1}{2}}$

58. $\dfrac{10\frac{3}{8}}{12\frac{2}{3}} = \dfrac{5\frac{3}{4}}{y}$

Skill Maintenance

In each of Exercises 59–66, fill in the blank with the correct term from the given list. Some of the choices may not be used.

59. A ratio is the _____ of two quantities.   [5.1a]

60. A number is divisible by 9 if the _____ of the digits is divisible by 9.   [2.2a]

61. To compute a(n) _____ of a set of numbers, add the numbers and then divide by the number of addends.   [1.9c], [3.7a]

62. To convert from _____ to _____, move the decimal point two places to the right and change the $ sign in front to the ¢ sign at the end.   [4.3b]

63. In the equation $103 - 13 = 90$, the _____ is 13.   [1.3a]

64. The number 0.125 is an example of a(n) _____ decimal.   [4.5a]

65. The sentence $\dfrac{2}{5} \cdot \dfrac{4}{9} = \dfrac{4}{9} \cdot \dfrac{2}{5}$ illustrates the _____ law of multiplication.   [1.4a]

66. To solve $\dfrac{x}{a} = \dfrac{c}{d}$ for x, equate the _____ and divide on both sides to get x alone.   [5.3b]

cross products
cents
dollars
terminating
repeating
sum
difference
minuend
subtrahend
associative
commutative
average
product
quotient

Synthesis

⊞ Solve.

67. $\dfrac{1728}{5643} = \dfrac{836.4}{x}$

68. $\dfrac{328.56}{627.48} = \dfrac{y}{127.66}$

69. *Strikeouts per Home Run.*   Baseball Hall-of-Famer Babe Ruth had 1330 strikeouts and 714 home runs in his career. Hall-of-Famer Mike Schmidt had 1883 strikeouts and 548 home runs in his career. Find the rate of strikeouts per home run for each player. (These rates were considered among the highest in the history of the game and yet each made the Hall of Fame.)

Mid-Chapter Review

Concept Reinforcement

Determine whether each statement is true or false.

_____ **1.** A ratio is the quotient of two quantities. [5.1a]

_____ **2.** A rate is a ratio. [5.2a]

_____ **3.** The largest size package of an item always has the lowest unit price. [5.2b]

_____ **4.** If $\dfrac{x}{t} = \dfrac{y}{s}$, then $xy = ts$. [5.3b]

Guided Solutions

5. What is the rate, or speed, in miles per hour? [5.2a]

120 mi, 2 hr

$$\frac{120\ \square}{\square\ \text{hr}} = \frac{120\ \square}{\square\ \text{hr}} = \square\ \text{mi/hr}$$

6. Solve: $\dfrac{x}{4} = \dfrac{3}{6}$. [5.3b]

$$\frac{x}{4} = \frac{3}{6}$$

$x \cdot \square = \square \cdot 3$ Equating cross products

$\dfrac{x \cdot 6}{\square} = \dfrac{4 \cdot 3}{\square}$ Dividing on both sides

$x = \square$ Simplifying

Mixed Review

Find fraction notation for each ratio. [5.1a]

7. 4 to 7

8. 313 to 199

9. 35 to 17

10. 59 to 101

Find the ratio of the first number to the second and simplify. [5.1b]

11. 8 to 12

12. 25 to 75

13. 32 to 28

14. 100 to 76

15. 112 to 56

16. 15 to 3

17. 2.4 to 8.4

18. 0.27 to 0.45

Find each rate, or speed, as a ratio of distance to time. Round to the nearest hundredth where appropriate. [5.2a]

19. 243 mi, 4 hr

20. 146 km, 3 hr

21. 65 m, 5 sec

22. 97 ft, 6 sec

23. *Record Snowfall.* The greatest recorded snowfall in a single storm occurred in the Mt. Shasta Ski Bowl in California, when 189 in. fell during a seven-day storm in 1959. What is the rate in inches per day? [5.2a]
Source: U.S. Army Corps of Engineers

24. *Free Throws.* During the 2007–2008 basketball season, Peja Stojakovic of the New Orleans Hornets attempted 140 free throws and made 130 of them. What is the rate in number of free throws made to number of free throws attempted? Round to the nearest thousandth. [5.2a]
Source: National Basketball Association

25. Jerome bought an 18-oz jar of grape jelly for $2.09. What is the unit price in cents per ounce? [5.2b]

26. Martha bought 12 oz of deli honey ham for $5.99. What is the unit price in cents per ounce? [5.2b]

Determine whether the two pairs of numbers are proportional. [5.3a]

27. 3, 7 and 15, 35

28. 9, 7 and 7, 5

29. 2.4, 1.5 and 3.2, 2.1

30. $1\frac{3}{4}, 1\frac{1}{3}$ and $8\frac{3}{4}, 6\frac{2}{3}$

Solve. [5.3b]

31. $\dfrac{9}{15} = \dfrac{x}{20}$

32. $\dfrac{x}{24} = \dfrac{30}{18}$

33. $\dfrac{12}{y} = \dfrac{20}{15}$

34. $\dfrac{2}{7} = \dfrac{10}{y}$

35. $\dfrac{y}{1.2} = \dfrac{1.1}{0.6}$

36. $\dfrac{0.24}{0.02} = \dfrac{y}{0.36}$

37. $\dfrac{\frac{1}{4}}{x} = \dfrac{\frac{1}{8}}{\frac{1}{4}}$

38. $\dfrac{1\frac{1}{2}}{3\frac{1}{4}} = \dfrac{7\frac{1}{2}}{x}$

Understanding Through Discussion and Writing

39. Can every ratio be written as the ratio of some number to 1? Why or why not? [5.1a]

40. What can be concluded about a rectangle's width if the ratio of length to perimeter is 1 to 3? Make some sketches and explain your reasoning. [5.1a, b]

41. Instead of equating cross products, a student solves $\frac{x}{7} = \frac{5}{3}$ by multiplying on both sides by the least common denominator, 21. Is his approach a good one? Why or why not? [5.3b]

42. An instructor predicts that a student's test grade will be proportional to the amount of time the student spends studying. What is meant by this? Write an example of a proportion that involves the grades of two students and their study times. [5.3b]

5.4 Applications of Proportions

a Applications and Problem Solving

OBJECTIVE

a Solve applied problems involving proportions.

Proportions have applications in such diverse fields as business, chemistry, health sciences, and home economics, as well as in many areas of daily life. Proportions are useful in making predictions.

EXAMPLE 1 *Predicting Total Distance.* Donna drives her delivery van 800 mi in 3 days. At this rate, how far will she drive in 15 days?

1. **Familiarize.** We let d = the distance traveled in 15 days.

2. **Translate.** We translate to a proportion. We make each side the ratio of distance to time, with distance in the numerator and time in the denominator.

$$\text{Distance in 15 days} \rightarrow \frac{d}{15} = \frac{800}{3} \leftarrow \text{Distance in 3 days}$$
$$\text{Time} \rightarrow \qquad\qquad \leftarrow \text{Time}$$

It may help to verbalize the proportion above as "the unknown distance d is to 15 days as the known distance 800 mi is to 3 days."

3. **Solve.** Next, we solve the proportion:

$$3 \cdot d = 15 \cdot 800 \qquad \text{Equating cross products}$$

$$\frac{3 \cdot d}{3} = \frac{15 \cdot 800}{3} \qquad \text{Dividing by 3 on both sides}$$

$$d = \frac{15 \cdot 800}{3}$$

$$d = 4000. \qquad \text{Multiplying and dividing}$$

4. **Check.** We substitute into the proportion and check cross products:

$$\frac{4000}{15} = \frac{800}{3};$$

$$4000 \cdot 3 = 12{,}000; \qquad 15 \cdot 800 = 12{,}000.$$

The cross products are the same.

5. **State.** Donna will drive 4000 mi in 15 days.

> Do Exercise 1.

Problems involving proportions can be translated in more than one way. For Example 1, any one of the following is also a correct translation:

$$\frac{15}{d} = \frac{3}{800}, \qquad \frac{15}{3} = \frac{d}{800}, \qquad \frac{800}{d} = \frac{3}{15}.$$

Equating the cross products in each proportion gives us the equation $3 \cdot d = 15 \cdot 800$, which is the equation we obtained in Example 1.

1. Burning Calories. The readout on Mary's treadmill indicates that she burns 108 calories when she walks for 24 min. How many calories will she burn if she walks at the same rate for 30 min?

Answer

1. 135 calories

EXAMPLE 2 *Recommended Dosage.* To control a fever, a doctor suggests that a child who weighs 28 kg be given 320 mg of a liquid pain reliever. If the dosage is proportional to the child's weight, how much of the medication is recommended for a child who weighs 35 kg?

1. **Familiarize.** We let t = the number of milligrams of the liquid pain reliever.

2. **Translate.** We translate to a proportion, keeping the amount of medication in the numerators.

$$\text{Medication suggested} \to \frac{320}{28} = \frac{t}{35} \leftarrow \text{Medication suggested}$$
$$\text{Child's weight} \to \qquad\qquad \leftarrow \text{Child's weight}$$

3. **Solve.** Next, we solve the proportion:

$$320 \cdot 35 = 28 \cdot t \qquad \text{Equating cross products}$$

$$\frac{320 \cdot 35}{28} = \frac{28 \cdot t}{28} \qquad \text{Dividing by 28 on both sides}$$

$$\frac{320 \cdot 35}{28} = t$$

$$400 = t. \qquad \text{Multiplying and dividing}$$

4. **Check.** We substitute into the proportion and check cross products:

$$\frac{320}{28} = \frac{400}{35};$$

$$320 \cdot 35 = 11{,}200; \qquad 28 \cdot 400 = 11{,}200.$$

The cross products are the same.

5. **State.** The dosage for a child who weighs 35 kg is 400 mg.

Do Exercise 2.

2. Determining Paint Needs.
Lowell and Chris run a summer painting company to pay for their college expenses. They can paint 1600 ft^2 of clapboard with 4 gal of paint. How much paint would be needed for a building with 6000 ft^2 of clapboard?

EXAMPLE 3 *Purchasing Tickets.* Carey bought 8 tickets to an international food festival for $52. How many tickets could she purchase with $90?

1. **Familiarize.** We let n = the number of tickets that can be purchased with $90.

2. **Translate.** We translate to a proportion, keeping the number of tickets in the numerators.

$$\text{Tickets} \to \frac{8}{52} = \frac{n}{90} \leftarrow \text{Tickets}$$
$$\text{Cost} \to \qquad\qquad \leftarrow \text{Cost}$$

Answer

2. 15 gal

3. **Solve.** Next, we solve the proportion:

$52 \cdot n = 8 \cdot 90$ Equating cross products

$\dfrac{52 \cdot n}{52} = \dfrac{8 \cdot 90}{52}$ Dividing by 52 on both sides

$n = \dfrac{8 \cdot 90}{52}$

$n \approx 13.8.$ Multiplying and dividing

Because it is impossible to buy a fractional part of a ticket, we must round our answer *down* to 13.

4. **Check.** As a check, we use a different approach: We find the cost per ticket and then divide $90 by that price. Since $52 \div 8 = 6.50$ and $90 \div 6.50 \approx 13.8$, we have a check.

5. **State.** Carey could purchase 13 tickets with $90.

Do Exercise 3.

EXAMPLE 4 *Waist-to-Hip Ratio.* To reduce the risk of heart disease, it is recommended that a man's waist-to-hip ratio be 0.9 or lower. Mac's hip measurement is 40 in. To meet the recommendation, what should his waist measurement be?

Source: Mayo Clinic

Waist measurement is the smallest measurement below the ribs but above the navel.

Hip measurement is the largest measurement around the widest part of the buttocks.

1. **Familiarize.** Note that $0.9 = \dfrac{9}{10}$. We let $w =$ Mac's waist measurement.

2. **Translate.** We translate to a proportion as follows:

Waist measurement $\rightarrow \dfrac{w}{40} = \dfrac{9}{10} \begin{array}{l} \leftarrow \text{Recommended} \\ \leftarrow \text{waist-to-hip ratio} \end{array}$
Hip measurement $\rightarrow$

3. **Solve.** Next, we solve the proportion:

$10 \cdot w = 40 \cdot 9$ Equating cross products

$\dfrac{10 \cdot w}{10} = \dfrac{40 \cdot 9}{10}$ Dividing by 10 on both sides

$w = \dfrac{40 \cdot 9}{10}$

$w = 36.$ Multiplying and dividing

4. Check. As a check, we divide 36 by 40: $36 \div 40 = 0.9$. This is the desired ratio.

5. State. Mac's recommended waist measurement is 36 in. or less.

Do Exercise 4.

EXAMPLE 5 *Construction Plans.* Architects make blueprints of projects to be constructed. These are scale drawings in which lengths are in proportion to actual sizes. The Hennesseys are adding a rectangular deck to their house. The architectural blueprints are rendered such that $\frac{3}{4}$ in. on the drawing is actually 2.25 ft on the deck. The width of the deck on the drawing is 4.3 in. How wide is the deck in reality?

28.5 ft
w
l
4.3 in.

1. Familiarize. We let w = the width of the deck.

2. Translate. Then we translate to a proportion, using 0.75 for $\frac{3}{4}$ in.

$$\text{Measure on drawing} \rightarrow \frac{0.75}{2.25} = \frac{4.3}{w} \leftarrow \text{Width on drawing} \atop \leftarrow \text{Width on deck}$$

Measure on deck

3. Solve. Next, we solve the proportion:

$$0.75 \times w = 2.25 \times 4.3 \qquad \text{Equating cross products}$$

$$\frac{0.75 \times w}{0.75} = \frac{2.25 \times 4.3}{0.75} \qquad \text{Dividing by 0.75 on both sides}$$

$$w = \frac{2.25 \times 4.3}{0.75}$$

$$w = 12.9.$$

4. Check. We substitute into the proportion and check cross products:

$$\frac{0.75}{2.25} = \frac{4.3}{12.9};$$

$$0.75 \times 12.9 = 9.675; \qquad 2.25 \times 4.3 = 9.675.$$

The cross products are the same.

5. State. The width of the deck is 12.9 ft.

Do Exercise 5.

Answers

4. 34 in. or less   **5.** 9.5 in.

EXAMPLE 6 *Estimating a Wildlife Population.* To determine the number of fish in a lake, a conservationist catches 225 fish, tags them, and throws them back into the lake. Later, 108 fish are caught, and it is found that 15 of them are tagged. Estimate how many fish are in the lake.

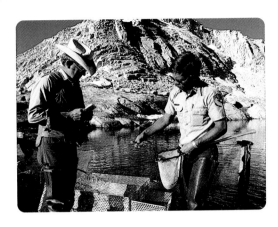

1. **Familiarize.** We let F = the number of fish in the lake. We assume that the ratio of the number of tagged fish to the total number of fish in the lake is the same as the ratio of the number of tagged fish caught later to the total number of fish caught later.

2. **Translate.** We translate to a proportion as follows:

Fish tagged originally $\rightarrow \dfrac{225}{F} = \dfrac{15}{108}.$ $\leftarrow$ Tagged fish caught later
Fish in lake $\rightarrow$ $\qquad\qquad\qquad \leftarrow$ Fish caught later

3. **Solve.** Next, we solve the proportion:

$225 \cdot 108 = F \cdot 15$ ⠀⠀ Equating cross products

$\dfrac{225 \cdot 108}{15} = \dfrac{F \cdot 15}{15}$ ⠀⠀ Dividing by 15 on both sides

$\dfrac{225 \cdot 108}{15} = F$

$1620 = F.$ ⠀⠀ Multiplying and dividing

4. **Check.** We substitute into the proportion and check cross products:

$\dfrac{225}{1620} = \dfrac{15}{108};$

$225 \cdot 108 = 24{,}300; \qquad 1620 \cdot 15 = 24{,}300.$

The cross products are the same.

5. **State.** We estimate that there are 1620 fish in the lake.

Do Exercise 6.

Do Exercise 6.

STUDY TIPS

WRITING ALL THE STEPS

Take the time to include all the steps when doing your homework. Doing so will help you organize your thinking and avoid computational errors. If you get a wrong answer, being able to see all the steps allows easier checking of your work. It will also give you complete, step-by-step solutions of the exercises that can be used to study for an exam.

6. Estimating a Deer Population. To determine the number of deer in a forest, a conservationist catches 153 deer, tags them, and releases them. Later, 62 deer are caught, and it is found that 18 of them are tagged. Estimate how many deer are in the forest.

Answer

6. 527 deer

Translating
for Success

1. *Calories in Cereal.* There are 140 calories in a $1\frac{1}{2}$-cup serving of Brand A cereal. How many calories are there in 6 cups of the cereal?

2. *Calories in Cereal.* There are 140 calories in 6 cups of Brand B cereal. How many calories are there in a $1\frac{1}{2}$-cup serving of the cereal?

3. *Gallons of Gasoline.* Jared's SUV traveled 310 mi on 15.5 gal of gasoline. At this rate, how many gallons would be needed to travel 465 mi?

4. *Gallons of Gasoline.* Elizabeth's new fuel-efficient car traveled 465 mi on 15.5 gal of gasoline. At this rate, how many gallons will be needed to travel 310 mi?

5. *Perimeter.* What is the perimeter of a rectangular field that measures 83.7 m by 62.4 m?

The goal of these matching questions is to practice step (2), *Translate*, of the five-step problem-solving process. Translate each word problem to an equation and select a correct translation from equations A–O.

A. $\dfrac{310}{15.5} = \dfrac{465}{x}$

B. $180 = 1\frac{1}{2} \cdot x$

C. $x = 71\frac{1}{8} - 76\frac{1}{2}$

D. $71\frac{1}{8} \cdot x = 74$

E. $74 \cdot 71\frac{1}{8} = x$

F. $x = 83.7 + 62.4$

G. $71\frac{1}{8} + x = 76\frac{1}{2}$

H. $x = 1\frac{2}{3} \cdot 180$

I. $\dfrac{140}{6} = \dfrac{x}{1\frac{1}{2}}$

J. $x = 2(83.7 + 62.4)$

K. $\dfrac{465}{15.5} = \dfrac{310}{x}$

L. $x = 83.7 \cdot 62.4$

M. $x = 180 \div 1\frac{2}{3}$

N. $\dfrac{140}{1\frac{1}{2}} = \dfrac{x}{6}$

O. $x = 1\frac{2}{3} \div 180$

Answers on page A-8

6. *Electric Bill.* Last month Todd's electric bills for his two rentals were $83.70 and $62.40. What was the total electric bill for the two properties?

7. *Package Tape.* A postal service center uses rolls of package tape that each contain 180 ft of tape. If it takes an average of $1\frac{2}{3}$ ft of tape per package, how many packages can be taped with one roll?

8. *Online Price.* Jane spent $180 for an area rug in a department store. Later, she saw the same rug for sale online and realized she had paid $1\frac{1}{2}$ times the online price. What was the online price?

9. *Heights of Sons.* Henry's three sons play basketball on three different college teams. Jeff's, Jason's, and Jared's heights are 74 in., $71\frac{1}{8}$ in., and $76\frac{1}{2}$ in., respectively. How much taller is Jared than Jason?

10. *Total Investment.* An investor bought 74 shares of stock at $71\frac{1}{8}$ per share. What was the total investment?

a Solve.

1. *Study Time and Test Grades.* An English instructor asserted that students' test grades are directly proportional to the amount of time spent studying. Lisa studies for 9 hr for a particular test and gets a score of 75. At this rate, for how many hours would she have had to study to get a score of 92?

2. *Study Time and Test Grades.* A mathematics instructor asserted that students' test grades are directly proportional to the amount of time spent studying. Brent studies for 15 hr for a final exam and gets a score of 75. At this rate, what score would he have received if he had studied for 18 hr?

3. *Lefties.* In a class of 40 students, on average, 6 will be left-handed. If a class includes 9 "lefties," how many students would you estimate are in the class?

4. *Sugaring.* When 20 gal of maple sap are boiled down, the result is $\frac{1}{2}$ gal of maple syrup. How much sap is needed to produce 9 gal of syrup?
Source: University of Maine

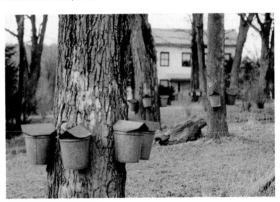

5. *Gas Mileage.* A 2008 Ford Mustang GT Convertible will travel 341 mi on 15.5 gal of gasoline in highway driving.
   a) How many gallons of gasoline will it take to drive 2690 mi from Boston to Phoenix?
   b) How far can the car be driven on 140 gal of gasoline?
   Source: Ford Motor Company

6. *Gas Mileage.* A 2008 Volkswagen Beetle will travel 462 mi on 16.5 gal of gasoline in highway driving.
   a) How many gallons of gasoline will it take to drive 1650 mi from Pittsburgh to Albuquerque?
   b) How far can the car be driven on 130 gal of gasoline?
   Source: Volkswagen of America, Inc.

7. *Overweight Americans.* A recent study determined that of every 100 Americans, 66 are overweight or obese. It is estimated that the U.S. population will be about 322 million in 2015. At the given rate, how many Americans would be considered overweight or obese in 2015?
   Source: U.S. Centers for Disease Control and Prevention

8. *Prevalence of Diabetes.* A recent study determined that of every 1000 Americans in the 65- to 74-year age group, 185 have been diagnosed with diabetes. It is estimated that there will be about 26.6 million Americans in this age group in 2015. At the given rate, how many in this age group will be diagnosed with diabetes in 2015?
   Source: U.S. Centers for Disease Control and Prevention

9. *Quality Control.* A quality-control inspector examined 100 lightbulbs and found 7 of them to be defective. At this rate, how many defective bulbs will there be in a lot of 2500?

10. *Grading.* A high school math teacher must grade 32 reports on famous mathematicians. She can grade 4 reports in 90 min. At this rate, how long will it take her to grade all 32 reports?

11. *Painting.* Fred uses 3 gal of paint to cover 1275 ft^2 of siding. How much siding can Fred paint with 7 gal of paint?

12. *Waterproofing.* Bonnie can waterproof 450 ft^2 of decking with 2 gal of sealant. How many gallons should Bonnie buy for a 1200-ft^2 deck?

13. *Publishing.* Every 6 pages of an author's manuscript corresponds to 5 published pages. How many published pages will a 540-page manuscript become?

14. *Turkey Servings.* An 8-lb turkey breast contains 36 servings of meat. How many pounds of turkey breast would be needed for 54 servings?

15. *Currency Exchange.* On 12 September 2008, 1 U.S. dollar was worth about 0.5613 British pound.
a) How much would 50 U.S. dollars be worth in British pounds?
b) How much would a car that costs 8640 British pounds cost in U.S. dollars?

16. *Currency Exchange.* On 12 September 2008, 1 U.S. dollar was worth about 1.133 Swiss francs.
a) How much would 360 U.S. dollars be worth in Swiss francs?
b) How much would a pair of jeans that costs 80 Swiss francs cost in U.S. dollars?

17. *Currency Exchange.* On 12 September 2008, 1 U.S. dollar was worth about 107.197 Japanese yen.
a) How much would 125 U.S. dollars be worth in Japanese yen?
b) While traveling in Japan, Dana bought a vase for 5120 Japanese yen. How much did it cost in U.S. dollars?

18. *Currency Exchange.* On 12 September 2008, 1 U.S. dollar was worth about 10.5859 Mexican pesos.
a) How much would 150 U.S. dollars be worth in Mexican pesos?
b) While traveling in Mexico, Jake bought a watch that cost 3600 Mexican pesos. How much did it cost in U.S. dollars?

19. *Mileage.*   Jean bought a new car. In the first 8 months, it was driven 9000 mi. At this rate, how many miles will the car be driven in 1 yr?

20. *Coffee Production.*   Coffee beans from 18 trees are required to produce enough coffee each year for a person who drinks 2 cups of coffee per day. Jared brews 15 cups of coffee each day for himself and his coworkers. How many coffee trees are required for this?

21. *Cap'n Crunch's Peanut Butter Crunch® Cereal.*   The nutritional chart on the side of a box of Quaker Cap'n Crunch's Peanut Butter Crunch® Cereal states that there are 110 calories in a $\frac{3}{4}$-cup serving. How many calories are there in 6 cups of the cereal?

Nutrition Facts

Serving Size 3/4 Cup (27g)

Amount Per Serving	Cereal Alone	with 1/2 Cup Vitamin A&D Fortified Skim Milk
Calories	110	150
Calories from Fat	25	25
	% Daily Value	
Total Fat 2.5g	4%	4%
Saturated Fat 1g	5%	6%
Trans Fat 0g		
Polyunsaturated Fat 0.5g		
Monounsaturated Fat 1g		
Cholesterol 0mg	0%	1%
Sodium 200mg	8%	10%
Potassium 65mg	2%	7%
Total Carbohydrate 21g	7%	9%
Dietary Fiber 1g	3%	3%
Sugars 9g		
Other Carbohydrate 11g		
Protein 2g		

22. *Rice Krispies® Cereal.*   The nutritional chart on the side of a box of Kellogg's Rice Krispies® Cereal states that there are 130 calories in a $1\frac{1}{4}$-cup serving. How many calories are there in 5 cups of the cereal?

Nutrition Facts

Serving Size $1\frac{1}{4}$ Cups (33g/1.2oz)

Amount Per Serving	Cereal	Cereal with 1/2 Cup Vitamins A&D Fat Free Milk
Calories	130	170
Calories from Fat	0	0
	% Daily Value	
Total Fat 0g	0%	0%
Saturated Fat 1g	0%	0%
Trans Fat 0g		
Cholesterol 0mg	0%	0%
Sodium 220mg	9%	12%
Potassium 30mg	1%	7%
Total Carbohydrate 29g	10%	11%
Dietary Fiber 0g	0%	0%
Sugars 4g		
Other Carbohydrate 25g		
Protein 2g		

23. *Metallurgy.*   In a metal alloy, the ratio of zinc to copper is 3 to 13. If there are 520 lb of copper, how many pounds of zinc are there?

24. *Class Size.*   A college advertises that its student-to-faculty ratio is 14 to 1. If 56 students register for Introductory Spanish, how many sections of the course would you expect to see offered?

25. *Painting.*   Helen can paint 950 ft² with 2 gal of paint. How many 1-gal cans does she need in order to paint a 30,000-ft² wall?

26. *Snow to Water.*   Under typical conditions, $1\frac{1}{2}$ ft of snow will melt to 2 in. of water. To how many inches of water will $5\frac{1}{2}$ ft of snow melt?

27. *Grass-Seed Coverage.* It takes 60 oz of grass seed to seed 3000 ft² of lawn. At this rate, how much would be needed for 5000 ft² of lawn?

28. *Grass-Seed Coverage.* In Exercise 27, how much seed would be needed for 7000 ft² of lawn?

29. *Estimating a Deer Population.* To determine the number of deer in a game preserve, a forest ranger catches 318 deer, tags them, and releases them. Later, 168 deer are caught, and it is found that 56 of them are tagged. Estimate how many deer are in the game preserve.

30. *Estimating a Trout Population.* To determine the number of trout in a lake, a conservationist catches 112 trout, tags them, and throws them back into the lake. Later, 82 trout are caught, and it is found that 32 of them are tagged. Estimate how many trout there are in the lake.

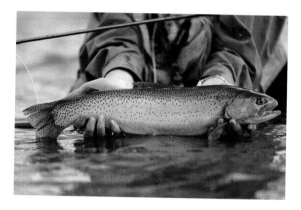

31. *Map Scaling.* On a road atlas map, 1 in. represents 16.6 mi. If two cities are 3.5 in. apart on the map, how far apart are they in reality?

32. *Map Scaling.* On a map, $\frac{1}{4}$ in. represents 50 mi. If two cities are $3\frac{1}{4}$ in. apart on the map, how far apart are they in reality?

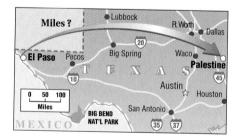

33. *Gasoline Mileage.*   Nancy's van traveled 84 mi on 6.5 gal of gasoline. At this rate, how many gallons would be needed to travel 126 mi?

34. *Bicycling.*   Roy bicycled 234 mi in 14 days. At this rate, how far would Roy bicycle in 42 days?

35. *Home Runs.*   After playing 147 games in the 2008 Major League Baseball season, Ryan Howard of the Philadelphia Phillies had hit 43 home runs.

  a) At this rate, how many games would it take him to hit 50 home runs?
  b) At this rate, how many home runs would Howard hit in the 162-game baseball season?

Source: Major League Baseball

36. *Hits.*   After playing 145 games in the 2008 Major League Baseball season, Ichiro Suzuki of the Seattle Mariners had 191 hits.

  a) At this rate, how many games would it take him to get 200 hits?
  b) At this rate, how many hits would Suzuki get in the 162-game baseball season?

Source: Major League Baseball

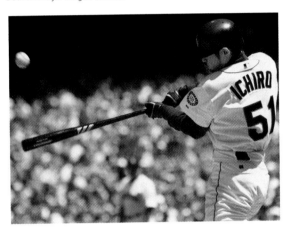

Skill Maintenance

Determine whether each number is prime, composite, or neither.   [2.1c]

37. 1   **38.** 83   **39.** 28   **40.** 93   **41.** 47

Find the prime factorization of each number.   [2.1d]

42. 808   **43.** 56   **44.** 866   **45.** 93   **46.** 2020

Synthesis

47. 🖩 Carney College is expanding from 850 to 1050 students. To avoid any rise in the student-to-faculty ratio, the faculty of 69 professors must also increase. How many new faculty positions should be created?

48. 🖩 In recognition of her outstanding work, Sheri's salary has been increased from $26,000 to $29,380. Tim is earning $23,000 and is requesting a proportional raise. How much more should he ask for?

49. *Baseball Statistics.*   Cy Young, one of the greatest baseball pitchers of all time, gave up an average of 2.63 earned runs every 9 innings. Young pitched 7356 innings, more than anyone else in the history of baseball. How many earned runs did he give up?

50. 🖩 *Real-Estate Values.*   According to Coldwell Banker Real Estate Corporation, a home selling for $189,000 in Austin, Texas, would sell for $437,850 in Denver, Colorado. How much would a $350,000 home in Denver sell for in Austin? Round to the nearest $1000.

Source: Coldwell Banker Real Estate Corporation

51. 🖩 The ratio 1 : 3 : 2 is used to estimate the relative costs of a CD player, receiver, and speakers when shopping for a sound system. That is, the receiver should cost three times the amount spent on the CD player and the speakers should cost twice the amount spent on the CD player. If you had $900 to spend, how would you allocate the money, using this ratio?

5.5

Geometric Applications

OBJECTIVES

a Find lengths of sides of similar triangles using proportions.

b Use proportions to find lengths in pairs of figures that differ only in size.

(a) Proportions and Similar Triangles

Look at the pair of triangles below. Note that they appear to have the same shape, but their sizes are different. These are examples of **similar triangles**. You can imagine using a magnifying glass to enlarge the smaller triangle and get the larger. This process works because the corresponding sides of each triangle have the same ratio. That is, the following proportion is true.

$$\frac{a}{d} = \frac{b}{e} = \frac{c}{f}$$

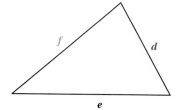

SIMILAR TRIANGLES

Similar triangles have the same shape. The lengths of their corresponding sides have the same ratio—that is, they are proportional.

EXAMPLE 1 The triangles below are similar triangles. Find the missing length x.

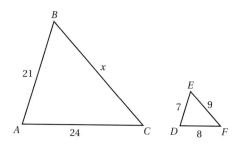

The ratio of x to 9 is the same as the ratio of 24 to 8 or 21 to 7. We get the proportions

$$\frac{x}{9} = \frac{24}{8} \quad \text{and} \quad \frac{x}{9} = \frac{21}{7}.$$

We can solve either one of these proportions. We use the first one:

$$\frac{x}{9} = \frac{24}{8}$$

$x \cdot 8 = 9 \cdot 24$ Equating cross products

$\dfrac{x \cdot 8}{8} = \dfrac{9 \cdot 24}{8}$ Dividing by 8 on both sides

$x = 27.$ Simplifying

The missing length x is 27. Other proportions could also be used.

Do Exercise 1.

---

1. This pair of triangles is similar. Find the missing length x.

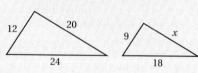

Answer

1. 15

Similar triangles and proportions can often be used to find lengths that would ordinarily be difficult to measure. For example, we could find the height of a flagpole without climbing it or the distance across a river without crossing it.

EXAMPLE 2 How high is a flagpole that casts a 56-ft shadow at the same time that a 6-ft man casts a 5-ft shadow?

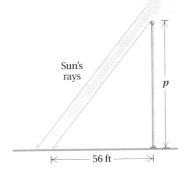

If we use the sun's rays to represent the third side of the triangle in our drawing of the situation, we see that we have similar triangles. Let p = the height of the flagpole. The ratio of 6 to p is the same as the ratio of 5 to 56. Thus we have the proportion

$$\text{Height of man} \longrightarrow \frac{6}{p} = \frac{5}{56}. \longleftarrow \text{Length of shadow of man}$$
$$\text{Height of pole} \longrightarrow p \qquad 56 \longleftarrow \text{Length of shadow of pole}$$

Solve: $6 \cdot 56 = 5 \cdot p$ Equating cross products

$$\frac{6 \cdot 56}{5} = \frac{5 \cdot p}{5}$$ Dividing by 5 on both sides

$$\frac{6 \cdot 56}{5} = p$$ Simplifying

$$67.2 = p.$$

The height of the flagpole is 67.2 ft.

> Do Exercise 2.

2. How high is a flagpole that casts a 45-ft shadow at the same time that a 5.5-ft woman casts a 10-ft shadow?

EXAMPLE 3 *Rafters of a House.* Carpenters use similar triangles to determine the lengths of rafters for a house. They first choose the pitch of the roof, or the ratio of the rise over the run. Then using a triangle with that ratio, they calculate the length of the rafter needed for the house. Loren is constructing rafters for a roof with a 6/12 pitch on a house that is 30 ft wide. Using a rafter guide, Loren knows that the rafter length corresponding to the 6/12 pitch is 13.4. Find the length x of the rafter of the house to the nearest tenth of a foot.

Answer

2. 24.75 ft

We have the proportion

Length of rafter
in 6/12 triangle → $\dfrac{13.4}{x} = \dfrac{12}{15}$. ← Run in 6/12 triangle
Length of rafter →
on the house ← Run in similar triangle on the house

Solve: $13.4 \cdot 15 = x \cdot 12$ Equating cross products

$$\dfrac{13.4 \cdot 15}{12} = \dfrac{x \cdot 12}{12}$$ Dividing by 12 on both sides

$$\dfrac{13.4 \cdot 15}{12} = x$$

$16.8 \text{ ft} \approx x.$ Rounding to the nearest tenth of a foot

The length x of the rafter of the house is about 16.8 ft.

Do Exercise 3.

3. Rafters of a House. Referring to Example 3, find the length y of the rise of the rafter of the house to the nearest tenth of a foot.

(b) Proportions and Other Geometric Shapes

When one geometric figure is a magnification of another, the figures are similar. Thus the corresponding lengths are proportional.

EXAMPLE 4 The sides in the photographs below are proportional. Find the width of the larger photograph.

2.5 cm |←—3.5 cm—→| x |←————— 10.5 cm —————→|

We let $x =$ the width of the photograph. Then we translate to a proportion.

Larger width → $\dfrac{x}{2.5} = \dfrac{10.5}{3.5}$ ← Larger length
Smaller width →   ← Smaller length

Solve: $3.5 \times x = 2.5 \times 10.5$ Equating cross products

$$\dfrac{3.5 \times x}{3.5} = \dfrac{2.5 \times 10.5}{3.5}$$ Dividing by 3.5 on both sides

$$x = \dfrac{2.5 \times 10.5}{3.5}$$ Simplifying

$$x = 7.5.$$

Thus the width of the larger photograph is 7.5 cm.

Do Exercise 4.

4. The sides in the photographs below are proportional. Find the width of the larger photograph.

6 cm |←10 cm→| x |←——— 35 cm ———→|

EXAMPLE 5 A scale model of an addition to an athletic facility is 12 cm wide at the base and rises to a height of 15 cm. If the actual base is to be 52 ft, what will the actual height of the addition be?

We let h = the height of the addition. Then we translate to a proportion.

$$\text{Width in model} \rightarrow \frac{12}{52} = \frac{15}{h} \begin{array}{l} \leftarrow \text{Height in model} \\ \leftarrow \text{Actual height} \end{array}$$
$$\text{Actual width} \rightarrow$$

Solve: $12 \cdot h = 52 \cdot 15$ Equating cross products

$\dfrac{12 \cdot h}{12} = \dfrac{52 \cdot 15}{12}$ Dividing by 12 on both sides

$h = \dfrac{52 \cdot 15}{12} = 65.$

Thus the height of the addition will be 65 ft.

Do Exercise 5.

5. Refer to the figures in Example 5. If a skylight on the model is 3 cm wide, how wide will the actual skylight be?

Answer

5. 13 ft

a The triangles in each exercise are similar. Find the missing lengths.

1.

2.

3.

4.

5.

6.

7.

8.

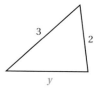

9. When a tree 8 m high casts a shadow 5 m long, how long a shadow is cast by a person 2 m tall?

10. How high is a flagpole that casts a 42-ft shadow at the same time that a $5\frac{1}{2}$-ft woman casts a 7-ft shadow?

11. How high is a tree that casts a 27-ft shadow at the same time that a 4-ft fence post casts a 3-ft shadow?

12. How high is a tree that casts a 32-ft shadow at the same time that an 8-ft light pole casts a 9-ft shadow?

13. Find the height h of the wall.

6 ft

6 ft

7 ft

h

14. Find the length L of the lake. Assume that the ratio of L to 120 yd is the same as the ratio of 720 yd to 30 yd.

120 yd

30 yd

720 yd

L

15. Find the distance across the river. Assume that the ratio of d to 25 ft is the same as the ratio of 40 ft to 10 ft.

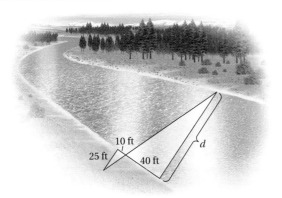

10 ft

25 ft

40 ft

d

16. To measure the height of a hill, a string is stretched from level ground to the top of the hill. A 3-ft stick is placed under the string, touching it at point P, a distance of 5 ft from point G, where the string touches the ground. The string is then detached and found to be 120 ft long. How high is the hill?

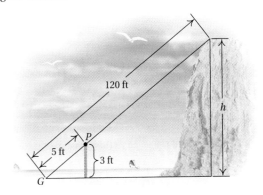

120 ft

P

5 ft

3 ft

G

h

b In each of Exercises 17–26, the sides in each pair of figures are proportional. Find the missing lengths.

17.

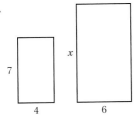

6

9

x

6

18.

5

7

x

14

19.

7

x

4

6

20.

11

x

4

3

21.

22.

23.

24.

25.

26.

27. A scale model of an addition to an athletic facility is 15 cm wide at the base and rises to a height of 19 cm. If the actual base is to be 120 ft, what will the actual height of the addition be?

28. Refer to the figures in Exercise 27. If a skylight on the model is 3 cm wide, how wide will the actual skylight be?

Skill Maintenance

29. *Expense Needs.* A student has $34.97 to spend for a book at $49.95, a CD at $14.88, and a sweatshirt at $29.95. How much more money does the student need to make these purchases? [4.7a]

30. Divide: $80.892 \div 8.4$. [4.4a]

Multiply. [4.3a]

31. 8.4×80.892

32. 0.01×274.568

33. 100×274.568

34. 0.002×274.568

Find decimal notation and round to the nearest thousandth, if appropriate. [4.5a, b]

35. $\dfrac{17}{20}$

36. $\dfrac{73}{40}$

37. $\dfrac{10}{11}$

38. $\dfrac{43}{51}$

Synthesis

Hockey Goals. An official hockey goal is 6 ft wide. To make scoring more difficult, goalies often locate themselves far in front of the goal to "cut down the angle." In Exercises 39 and 40, suppose that a slapshot from point A is attempted and that the goalie is 2.7 ft wide. Determine how far from the goal the goalie should be located if point A is the given distance from the goal. (*Hint*: First find how far the goalie should be from point A.)

39. 25 ft

40. 35 ft

41. A miniature basketball hoop is built for the model referred to in Exercise 27. An actual hoop is 10 ft high. How high should the model hoop be?

 Solve. Round the answer to the nearest thousandth.

42. $\dfrac{8664.3}{10{,}344.8} = \dfrac{x}{9776.2}$

43. $\dfrac{12.0078}{56.0115} = \dfrac{789.23}{y}$

The triangles in each exercise are similar triangles. Find the lengths not given.

44.

45.

Key Terms

ratio, p. 296
rate, p. 303

unit price, p. 304
unit rate, p. 304

proportion, p. 310
similar triangles, p. 330

Concept Reinforcement

Determine whether each statement is true or false.

_____ **1.** When we simplify a ratio like $\frac{8}{12}$, we find two other numbers in the same ratio.   [5.1b]

_____ **2.** The proportion $\frac{a}{b} = \frac{c}{d}$ can also be written as $\frac{c}{a} = \frac{d}{b}$.   [5.3a]

_____ **3.** Similar triangles must be the same size.   [5.5a]

_____ **4.** Lengths of corresponding sides of similar figures have the same ratio.   [5.5b]

Important Concepts

Objective 5.1a   Find fraction notation for ratios.

Example   Find the ratio of 7 to 18.

Write a fraction with a numerator of 7 and a denominator of 18: $\frac{7}{18}$.

Practice Exercise

1. Find the ratio of 17 to 3.

Objective 5.1b   Simplify ratios.

Example   Simplify the ratio 8 to 2.5.

We first write the ratio in fraction notation. Next, we multiply by 1 to clear the decimal from the denominator. Then we simplify.

$$\frac{8}{2.5} = \frac{8}{2.5} \cdot \frac{10}{10} = \frac{80}{25} = \frac{5 \cdot 16}{5 \cdot 5}$$

$$= \frac{5}{5} \cdot \frac{16}{5} = \frac{16}{5}$$

Practice Exercise

2. Simplify the ratio 3.2 to 2.8.

Objective 5.2a   Give the ratio of two different measures as a rate.

Example   A driver travels 156 mi on 6.5 gal of gas. What is the rate in miles per gallon?

$$\frac{156 \text{ mi}}{6.5 \text{ gal}} = \frac{156}{6.5} \frac{\text{mi}}{\text{gal}} = 24 \frac{\text{mi}}{\text{gal}}, \text{ or } 24 \text{ mpg}$$

Practice Exercise

3. A student earned $120 for working 16 hr. What was the rate of pay per hour?

Objective 5.2b Find unit prices and use them to compare purchases.

Example A 16-oz can of Garcia diced tomatoes costs $1.00. A 20-oz can of Del's diced tomatoes costs $1.23. Which has the lower unit price?

We find each unit price:

For Garcia: $\dfrac{\$1.00}{16\,oz} = \dfrac{100\text{ cents}}{16\,oz} = 6.25\,\dfrac{\text{cents}}{oz}$;

For Del's: $\dfrac{\$1.23}{20\,oz} = \dfrac{123\text{ cents}}{20\,oz} = 6.15\,\dfrac{\text{cents}}{oz}$.

Thus Del's has the lower unit price.

Practice Exercise

4. A 28-oz jar of Brand A spaghetti sauce costs $2.79. A 32-oz jar of Brand B spaghetti sauce costs $3.29. Find the unit price of each brand and determine which is a better buy based on unit price alone.

Objective 5.3a Determine whether two pairs of numbers are proportional.

Example Determine whether 3, 4 and 7, 9 are proportional.

We have

$$3 \cdot 9 = 27 \qquad \dfrac{3}{4} \overset{?}{=} \dfrac{7}{9} \qquad 4 \cdot 7 = 28.$$

Since the cross products are not the same ($27 \ne 28$), $\dfrac{3}{4} \ne \dfrac{7}{9}$ and the numbers are not proportional.

Practice Exercise

5. Determine whether 7, 9 and 21, 27 are proportional.

Objective 5.3b Solve proportions.

Example Solve: $\dfrac{3}{4} = \dfrac{y}{7}$.

$$\dfrac{3}{4} = \dfrac{y}{7}$$

$3 \cdot 7 = 4 \cdot y$ Equating cross products

$\dfrac{3 \cdot 7}{4} = \dfrac{4 \cdot y}{4}$ Dividing by 4 on both sides

$\dfrac{21}{4} = y$

The solution is $\dfrac{21}{4}$.

Practice Exercise

6. Solve: $\dfrac{9}{x} = \dfrac{8}{3}$.

Objective 5.4a Solve applied problems involving proportions.

Example Martina bought 3 tickets to a campus theater production for $16.50. How much would 8 tickets cost?

We translate to a proportion.

$$\begin{array}{c} Tickets \rightarrow \\ Cost \rightarrow \end{array} \dfrac{3}{16.50} = \dfrac{8}{c} \begin{array}{c} \leftarrow Tickets \\ \leftarrow Cost \end{array}$$

$3 \cdot c = 16.50 \cdot 8$ Equating cross products

$c = \dfrac{16.50 \cdot 8}{3}$

$c = 44$

Eight tickets would cost $44.

Practice Exercise

7. On a map, $\frac{1}{2}$ in. represents 50 mi. If two cities are $1\frac{3}{4}$ in. apart on the map, how far apart are they in reality?

Objective 5.5a Find lengths of sides of similar triangles using proportions.

Example The triangles below are similar. Find the missing length *x*.

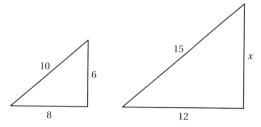

The ratio of 6 to *x* is the same as the ratio of 8 to 12 (and also as the ratio of 10 to 15). We write and solve a proportion:

$$\frac{6}{x} = \frac{8}{12}$$

$$6 \cdot 12 = x \cdot 8$$

$$\frac{6 \cdot 12}{8} = \frac{x \cdot 8}{8}$$

$$9 = x.$$

The missing length is 9. (We could also have used other proportions, including $\frac{6}{x} = \frac{10}{15}$, to find *x*.)

Practice Exercise

8. The triangles below are similar. Find the missing length *y*.

Review Exercises

Write fraction notation for each ratio. Do not simplify. [5.1a]

1. 47 to 84

2. 46 to 1.27

3. 83 to 100

4. 0.72 to 197

5. At Preston Seafood Market, 12,480 lb of tuna and 16,640 lb of salmon were sold one year. [5.1a, b]

   a) Write fraction notation for the ratio of tuna sold to salmon sold.

   b) Write fraction notation for the ratio of salmon sold to the total number of pounds of both kinds of fish sold.

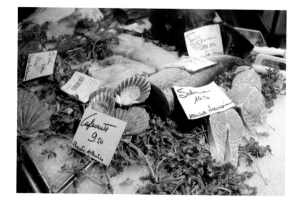

Find the ratio of the first number to the second number and simplify. [5.1b]

6. 9 to 12

7. 3.6 to 6.4

8. *Gas Mileage.* The Chrysler PT Cruiser will travel 377 mi on 14.5 gal of gasoline in highway driving. What is the rate in miles per gallon? [5.2a]
   Source: Chrysler Motor Corporation

9. *Flywheel Revolutions.* A certain flywheel makes 472,500 revolutions in 75 min. What is the rate of spin in revolutions per minute? [5.2a]

10. A lawn requires 319 gal of water for every 500 ft². What is the rate in gallons per square foot? [5.2a]

11. *Calcium Supplement.* The price for a particular calcium supplement is $18.99 for 300 tablets. Find the unit price in cents per tablet. [5.2b]

12. Nola bought a 24-oz loaf of 12-grain bread for $2.69. Find the unit price in cents per ounce. [5.2b]

13. *Vegetable Oil.* Find the unit price. Then determine which size has the lowest unit price. [5.2b]

PACKAGE	PRICE	UNIT PRICE
32 oz	$4.79	
48 oz	$5.99	
64 oz	$9.99	

Determine whether the two pairs of numbers are proportional. [5.3a]

14. 9, 15 and 36, 60

15. 24, 37 and 40, 46.25

Solve. [5.3b]

16. $\dfrac{8}{9} = \dfrac{x}{36}$

17. $\dfrac{6}{x} = \dfrac{48}{56}$

18. $\dfrac{120}{\frac{3}{7}} = \dfrac{7}{x}$

19. $\dfrac{4.5}{120} = \dfrac{0.9}{x}$

Solve. [5.4a]

20. *Quality Control.* A factory manufacturing computer circuits found 3 defective circuits in a lot of 65 circuits. At this rate, how many defective circuits can be expected in a lot of 585 circuits?

21. *Exchanging Money.* On 16 September 2008, 1 U.S. dollar was worth about 1.068 Canadian dollars.

  a) How much would 250 U.S. dollars be worth in Canada?

  b) While traveling in Canada, Jamal saw a sweatshirt that cost 50 Canadian dollars. How much would it cost in U.S. dollars?

22. A train travels 448 mi in 7 hr. At this rate, how far will it travel in 13 hr?

23. Fifteen acres of land is required to produce 54 bushels of tomatoes. At this rate, how many acres are required to produce 97.2 bushels of tomatoes?

24. *Trash Production.* A study shows that 5 people generate 23 lb of trash each day. The population of Austin, Texas, is 743,074. How many pounds of trash are produced in Austin in one day?

Sources: U.S. Environmental Protection Agency; U.S. Census Bureau

25. *Snow to Water.* Under typical conditions, $1\frac{1}{2}$ ft of snow will melt to 2 in. of water. To how many inches of water will $4\frac{1}{2}$ ft of snow melt?

26. *Lawyers in Chicago.* In Illinois, there are about 4.8 lawyers for every 1000 people. The population of Chicago is 2,842,518. How many lawyers would you expect there to be in Chicago?

Sources: American Bar Association; U.S. Census Bureau

Each pair of triangles in Exercises 27 and 28 is similar. Find the missing length(s). [5.5a]

27.

28.

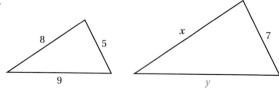

29. How high is a billboard that casts a 25-ft shadow at the same time that an 8-ft sapling casts a 5-ft shadow?   [5.5a]

30. The lengths in the figures below are proportional. Find the missing lengths.   [5.5b]

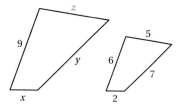

31. *Turkey Servings.*   A 25-lb turkey serves 18 people. Find the rate in servings per pound.   [5.2a]

  A. 0.36 serving/lb   **B.** 0.72 serving/lb
  C. 0.98 serving/lb   **D.** 1.39 servings/lb

32. If 3 dozen eggs cost $5.04, how much will 5 dozen eggs cost?   [5.4a]

  A. $6.72   **B.** $6.96
  C. $8.40   **D.** $10.08

Synthesis

33. *Paper Towels.*   Find the unit price and determine which item would be the best buy based on unit price alone.   [5.2b]

PACKAGE	PRICE	UNIT PRICE PER SHEET
8 rolls, 60 (2 ply) sheets per roll	$6.38	
15 rolls, 60 (2 ply) sheets per roll	$13.99	
6 big rolls, 165 (2 ply) sheets per roll	$10.99	

34. 🖩 The following triangles are similar. Find the missing lengths.   [5.5a]

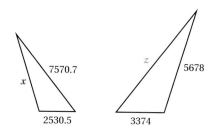

35. Shine-and-Glo Painters uses 2 gal of finishing paint for every 3 gal of primer. Each gallon of finishing paint covers 450 ft^2. If a surface of 4950 ft^2 needs both primer and finishing paint, how many gallons of each should be purchased?   [5.4a]

Understanding Through Discussion and Writing

1. If you were a college president, which would you prefer: a low or high faculty-to-student ratio? Why?   [5.1a]

2. Can unit prices be used to solve proportions that involve money? Explain why or why not.   [5.2b], [5.4a]

3. Write a proportion problem for a classmate to solve. Design the problem so that the solution is "Leslie would need 16 gal of gasoline in order to travel 368 mi."   [5.4a]

4. Is it possible for two triangles to have two pairs of sides that are proportional without the triangles being similar? Why or why not?   [5.5a]

CHAPTER
5

Test   For Extra Help
CHAPTER
Test Prep
VIDEOS

Step-by-step test solutions are found on the Chapter Test Prep Videos available via the Video Resources on DVD, in *MyMathLab*, and on YouTube (search "BittingerBasicMath" and click on "Channels").

Write fraction notation for each ratio. Do not simplify.

1. 85 to 97

2. 0.34 to 124

Find the ratio of the first number to the second number and simplify.

3. 18 to 20

4. 0.75 to 0.96

5. What is the rate in feet per second?
   10 feet, 16 seconds

6. *Ham Servings.*   A 12-lb shankless ham contains 16 servings. What is the rate in servings per pound?

7. *Gas Mileage.*   The 2008 Chevrolet Malibu LTZ will travel 464 mi on 14.5 gal of gasoline in highway driving. What is the rate in miles per gallon?
   Source: General Motors Corporation

8. *Bagged Salad Greens.*   Ron bought a 16-oz bag of salad greens for $2.49. Find the unit price in cents per ounce.

9. The table below lists prices for concentrated liquid laundry detergent. Find the unit price of each size in cents per ounce. Then determine which has the lowest unit price.

SIZE	PRICE	UNIT PRICE
40 oz	$6.59	
50 oz	$6.99	
100 oz	$11.49	
150 oz	$24.99	

Determine whether the two pairs of numbers are proportional.

10. 7, 8 and 63, 72

11. 1.3, 3.4 and 5.6, 15.2

Solve.

12. $\dfrac{9}{4} = \dfrac{27}{x}$

13. $\dfrac{150}{2.5} = \dfrac{x}{6}$

14. $\dfrac{x}{100} = \dfrac{27}{64}$

15. $\dfrac{68}{y} = \dfrac{17}{25}$

Solve.

16. *Distance Traveled.*   An ocean liner traveled 432 km in 12 hr. At this rate, how far would the boat travel in 42 hr?

17. *Time Loss.*   A watch loses 2 min in 10 hr. At this rate, how much will it lose in 24 hr?

18. *Map Scaling.* On a map, 3 in. represents 225 mi. If two cities are 7 in. apart on the map, how far are they apart in reality?

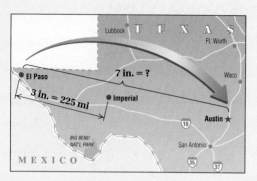

19. *Tower Height.* A birdhouse built on a pole that is 3 m high casts a shadow 5 m long. At the same time, the shadow of a tower is 110 m long. How high is the tower?

20. *Exchanging Money.* On 18 September 2008, 1 U.S. dollar was worth about 7.781 Hong Kong dollars.

  a) How much would 450 U.S. dollars be worth in Hong Kong dollars?

  b) While traveling in Hong Kong, Mitchell saw a DVD player that cost 795 Hong Kong dollars. How much would it cost in U.S. dollars?

21. *Thanksgiving Dinner.* A traditional turkey dinner for 8 people cost about $33.81 in a recent year. How much would it cost to serve a turkey dinner for 14 people?

Source: American Farm Bureau Federation

The sides in each pair of figures are proportional. Find the missing lengths.

22.

23.

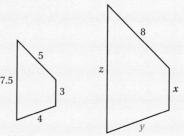

24. Lucita walks $4\frac{1}{2}$ mi in $1\frac{1}{2}$ hr. What is her rate in miles per hour?

  A. $\frac{1}{3}$ mph

  B. $1\frac{1}{2}$ mph

  C. 3 mph

  D. $4\frac{1}{2}$ mph

Synthesis

25. Nancy wants to win a gift card from the campus bookstore by guessing the number of marbles in an 8-gal jar. She knows that there are 128 oz in a gallon. She goes home and fills an 8-oz jar with 46 marbles. How many marbles should she guess are in the 8-gal jar?

Cumulative Review

Calculate and simplify.

1.
```
    2 7.6 8
       3.0 1 9
 + 4 8 3.2 9 7
```

2.
$$2\tfrac{1}{3}$$
$$+ 4\tfrac{5}{12}$$

3. $\dfrac{6}{35} + \dfrac{5}{28}$

4.
```
    4 0.2
 −    9.7 0 9
```

5. $73.82 - 0.908$

6. $\dfrac{4}{15} - \dfrac{3}{20}$

7.
```
   3 7.6 4
 ×     5.9
```

8. 5.678×100

9. $2\tfrac{1}{3} \cdot 1\tfrac{2}{7}$

10. $2.3\overline{)9\,8.9}$

11. $5\,4\overline{)4\,8{,}5\,4\,6}$

12. $\dfrac{7}{11} \div \dfrac{14}{33}$

13. Write expanded notation: 30,074.

14. Write a word name for 120.07.

Which number is larger?

15. 0.7, 0.698

16. 0.799, 0.8

17. Find the prime factorization of 144.

18. Find the LCM of 18 and 30.

19. What part is shaded?

20. Simplify: $\dfrac{90}{144}$.

Calculate.

21. $\dfrac{3}{5} \times 9.53$

22. $\dfrac{1}{3} \times 0.645 - \dfrac{3}{4} \times 0.048$

23. Write fraction notation for the ratio 0.3 to 15. Do not simplify.

24. Determine whether the pairs 3, 9 and 25, 75 are proportional.

25. What is the rate in meters per second?

   660 meters, 12 seconds

26. *Unit Prices.* An 8-oz can of pineapple chunks costs $0.99. A 24.5-oz jar of pineapple chunks costs $3.29. Which has the lower unit price?

Solve.

27. $\dfrac{14}{25} = \dfrac{x}{54}$

28. $423 = 16 \cdot t$

29. $\dfrac{2}{3} \cdot y = \dfrac{16}{27}$

30. $\dfrac{7}{16} = \dfrac{56}{x}$

31. $34.56 + n = 67.9$

32. $t + \dfrac{7}{25} = \dfrac{5}{7}$

33. Ramona's recipe for fettuccini Alfredo has 520 calories in 1 cup. How many calories are there in $\frac{3}{4}$ cup?

34. A machine can stamp out 925 washers in 5 min. An order is placed for 1295 washers. How long will it take to stamp them out?

35. A 46-oz juice can contains $5\frac{3}{4}$ cups of juice. A recipe calls for $3\frac{1}{2}$ cups of juice. How many cups are left over?

36. It takes a carpenter $\frac{2}{3}$ hr to hang a door. How many doors can the carpenter hang in 8 hr?

37. *Car Travel.* A car travels 337.62 mi in 8 hr. How far does it travel in 1 hr?

38. *Shuttle Orbits.* A space shuttle made 16 orbits a day during an 8.25-day mission. How many orbits were made during the entire mission?

39. How many even prime numbers are there?
   A. 5　　　　　**B.** 3　　　　　**C.** 2
   D. 1　　　　　**E.** None

40. The gas mileage of a car is 28.16 mpg. How many gallons per mile is this?
   A. $\dfrac{704}{25}$　　　　**B.** $\dfrac{25}{704}$　　　　**C.** $\dfrac{2816}{100}$
   D. $\dfrac{250}{704}$　　　　**E.** None

Synthesis

41. A soccer goalie wishing to block an opponent's shot moves toward the shooter to reduce the shooter's view of the goal. If the goalie can only defend a region 10 ft wide, how far in front of the goal should the goalie be? (See the figure at right.)

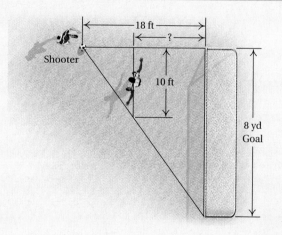

Percent Notation

Real-World Application

The Dow Jones Industrial Average (DJIA) plunged from 11,143 to 10,365 on September 29, 2008. This was the largest one-day drop in its history. What was the percent of decrease?

Sources: Nightly Business Reports, September 29, 2008; DJIA

This problem appears as Example 3 in Section 6.5.

6.1

Percent Notation

OBJECTIVES

a Write three kinds of notation for a percent.

b Convert between percent notation and decimal notation.

SKILL TO REVIEW

Objective 4.3a: Multiply using decimal notation.

Multiply.

1. 68.3×0.01

2. 3013×2.4

a Understanding Percent Notation

Of all the surface area of the earth, 70% is covered by water. What does this mean? It means that of every 100 square miles of the earth's surface area, 70 square miles are covered by water. Thus, 70% is a ratio of 70 to 100, or $\frac{70}{100}$.

Source: *The Handy Geography Answer Book*

70 of 100 squares are shaded.

70% or $\frac{70}{100}$ or 0.70 of the large square is shaded.

Percent notation is used extensively in our everyday lives. Here are some examples.

25.2% of the adult population in Washington, D.C., has a graduate degree.

Almost 22% of the glass that is produced is recycled.

A blood alcohol level of 0.08% is the standard used by most states as the legal limit for drunk driving.

71% of people in the United States use the Internet; 17% of people in China use the Internet.

In California, 42.5% of people age 5 and older speak a language other than English at home. In Kansas, the percentage is 10.3%.

Percent notation is often represented using a circle graph, or pie chart, to show how the parts of a quantity are related. For example, the circle graph at left illustrates the percentage of people in the United States with each of the four blood types.

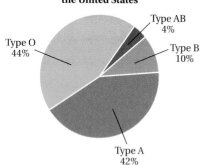

Blood Types in the United States

Type O 44%

Type AB 4%

Type B 10%

Type A 42%

SOURCE: bloodcenter.stanford.edu/ about_blood/blood_types.html

PERCENT NOTATION

The notation **n%** means "*n* per hundred."

This definition leads us to the following equivalent ways of defining percent notation.

> **NOTATION FOR _n_%**
>
> **Percent notation, _n_%,** can be expressed using:
>
> ratio $\longrightarrow$ $n\% = $ the ratio of n to $100 = \dfrac{n}{100}$,
>
> fraction notation $\longrightarrow$ $n\% = n \times \dfrac{1}{100}$, or
>
> decimal notation $\longrightarrow$ $n\% = n \times 0.01$.

EXAMPLE 1 Write three kinds of notation for 35%.

Using ratio: $\qquad 35\% = \dfrac{35}{100}$ A ratio of 35 to 100

Using fraction notation: $\quad 35\% = 35 \times \dfrac{1}{100}$ Replacing % with $\times \dfrac{1}{100}$

Using decimal notation: $\quad 35\% = 35 \times 0.01$ Replacing % with $\times 0.01$

EXAMPLE 2 Write three kinds of notation for 67.8%.

Using ratio: $\qquad 67.8\% = \dfrac{67.8}{100}$ A ratio of 67.8 to 100

Using fraction notation: $\quad 67.8\% = 67.8 \times \dfrac{1}{100}$ Replacing % with $\times \dfrac{1}{100}$

Using decimal notation: $\quad 67.8\% = 67.8 \times 0.01$ Replacing % with $\times 0.01$

Do Exercises 1–4.

At the airport in Oslo, Norway, 64% of all passengers arrive and leave using trains and buses. At the airport in San Francisco, the percentage is 23%.
SOURCE: Transportation Research Board

Write three kinds of notation as in Examples 1 and 2.

1. 70%   **2.** 23.4%

3. 100%   **4.** 0.6%

(b) Converting Between Percent Notation and Decimal Notation

Consider 78%. To convert to decimal notation, we can think of percent notation as a ratio and write

$$78\% = \dfrac{78}{100} \qquad \text{Using the definition of percent as a ratio}$$
$$= 0.78. \qquad \text{Dividing}$$

Similarly,

$$4.9\% = \dfrac{4.9}{100} \qquad \text{Using the definition of percent as a ratio}$$
$$= 0.049. \qquad \text{Dividing}$$

We could also convert 78% to decimal notation by replacing "%" with "$\times 0.01$" and write

$$78\% = 78 \times 0.01 \qquad \text{Replacing % with } \times 0.01$$
$$= 0.78. \qquad \text{Multiplying}$$

Answers

1. $\dfrac{70}{100}$; $70 \times \dfrac{1}{100}$; 70×0.01

2. $\dfrac{23.4}{100}$; $23.4 \times \dfrac{1}{100}$; 23.4×0.01

3. $\dfrac{100}{100}$; $100 \times \dfrac{1}{100}$; 100×0.01

4. $\dfrac{0.6}{100}$; $0.6 \times \dfrac{1}{100}$; 0.6×0.01

Find decimal notation.

5. 34%   **6.** 78.9%

Find decimal notation for the percent notation in each sentence.

7. Working Online.   Of all American adults who use the Internet, 42% have gone online to work from home.
Sources: U.S. Bureau of Labor Statistics; Nielsen Home Technology Report

8. Blood Alcohol Level.   A blood alcohol level of 0.08% is the standard used by most states as the legal limit for drunk driving.

Similarly,

$$4.9\% = 4.9 \times 0.01 \qquad \text{Replacing \% with } \times 0.01$$
$$= 0.049. \qquad \text{Multiplying}$$

Dividing by 100 amounts to moving the decimal point two places to the left, which is the same as multiplying by 0.01. This leads us to a quick way to convert from percent notation to decimal notation: We drop the percent symbol and move the decimal point two places to the left.

To convert from percent notation to decimal notation,	36.5%
a) replace the percent symbol % with $\times$ 0.01, and	36.5 $\times$ 0.01
b) multiply by 0.01, which means move the decimal point two places to the left.	0.36.5 Move 2 places to the left.
	36.5% = 0.365

EXAMPLE 3   Find decimal notation for 99.44%.

a) Replace the percent symbol with $\times$ 0.01.   99.44 $\times$ 0.01

b) Move the decimal point two places to the left.   0.99.44

Thus, 99.44% = 0.9944.

EXAMPLE 4   The interest rate on a $2\frac{1}{2}$-year certificate of deposit is $6\frac{3}{8}\%$. Find decimal notation for $6\frac{3}{8}\%$.

a) Convert $6\frac{3}{8}$ to decimal notation and replace the percent symbol with $\times$ 0.01. $6\frac{3}{8}\%$
    6.375 $\times$ 0.01

b) Move the decimal point two places to the left.   0.06.375

Thus, $6\frac{3}{8}\% = 0.06375$.

Do Exercises 5–8.

To convert 0.38 to percent notation, we can first write fraction notation, as follows:

$$0.38 = \frac{38}{100} \qquad \text{Converting to fraction notation}$$
$$= 38\%. \qquad \text{Using the definition of percent as a ratio}$$

Note that $100\% = 100 \times 0.01 = 1$. Thus to convert 0.38 to percent notation, we can multiply by 1, using 100% as a symbol for 1.

$$0.38 = 0.38 \times 1$$
$$= 0.38 \times 100\%$$
$$= 0.38 \times 100 \times 0.01 \qquad \text{Replacing 100\% with } 100 \times 0.01$$
$$= (0.38 \times 100) \times 0.01 \qquad \text{Using the associative law of multiplication}$$
$$= 38 \times 0.01$$
$$= 38\% \qquad \text{Replacing } \times 0.01 \text{ with \%}$$

Answers
5. 0.34   **6.** 0.789
7. 0.42   **8.** 0.0008

Even more quickly, since $0.38 = 0.38 \times 100\%$, we can simply multiply 0.38 by 100 and write the % symbol.

To convert from decimal notation to percent notation, we multiply by 100%. That is, we move the decimal point two places to the right and write a percent symbol.

To convert from decimal notation to percent notation, multiply by 100%. That is,	$0.675 = 0.675 \times 100\%$
a) move the decimal point two places to the right, and	0.67.5 Move 2 places to the right.
b) write a % symbol.	67.5% $0.675 = 67.5\%$

EXAMPLE 5 Find percent notation for 1.27.

a) Move the decimal point two places to the right.   1.27.

b) Write a % symbol.   127%

Thus, $1.27 = 127\%$.

EXAMPLE 6 Of the time that people declare as sick leave, 0.21 is actually used for family issues. Find percent notation for 0.21.

Source: CCH Inc.

a) Move the decimal point two places to the right.   0.21.

b) Write a % symbol.   21%

Thus, $0.21 = 21\%$.

EXAMPLE 7 Find percent notation for 5.6.

a) Move the decimal point two places to the right, adding an extra zero.   5.60.

b) Write a % symbol.   560%

Thus, $5.6 = 560\%$.

EXAMPLE 8 Of those who play golf, 0.149 play 8–24 rounds per year. Find percent notation for 0.149.

Source: U.S. Golf Association

a) Move the decimal point two places to the right.   0.14.9

b) Write a % symbol.   14.9%

Thus, $0.149 = 14.9\%$.

Do Exercises 9–14.

Find percent notation.

9. 0.24   **10.** 3.47

11. 1   **12.** 0.05

Find percent notation for the decimal notation in each sentence.

13. Women in Congress. In 2008, 0.16 of the members of the United States Congress were women.

Source: Center for American Women and Politics at Rutgers University

14. Soccer. For Americans in the 18–24 age group, 0.311 have played soccer; for the 12–17 age group, 0.396 have played.

Source: ESPN Sports Poll, a service of TNS Sport

Answers

9. 24%   **10.** 347%   **11.** 100%   **12.** 5%
13. 16%   **14.** 31.1%; 39.6%

a Write three kinds of notation as in Examples 1 and 2 on p. 349.

1. 90%

2. 58.7%

3. 12.5%

4. 130%

b Find decimal notation.

5. 67%

6. 17%

7. 45.6%

8. 76.3%

9. 59.01%

10. 30.02%

11. 10%

12. 80%

13. 1%

14. 100%

15. 200%

16. 300%

17. 0.1%

18. 0.4%

19. 0.09%

20. 0.12%

21. 0.18%

22. 5.5%

23. 23.19%

24. 87.99%

25. $14\frac{7}{8}\%$

26. $93\frac{1}{8}\%$

27. $56\frac{1}{2}\%$

28. $61\frac{3}{4}\%$

Find decimal notation for the percent notation(s) in each sentence.

29. *Video Games.* According to a recent survey, 97% of the 12–17 age group play video games.

Sources: Pew Survey; *Time*, September 29, 2008

30. *Female Astronauts.* Of the 466 astronauts who have flown in space, 10.52% are female.

Source: *Encyclopedia Astronautica*

31. *Fuel Efficiency.*   Speeding up by only 5 mph on the highway cuts fuel efficiency by approximately 7% to 8%.

Source: *Wall Street Journal*, "Pain Relief," by A. J. Miranda, September 15, 2008

32. *Foreign-Born Population.*   In 2008, the U.S. foreign-born population was 12.6%, the highest since 1920.

Source: U.S. Census Bureau

33. *High School Sports.*   During the 2007–2008 academic year, 54.8% of all high school students were involved in high school sports.

Source: National Federation of State High School Associations

34. *Eating Out.*   On a given day, 58% of all Americans eat meals and snacks away from home.

Source: U.S. Department of Agriculture

Find percent notation.

35. 0.47

36. 0.87

37. 0.03

38. 0.01

39. 8.7

40. 4

41. 0.334

42. 0.889

43. 0.75

44. 0.99

45. 0.4

46. 0.5

47. 0.006

48. 0.008

49. 0.017

50. 0.024

51. 0.2718

52. 0.8911

53. 0.0239

54. 0.00073

Find percent notation for the decimal notation(s) in each sentence.

55. *Wasting Food.*   Americans waste an estimated 0.27 of the food available for consumption. The waste occurs in restaurants, supermarkets, cafeterias, and household kitchens.

Source: *New York Times*, "One Country's Table Scraps, Another Country's Meal," by Andrew Martin, May 18, 2008

56. *Recycling Newspapers.*   Over 0.73 of all newspapers are recycled.

Source: Newspaper Association of America

57. *Age 65 and Older.*   In Alaska, 0.057 of the residents are age 65 and older. In Florida, 0.176 are age 65 and older.

Source: U.S. Census Bureau

58. *Dining Together.*   In 2008, 0.2 of families dined together every evening. This rate declined from 0.59 in 1987.

Source: Online polls at USATODAY.com

59. *Cancer Survival.* In 2005, the estimated 10-yr survival rate was 0.906 for children diagnosed with non-Hodgkin's lymphoma (NHL) and 0.88 for those diagnosed with acute lymphoblastic leukemia (ALL).

Source: *Journal of the National Cancer Institute,* news release, September 8, 2008

60. *Postsecondary Degrees.* In 2007, 0.296 of those 25 to 29 years old in the United States had attained a bachelor's degree or higher.

Source: National Center for Education Statistics, U.S. Department of Commerce, U.S. Census Bureau, Current Population Survey, March Supplement 1971–2007

Find decimal notation for each percent notation in the graph.

61.

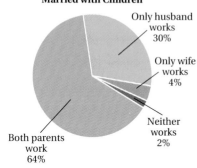

U.S. Households:
Married with Children

Only husband works 30%

Only wife works 4%

Neither works 2%

Both parents work 64%

SOURCE: U.S. Census Bureau;
U.S. Bureau of Labor Statistics

62.

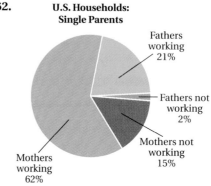

U.S. Households:
Single Parents

Fathers working 21%

Fathers not working 2%

Mothers not working 15%

Mothers working 62%

SOURCE: U.S. Census Bureau;
U.S. Bureau of Labor Statistics

Skill Maintenance

Convert to a mixed numeral.   [3.4a]

63. $\dfrac{100}{3}$

64. $\dfrac{75}{2}$

65. $\dfrac{75}{8}$

66. $\dfrac{297}{16}$

67. $\dfrac{567}{98}$

68. $\dfrac{2345}{21}$

Convert to decimal notation.   [4.5a]

69. $\dfrac{2}{3}$

70. $\dfrac{1}{3}$

71. $\dfrac{5}{6}$

72. $\dfrac{17}{12}$

73. $\dfrac{8}{3}$

74. $\dfrac{15}{16}$

Synthesis

Find percent notation. (*Hint*: Multiply by a form of 1 and obtain a denominator of 100.)

75. $\dfrac{1}{2}$

76. $\dfrac{3}{4}$

77. $\dfrac{7}{10}$

78. $\dfrac{2}{5}$

Find percent notation for each shaded area.

79.

80.

6.2 Percent Notation and Fraction Notation

a Convert from Fraction Notation to Percent Notation

OBJECTIVES

a Convert from fraction notation to percent notation.

b Convert from percent notation to fraction notation.

Consider the fraction notation $\frac{7}{8}$. To convert to percent notation, we use two skills that we already have. We first find decimal notation by dividing: $7 \div 8$.

$$
\begin{array}{r}
0.8\,7\,5 \\
8\,\overline{)\,7.0\,0\,0} \\
\underline{6\,4} \\
6\,0 \\
\underline{5\,6} \\
4\,0 \\
\underline{4\,0} \\
0
\end{array}
\qquad \frac{7}{8} = 0.875
$$

Then we convert the decimal notation to percent notation. We move the decimal point two places to the right

$$0.8\,7.5$$

and write a % symbol:

$$\frac{7}{8} = 87.5\%, \text{ or } 87\tfrac{1}{2}\%. \qquad 0.5 = \tfrac{1}{2}$$

SKILL TO REVIEW

Objective 4.5a: Convert from fraction notation to decimal notation.

Find decimal notation.

1. $\dfrac{11}{16}$ **2.** $\dfrac{5}{9}$

To convert from fraction notation to percent notation,

a) find decimal notation by division, and

b) convert the decimal notation to percent notation.

$$\frac{3}{5} \qquad \text{Fraction notation}$$

$$
\begin{array}{r}
0.6 \\
5\,\overline{)\,3.0} \\
\underline{3\,0} \\
0
\end{array}
$$

$$0.6 = 0.60 = 60\% \qquad \text{Percent notation}$$

$$\frac{3}{5} = 60\%$$

EXAMPLE 1 Find percent notation for $\frac{9}{16}$.

a) We first find decimal notation by division.

$$
\begin{array}{r}
0.5\,6\,2\,5 \\
1\,6\,\overline{)\,9.0\,0\,0\,0} \\
\underline{8\,0} \\
1\,0\,0 \\
\underline{9\,6} \\
4\,0 \\
\underline{3\,2} \\
8\,0 \\
\underline{8\,0} \\
0
\end{array}
\qquad \frac{9}{16} = 0.5625
$$

Calculator Corner

Converting from Fraction Notation to Percent Notation A calculator can be used to convert from fraction notation to percent notation. We simply perform the division on the calculator and then use the percent key. To convert $\frac{17}{40}$ to percent notation, for example, we press

[1] [7] [÷] [4] [0] [2nd] [%] , or

[1] [7] [÷] [4] [0] [SHIFT] [%] .

The display reads [42.5] , so $\frac{17}{40} = 42.5\%$.

Exercises: Use a calculator to find percent notation. Round to the nearest hundredth of a percent.

1. $\dfrac{13}{25}$ **2.** $\dfrac{5}{13}$

3. $\dfrac{43}{39}$ **4.** $\dfrac{12}{7}$

5. $\dfrac{217}{364}$ **6.** $\dfrac{2378}{8401}$

Answers

Skill to Review:
1. 0.6875   2. $0.\overline{5}$

b) Next, we convert the decimal notation to percent notation. We move the decimal point two places to the right and write a % symbol.

0.56.25

$\dfrac{9}{16} = 56.25\%$, or $56\tfrac{1}{4}\%$ $\qquad 0.25 = \tfrac{1}{4}$

Don't forget the % symbol.

Do Exercises 1 and 2.

Fractions named by repeating decimals also can be converted to percent notation.

EXAMPLE 2 *Without Health Insurance.* Approximately $\tfrac{1}{6}$ of all people in the United States are without health insurance. Find percent notation for $\tfrac{1}{6}$.
Source: U.S. Census Bureau, *Current Population Survey*, March 2003

a) Find decimal notation by division.

$$
\begin{array}{r}
0.1\ 6\ 6 \\
6\overline{)\,1.0\ 0\ 0} \\
\underline{6} \\
4\ 0 \\
\underline{3\ 6} \\
4\ 0 \\
\underline{3\ 6} \\
4
\end{array}
$$

We get a repeating decimal: $0.16\overline{6}$.

b) Convert the answer to percent notation.

0.16.$\overline{6}$

$\dfrac{1}{6} = 16.\overline{6}\%$, or $16\tfrac{2}{3}\%$ $\qquad 0.\overline{6} = \tfrac{2}{3}$

Do Exercises 3 and 4.

In some cases, division is not the fastest way to convert. The following are some optional ways in which conversion might be done.

EXAMPLE 3   Find percent notation for $\frac{69}{100}$.

We use the definition of percent as a ratio.

$$\frac{69}{100} = 69\%$$

EXAMPLE 4   Find percent notation for $\frac{17}{20}$.

We want to multiply by 1 to get 100 in the denominator. We think of what we must multiply 20 by in order to get 100. That number is 5, so we multiply by 1 using $\frac{5}{5}$.

$$\frac{17}{20} \cdot \frac{5}{5} = \frac{85}{100} = 85\%$$

Note that this shortcut works only when the denominator is a factor of 100.

EXAMPLE 5   Find percent notation for $\frac{18}{25}$.

$$\frac{18}{25} = \frac{18}{25} \cdot \frac{4}{4} = \frac{72}{100} = 72\%$$

Do Exercises 5–8.

Find percent notation.

5. $\frac{57}{100}$

6. $\frac{19}{25}$

7. $\frac{7}{10}$

8. $\frac{1}{4}$

b Converting from Percent Notation to Fraction Notation

> To convert from percent notation to fraction notation,   30%   Percent notation
>
> **a)** use the definition of percent as a ratio, and $\frac{30}{100}$
>
> **b)** simplify, if possible. $\frac{3}{10}$ Fraction notation

EXAMPLE 6   Find fraction notation for 75%.

$$75\% = \frac{75}{100} \qquad \text{Using the definition of percent}$$

$$= \frac{3 \cdot 25}{4 \cdot 25} = \frac{3}{4} \cdot \frac{25}{25} \Bigg\} \quad \text{Simplifying}$$

$$= \frac{3}{4}$$

EXAMPLE 7 Find fraction notation for 62.5%.

$$62.5\% = \frac{62.5}{100} \qquad \text{Using the definition of percent}$$

$$= \frac{62.5}{100} \times \frac{10}{10} \qquad \text{Multiplying by 1 to eliminate the decimal point in the numerator}$$

$$= \frac{625}{1000}$$

$$= \frac{5 \cdot 125}{8 \cdot 125} = \frac{5}{8} \cdot \frac{125}{125} \Big\}$$

$$= \frac{5}{8} \qquad\qquad\qquad \text{Simplifying}$$

EXAMPLE 8 Find fraction notation for $16\frac{2}{3}\%$.

$$16\frac{2}{3}\% = \frac{50}{3}\% \qquad \text{Converting from the mixed numeral to fraction notation}$$

$$= \frac{50}{3} \times \frac{1}{100} \qquad \text{Using the definition of percent}$$

$$= \frac{50 \cdot 1}{3 \cdot 50 \cdot 2} = \frac{1}{3 \cdot 2} \cdot \frac{50}{50} \Big\}$$

$$= \frac{1}{6} \qquad\qquad\qquad \text{Simplifying}$$

Do Exercises 9–12.

Find fraction notation.

9. 60%

10. 3.25%

11. $66\frac{2}{3}\%$

12. $12\frac{1}{2}\%$

The table below lists fraction, decimal, and percent equivalents used so often that it would speed up your work if you memorized them. For example, $\frac{1}{3} = 0.\overline{3}$, so we say that the **decimal equivalent** of $\frac{1}{3}$ is $0.\overline{3}$, or that $0.\overline{3}$ has the **fraction equivalent** $\frac{1}{3}$. This table also appears on the inside back cover.

FRACTION, DECIMAL, AND PERCENT EQUIVALENTS

FRACTION NOTATION	$\frac{1}{10}$	$\frac{1}{8}$	$\frac{1}{6}$	$\frac{1}{5}$	$\frac{1}{4}$	$\frac{3}{10}$	$\frac{1}{3}$	$\frac{3}{8}$	$\frac{2}{5}$	$\frac{1}{2}$	$\frac{3}{5}$	$\frac{5}{8}$	$\frac{2}{3}$	$\frac{7}{10}$	$\frac{3}{4}$	$\frac{4}{5}$	$\frac{5}{6}$	$\frac{7}{8}$	$\frac{9}{10}$	$\frac{1}{1}$
DECIMAL NOTATION	0.1	0.125	$0.16\overline{6}$	0.2	0.25	0.3	$0.33\overline{3}$	0.375	0.4	0.5	0.6	0.625	$0.66\overline{6}$	0.7	0.75	0.8	$0.83\overline{3}$	0.875	0.9	1
PERCENT NOTATION	10%	12.5%, or $12\frac{1}{2}\%$	$16.\overline{6}\%$, or $16\frac{2}{3}\%$	20%	25%	30%	$33.\overline{3}\%$, or $33\frac{1}{3}\%$	37.5%, or $37\frac{1}{2}\%$	40%	50%	60%	62.5%, or $62\frac{1}{2}\%$	$66.\overline{6}\%$, or $66\frac{2}{3}\%$	70%	75%	80%	$83.\overline{3}\%$, or $83\frac{1}{3}\%$	87.5%, or $87\frac{1}{2}\%$	90%	100%

EXAMPLE 9 Find fraction notation for $16.\overline{6}\%$.

We can use the table above or recall that $16.\overline{6}\% = 16\frac{2}{3}\% = \frac{1}{6}$. We can also recall from our work with repeating decimals in Chapter 4 that $0.\overline{6} = \frac{2}{3}$. Then we have $16.\overline{6}\% = 16\frac{2}{3}\%$ and can proceed as in Example 8.

Do Exercises 13 and 14.

Find fraction notation.

13. $33.\overline{3}\%$

14. $83.\overline{3}\%$

Answers

9. $\frac{3}{5}$ 10. $\frac{13}{400}$ 11. $\frac{2}{3}$

12. $\frac{1}{8}$ 13. $\frac{1}{3}$ 14. $\frac{5}{6}$

a Find percent notation.

1. $\dfrac{41}{100}$

2. $\dfrac{36}{100}$

3. $\dfrac{5}{100}$

4. $\dfrac{1}{100}$

5. $\dfrac{2}{10}$

6. $\dfrac{7}{10}$

7. $\dfrac{3}{10}$

8. $\dfrac{9}{10}$

9. $\dfrac{1}{2}$

10. $\dfrac{3}{4}$

11. $\dfrac{7}{8}$

12. $\dfrac{1}{8}$

13. $\dfrac{4}{5}$

14. $\dfrac{2}{5}$

15. $\dfrac{2}{3}$

16. $\dfrac{1}{3}$

17. $\dfrac{1}{6}$

18. $\dfrac{5}{6}$

19. $\dfrac{3}{16}$

20. $\dfrac{11}{16}$

21. $\dfrac{13}{16}$

22. $\dfrac{7}{16}$

23. $\dfrac{4}{25}$

24. $\dfrac{17}{25}$

25. $\dfrac{1}{20}$

26. $\dfrac{31}{50}$

27. $\dfrac{17}{50}$

28. $\dfrac{3}{20}$

Find percent notation for the fraction notation in each sentence.

29. *Heart Transplants.* In the United States in 2006, $\dfrac{2}{25}$ of the organ transplants were heart transplants and $\dfrac{59}{100}$ were kidney transplants.

   Source: 2007 OPTN/SRTR Annual Report, Table 1.7

30. *Car Colors.* The four most popular colors for 2006 compact/sports cars were silver, gray, black, and red. Of all cars in this category, $\dfrac{9}{50}$ were silver, $\dfrac{3}{20}$ gray, $\dfrac{3}{20}$ black, and $\dfrac{3}{20}$ red.

   Sources: Ward's Automotive Group; DuPont Automotive Products

In Exercises 31–36, write percent notation for the fractions in the pie chart below.

How Food Dollars Are Spent

Dairy $\dfrac{3}{25}$

Beverages (nonalcoholic) $\dfrac{3}{25}$

Sugar $\dfrac{1}{25}$

Cereal and baked goods $\dfrac{13}{100}$

Fats and oils $\dfrac{3}{100}$

Fruits and vegetables $\dfrac{3}{20}$

Other $\dfrac{9}{50}$

Eggs $\dfrac{1}{50}$

Meat, poultry, and fish $\dfrac{11}{50}$

SOURCES: U.S. Bureau of Labor Statistics; Consumer Price Index; *The Hoosier Farmer*, Summer 2008

31. $\dfrac{11}{50}$

32. $\dfrac{9}{50}$

33. $\dfrac{3}{25}$

34. $\dfrac{1}{25}$

35. $\dfrac{3}{20}$

36. $\dfrac{13}{100}$

b Find fraction notation. Simplify.

37. 85%

38. 55%

39. 62.5%

40. 12.5%

41. $33\frac{1}{3}\%$

42. $83\frac{1}{3}\%$

43. $16.\overline{6}\%$

44. $66.\overline{6}\%$

45. 7.25%

46. 4.85%

47. 0.8%

48. 0.2%

49. $25\frac{3}{8}\%$

50. $48\frac{7}{8}\%$

51. $78\frac{2}{9}\%$

52. $16\frac{5}{9}\%$

53. $64\frac{7}{11}\%$

54. $73\frac{3}{11}\%$

55. 150%

56. 110%

57. 0.0325%

58. 0.419%

59. $33.\overline{3}\%$

60. $83.\overline{3}\%$

In Exercises 61–66, find fraction notation for the percent notations in the table below.

U.S. POPULATION BY SELECTED AGE CATEGORIES (Data have been rounded to the nearest percent.)

AGE CATEGORY	PERCENT OF POPULATION
5–17 years	18%
18–24 years	10
15–44 years	42
18 years and older	75
65 years and older	12
75 years and older	6

SOURCES: U.S. Census Bureau; 2006 American Community Survey

61. 6%

62. 18%

63. 12%

64. 42%

65. 75%

66. 10%

Find fraction notation for the percent notation in each sentence.

67. A $\frac{3}{4}$-cup serving of Post Selects Great Grains cereal with $\frac{1}{2}$ cup fat-free milk satisfies 15% of the minimum daily requirement for calcium.
Source: Kraft Foods Global, Inc.

68. A 1.8-oz serving of Frosted Mini-Wheat Muffin, with $\frac{1}{2}$ cup fat-free milk satisfies minimum daily requirement for Vitamin B$_{12}$.
Source: Kellogg, Inc.

69. In 2006, 20.9% of Americans age 18 and older smoked cigarettes.
Sources: *Washington Post,* March 9, 2006; U.S. Centers for Disease Control and Prevention

70. In 2006, 14.9% of the adults age 18 and older in California smoked cigarettes.
Sources: U.S. Centers for Disease Control and Prevention; *AARP Magazine,* March 2008

Complete each table.

71.

FRACTION NOTATION	DECIMAL NOTATION	PERCENT NOTATION
$\frac{1}{8}$		12.5%, or $12\frac{1}{2}$%
$\frac{1}{6}$		
		20%
	0.25	
		33.$\overline{3}$%, or $33\frac{1}{3}$%
		37.5%, or $37\frac{1}{2}$%
		40%
$\frac{1}{2}$		

72.

FRACTION NOTATION	DECIMAL NOTATION	PERCENT NOTATION
$\frac{3}{5}$		
	0.625	
$\frac{2}{3}$		
	0.75	75%
$\frac{4}{5}$		
$\frac{5}{6}$		83.$\overline{3}$%, or $83\frac{1}{3}$%
$\frac{7}{8}$		87.5%, or $87\frac{1}{2}$%
		100%

73.

FRACTION NOTATION	DECIMAL NOTATION	PERCENT NOTATION
	0.5	
$\frac{1}{3}$		
		25%
		16.$\overline{6}$%, or $16\frac{2}{3}$%
	0.125	
$\frac{3}{4}$		
	0.8$\overline{3}$	
$\frac{3}{8}$		

74.

FRACTION NOTATION	DECIMAL NOTATION	PERCENT NOTATION
		40%
		62.5%, or $62\frac{1}{2}$%
	0.875	
$\frac{1}{1}$		
	0.6	
	0.$\overline{6}$	
$\frac{1}{5}$		

Solve.

75. $13 \cdot x = 910$ [1.7b]

76. $15 \cdot y = 75$ [1.7b]

77. $0.05 \times b = 20$ [4.4b]

78. $3 = 0.16 \times b$ [4.4b]

79. $\dfrac{24}{37} = \dfrac{15}{x}$ [5.3b]

80. $\dfrac{17}{18} = \dfrac{x}{27}$ [5.3b]

81. $\dfrac{9}{10} = \dfrac{x}{5}$ [5.3b]

82. $\dfrac{7}{x} = \dfrac{4}{5}$ [5.3b]

Convert to a mixed numeral. [3.4a]

83. $\dfrac{100}{3}$

84. $\dfrac{75}{2}$

85. $\dfrac{250}{3}$

86. $\dfrac{123}{6}$

87. $\dfrac{345}{8}$

88. $\dfrac{373}{6}$

89. $\dfrac{75}{4}$

90. $\dfrac{67}{9}$

Convert from a mixed numeral to fraction notation. [3.4a]

91. $1\dfrac{1}{17}$

92. $20\dfrac{9}{10}$

93. $101\dfrac{1}{2}$

94. $32\dfrac{3}{8}$

Synthesis

Write percent notation.

95. $\dfrac{41}{369}$

96. $\dfrac{54}{999}$

97. $2.5\overline{74631}$

98. $3.2\overline{93847}$

Write decimal notation.

99. $\dfrac{14}{9}\%$

100. $\dfrac{19}{12}\%$

101. $\dfrac{729}{7}\%$

102. $\dfrac{637}{6}\%$

103. Arrange the following numbers from smallest to largest.

$$16\dfrac{1}{6}\%, \ 1.6, \ \dfrac{1}{6}\%, \ \dfrac{1}{2}, \ 0.2, \ 1.6\%, \ 1\dfrac{1}{6}\%, \ 0.5\%, \ \dfrac{2}{7}\%, \ 0.\overline{54}$$

6.3 Solving Percent Problems Using Percent Equations

a Translating to Equations

To solve a problem involving percents, it is helpful to translate first to an equation. To distinguish the method discussed in this section from that of Section 6.4, we will call these *percent equations*.

> **KEY WORDS IN PERCENT TRANSLATIONS**
>
> "**Of**" translates to "·", or "×".   "**Is**" translates to "=".
> "**What**" translates to any letter.   "**%**" translates to "× $\frac{1}{100}$" or "× 0.01".

EXAMPLES   Translate each of the following.

1. 23%   of   5   is   what?
   23%   ·   5   =   a   This is a *percent equation*.

2. What   is   11%   of   49?
   a   =   11%   ·   49   Any letter can be used.

Do Margin Exercises 1 and 2.

EXAMPLES   Translate each of the following.

3. 3   is   10%   of   what?
   3   =   10%   ·   b

4. 45%   of   what   is   23?
   45%   ×   b   =   23

Do Exercises 3 and 4.

EXAMPLES   Translate each of the following.

5. 10   is   what percent   of   20?
   10   =   p   ×   20

6. What percent   of   50   is   7?
   p   ·   50   =   7

Do Exercises 5 and 6.

OBJECTIVES

a Translate percent problems to percent equations.

b Solve basic percent problems.

SKILL TO REVIEW
Objective 4.4b: Solve equations of the type $a \cdot x = b$, where a and b may be in decimal notation.

Solve.

1. $0.05 \cdot x = 830$

2. $8 \cdot y = 40.648$

Translate to an equation. Do not solve.

1. 12% of 50 is what?

2. What is 40% of 60?

Translate to an equation. Do not solve.

3. 45 is 20% of what?

4. 120% of what is 60?

Translate to an equation. Do not solve.

5. 16 is what percent of 40?

6. What percent of 84 is 10.5?

Answers

Skill to Review:
1. 16,600   **2.** 5.081

Margin Exercises:
1. 12% × 50 = a   **2.** a = 40% × 60
3. 45 = 20% × b   **4.** 120% × b = 60
5. 16 = p × 40   **6.** p × 84 = 10.5

b Solving Percent Problems

In solving percent problems, we use the *Translate* and *Solve* steps in the problem-solving strategy used throughout this text.

Percent problems are actually of three different types. Although the method we present does *not* require that you be able to identify which type you are solving, it is helpful to know them. Each of the three types of percent problems depends on which of the three pieces of information is missing.

1. **Finding the *amount* (the result of taking the percent)**

   Example: **What is 25% of 60?**

   Translation: *a* = 25% · 60

2. **Finding the *base* (the number you are taking the percent of)**

   Example: 15 is 25% of what?

   Translation: 15 = 25% · *b*

3. **Finding the *percent number* (the percent itself)**

   Example: 15 is what percent of 60?

   Translation: 15 = *p* · 60

Finding the Amount

EXAMPLE 7 What is 23.9% of $40,000,000,000?

Translate: $a = 23.9\% \times 40{,}000{,}000{,}000$.

Solve: The letter is by itself. To solve the equation, we just convert 23.9% to decimal notation and multiply:

$$a = 23.9\% \times 40{,}000{,}000{,}000$$
$$a = 0.239 \times 40{,}000{,}000{,}000 = 9{,}560{,}000{,}000.$$

Thus, $9,560,000,000 is 23.9% of $40,000,000,000. The answer is $9,560,000,000.

Do Exercise 7.

EXAMPLE 8 120% of 42 is what?

Translate: $120\% \times 42 = a$.

Solve: The letter is by itself. To solve the equation, we carry out the calculation:

$$a = 120\% \times 42$$
$$a = 1.2 \times 42 \qquad 120\% = 1.2$$
$$a = 50.4.$$

Thus, 120% of 42 is 50.4. The answer is 50.4.

Do Exercise 8.

Each year, Americans spend about $40 billion on pets; approximately 23.9% of that amount is spent on veterinary care. What is spent per year on veterinary care? (See Example 7.)

SOURCE: American Pet Products Manufacturers Association

7. Solve:

What is 12% of $50?

8. Solve:

64% of 55 is what?

Finding the Base

EXAMPLE 9 5% of what is 20?

Translate: $5\% \times b = 20$.

Solve: This time the letter is *not* by itself. To solve the equation, we divide by 5% on both sides:

$$\frac{5\% \times b}{5\%} = \frac{20}{5\%} \qquad \text{Dividing by 5\% on both sides}$$

$$b = \frac{20}{0.05} \qquad 5\% = 0.05$$

$$b = 400.$$

Thus, 5% of 400 is 20. The answer is 400.

EXAMPLE 10 $3 is 16% of what?

Translate:
$$\begin{array}{ccccc} \$3 & \text{is} & 16\% & \text{of} & \text{what?} \\ \downarrow & \downarrow & \downarrow & \downarrow & \downarrow \\ 3 & = & 16\% & \times & b \end{array}$$

Solve: To solve the equation, we divide by 16% on both sides:

$$\frac{3}{16\%} = \frac{16\% \times b}{16\%} \qquad \text{Dividing by 16\% on both sides}$$

$$\frac{3}{0.16} = b \qquad 16\% = 0.16$$

$$18.75 = b.$$

Thus, $3 is 16% of $18.75. The answer is $18.75.

Do Exercises 9 and 10.

Finding the Percent Number

In solving these problems, you *must* remember to convert to percent notation after you have solved the equation.

EXAMPLE 11 2100 is what percent of 30,000?

Translate:
$$\begin{array}{cccccc} 2100 & \text{is} & \text{what percent} & \text{of} & 30,000? \\ \downarrow & \downarrow & \downarrow & \downarrow & \downarrow \\ 2100 & = & p & \times & 30,000 \end{array}$$

Solve: To solve the equation, we divide by 30,000 on both sides and convert the result to percent notation:

$$p \times 30,000 = 2100$$

$$\frac{p \times 30,000}{30,000} = \frac{2100}{30,000} \qquad \text{Dividing by 30,000 on both sides}$$

$$p = 0.07 \qquad \text{Converting to decimal notation}$$

$$p = 7\%. \qquad \text{Converting to percent notation}$$

Thus, 2100 is 7% of 30,000. The answer is 7%.

In a survey of a group of people, it was found that 5%, or 20 people, chose strawberry as their favorite ice cream flavor. How many people were surveyed? (See Example 9.)
SOURCE: International Ice Cream Association

Solve.

9. 20% of what is 45?

10. $60 is 120% of what?

In 2007, there were about 30,000 earthquakes worldwide. Of this number, 2100 earthquakes had magnitudes greater than 5.0. What percent of the 30,000 earthquakes had magnitudes greater than 5.0? (See Example 11.)
SOURCE: National Earthquake Information Center, U.S. Geological Survey

Answers
9. 225 **10.** $50

EXAMPLE 12 What percent of $50 is $16?

Translate: What percent of $50 is $16?

$$p \times 50 = 16$$

Solve: To solve the equation, we divide by 50 on both sides and convert the answer to percent notation:

$$\frac{p \times 50}{50} = \frac{16}{50} \qquad \text{Dividing by 50 on both sides}$$

$$p = \frac{16}{50}$$

$$p = 0.32$$

$$p = 32\%. \qquad \text{Converting to percent notation}$$

Thus, 32% of $50 is $16. The answer is 32%.

Do Exercises 11 and 12.

Solve:

11. 16 is what percent of 40?

12. What percent of $84 is $10.50?

-- *Caution!* --

When a question asks "what percent?", be sure to give the answer in percent notation.

Calculator Corner

Using Percents in Computations Many calculators have a %⃞ key that can be used in computations. (See the Calculator Corner on page 350.) For example, to find 11% of 49, we press 1⃞ 1⃞ 2nd⃞ %⃞ ×⃞ 4⃞ 9⃞ =⃞, or 4⃞ 9⃞ ×⃞ 1⃞ 1⃞ SHIFT⃞ %⃞. The display reads ⎡ 5.39 ⎤, so 11% of 49 is 5.39.

In Example 9, we perform the computation 20/5%. To use the %⃞ key in this computation, we press 2⃞ 0⃞ ÷⃞ 5⃞ 2nd⃞ %⃞ =⃞, or 2⃞ 0⃞ ÷⃞ 5⃞ SHIFT⃞ %⃞. The result is 400.

We can also use the %⃞ key to find the percent number in a problem. In Example 11, for instance, we answer the question "2100 is what percent of 30,000?" On a calculator, we press 2⃞ 1⃞ 0⃞ 0⃞ ÷⃞ 3⃞ 0⃞ 0⃞ 0⃞ 0⃞ 2nd⃞ %⃞ =⃞, or 2⃞ 1⃞ 0⃞ 0⃞ ÷⃞ 3⃞ 0⃞ 0⃞ 0⃞ 0⃞ SHIFT⃞ %⃞. The result is 7, so 2100 is 7% of 30,000.

Exercises: Use a calculator to find each of the following.

1. What is 12.6% of $40?

2. 0.04% of 28 is what?

3. 8% of what is 36?

4. $45 is 4.5% of what?

5. 23 is what percent of 920?

6. What percent of $442 is $53.04?

Answers

11. 40%   **12.** 12.5%

a Translate to an equation. Do not solve.

1. What is 32% of 78?

2. 98% of 57 is what?

3. 89 is what percent of 99?

4. What percent of 25 is 8?

5. 13 is 25% of what?

6. 21.4% of what is 20?

b Translate to an equation and solve.

7. What is 85% of 276?

8. What is 74% of 53?

9. 150% of 30 is what?

10. 100% of 13 is what?

11. What is 6% of $300?

12. What is 4% of $45?

13. 3.8% of 50 is what?

14. $33\frac{1}{3}$% of 480 is what?
$\left(Hint: 33\frac{1}{3}\% = \frac{1}{3}.\right)$

15. $39 is what percent of $50?

16. $16 is what percent of $90?

17. 20 is what percent of 10?

18. 60 is what percent of 20?

19. What percent of $300 is $150?

20. What percent of $50 is $40?

21. What percent of 80 is 100?

22. What percent of 60 is 15?

23. 20 is 50% of what?

24. 57 is 20% of what?

25. 40% of what is $16?

26. 100% of what is $74?

27. 56.32 is 64% of what?

28. 71.04 is 96% of what?

29. 70% of what is 14?

30. 70% of what is 35?

31. What is $62\frac{1}{2}$% of 10?

32. What is $35\frac{1}{4}$% of 1200?

33. What is 8.3% of $10,200?

34. What is 9.2% of $5600?

35. 2.5% of what is 30.4?

36. 8.2% of what is 328?

Skill Maintenance

Write fraction notation. [4.1b]

37. 0.09

38. 1.79

39. 0.875

40. 0.125

41. 0.9375

42. 0.6875

Write decimal notation. [4.1b]

43. $\dfrac{89}{100}$

44. $\dfrac{7}{100}$

45. $\dfrac{3}{10}$

46. $\dfrac{17}{1000}$

Synthesis

Solve.

47. What is 7.75% of $10,880?

Estimate _____

Calculate _____

48. 50,951.775 is what percent of 78,995?

Estimate _____

Calculate _____

49. $2496 is 24% of what amount?

Estimate _____

Calculate _____

50. What is 38.2% of $52,345.79?

Estimate _____

Calculate _____

51. 40% of $18\frac{3}{4}$% of $25,000 is what?

6.4 Solving Percent Problems Using Proportions*

a Translating to Proportions

A percent is a ratio of some number to 100. For example, 46% is the ratio $\frac{46}{100}$. The numbers 67,620,000 and 147,000,000 have the same ratio as 46 and 100.

$$\frac{46}{100} = \frac{67,620,000}{147,000,000}$$

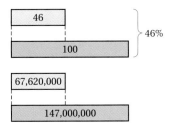

To solve a percent problem using a proportion, we translate as follows:

$$\text{Number} \rightarrow \frac{N}{100} = \frac{a}{b} \begin{matrix}\leftarrow \text{Amount} \\ \leftarrow \text{Base}\end{matrix}$$
$$100 \rightarrow$$

You might find it helpful to read this as "part is to whole as part is to whole."

For example, 60% of 25 is 15 translates to

$$\frac{60}{100} = \frac{15}{25} \begin{matrix}\leftarrow \text{Amount} \\ \leftarrow \text{Base}\end{matrix}$$

A clue in translating is that the base, *b*, corresponds to 100 and usually follows the wording "percent of." Also, *N*% always translates to *N*/100. Another aid in translating is to make a comparison drawing. To do this, we start with the percent side and list 0% at the top and 100% near the bottom. Then we estimate where the specified percent—in this case, 60%—is located. The corresponding quantities are then filled in. The base—in this case, 25—always corresponds to 100%, and the amount—in this case, 15—corresponds to the specified percent.

Percents	Quantities	Percents	Quantities	Percents	Quantities
0%	0	0%	0	0%	0
		60%		60%	15
100%		100%		100%	25

The proportion can then be read easily from the drawing: $\frac{60}{100} = \frac{15}{25}$.

---

*Note: This section presents an alternative method for solving basic percent problems. You can use either equations or proportions to solve percent problems, but you might prefer one method over the other, or your instructor may direct you to use one method over the other.

In the United States, 46% of the labor force is women. In 2007, there were approximately 147,000,000 people in the labor force. This means that about 67,620,000 were women.

SOURCES: U.S. Department of Labor; U.S. Bureau of Labor Statistics

Answers

Skill to Review:
1. 900   **2.** 21.5

EXAMPLE 1 Translate to a proportion.

23% of 5 is what?

$$\frac{23}{100} = \frac{a}{5}$$

EXAMPLE 2 Translate to a proportion.

What is 124% of 49?

$$\frac{124}{100} = \frac{a}{49}$$

Translate to a proportion. Do not solve.

1. 12% of 50 is what?

2. What is 40% of 60?

3. 130% of 72 is what?

Do Exercises 1–3.

EXAMPLE 3 Translate to a proportion.

3 is 10% of what?

$$\frac{10}{100} = \frac{3}{b}$$

EXAMPLE 4 Translate to a proportion.

45% of what is 23?

$$\frac{45}{100} = \frac{23}{b}$$

Translate to a proportion. Do not solve.

4. 45 is 20% of what?

5. 120% of what is 60?

Do Exercises 4 and 5.

EXAMPLE 5 Translate to a proportion.

10 is what percent of 20?

$$\frac{N}{100} = \frac{10}{20}$$

Answers

1. $\frac{12}{100} = \frac{a}{50}$ **2.** $\frac{40}{100} = \frac{a}{60}$ **3.** $\frac{130}{100} = \frac{a}{72}$

4. $\frac{20}{100} = \frac{45}{b}$ **5.** $\frac{120}{100} = \frac{60}{b}$

EXAMPLE 6 Translate to a proportion.

What percent of 50 is 7?

$$\frac{N}{100} = \frac{7}{50}$$

Do Exercises 6 and 7.

Translate to a proportion. Do not solve.

6. 16 is what percent of 40?

7. What percent of 84 is 10.5?

b Solving Percent Problems

After a percent problem has been translated to a proportion, we solve as in Section 5.3.

EXAMPLE 7 5% of what is $20?

Translate: $\dfrac{5}{100} = \dfrac{20}{b}$

Solve: $5 \cdot b = 100 \cdot 20$ Equating cross products

$\dfrac{5 \cdot b}{5} = \dfrac{100 \cdot 20}{5}$ Dividing by 5

$b = \dfrac{2000}{5}$

$b = 400$ Simplifying

Thus, 5% of $400 is $20. The answer is $400.

Do Exercise 8.

8. Solve:

20% of what is $45?

EXAMPLE 8 120% of 42 is what?

Translate: $\dfrac{120}{100} = \dfrac{a}{42}$

Solve: $120 \cdot 42 = 100 \cdot a$ Equating cross products

$\dfrac{120 \cdot 42}{100} = \dfrac{100 \cdot a}{100}$ Dividing by 100

$\dfrac{5040}{100} = a$

$50.4 = a$ Simplifying

Thus, 120% of 42 is 50.4. The answer is 50.4.

Do Exercises 9 and 10.

Solve.

9. 64% of 55 is what?

10. What is 12% of 50?

Answers

6. $\dfrac{N}{100} = \dfrac{16}{40}$ **7.** $\dfrac{N}{100} = \dfrac{10.5}{84}$ **8.** $225

9. 35.2   **10.** 6

EXAMPLE 9 210 is $10\frac{1}{2}$% of what?

Translate: $\dfrac{210}{b} = \dfrac{10.5}{100}$ $10\frac{1}{2}\% = 10.5\%$

Solve: $210 \cdot 100 = b \cdot 10.5$ Equating cross products

$\dfrac{210 \cdot 100}{10.5} = \dfrac{b \cdot 10.5}{10.5}$ Dividing by 10.5

$\dfrac{21,000}{10.5} = b$ Multiplying and simplifying

$2000 = b$ Dividing

Thus, 210 is $10\frac{1}{2}$% of 2000. The answer is 2000.

11. Solve:

60 is 120% of what?

> Do Exercise 11.

EXAMPLE 10 $10 is what percent of $20?

Translate: $\dfrac{10}{20} = \dfrac{N}{100}$

Solve: $10 \cdot 100 = 20 \cdot N$ Equating cross products

$\dfrac{10 \cdot 100}{20} = \dfrac{20 \cdot N}{20}$ Dividing by 20

$\dfrac{1000}{20} = N$ Multiplying and simplifying

$50 = N$ Dividing

> Note when solving percent problems using proportions that N is a percent and need not be converted.

Thus, $10 is 50% of $20. The answer is 50%.

12. Solve:

$12 is what percent of $40?

> Do Exercise 12.

EXAMPLE 11 What percent of 50 is 16?

Translate: $\dfrac{N}{100} = \dfrac{16}{50}$

Solve: $50 \cdot N = 100 \cdot 16$ Equating cross products

$\dfrac{50 \cdot N}{50} = \dfrac{100 \cdot 16}{50}$ Dividing by 50

$N = \dfrac{1600}{50}$ Multiplying and simplifying

$N = 32$ Dividing

Thus, 32% of 50 is 16. The answer is 32%.

13. Solve:

What percent of 84 is 10.5?

> Do Exercise 13.

Answers

11. 50   **12.** 30%   **13.** 12.5%

a Translate to a proportion. Do not solve.

1. What is 37% of 74?

2. 66% of 74 is what?

3. 4.3 is what percent of 5.9?

4. What percent of 6.8 is 5.3?

5. 14 is 25% of what?

6. 133% of what is 40?

b Translate to a proportion and solve.

7. What is 76% of 90?

8. What is 32% of 70?

9. 70% of 660 is what?

10. 80% of 920 is what?

11. What is 4% of 1000?

12. What is 6% of 2000?

13. 4.8% of 60 is what?

14. 63.1% of 80 is what?

15. $24 is what percent of $96?

16. $14 is what percent of $70?

17. 102 is what percent of 100?

18. 103 is what percent of 100?

19. What percent of $480 is $120?

20. What percent of $80 is $60?

21. What percent of 160 is 150?

22. What percent of 33 is 11?

23. $18 is 25% of what?

24. $75 is 20% of what?

25. 60% of what is 54?

26. 80% of what is 96?

27. 65.12 is 74% of what?

28. 63.7 is 65% of what?

29. 80% of what is 16?

30. 80% of what is 10?

31. What is $62\frac{1}{2}$% of 40?

32. What is $43\frac{1}{4}$% of 2600?

33. What is 9.4% of $8300?

34. What is 8.7% of $76,000?

35. 80.8 is $40\frac{2}{5}$% of what?

36. 66.3 is $10\frac{1}{5}$% of what?

Skill Maintenance

Solve. [5.3b]

37. $\dfrac{x}{188} = \dfrac{2}{47}$

38. $\dfrac{15}{x} = \dfrac{3}{800}$

39. $\dfrac{4}{7} = \dfrac{x}{14}$

40. $\dfrac{612}{t} = \dfrac{72}{244}$

41. $\dfrac{5000}{t} = \dfrac{3000}{60}$

42. $\dfrac{75}{100} = \dfrac{n}{20}$

43. $\dfrac{x}{1.2} = \dfrac{36.2}{5.4}$

44. $\dfrac{y}{1\frac{1}{2}} = \dfrac{2\frac{3}{4}}{22}$

Solve.

45. A recipe for muffins calls for $\frac{1}{2}$ qt of buttermilk, $\frac{1}{3}$ qt of skim milk, and $\frac{1}{16}$ qt of oil. How many quarts of liquid ingredients does the recipe call for? [3.2b]

46. The Ferristown School District purchased $\frac{3}{4}$ ton (T) of clay. If the clay is to be shared equally among the district's 6 art departments, how much will each art department receive? [2.7d]

Synthesis

Solve.

47. ▥ What is 8.85% of $12,640?

Estimate _____

Calculate _____

48. ▥ 78.8% of what is 9809.024?

Estimate _____

Calculate _____

Mid-Chapter Review

Concept Reinforcement

Determine whether each statement is true or false.

_____ **1.** When converting decimal notation to percent notation, move the decimal point two places to the right and write a percent symbol.   [6.1b]

_____ **2.** The symbol % is equivalent to $\times\ 0.10$.   [6.1a]

_____ **3.** Of the numbers $\frac{1}{10}$, 1%, 0.1%, 10%, and $\frac{1}{100}$, the smallest number is 0.1%.   [6.1b], [6.2a, b]

Guided Solutions

Fill in each blank with the number that creates a correct statement or solution.   [6.1b], [6.2a, b]

4. $\dfrac{1}{2}\% = \dfrac{1}{2} \cdot \dfrac{1}{\square} = \dfrac{1}{\square}$

5. $\dfrac{80}{1000} = \dfrac{\square}{100} = \square\ \%$

6. $5.5\% = \dfrac{\square}{100} = \dfrac{\square}{1000} = \dfrac{11}{\square}$

7. $0.375 = \dfrac{\square}{1000} = \dfrac{\square}{100} = \square\ \%$

8. Solve: 15 is what percent of 80?   [6.3b]

$15 = p \times \square$ $\qquad$ Translating

$\dfrac{15}{\square} = \dfrac{p \times \square}{\square}$ $\qquad$ Dividing on both sides

$\dfrac{15}{\square} = p$ $\qquad$ Simplifying

$\square = p$ $\qquad$ Dividing

$\square\% = p$ $\qquad$ Converting to percent notation

Mixed Review

Find decimal notation.   [6.1b]

9. 28%

10. 0.15%

11. $5\dfrac{3}{8}\%$

12. 240%

Find percent notation.   [6.1b], [6.2a]

13. 0.71

14. $\dfrac{9}{100}$

15. 0.3891

16. $\dfrac{3}{16}$

17. 0.005

18. $\dfrac{37}{50}$

19. 6

20. $\dfrac{5}{6}$

Find fraction notation. Simplify. [6.2b]

21. 85%

22. 0.048%

23. $22\frac{3}{4}\%$

24. $16.\overline{6}\%$

Write percent notation for the shaded area. [6.2a]

25.

26.

Solve. [6.3b], [6.4b]

27. 25% of what is 14.5?

28. 220 is what percent of 1320?

29. What is 3.2% of 80,000?

30. $17.50 is 35% of what?

31. What percent of $800 is $160?

32. 130% of $350 is what?

33. Arrange the following numbers from smallest to largest. [6.1b], [6.2a, b]

$$\frac{1}{2}\%, \ 5\%, \ 0.275, \ \frac{13}{100}, \ 1\%, \ 0.1\%, \ 0.05\%, \ \frac{3}{10}, \ \frac{7}{20}, \ 10\%$$

34. Solve: 8.5 is $2\frac{1}{2}\%$ of what? [6.3b], [6.4b]

  A. 3.4   **B.** 21.25
  C. 0.2125   **D.** 340

35. Solve: $102,000 is what percent of $3.6 million?
    [6.3b], [6.4b]

  A. $2.8\overline{3}$ million   **B.** $2\frac{5}{6}\%$
  C. $0.028\overline{3}\%$ **D.** $28.\overline{3}$

Understanding Through Discussion and Writing

36. Is it always best to convert from fraction notation to percent notation by first finding decimal notation? Why or why not? [6.2a]

37. Suppose we know that 40% of 92 is 36.8. What is a quick way to find 4% of 92? 400% of 92? Explain. [6.3b]

38. In solving Example 10 in Section 6.4 a student simplifies $\frac{10}{20}$ before solving. Is this a good idea? Why or why not? [6.4b]

39. What do the following have in common? Explain. [6.1b], [6.2a, b]

$$\frac{23}{16}, \ 1\frac{875}{2000}, \ 1.4375, \ \frac{207}{144}, \ 1\frac{7}{16}, \ 143.75\%, \ 1\frac{4375}{10,000}$$

6.5

Applications of Percent

a Applied Problems Involving Percent

Applied problems involving percent are not always stated in a manner easily translated to an equation. In such cases, it is helpful to rephrase the problem before translating. Sometimes it also helps to make a drawing.

EXAMPLE 1 *Extinction of Mammals.* According to a study conducted for the International Union for the Conservation of Nature (IUCN), the world's mammals are in danger of an extinction crisis. Of the 5487 species of mammals on Earth, 1141 are on the IUCN Red List of Threatened Species. Six of those mammals are shown below. What percent of all mammals are threatened with extinction?

Sources: Environment News Service, October 6, 2008; IUCN

OBJECTIVES

a Solve applied problems involving percent.

b Solve applied problems involving percent of increase or percent of decrease.

SKILL TO REVIEW
Objective 4.4a: Divide using decimal notation.

Divide.
  1. $345 \div 57.5$
  2. $111.87 \div 9.9$

Top row, left to right: Tasmanian devils, Père David's deer, African elephant. *Bottom row, left to right:* Iberian lynx, black-footed ferret, giant panda.

1. **Familiarize.** The question asks for a percent of the world's mammals that are in danger of extinction. We note that 5487 is approximately 5500 and 1141 is approximately 1100. Since 1100 is $\frac{1100}{5500}$, or $\frac{1}{5}$, or 20% of 5500, our answer is close to 20%. We let p = the percent of mammals that are in danger of extinction.

2. **Translate.** There are two ways in which we can translate this problem.

Percent equation (see Section 6.3):

$$1141 \text{ is what percent of } 5487?$$

$$1141 = p \cdot 5487$$

Proportion (see Section 6.4):

$$\frac{N}{100} = \frac{1141}{5487}$$

For proportions, $N\% = p$.

Percents	Quantities
0%	0
$N\%$	1141
100%	5487

Answers

Skill to Review:
1. 6   2. 11.3

3. **Solve.** We now have two ways in which to solve this problem.

Percent equation (see Section 6.3):

$$1141 = p \cdot 5487$$

$$\frac{1141}{5487} = \frac{p \cdot 5487}{5487} \qquad \text{Dividing by 5487 on both sides}$$

$$\frac{1141}{5487} = p$$

$$0.208 \approx p \qquad \text{Finding decimal notation and rounding to the nearest thousandth}$$

$$20.8\% \approx p \qquad \text{Remember to find percent notation.}$$

Note here that the solution, p, includes the % symbol.

Proportion (see Section 6.4):

$$\frac{N}{100} = \frac{1141}{5487}$$

$$N \cdot 5487 = 100 \cdot 1141 \qquad \text{Equating cross products}$$

$$\frac{N \cdot 5487}{5487} = \frac{114{,}100}{5487} \qquad \text{Dividing by 5487 on both sides}$$

$$N = \frac{114{,}100}{5487}$$

$$N = 20.8 \qquad \text{Dividing and rounding to the nearest tenth}$$

We use the solution of the proportion to express the answer to the problem as 20.8%. Note that in the proportion method, $N\% = p$.

4. **Check.** To check, we note that the answer 20.8% is close to 20%, as estimated in the *Familiarize* step.

5. **State.** About 20.8% of the world's mammals are threatened with extinction.

1. **Presidential Assassinations in Office.** Of the 43 different U.S. presidents, 4 have been assassinated in office. These were James A. Garfield, William McKinley, Abraham Lincoln, and John F. Kennedy. What percent have been assassinated in office?

Do Exercise 1.

EXAMPLE 2 *Transportation to Work.* In the United States, there are about 147,000,000 workers 16 years and older. Approximately 77% drive to work alone. How many workers drive to work alone?

Transportation to Work in the United States

Drive alone 76.9%
Taxi 0.1%
Motorcycle 0.2%
Bicycle 0.4%
Other means 0.9%
Walk 2.5%
Work from home 3.6%
Public transportation 4.7%
Carpool 10.7%

SOURCES: U.S. Census Bureau; American Community Survey

Answer

1. About 9.3%

1. **Familiarize.** We can simplify the pie chart shown on the preceding page to help familiarize ourselves with the problem. We let a = the total number of workers who drive to work alone.

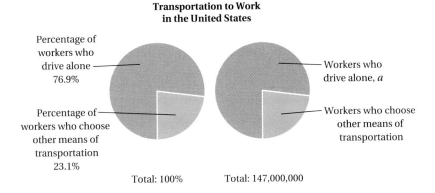

Transportation to Work in the United States

Percentage of workers who drive alone 76.9%

Percentage of workers who choose other means of transportation 23.1%

Total: 100%

Workers who drive alone, a

Workers who choose other means of transportation

Total: 147,000,000

2. **Translate.** There are two ways in which we can translate this problem.

Percent equation:

What number is 76.9% of 147,000,000?

$$a = 76.9\% \cdot 147,000,000$$

Proportion:

$$\frac{76.9}{100} = \frac{a}{147,000,000}$$

3. **Solve.** We now have two ways in which to solve this problem.

Percent equation:

$$a = 76.9\% \cdot 147,000,000$$

We convert 76.9% to decimal notation and multiply:

$$a = 0.769 \times 147,000,000 = 113,043,000.$$

Proportion:

$$\frac{76.9}{100} = \frac{a}{147,000,000}$$

$$76.9 \times 147,000,000 = 100 \cdot a \quad \text{Equating cross products}$$

$$\frac{76.9 \cdot 147,000,000}{100} = \frac{100 \cdot a}{100} \quad \text{Dividing by 100}$$

$$\frac{11,304,300,000}{100} = a$$

$$113,043,000 = a \quad \text{Simplifying}$$

4. **Check.** To check, we can repeat the calculations. We can also do a partial check by estimating. Since 76.9% is about 80%, or $\frac{4}{5}$, and $\frac{4}{5}$ of 147,000,000 is 117,600,000 and 117,600,000 is close to 113,043,000, our answer is reasonable.

5. **State.** The number of workers who drive to work alone is 113,043,000.

Do Exercise 2.

2. **Transportation to Work.** There are about 147,000,000 workers 16 years or older in the United States. Approximately 10.7% carpool to work. How many workers carpool to work?
Sources: U.S. Census Bureau; American Community Survey

Answer

2. 15,729,000 workers

Percent of Increase or Decrease

Percent is often used to state increase or decrease. Let's consider an example of each, using the price of a car as the original number.

Percent of Increase

One year a car sold for $20,455. The manufacturer decides to raise the price of the following year's model by 6%. The increase is 0.06 × $20,455, or $1227.30. The new price is $20,455 + $1227.30, or $21,682.30. Note that the new price is 106% of the *former* price.

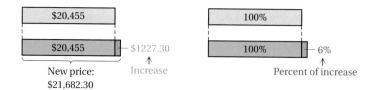

The increase, $1227.30, is 6% of the *former* price, $20,455. The *percent of increase* is 6%.

Percent of Decrease

Abigail buys the car listed above for $20,455. After one year, the car depreciates in value by 25%. The decrease is 0.25 × $20,455, or $5113.75. This lowers the value of the car to $20,455 − $5113.75, or $15,341.25. Note that the new value is 75% of the original price. If Abigail decides to sell the car after one year, $15,341.25 might be the most she could expect to get for it.

The decrease, $5113.75, is 25% of the *original* price, $20,455. The *percent of decrease* is 25%.

> Do Exercises 3 and 4.

When a quantity is decreased by a certain percent, we say that we have **percent of decrease**.

EXAMPLE 3   *Dow Jones Industrial Average.*   The Dow Jones Industrial Average (DJIA) plunged from 11,143 to 10,365 on September 29, 2008. This was the largest one-day drop in its history. What was the percent of decrease?

Sources: *Nightly Business Reports,* September 29, 2008; DJIA

New price: $21,682.30

Increase ⬆

Former price: $20,455

Original price: $20,455

Decrease ⬇

New value: $15,341.25

---

3. Percent of Increase.   The price of a car is $36,875. The price is increased by 4%.

a) How much is the increase?

b) What is the new price?

4. Percent of Decrease.   The value of a car is $36,875. The car depreciates in value by 25% after one year.

a) How much is the decrease?

b) What is the depreciated value of the car?

Answers

3. (a) $1475; (b) $38,350
4. (a) $9218.75; (b) $27,656.25

1. **Familiarize.** We first determine the amount of decrease and then make a drawing.

$$
\begin{array}{rl}
11{,}143 & \text{Opening average} \\
-10{,}365 & \text{Closing average} \\
\hline
778 & \text{Decrease}
\end{array}
$$

We are asking this question: The decrease is what percent of the opening average? We let $p = $ the percent of decrease.

2. **Translate.** There are two ways in which we can translate this problem.

Percent equation:

$$
\begin{array}{ccccc}
778 & \text{is} & \text{what percent} & \text{of} & 11{,}143? \\
\downarrow & \downarrow & \downarrow & \downarrow & \downarrow \\
778 & = & p & \times & 11{,}143
\end{array}
$$

Proportion:

$$
\frac{N}{100} = \frac{778}{11{,}143}
$$

For proportions, $N\% = p$.

3. **Solve.** We have two ways in which to solve this problem.

Percent equation:

$$
778 = p \times 11{,}143
$$

$$
\frac{778}{11{,}143} = \frac{p \times 11{,}143}{11{,}143} \qquad \text{Dividing by 11,143 on both sides}
$$

$$
\frac{778}{11{,}143} = p
$$

$$
0.07 \approx p
$$

$$
7\% \approx p \qquad \text{Converting to percent notation}
$$

Proportion:

$$
\frac{N}{100} = \frac{778}{11{,}143}
$$

$$
11{,}143 \times N = 100 \times 778 \qquad \text{Equating cross products}
$$

$$
\frac{11{,}143 \times N}{11{,}143} = \frac{100 \times 778}{11{,}143} \qquad \text{Dividing by 11,143 on both sides}
$$

$$
N = \frac{77{,}800}{11{,}143}
$$

$$
N \approx 7
$$

We use the solution of the proportion to express the answer to the problem as 7%.

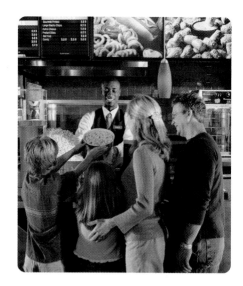

5. Volume of Mail. The volume of U.S. mail decreased from about 213,138 million pieces of mail in 2006 to 212,234 million pieces in 2007. What was the percent of decrease?

Source: U.S. Postal Service

4. Check. To check, we note that, with a 7% decrease, the closing Dow average should be 93% of the opening average. Since

$$93\% \times 11{,}143 = 0.93 \times 11{,}143 \approx 10{,}363,$$

and 10,363 is close to 10,365, our answer checks. (Remember that we rounded to get 7%.)

5. State. The percent of decrease in the Dow average was approximately 7%.

> Do Exercise 5.

When a quantity is increased by a certain percent, we say we have **percent of increase**.

EXAMPLE 4 *Costs for Moviegoers.* The average cost of movie tickets for a family of four was $16.56 in 1993. The cost rose to $28.32 in 2008. What was the percent of increase in the cost for a family of four to attend a movie?

Source: Motion Picture Association of America

1. Familiarize. We first determine the increase in the cost and then make a drawing.

$$
\begin{array}{rl}
\$28.32 & \text{Cost in 2008} \\
-\ \ 16.56 & \text{Cost in 1993} \\
\hline
\$11.76 & \text{Increase}
\end{array}
$$

We are asking this question: The increase is what percent of the *original* cost? We let $p =$ the percent of increase.

2. Translate. There are two ways in which we can translate this problem.

Percent equation:

11.76   is   what percent   of   16.56?

$$11.76 = p \cdot 16.56$$

Proportion:

$$\frac{N}{100} = \frac{11.76}{16.56}$$

For proportions, $N\% = p$.

Percents	Quantities
0%	0
$N\%$	11.76
100%	16.56

Answer

5. About 0.42%

3. Solve. We have two ways in which to solve this problem.

Percent equation:

$$11.76 = p \times 16.56$$

$$\frac{11.76}{16.56} = \frac{p \times 16.56}{16.56} \qquad \text{Dividing by 16.56 on both sides}$$

$$\frac{11.76}{16.56} = p$$

$$0.71 \approx p$$

$$71\% \approx p \qquad \text{Converting to percent notation}$$

Proportion:

$$\frac{N}{100} = \frac{11.76}{16.56}$$

$$16.56 \times N = 100 \times 11.76 \qquad \text{Equating cross products}$$

$$\frac{16.56 \times N}{16.56} = \frac{100 \times 11.76}{16.56} \qquad \text{Dividing by 16.56 on both sides}$$

$$N = \frac{1176}{16.56}$$

$$N \approx 71$$

We use the solution of the proportion to express the answer to the problem as 71%.

4. Check. To check, we take 71% of 16.56:

$$71\% \times 16.56 = 0.71 \times 16.56 = 11.7576.$$

The approximation 11.7576 is close enough to 11.76 to be an excellent check. (Remember that we rounded to get 71%.)

5. State. The percent of increase in the cost for a family of four to attend a movie is 71%.

Do Exercise 6.

6. Shipping Costs. The total cost of shipping a 40-ft container from Shanghai to New York rose from $5100 in 2005 to $8350 in 2008. Find the percent of increase.

Source: CIBC World Markets, IMF

Answer

6. About 64%

Translating for Success

1. *Distance Walked.* After knee replacement, Alex walked $\frac{1}{8}$ mi each morning and $\frac{1}{5}$ mi each afternoon. How much farther did he walk in the afternoon?

2. *Stock Prices.* A stock sold for $5 per share on Monday and only $2.125 per share on Friday. What was the percent of decrease from Monday to Friday?

3. *SAT Score.* After attending a class titled *Improving Your SAT Scores*, Jacob raised his total score from 884 to 1040. What was the percent of increase?

4. *Change in Population.* The population of a small farming community decreased from 1040 to 884. What was the percent of decrease?

5. *Lawn Mowing.* During the summer, brothers Steve and Rob earned money for college by mowing lawns. The largest lawn that they mowed was $2\frac{1}{8}$ acres. Steve can mow $\frac{1}{5}$ acre per hour, and Rob can mow only $\frac{1}{8}$ acre per hour. Working together, how many acres did they mow per hour?

The goal of these matching questions is to practice step (2), *Translate*, of the five-step problem-solving process. Translate each word problem to an equation and select a correct translation from equations A–O.

A. $x + \dfrac{1}{5} = \dfrac{1}{8}$

B. $250 = x \cdot 1040$

C. $884 = x \cdot 1040$

D. $\dfrac{250}{16.25} = \dfrac{1000}{x}$

E. $156 = x \cdot 1040$

F. $16.25 = 250 \cdot x$

G. $\dfrac{1}{5} + \dfrac{1}{8} = x$

H. $2\dfrac{1}{8} = x \cdot 5$

I. $5 = 2.875 \cdot x$

J. $\dfrac{1}{8} + x = \dfrac{1}{5}$

K. $1040 = x \cdot 884$

L. $\dfrac{250}{16.25} = \dfrac{x}{1000}$

M. $2.875 = x \cdot 5$

N. $x \cdot 884 = 156$

O. $x = 16.25 \cdot 250$

Answers on page A-10

---

6. *Land Sale.* Cole sold $2\frac{1}{8}$ acres from the 5 acres he inherited from his uncle. What percent did he sell?

7. *Travel Expenses.* A magazine photographer is reimbursed 16.25¢ per mile for business travel up to 1000 mi per week. In a recent week, he traveled 250 mi. What was the total reimbursement for travel?

8. *Trip Expenses.* The total expenses for Claire's recent business trip were $1040. She put $884 on her credit card and paid the balance in cash. What percent did she place on her credit card?

9. *Cost of Copies.* During the first summer session at a community college, the campus copy center advertised 250 copies for $16.25. At this rate, what is the cost of 1000 copies?

10. *Cost of Insurance.* Following a raise in the cost of health insurance, 250 of a company's 1040 employees dropped their health coverage. What percent of the employees canceled their insurance?

a Solve.

1. *Foreign Students.* In the 2006–2007 school year, of the 15 million college students in the United States, 583,000 were from foreign countries. Approximately 10.70% of the foreign students were from South Korea and 6.05% were from Japan. How many foreign students were from South Korea? from Japan?
Source: Institute of International Education

2. *Mississippi River.* The Mississippi River, which extends from its source, at Lake Itasca in Minnesota, to the Gulf of Mexico, is 2348 mi long. Approximately 77% of the river is navigable. How many miles of the river are navigable?
Source: National Oceanic and Atmospheric Administration

Mississippi River

3. A person earns $43,200 one year and receives an 8% raise in salary. What is the new salary?

4. A person earns $28,600 one year and receives a 5% raise in salary. What is the new salary?

5. *Test Results.* On a test, Juan got 85%, or 119, of the items correct. How many items were on the test?

6. *Test Results.* On a test, Maj Ling got 86%, or 81.7, of the items correct. (There was partial credit on some items.) How many items were on the test?

7. *Farmland.* In Kansas, 47,000,000 acres are farmland. About 5% of all the farm acreage in the United States is in Kansas. What is the total number of acres of farmland in the United States?
Sources: U.S. Department of Agriculture; National Agricultural Statistics Service

8. World Population. World population is increasing by 1.2% each year. In 2008, it was 6.68 billion. What will the population be in 2015?
Sources: U.S. Census Bureau; International Data Base

ion. A car generally depreciates 25% of e in the first year. A car is worth $27,300 ar. What was its original cost?

10. *Car Depreciation.* Given normal use, an American-made car will depreciate 25% of its original cost the first year and 14% of its remaining value in the second year. What is the value of a car at the end of the second year if its original cost was $36,400? $28,400? $26,800?

11. *Test Results.* On a test of 80 items, Pedro got 93% correct. (There was partial credit on some items.) How many items did he get correct? incorrect?

12. *Test Results.* On a test of 40 items, Christina got 91% correct. (There was partial credit on some items.) How many items did she get correct? incorrect?

13. *Under 15 Years Old.* In Egypt, 32.6% of the population is under 15 years old. In the United States, 20.4% of the population is under 15. The population of Egypt is 75,449,000, and the population of the United States is 305,468,000. How many are under 15 years old in Egypt? in the United States?

Sources: U.S. Census Bureau; International Data Base

14. *Age 65 and Older.* In Egypt, 4.5% of the population is age 65 and older. In the United States, 12.5% of the population is age 65 and older. The population of Egypt is 75,449,000, and the population of the United States is 305,468,000. How many are age 65 and older in Egypt? in the United States?

Sources: U.S. Census Bureau; International Data Base

15. *Transplant Waiting List.* The total number of patients waiting for transplants as of September 27, 2007, was 96,749. Of this number, 16,737 were waiting for a liver transplant. What percent were waiting for a liver transplant?

Source: United Network for Organ Sharing

16. *Doctors' Salaries.* In 2007, the starting salary for a neurologist was approximately 64.5% of that of an anesthesiologist. The beginning salary for an anesthesiologist was $275,000. What was the beginning salary for a neurologist?

Source: *Journal of the American Medical Association*

17. *Tipping.* For a party of 8 or more, some restaurants add an 18% tip to the bill. What is the total amount charged for a party of 10 if the cost of the meal, without tip, is $195?

18. *Tipping.* Diners frequently add a 15% tip when charging a meal to a credit card. What is the total amount charged if the cost of the meal, without tip, is $18? $34? $49?

19. *Fast-Food Cooks.* The United States has 392,850 full-time farmers. This number is about 64.2% of the number of fast-food cooks. How many fast-food cooks are there in the United States?

Source: U.S. Department of Agriculture

20. *Spending in Restaurants.* America$ $364 billion in grocery stores in 2007 is about $93\frac{1}{3}$% of the amount spent in How much was spent in restaurants?

Source: U.S. Department of Agriculture

21. A lab technician has 540 mL of a solution of alcohol and water; 8% is alcohol. How many milliliters are alcohol? water?

22. A lab technician has 680 mL of a solution of water and acid; 3% is acid. How many milliliters are acid? water?

23. *U.S. Armed Forces.* There were 1,385,000 people in the United States in active military service in 2006. The numbers in the four armed services are listed in the table below. What percent of the total does each branch represent? Round the answers to the nearest tenth of a percent.

U.S. ARMED FORCES: 2006

TOTAL	1,385,000*
AIR FORCE	349,000
ARMY	505,000
NAVY	350,000
MARINES	180,000

*Includes National Guard, Reserve, and retired regular personnel on extended or continuous active duty. Excludes Coast Guard.

SOURCES: U.S. Department of Defense; U.S. Census Bureau

24. *Living Veterans.* There were 23,977,000 living veterans in the United States in 2006. Numbers in selected age groups are listed in the table below. What percent of the total does each age group represent? Round the answers to the nearest tenth of a percent.

LIVING VETERANS BY AGE: 2006

TOTAL	23,977,000
UNDER 35 YEARS OLD	1,949,000
35–44 YEARS OLD	2,901,000
45–54 YEARS OLD	3,846,000
55–64 YEARS OLD	6,081,000
65 YEARS OLD AND OLDER	9,200,000

SOURCES: U.S. Department of Defense; U.S. Census Bureau

 Solve.

25. *Mortgage Payment Increase.* A monthly mortgage payment increases from $840 to $882. What is the percent of increase?

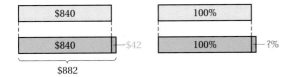

26. *Savings Increase.* The amount in a savings account increased from $200 to $216. What was the percent of increase?

on a diet goes from a weight of 160 lb to a
ght of 136 lb. What is the percent of decrease?

160 lb

100%

136 lb — 24 lb

?%

28. During a sale, a dress decreased in price from $90 to $72. What was the percent of decrease?

$90

100%

$72 — $18

?%

29. *Insulation.* A 15" roll of unfaced fiberglass insulation has a retail price of $23.43. For two weeks, it is on sale for $15.31. What is the percent of decrease?

30. *Set of Weights.* A 300-lb weight set retails for $199.95. For its grand opening, a sporting goods store reduced the price to $154.95. What is the percent of decrease?

31. *Mass Transit.* From April through June of 2008, people took 2.8 billion rides on public transit in the United States. The ridership was up from 2.1 billion rides in 1998. What is the percent of increase?

Source: American Public Transportation Association

32. *Miles of Railroad Track.* The greatest combined length of U.S.-owned operating railroad track was 254,037 mi in 1916, when industrial activity increased during World War I. The total length has decreased ever since. By 2006, the number of miles of track had decreased to 140,490 mi. What is the percent of decrease from 1916 to 2006?

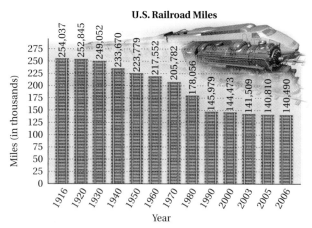

NOTE: The lengths exclude yard tracks, sidings, and parallel tracks.
SOURCE: Association of American Railroads

33. *Overdraft Fees.* Consumers are paying record amounts of fees for overdrawing their bank accounts. In 2007, banks, thrifts, and credit unions collected $45.6 billion in overdraft fees, which is $15.1 billion more than in 2001. What is the percent of increase?

Source: Moebs Services

34. *Credit-Card Debt.* In 2007, the average credit-card debt per household in the United States was $9840. In 1990, the average credit-card debt was $2966. What is the percent of increase?

Sources: CardWeb.com; *USA TODAY*

35. *Immigrant Applications.* From January through June of 2008, 46,866 immigrants applied for citizenship each month. During this same period in 2007, 114,469 immigrants applied each month. What is the percent of decrease?

Source: U.S. Citizenship and Immigration Services (USCIS)

36. *Pharmacists.* It is projected that there will be 296,000 people employed as pharmacists in 2016. In 2006, 243,000 members of the labor force were pharmacists. What is the percent of increase?

Source: EarnMyDegree.com

37. *Patents Issued.* The U.S. Patent and Trademark Office (USPTO) issued a total of 157,284 utility patents in 2007. This number of patents is down from 173,794 in 2006. What is the percent of decrease?

Source: IFI Patent Intelligence

38. *Highway Fatalities.* In 2007, there were 41,059 highway fatalities, which was 1649 fewer deaths than in 2006. What is the percent of decrease?

Source: National Highway Traffic Safety Administration

39. *Two-by-Four.* A cross-section of a standard or nominal "two-by-four" board actually measures $1\frac{1}{2}$ in. by $3\frac{1}{2}$ in. The rough board is 2 in. by 4 in. but is planed and dried to the finished size. What percent of the wood is removed in planing and drying?

40. *Strike Zone.* In baseball, the *strike zone* is normally a 17-in. by 30-in. rectangle. Some batters give the pitcher an advantage by swinging at pitches thrown out of the strike zone. By what percent is the area of the strike zone increased if a 2-in. border is added to the outside?

Source: Major League Baseball

Population Increase. The table below provides data showing how the populations of various states increased from 2000 to 2006. Complete the table by filling in the missing numbers. Round percents to the nearest tenth of a percent.

	STATE	POPULATION IN 2000	POPULATION IN 2006	CHANGE	PERCENT CHANGE
41.	Vermont	608,827	623,908		
42.	Wisconsin	5,363,675		192,831	
43.	Arizona		6,166,318	1,035,686	
44.	Virginia		7,642,884	564,369	
45.	Idaho	1,293,953		172,512	
46.	Georgia	8,186,453	9,363,941		

SOURCE: U.S. Census Bureau

47. *Decrease in Population.* Between 2000 and 2006, the population of North Dakota decreased from 642,200 to 635,867. What was the percent of decrease?
Sources: U.S. Census Bureau; U.S. Department of Commerce

48. *Decrease in Population.* Between 2000 and 2006, the population of Louisiana decreased from 4,468,976 to 4,287,768. What was the percent of decrease?
Sources: U.S. Census Bureau; U.S. Department of Commerce

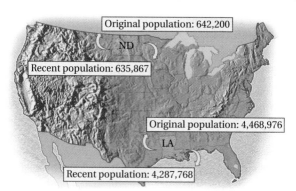

Original population: 642,200
ND
Recent population: 635,867
Original population: 4,468,976
LA
Recent population: 4,287,768

Skill Maintenance

Convert to decimal notation. [4.1b], [4.5a]

49. $\dfrac{25}{11}$

50. $\dfrac{11}{25}$

51. $\dfrac{27}{8}$

52. $\dfrac{43}{9}$

53. $\dfrac{23}{25}$

54. $\dfrac{20}{24}$

55. $\dfrac{14}{32}$

56. $\dfrac{2317}{1000}$

57. $\dfrac{34,809}{10,000}$

58. $\dfrac{27}{40}$

Synthesis

59. A coupon allows a couple to have dinner and then have $10 subtracted from the bill. Before subtracting $10, however, the restaurant adds a tip of 20%. If the couple is presented with a bill for $40.40, how much would the dinner (without tip) have cost without the coupon?

60. If p is 120% of q, then q is what percent of p?

6.6 Sales Tax, Commission, and Discount

a Sales Tax

Sales tax computations represent a special type of percent of increase problem. The sales tax rate in Pennsylvania is 6%. This means that the tax is 6% of the purchase price. Suppose the purchase price of a guitar is $839.95. The sales tax is then 6% of $839.95, or 0.06 × $839.95, or $50.397, or about $50.40.

PENNSYLVANIA

$839.95
+ 6% sales tax

BILL:

Purchase price	= $839.95
Sales tax (6% of $839.95)	= + 50.40
Total price	$890.35

The total that you pay is the purchase price plus the sales tax:

$839.95 + $50.40,   or $890.35.

SALES TAX

Sales tax = Sales tax rate × Purchase price

Total price = Purchase price + Sales tax

EXAMPLE 1 *Wisconsin Sales Tax.*   The sales tax rate in Wisconsin is 5%. How much tax is charged on the purchase of 3 gal of paint at $42.99 each? What is the total price?

a) We first find the cost of the paint. It is

3 × $42.99 = $128.97.

b) The sales tax on items costing $128.97 is

$$\underbrace{\text{Sales tax rate}}_{5\%} \times \underbrace{\text{Purchase price}}_{\$128.97},$$

or 0.05 × 128.97, or 6.4485. Thus the tax is $6.45 (rounded to the nearest cent).

$42.99 each plus 5% sales tax

TRUE BLUE

WISCONSIN

c) The total price is given by the purchase price plus the sales tax:

$128.97 + $6.45,   or $135.42.

To check, note that the total price is the purchase price plus 5% of the purchase price. Thus the total price is 105% of the purchase price. Since 1.05 × 128.97 ≈ 135.42, we have a check. The sales tax is $6.45, and the total price is $135.42.

1. **Texas Sales Tax.** The sales tax rate in Texas is 6.25%. In Texas, how much tax is charged on the purchase of an ultrasound toothbrush that sells for $139.95? What is the total price?

2. **Wyoming Sales Tax.** Samantha buys 7 copies of *The Last Lecture* by Randy Pausch with Jeffrey Zaslow for $14.95 each. The sales tax rate in Wyoming is 4%. In Wyoming, how much sales tax will be charged? What is the total price?

3. The sales tax on the purchase of a set of holiday dishes that costs $449 is $26.94. What is the sales tax rate?

Price: ?
$12.74 tax @ 8%

4. The sales tax on the purchase of a pair of designer jeans is $4.84 and the sales tax rate is 5.5%. Find the purchase price (the price before taxes are added).

Answers

1. $8.75; $148.70   2. $4.19; $108.84
3. 6%   4. $88

Do Exercises 1 and 2.

EXAMPLE 2 The sales tax on the purchase of this GPS navigator, which costs $799, is $55.93. What is the sales tax rate?

GPS Navigator 4.3"-screen with speech recognition
$799
+ $55.93 sales tax

We rephrase and translate as follows:

Rephrase: Sales tax | is | what percent | of | purchase price?
Translate: 55.93 | = | r | $\times$ | 799.

To solve the equation, we divide by 799 on both sides:

$$\frac{55.93}{799} = \frac{r \times 799}{799}$$

$$\frac{55.93}{799} = r$$

$$0.07 = r$$

$$7\% = r.$$

The sales tax rate is 7%.

Do Exercise 3.

EXAMPLE 3 The sales tax on the purchase of a stone-top firepit is $12.74 and the sales tax rate is 8%. Find the purchase price (the price before taxes are added).

We rephrase and translate as follows:

Rephrase: Sales tax | is | 8% | of | what?
Translate: 12.74 | = | 8% | $\times$ | b, or
12.74 = 0.08 $\times$ b.

To solve, we divide by 0.08 on both sides:

$$\frac{12.74}{0.08} = \frac{0.08 \times b}{0.08}$$

$$\frac{12.74}{0.08} = b$$

$$159.25 = b.$$

The purchase price is $159. 25.

Do Exercise 4.

b) Commission

When you work for a **salary**, you receive the same amount of money each week or month. When you work for a **commission**, you are paid a percentage of the total sales for which you are responsible.

> ## COMMISSION
>
> **Commission** = Commission rate × Sales

EXAMPLE 4 *Appliance Sales.* A salesperson's commission rate is 3%. What is the commission from the sale of $8300 worth of appliances?

$$Commission = Commission\ rate \times Sales$$
$$C = 3\% \times 8300$$
$$C = 0.03 \times 8300$$
$$C = 249$$

The commission is $249.

Do Exercise 5.

Do Exercise 5.

EXAMPLE 5 *Earth-Moving Equipment Sales.* Gavin earns a commission of $20,800 selling $320,000 worth of earth-moving equipment. What is the commission rate?

$$Commission = Commission\ rate \times Sales$$
$$20,800 = r \times 320,000$$

5. Hailey's commission rate is 15%. What is the commission from the sale of $9260 worth of exercise equipment?

Answer

5. $1389

To solve this equation, we divide by 320,000 on both sides:

$$\frac{20,800}{320,000} = \frac{r \times 320,000}{320,000}$$

$$0.065 = r$$

$$6.5\% = r.$$

The commission rate is 6.5%.

> Do Exercise 6.

6. William earns a commission of $2040 selling $17,000 worth of concert tickets. What is the commission rate?

EXAMPLE 6 *Cruise Vacations.* Mia's commission rate is 5.6%. She received a commission of $2457 on cruise vacation packages that she sold in November. How many dollars worth of cruise vacations did she sell?

$$Commission = Commission\ rate \times Sales$$
$$2457 = 5.6\% \times S,\ \text{or}$$
$$2457 = 0.056 \times S$$

To solve this equation, we divide by 0.056 on both sides:

$$\frac{2457}{0.056} = \frac{0.056 \times S}{0.056}$$

$$\frac{2457}{0.056} = S$$

$$43,875 = S.$$

Mia sold $43,875 worth of cruise vacation packages.

> Do Exercise 7.

7. Dylan's commission rate is 7.5%. He receives a commission of $2970 from the sale of winter ski passes. How many dollars worth of ski passes did he sell?

c Discount

Suppose that the regular price of a rug is $60, and the rug is on sale at 25% off. Since 25% of $60 is $15, the sale price is $60 − $15, or $45. We call $60 the **original**, or **marked**, **price**, 25% the **rate of discount**, $15 the **discount**, and $45 the **sale price**. Note that discount problems are a type of percent of decrease problem.

> **DISCOUNT AND SALE PRICE**
>
> **Discount** = Rate of discount × Original price
> **Sale price** = Original price − Discount

Answers

6. 12% **7.** $39,600

EXAMPLE 7 A leather sofa marked $2379 is on sale at $33\frac{1}{3}\%$ off. What is the discount? the sale price?

Leather sofa
$2379 original price
Save $33\frac{1}{3}\%$

a) *Discount* = *Rate of discount* × *Original price*

$$D = 33\frac{1}{3}\% \times 2379$$

$$D = \frac{1}{3} \times 2379$$

$$D = \frac{2379}{3} = 793$$

b) *Sale price* = *Original price* − *Discount*

$$S = 2379 - 793$$

$$S = 1586$$

The discount is $793, and the sale price is $1586.

8. A computer marked $660 is on sale at $16\frac{2}{3}\%$ off. What is the discount? the sale price?

Do Exercise 8.

EXAMPLE 8 The price of a snowblower is marked down from $950 to $779. What is the rate of discount?

We first find the discount by subtracting the sale price from the original price:

$$950 - 779 = 171.$$

The discount is $171.

Next, we use the equation for discount:

Discount = *Rate of discount* × *Original price*

$$171 = r \times 950.$$

To solve, we divide by 950 on both sides:

$$\frac{171}{950} = \frac{r \times 950}{950}$$

$$\frac{171}{950} = r$$

$$0.18 = r$$

$$18\% = r.$$

The discount rate is 18%.

To check, note that an 18% discount rate means that 82% of the original price is paid:

$$0.82 \times \$950 = \$779.$$

9. The price of a winter coat is reduced from $75 to $60. Find the rate of discount.

Do Exercise 9.

Answers

8. $110; $550   **9.** 20%

rcise Set

For Extra Help

MyMathLab

Math XL
PRACTICE

WATCH

DOWNLOAD

READ

REVIEW

sales tax rate in Wyoming is
...ould be charged on a fireplace
...doors that costs $239?

2. *Kansas Sales Tax.*   The sales tax rate in Kansas is 5.3%. How much sales tax would be charged on a fireplace screen with doors that costs $239?

3. *Ohio Sales Tax.*   The sales tax rate in Ohio is 5.5%. How much sales tax would be charged on a dog jacket that sells for $29.50?

4. *New Mexico Sales Tax.*   The sales tax rate in New Mexico is 5%. How much sales tax would be charged on a pair of sunglasses that sells for $129.95?

5. *California Sales Tax.*   The sales tax rate in California is 7.25%. How much sales tax is charged on a purchase of 4 travel contour foam pillows at $39.95 each? What is the total price?

6. *Illinois Sales Tax.*   The sales tax rate in Illinois is 6.25%. How much sales tax is charged on a purchase of 3 wet–dry vacs at $60.99 each? What is the total price?

7. The sales tax is $30 on the purchase of a diamond ring that sells for $750. What is the sales tax rate?

8. The sales tax is $48 on the purchase of a dining room set that sells for $960. What is the sales tax rate?

9. The sales tax is $9.12 on the purchase of a patio set that sells for $456. What is the sales tax rate?

10. The sales tax is $35.80 on the purchase of a refrigerator–freezer that sells for $895. What is the sales tax rate?

11. The sales tax on the purchase of a new fishing boat is $112 and the sales tax rate is 2%. What is the purchase price (the price before tax is added)?

12. The sales tax on the purchase of a used car is $100 and the sales tax rate is 5%. What is the purchase price?

13. The sales tax rate in New York City, New York, is 4.375% for the city and 4% for the state. Find the total amount paid for 6 boxes of chocolates at $17.95 each.

14. The sales tax rate in Nashville, Te Davidson County and 7% for the amount paid for 2 ladders at $39

25. A real es sion fr

15. The sales tax rate in Seattle, Washington, is 2.5% for King County and 6.5% for the state. Find the total amount paid for 3 ceiling fans at $84.49 each.

16. The sales tax rate in Miami, Florida, is 1% for Dade County and 6% for the state. Find the total amount paid for 2 tires at $49.95 each.

17. The sales tax rate in Atlanta, Georgia, is 1% for the city, 3% for Fulton County, and 4% for the state. Find the total amount paid for 6 basketballs at $29.95 each.

18. The sales tax rate in Dallas, Texas, is 1% for the city, 1% for Dallas County, and 6.25% for the state. Find the total amount paid for 5 shrubs at $19.95 each.

b Solve.

19. Jose's commission rate is 21%. What is the commission from the sale of $12,500 worth of windows?

20. Jasmine's commission rate is 6%. What is the commission from the sale of $45,000 worth of lawn irrigation systems?

21. Olivia earns $408 selling $3400 worth of shoes. What is the commission rate?

22. Mitchell earns $120 selling $2400 worth of television sets. What is the commission rate?

23. *Real Estate Commission.* A real estate agent's commission rate is 7%. She receives a commission of $12,950 from the sale of a home. How much did the home sell for?

24. *Clothing Consignment Commission.* A clothing consignment shop's commission rate is 40%. The shop receives a commission of $552. How many dollars worth of clothing were sold?

ate commission rate is 8%. What is the commis-
om the sale of a piece of land for $68,000?

26. A real estate commission rate is 6%. What is the commission from the sale of a $98,000 home?

27. David earns $1147.50 selling $7650 worth of car parts. What is the commission rate?

28. Isabella earns $280.80 selling $2340 worth of tee shirts. What is the commission rate?

29. Sabrina's commission is increased according to how much she sells. She receives a commission of 4% for the first $1000 of sales and 7% for the amount over $1000. What is the total commission on sales of $5500?

30. Miguel's commission is increased according to how much he sells. He receives a commission of 5% for the first $2000 of sales and 8% for the amount over $2000. What is the total commission on sales of $6200?

c Complete the table below by filling in the missing numbers.

	MARKED PRICE	RATE OF DISCOUNT	DISCOUNT	SALE PRICE
31.	$300	10%		
32.	$2000	40%		
33.	$17	15%		
34.	$20	25%		
35.		10%	$12.50	
36.		15%	$65.70	
37.	$600		$240	
38.	$12,800		$1920	

39. Find the marked price and the rate of discount for the steel log rack in this ad.

40. Find the marked price and the rate of discount for the digital photo frame in this ad.

41. Find the discount and the rate of discount for the pinball machine in this ad.

42. Find the discount and the rate of discount for the amaryllis in this ad.

Skill Maintenance

Solve. [5.3b]

43. $\dfrac{x}{12} = \dfrac{24}{16}$

44. $\dfrac{7}{2} = \dfrac{11}{x}$

Solve. [4.4b]

45. $0.64 \cdot x = 170$

46. $29.44 = 25.6 \times y$

Find decimal notation. [4.5a]

47. $\dfrac{5}{9}$

48. $\dfrac{23}{11}$

49. $\dfrac{11}{12}$

50. $\dfrac{13}{7}$

51. $\dfrac{15}{7}$

52. $\dfrac{19}{12}$

Convert to standard notation. [4.3b]

53. 4.03 trillion

54. 5.8 million

55. 42.7 million

56. 6.09 trillion

Synthesis

57. *Magazine Subscriptions.* In a recent subscription drive, *Car and Driver* offered a 1-year subscription of 12 issues for the price of $1.50 per issue. They advertised that this was a savings of 69.94% off the cover price. What was the cover price?
Source: *Car and Driver*

58. Elijah collects baseball memorabilia. He bought two autographed plaques, but became short of funds and had to sell them quickly for $200 each. On one, he made a 20% profit, and on the other, he lost 20%. Did he make or lose money on the sale?

59. 🖩 John receives a 10% commission on the first $5000 in sales and 15% on all sales beyond $5000. If John receives a commission of $2405, how much did he sell? Use a calculator and trial and error if you wish.

60. Tee shirts are being sold at the mall for $5 each, or 3 for $10. If you buy three tee shirts, what is the rate of discount?

6.7

Simple Interest and Compound Interest; Credit Cards

OBJECTIVES

a Solve applied problems involving simple interest.

b Solve applied problems involving compound interest.

c Solve applied problems involving interest rates on credit cards.

SKILL TO REVIEW
Objective 6.1b: Convert between percent notation and decimal notation.

Find decimal notation.

1. $34\frac{5}{8}\%$ **2.** $5\frac{1}{4}\%$

a Simple Interest

Suppose you put $1000 into an investment for 1 year. The $1000 is called the **principal**. If the **interest rate** is 5%, in addition to the principal, you get back 5% of the principal, which is

5% of $1000,   or   0.05 × $1000,   or $50.00.

The $50.00 is called **simple interest**. It is, in effect, the price that a financial institution pays for the use of the money over time.

SIMPLE INTEREST FORMULA

The **simple interest** I on principal P, invested for t years at interest rate r, is given by

$$I = P \cdot r \cdot t.$$

EXAMPLE 1 What is the simple interest on $2500 invested at an interest rate of 6% for 1 year?

We use the formula $I = P \cdot r \cdot t$:

$$I = P \cdot r \cdot t = \$2500 \times 6\% \times 1$$
$$= \$2500 \times 0.06$$
$$= \$150.$$

The simple interest for 1 year is $150.

Do Margin Exercise 1.

1. What is the simple interest on $4300 invested at an interest rate of 4% for 1 year?

EXAMPLE 2 What is the simple interest on a principal of $2500 invested at an interest rate of 6% for 3 months?

We use the formula $I = P \cdot r \cdot t$ and express 3 months as a fraction of a year:

$$I = P \cdot r \cdot t = \$2500 \times 6\% \times \frac{3}{12} = \$2500 \times 6\% \times \frac{1}{4}$$
$$= \frac{\$2500 \times 0.06}{4} = \$37.50.$$

The simple interest for 3 months is $37.50.

Do Margin Exercise 2.

When time is given in days, we generally divide it by 365 to express the time as a fractional part of a year.

2. What is the simple interest on a principal of $4300 invested at an interest rate of 4% for 9 months?

EXAMPLE 3 To pay for a shipment of lawn furniture, Patio by Design borrows $8000 at $9\frac{3}{4}\%$ for 60 days. Find **(a)** the amount of simple interest that is due and **(b)** the total amount that must be paid after 60 days.

Answers

Skill to Review:
1. 0.34625 **2.** 0.0525

Margin Exercises:
1. $172 **2.** $129

a) We express 60 days as a fractional part of a year:

$$I = P \cdot r \cdot t = \$8000 \times 9\frac{3}{4}\% \times \frac{60}{365}$$

$$= \$8000 \times 0.0975 \times \frac{60}{365}$$

$$\approx \$128.22.$$

The interest due for 60 days is $128.22.

b) The total amount to be paid after 60 days is the principal plus the interest:

$$\$8000 + \$128.22 = \$8128.22.$$

The total amount due is $8128.22.

Do Exercise 3.

3. The Glass Nook borrows $4800 at $8\frac{1}{2}\%$ for 30 days. Find **(a)** the amount of simple interest due and **(b)** the total amount that must be paid after 30 days.

(b) Compound Interest

When interest is paid *on interest*, we call it **compound interest**. This is the type of interest usually paid on investments. Suppose you have $5000 in a savings account at 6%. In 1 year, the account will contain the original $5000 plus 6% of $5000. Thus the total in the account after 1 year will be

106% of $5000, or 1.06 × $5000, or $5300.

Now suppose that the total of $5300 remains in the account for another year. At the end of this second year, the account will contain the $5300 plus 6% of $5300. The total in the account would thus be

106% of $5300, or 1.06 × $5300, or $5618.

Note that in the second year, interest is also earned on the first year's interest. When this happens, we say that interest is **compounded annually**.

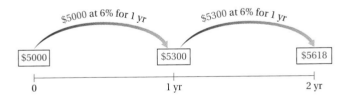

EXAMPLE 4 Find the amount in an account if $2000 is invested at 8%, compounded annually, for 2 years.

a) After 1 year, the account will contain 108% of $2000:

1.08 × $2000 = $2160.

b) At the end of the second year, the account will contain 108% of $2160:

1.08 × $2160 = $2332.80.

The amount in the account after 2 years is $2332.80.

Do Exercise 4.

4. Find the amount in an account if $2000 is invested at 9%, compounded annually, for 2 years.

Suppose that the interest in Example 4 were **compounded semi-annually**—that is, every half year. Interest would then be calculated twice a year at a rate of 8% ÷ 2, or 4% each time. The approach used in Example 4 can then be adapted, as follows.

After the first $\frac{1}{2}$ year, the account will contain 104% of $2000:

$$1.04 \times \$2000 = \$2080.$$

After a second $\frac{1}{2}$ year (1 full year), the account will contain 104% of $2080:

$$1.04 \times \$2080 = \$2163.20.$$

After a third $\frac{1}{2}$ year $\left(1\frac{1}{2} \text{ full years}\right)$, the account will contain 104% of $2163.20:

$$1.04 \times \$2163.20 = \$2249.728$$
$$\approx \$2249.73. \qquad \text{Rounding to the nearest cent}$$

Finally, after a fourth $\frac{1}{2}$ year (2 full years), the account will contain 104% of $2249.73:

$$1.04 \times \$2249.73 = \$2339.7192$$
$$\approx \$2339.72. \qquad \text{Rounding to the nearest cent}$$

Let's summarize our results and look at them another way:

End of 1st $\frac{1}{2}$ year $\rightarrow 1.04 \times 2000 = 2000 \times (1.04)^1$;
End of 2nd $\frac{1}{2}$ year $\rightarrow 1.04 \times (1.04 \times 2000) = 2000 \times (1.04)^2$;
End of 3rd $\frac{1}{2}$ year $\rightarrow 1.04 \times (1.04 \times 1.04 \times 2000) = 2000 \times (1.04)^3$;
End of 4th $\frac{1}{2}$ year $\rightarrow 1.04 \times (1.04 \times 1.04 \times 1.04 \times 2000) = 2000 \times (1.04)^4$.

Note that each multiplication was by 1.04 and that

$$\$2000 \times 1.04^4 \approx \$2339.72. \qquad \text{Using a calculator and rounding to the nearest cent}$$

We have illustrated the following result.

COMPOUND INTEREST FORMULA

If a principal P has been invested at interest rate r, compounded n times a year, in t years it will grow to an amount A given by

$$A = P \cdot \left(1 + \frac{r}{n}\right)^{n \cdot t}.$$

Let's apply this formula to confirm our preceding discussion, where the amount invested is $P = \$2000$, the number of years is $t = 2$, and the number of compounding periods each year is $n = 2$. Substituting into the compound interest formula, we have

$$A = P \cdot \left(1 + \frac{r}{n}\right)^{n \cdot t} = 2000 \cdot \left(1 + \frac{8\%}{2}\right)^{2 \cdot 2}$$
$$= \$2000 \cdot \left(1 + \frac{0.08}{2}\right)^4 = \$2000(1.04)^4$$
$$= \$2000 \times 1.16985856 \approx \$2339.72.$$

If you were using a calculator, you could perform this computation in one step.

EXAMPLE 5 The Ibsens invest $4000 in an account paying $5\frac{5}{8}\%$, compounded quarterly. Find the amount in the account after $2\frac{1}{2}$ years.

The compounding is quarterly, so n is 4. We substitute $4000 for P, $5\frac{5}{8}\%$, or 0.05625, for r, 4 for n, and $2\frac{1}{2}$, or $\frac{5}{2}$, for t and compute A:

$$A = P \cdot \left(1 + \frac{r}{n}\right)^{n \cdot t} = \$4000 \cdot \left(1 + \frac{5\frac{5}{8}\%}{4}\right)^{4 \cdot 5/2}$$

$$= \$4000 \cdot \left(1 + \frac{0.05625}{4}\right)^{10}$$

$$= \$4000(1.0140625)^{10}$$

$$\approx \$4599.46.$$

The amount in the account after $2\frac{1}{2}$ years is $4599.46.

Do Exercise 5.

> **5.** A couple invests $7000 in an account paying $6\frac{3}{8}\%$, compounded semiannually. Find the amount in the account after $1\frac{1}{2}$ years.

Calculator Corner

Compound Interest A calculator is useful in computing compound interest. Not only does it perform computations quickly but it also eliminates the need to round until the computation is completed. This minimizes "round-off errors" that occur when rounding is done at each stage of the computation. We must keep order of operations in mind when computing compound interest.

To find the amount due on a $20,000 loan made for 25 days at 11% interest, compounded daily, we would compute

$20,000\left(1 + \dfrac{0.11}{365}\right)^{25}$. To do this on a calculator, we press 2 0 0 0 0 × ((1 + . 1 1 ÷ 3 6 5)) y^x

(or ∧) 2 5 =. Using a calculator that does not have parentheses, we would first find $1 + \dfrac{0.11}{365}$, raise this result to the 25th power,

and then multiply by 20,000. To do this, we press 1 + . 1 1 ÷ 3 6 5 = y^x (or ∧) 2 5 = × 2 0 0 0 0. In either case, the result is 20,151.23, rounded to the nearest cent.

Some calculators have business keys that allow such computations to be done more quickly.

Exercises:

1. Find the amount due on a $16,000 loan made for 62 days at 13% interest, compounded daily.

2. An investment of $12,500 is made for 90 days at 8.5% interest, compounded daily. How much is the investment worth after 90 days?

---

c Credit Cards

According to CardWeb.com, the average credit-card debt per household in the United States was $9840 in 2007. According to a 2007 survey by student lender Nellie Mae, the average college graduate has a $2748 credit-card debt.

The money you obtain through the use of a credit card is not "free" money. There is a price (interest) to be paid for the privilege. A balance carried on a credit card is a type of loan. Comparing interest rates is essential if one is to become financially responsible. A small change in an interest rate can make a large difference in the cost of a loan. When you make a payment on a credit card, do you know how much of that payment is interest and how much is applied to reducing the principal?

Answer

5. $7690.94

EXAMPLE 6 *Credit Cards.* After the holidays, Sarah has a balance of $3216.28 on a credit card with an annual percentage rate (APR) of 19.7%. She decides not to make additional purchases with this card until she has paid off the balance.

a) Many credit cards require a minimum monthly payment of 2% of the balance. At this rate, what is Sarah's minimum payment on a balance of $3216.28? Round the answer to the nearest dollar.

b) Find the amount of interest and the amount applied to reduce the principal in the minimum payment found in part (a).

c) If Sarah had transferred her balance to a card with an APR of 12.5%, how much of her first payment would be interest and how much would be applied to reduce the principal?

d) Compare the amounts for 12.5% from part (c) with the amounts for 19.7% from part (b).

We solve as follows.

a) We multiply the balance of $3216.28 by 2%:

$$0.02 \times \$3216.28 = \$64.3256.$$ Sarah's minimum payment, rounded to the nearest dollar, is $64.

b) The amount of interest on $3216.28 at 19.7% for one month* is given by

$$I = P \cdot r \cdot t = \$3216.28 \times 0.197 \times \frac{1}{12} \approx \$52.80.$$

We subtract to find the amount applied to reduce the principal in the first payment:

$$\text{Amount applied to reduce the principal} = \text{Minimum payment} - \text{Interest for the month}$$

$$= \$64 - \$52.80$$

$$= \$11.20.$$

Thus the principal of $3216.28 is decreased by only $11.20 with the first payment. (Sarah still owes $3205.08.)

c) The amount of interest on $3216.28 at 12.5% for one month is

$$I = P \cdot r \cdot t = \$3216.28 \times 0.125 \times \frac{1}{12} \approx \$33.50.$$

We subtract to find the amount applied to reduce the principal in the first payment:

$$\text{Amount applied to reduce the principal} = \text{Minimum payment} - \text{Interest for the month}$$

$$= \$64 - \$33.50$$

$$= \$30.50.$$

Thus the principal of $3216.28 is decreased by $30.50 with the first payment. (Sarah still owes $3185.78.)

---

6. Credit Cards. After the holidays, Jamal has a balance of $4867.59 on a credit card with an annual percentage rate (APR) of 21.3%. He decides not to make additional purchases with this card until he has paid off the balance.

a) Many credit cards require a minimum monthly payment of 2% of the balance. What is Jamal's minimum payment on a balance of $4867.59? Round the answer to the nearest dollar.

b) Find the amount of interest and the amount applied to reduce the principal in the minimum payment found in part (a).

c) If Jamal had transferred his balance to a card with an APR of 13.6%, how much of his first payment would be interest and how much would be applied to reduce the principal?

d) Compare the amounts for 13.6% from part (c) with the amounts for 21.3% from part (b).

Answers

6. (a) $97; **(b)** interest: $86.40; amount applied to principal: $10.60; **(c)** interest: $55.17; amount applied to principal: $41.83; **(d)** At 13.6%, the principal was reduced by $31.23 more than at the 21.3% rate. The interest at 13.6% is $31.23 less than at 21.3%.

---

*Actually, the interest on a credit card is computed daily with a rate called a daily percentage rate (DPR). The DPR for Example 6 would be 19.7%/365 = 0.054%. When no payments or additional purchases are made during the month, the difference in total interest for the month is minimal and we will not deal with it here.

d) Let's organize the information for both rates in the following table.

BALANCE BEFORE FIRST PAYMENT	FIRST MONTH'S PAYMENT	% APR	AMOUNT OF INTEREST	AMOUNT APPLIED TO PRINCIPAL	BALANCE AFTER FIRST PAYMENT
$3216.28	$64	19.7%	$52.80	$11.20	$3205.08
3216.28	64	12.5	33.50	30.50	3185.78

Difference in balance after first payment → $19.30

At 19.7%, the interest is $52.80 and the principal is decreased by $11.20. At 12.5%, the interest is $33.50 and the principal is decreased by $30.50. Thus the interest at 19.7% is $52.80 − $33.50, or $19.30, greater than the interest at 12.5%. Thus the principal is decreased by $30.50 − $11.20, or $19.30, more with the 12.5% rate than with the 19.7% rate.

Do Exercise 6 on the preceding page.

Even though the mathematics of the information in the table below is beyond the scope of this text, it is interesting to compare how long it takes to pay off the balance of Example 6 if Sarah continues to pay $64 for each payment with how long it takes if she pays double that amount, $128, for each payment. Financial consultants frequently tell clients that if they want to take control of their debt, they should pay double the minimum payment.

RATE	PAYMENT PER MONTH	NUMBER OF PAYMENTS TO PAY OFF DEBT	TOTAL PAID BACK	ADDITIONAL COST OF PURCHASES
19.7%	$64	107, or 8 yr 11 mo	$6848	$3631.72
19.7	128	33, or 2 yr 9 mo	4224	1007.72
12.5	64	72, or 6 yr	4608	1391.72
12.5	128	29, or 2 yr 5 mo	3712	495.72

As with most loans, if you pay an extra amount toward the principal with each payment, the length of the loan can be greatly reduced. Note that at the rate of 19.7%, it will take Sarah almost 9 yr to pay off her debt if she pays only $64 per month and does not make additional purchases. If she transfers her balance to a card with a 12.5% rate and pays $128 per month, she can eliminate her debt in approximately $2\frac{1}{2}$ yr. You can see how debt can get out of control if you continue to make purchases and pay only the minimum payment each month. The debt will never be eliminated.

a   Find the simple interest.

	PRINCIPAL	RATE OF INTEREST	TIME	SIMPLE INTEREST
1.	$200	4%	1 year	
2.	$200	7.7%	$\frac{1}{2}$ year	
3.	$4300	10.56%	$\frac{1}{4}$ year	
4.	$80,000	$6\frac{3}{4}\%$	$\frac{1}{12}$ year	
5.	$20,000	$4\frac{5}{8}\%$	1 year	
6.	$8000	9.42%	2 months	
7.	$50,000	$5\frac{3}{8}\%$	3 months	
8.	$100,000	$3\frac{1}{4}\%$	1 year	

Solve. Assume that simple interest is being calculated in each case.

9. CopiPix, Inc. borrows $10,000 at 9% for 60 days. Find **(a)** the amount of interest due and **(b)** the total amount that must be paid after 60 days.

10. Sal's Laundry borrows $8000 at 10% for 90 days. Find **(a)** the amount of interest due and **(b)** the total amount that must be paid after 90 days.

11. Animal Instinct, a pet supply shop, borrows $6500 at $5\frac{1}{4}\%$ for 90 days. Find **(a)** the amount of interest due and **(b)** the total amount that must be paid after 90 days.

12. Andante's Cafe borrows $4500 at $12\frac{1}{2}\%$ for 60 days. Find **(a)** the amount of interest due and **(b)** the total amount that must be paid after 60 days.

13. Jean's Garage borrows $5600 at 10% for 30 days. Find **(a)** the amount of interest due and **(b)** the total amount that must be paid after 30 days.

14. Shear Delights Hair Salon borrows $3600 at 4% for 30 days. Find **(a)** the amount of interest due and **(b)** the total amount that must be paid after 30 days.

b   Interest is compounded annually. Find the amount in the account after the given length of time. Round to the nearest cent.

	PRINCIPAL	RATE OF INTEREST	TIME	AMOUNT IN THE ACCOUNT
15.	$400	5%	2 years	
16.	$450	4%	2 years	
17.	$2000	8.8%	4 years	
18.	$4000	7.7%	4 years	
19.	$4300	10.56%	6 years	
20.	$8000	9.42%	6 years	
21.	$20,000	$6\frac{5}{8}\%$	25 years	
22.	$100,000	$5\frac{7}{8}\%$	30 years	

Interest is compounded semiannually. Find the amount in the account after the given length of time. Round to the nearest cent.

	PRINCIPAL	RATE OF INTEREST	TIME	AMOUNT IN THE ACCOUNT
23.	$4000	6%	1 year	
24.	$1000	5%	1 year	
25.	$20,000	8.8%	4 years	
26.	$40,000	7.7%	4 years	
27.	$5000	10.56%	6 years	
28.	$8000	9.42%	8 years	
29.	$20,000	$7\frac{5}{8}\%$	25 years	
30.	$100,000	$4\frac{7}{8}\%$	30 years	

Solve.

31. A family invests $4000 in an account paying 6%, compounded monthly. How much is in the account after 5 months?

32. A couple invests $2500 in an account paying 3%, compounded monthly. How much is in the account after 6 months?

33. A couple invests $1200 in an account paying 10%, compounded quarterly. How much is in the account after 1 year?

34. The O'Hares invest $6000 in an account paying 8%, compounded quarterly. How much is in the account after 18 months?

c Solve.

35. *Credit Cards.* Amelia has a balance of $1278.56 on a credit card with an annual percentage rate (APR) of 19.6%. The minimum payment required in the current statement is $25.57. Find the amount of interest and the amount applied to reduce the principal in this payment and the balance after this payment.

36. *Credit Cards.* Lawson has a balance of $1834.90 on a credit card with an annual percentage rate (APR) of 22.4%. The minimum payment required in the current statement is $36.70. Find the amount of interest and the amount applied to reduce the principal in this payment and the balance after this payment.

37. *Credit Cards.* Antonio has a balance of $4876.54 on a credit card with an annual percentage rate (APR) of 21.3%.

a) Many credit cards require a minimum monthly payment of 2% of the balance. What is Antonio's minimum payment on a balance of $4876.54? Round the answer to the nearest dollar.

b) Find the amount of interest and the amount applied to reduce the principal in the minimum payment found in part (a).

c) If Antonio had transferred his balance to a card with an APR of 12.6%, how much of his payment would be interest and how much would be applied to reduce the principal?

d) Compare the amounts for 12.6% from part (c) with the amounts for 21.3% from part (b).

38. *Credit Cards.* Becky has a balance of $5328.88 on a credit card with an annual percentage rate (APR) of 18.7%.

a) Many credit cards require a minimum monthly payment of 2% of the balance. What is Becky's minimum payment on a balance of $5328.88? Round the answer to the nearest dollar.

b) Find the amount of interest and the amount applied to reduce the principal in the minimum payment found in part (a).

c) If Becky had transferred her balance to a card with an APR of 13.2%, how much of her payment would be interest and how much would be applied to reduce the principal?

d) Compare the amounts for 13.2% from part (c) with the amounts for 18.7% from part (b).

Skill Maintenance

In each of Exercises 39–46, fill in the blank with the correct term from the given list. Some of the choices may not be used.

39. If the product of two numbers is 1, they are _____ of each other. [2.7a]

40. A number is _____ if its ones digit is even and the sum of its digits is divisible by 3. [2.2a]

41. The number 0 is the _____ identity. [1.2a]

42. A(n) _____ is the ratio of price to the number of units. [5.2b]

43. The distance around an object is its _____. [1.2b]

44. A number is _____ if the sum of its digits is divisible by 3. [2.2a]

45. A natural number that has exactly two different factors, only itself and 1, is called a _____ number. [2.1c]

46. When two pairs of numbers have the same ratio, they are _____. [5.3a]

divisible by 3

divisible by 4

divisible by 6

divisible by 9

perimeter

area

unit price

reciprocals

proportional

composite

prime

additive

multiplicative

Synthesis

Effective Yield. The *effective yield* is the yearly rate of simple interest that corresponds to a rate for which interest is compounded two or more times a year. For example, if P is invested at 12%, compounded quarterly, we would multiply P by $(1 + 0.12/4)^4$, or 1.03^4. Since $1.03^4 \approx 1.126$, the 12% compounded quarterly corresponds to an effective yield of approximately 12.6%. In Exercises 47 and 48, find the effective yield for the indicated account.

47. 🖩 The account pays 9% compounded monthly.

48. 🖩 The account pays 10% compounded daily.

Summary and Review

Key Terms and Formulas

percent notation, $n\%$, p. 348
percent of decrease, p. 380
percent of increase, p. 382
purchase price, p. 391
sales tax, p. 391
total price, p. 391

commission, p. 393
original price, p. 394
marked price, p. 394
rate of discount, p. 394
discount, p. 394
sale price, p. 394

principal, p. 400
interest rate, p. 400
simple interest, p. 400
compound interest, p. 401
compounded annually, p. 401
compounded semi-annually, p. 402

Commission = Commission rate × Sales

Discount = Rate of discount × Original price

Sale price = Original price − Discount

Simple Interest: $\quad I = P \cdot r \cdot t$

Compound Interest: $\quad A = P \cdot \left(1 + \dfrac{r}{n}\right)^{n \cdot t}$

Concept Reinforcement

Determine whether each statement is true or false.

_____ 1. A fixed principal invested for 4 years will earn more interest when interest is compounded quarterly than when interest is compounded semiannually. [6.7b]

_____ 2. Of the numbers 0.5%, $\dfrac{5}{1000}\%$, $\dfrac{1}{2}\%$, and $\dfrac{1}{5}$, and $0.\overline{1}$, the largest number is $0.\overline{1}$. [6.1b], [6.2a, b]

_____ 3. If principal A equals principal B and principal A is invested for 2 years at 4%, compounded quarterly, while principal B is invested for 4 years at 2%, compounded semiannually, the interest earned from each investment is the same. [6.7b]

Important Concepts

Objective 6.1b Convert between percent notation and decimal notation.

Example Find decimal notation for $12\dfrac{3}{4}\%$.

$$12\dfrac{3}{4}\% = 12.75 \times 0.01 = 0.1275$$

Practice Exercise

1. Find decimal notation for $62\dfrac{5}{8}\%$.

Objective 6.2a Convert from fraction notation to percent notation.

Example Find percent notation for $\dfrac{5}{12}$.

$$\begin{array}{r} 0.4\ 1\ 6 \\ 1\ 2\ \overline{)\ 5.0\ 0\ 0} \\ \underline{4\ 8} \\ 2\ 0 \\ \underline{1\ 2} \\ 8\ 0 \\ \underline{7\ 2} \\ 8 \end{array}$$

$\dfrac{5}{12} = 0.41\overline{6} = 41.\overline{6}\%$, or $41\dfrac{2}{3}\%$

Practice Exercise

2. Find percent notation for $\dfrac{7}{11}$.

Objective 6.2b Convert from percent notation to fraction notation.

Example Find fraction notation for 9.5%.

$$9.5\% = \frac{9.5}{100} = \frac{95}{1000} = \frac{5 \cdot 19}{5 \cdot 200}$$

$$= \frac{5}{5} \cdot \frac{19}{200} = \frac{19}{200}$$

Practice Exercise

3. Find fraction notation for 6.8%.

Objective 6.3b Solve basic percent problems.

Example 165 is what percent of 3300?

We have

$$165 = p \cdot 3300 \qquad \text{Translating to a percent equation}$$

$$\frac{165}{3300} = \frac{p \cdot 3300}{3300}$$

$$\frac{165}{3300} = p$$

$$0.05 = p$$

$$5\% = p.$$

Thus, 165 is 5% of 3300.

Practice Exercise

4. 12 is what percent of 288?

Objective 6.4b Solve basic percent problems.

Example 18% of what is 1296?

$$\frac{18}{100} = \frac{1296}{b} \qquad \text{Translating to a proportion}$$

$$18 \cdot b = 100 \cdot 1296$$

$$\frac{18 \cdot b}{18} = \frac{129{,}600}{18}$$

$$b = 7200.$$

Thus, 18% of 7200 is 1296.

Practice Exercise

5. 3% of what is 300?

Objective 6.5b Solve applied problems involving percent of increase or percent of decrease.

Example The total cost for 16 basic grocery items in the second quarter of 2008 averaged $46.67 nationally. The total cost of these 16 items in the second quarter of 2007 averaged $42.95. What was the percent of increase?

Source: American Farm Bureau Federation

We first determine the amount of increase: $46.67 − $42.95 = $3.72. Then we translate to a percent equation or a proportion and solve.

Rewording: $3.72 is what percent of $42.95?

Practice Exercise

6. In Indiana, the cost for 16 basic grocery items increased from $40.07 in the second quarter of 2007 to $46.20 in the second quarter of 2008. What was the percent of increase from 2007 to 2008?

Objective 6.5b *(continued)*

Percent Equation:	*Proportion:*

$$3.72 = p \cdot 42.95$$

$$\frac{N}{100} = \frac{3.72}{42.95}$$

$$\frac{3.72}{42.95} = \frac{p \cdot 42.95}{42.95}$$

$$42.95N = 100 \cdot 3.72$$

$$\frac{3.72}{42.95} = p$$

$$\frac{42.95N}{42.95} = \frac{100 \cdot 3.72}{42.95}$$

$$0.087 \approx p$$

$$N = \frac{372}{42.95}$$

$$8.7\% \approx p$$

$$N \approx 8.7$$

The percent of increase was 8.7%.

Objective 6.6a Solve applied problems involving sales tax and percent.

Example The sales tax is $34.23 on the purchase of a flat-screen high-definition television that costs $489. What is the sales tax rate?

Rephrase: Sales tax is what percent of purchase price?

Translate: $34.23 = r \times 489$

Solve:

$$\frac{34.23}{489} = \frac{r \times 489}{489}$$

$$\frac{34.23}{489} = r$$

$$0.07 = r$$

$$7\% = r$$

The sales tax rate is 7%.

Practice Exercise

7. The sales tax is $1102.20 on the purchase of a new car that costs $18,370. What is the sales tax rate?

Objective 6.6b Solve applied problems involving commission and percent.

Example A real estate agent's commission rate is $6\frac{1}{2}\%$. She received a commission of $17,160 on the sale of a home. For how much did the home sell?

Rephrase: Commission is $6\frac{1}{2}\%$ of what selling price?

Translate: $17,160 = 6\frac{1}{2}\% \times S$

Solve: $17,160 = 0.065 \times S$

$$\frac{17,160}{0.065} = \frac{0.065 \times S}{0.065}$$

$$264,000 = S$$

The home sold for $264,000.

Practice Exercise

8. A real estate agent's commission rate is 7%. He received a commission of $12,950 on the sale of a home. For how much did the home sell?

Objective 6.7a   Solve applied problems involving simple interest.

Example   To meet its payroll, a business borrows $5200 at $4\frac{1}{4}\%$ for 90 days. Find the amount of simple interest that is due and the total amount that must be paid after 90 days.

$$I = P \cdot r \cdot t = \$5200 \times 4\tfrac{1}{4}\% \times \frac{90}{365}$$

$$= \$5200 \times 0.0425 \times \frac{90}{365}$$

$$\approx \$54.49$$

The interest due for 90 days $= \$54.49$.

The total amount due $= \$5200 + \$54.49 = \$5254.49$.

Practice Exercise

9. A student borrows $2500 for tuition at $5\frac{1}{2}\%$ for 60 days. Find the amount of simple interest that is due and the total amount that must be paid after 60 days.

Objective 6.7b   Solve applied problems involving compound interest.

Example   Find the amount in an account if $3200 is invested at 5%, compounded semiannually, for $1\frac{1}{2}$ years.

$$A = P \cdot \left(1 + \frac{r}{n}\right)^{n \cdot t}$$

$$= \$3200\left(1 + \frac{0.05}{2}\right)^{2 \cdot \frac{3}{2}}$$

$$= \$3200(1.025)^3$$

$$= \$3446.05$$

The amount in the account after $1\frac{1}{2}$ years is $3446.05.

Practice Exercise

10. Find the amount in an account if $6000 is invested at $4\frac{3}{4}\%$, compounded quarterly, for 2 years.

Review Exercises

Find decimal notation for the percent notations in each sentence.   [6.1b]

1. In the 2006–2007 school year, about 4% of the 15 million college students in the United States were foreign students. Approximately 14.4% of the foreign students were from India.
   Source: Institute of International Education

2. Poland is 62.1% urban; Sweden is 84.2% urban.
   Source: The World Almanac, 2008

Find percent notation   [6.2a]

3. $\dfrac{3}{8}$

4. $\dfrac{1}{3}$

Find percent notation.   [6.1b]

5. 1.7

6. 0.065

Find fraction notation.   [6.2b]

7. 24%

8. 6.3%

Translate to a percent equation. Then solve.   [6.3a, b]

9. 30.6 is what percent of 90?

10. 63 is 84% of what?

11. What is $38\frac{1}{2}\%$ of 168?

Translate to a proportion. Then solve.   [6.4a, b]

12. 24 percent of what is 16.8?

13. 42 is what percent of 30?

14. What is 10.5% of 84?

Solve.   [6.5a, b]

15. *Favorite Ice Creams.*   According to a survey, 8.9% of those interviewed chose chocolate as their favorite ice cream flavor and 4.2% chose butter pecan. At this rate, of the 2000 students in a freshman class, how many would choose chocolate as their favorite ice cream? butter pecan?
Source: International Ice Cream Association

16. *Prescriptions.*   Of the 305 million people in the United States, 140.3 million take at least one type of prescription drug per day. What percent take at least one type of prescription drug per day?
Source: William N. Kelly, *Pharmacy: What It Is and How It Works*, 2nd ed., CRC Press Pharmaceutical Education, 2006

17. *Water Output.*   The average person expels 200 mL of water per day by sweating. This is 8% of the total output of water from the body. How much is the total output of water?
Source: Elaine N. Marieb, *Essentials of Human Anatomy and Physiology*, 6th ed. Boston: Addison Wesley Longman, Inc., 2000

18. *Test Scores.*   After Sheila got a 75 on a math test, she was allowed to go to the math lab and take a retest. She increased her score to 84. What was the percent of increase?

19. *Test Scores.*   James got an 80 on a math test. By taking a retest in the math lab, he increased his score by 15%. What was his new score?

Solve.   [6.6a, b, c]

20. A state charges a meals tax of $7\frac{1}{2}$%. What is the meals tax charged on a dinner party costing $320?

21. In a certain state, a sales tax of $453.60 is collected on the purchase of a used car for $7560. What is the sales tax rate?

22. Kim earns $753.50 selling $6850 worth of televisions. What is the commission rate?

23. An air conditioner has a marked price of $350. It is placed on sale at 12% off. What are the discount and the sale price?

24. The price of a fax machine is marked down from $305 to $262.30. What is the rate of discount?

25. An insurance salesperson receives a 7% commission. If $42,000 worth of life insurance is sold, what is the commission?

26. What is the rate of discount of this stepladder?

SPECIAL VALUE!
Now $67   Was $82
8' Aluminum Stepladder

Solve.   [6.7a, b, c]

27. What is the simple interest on $1800 at 6% for $\frac{1}{3}$ year?

28. The Dress Shack borrows $24,000 at 10% simple interest for 60 days. Find **(a)** the amount of interest due and **(b)** the total amount that must be paid after 60 days.

29. What is the simple interest on $2200 principal at an interest rate of 5.5% for 1 year?

30. The Armstrongs invest $7500 in an investment account paying an annual interest rate of 4%, compounded monthly. How much is in the account after 3 months?

31. Find the amount in an investment account if $8000 is invested at 9%, compounded annually, for 2 years.

32. *Credit Cards.* At the end of her junior year of college, Kasha has a balance of $6428.74 on a credit card with an annual percentage rate (APR) of 18.7%. She decides not to make additional purchases with this card until she has paid off the balance.

a) Many credit cards require a minimum payment of 2% of the balance. At this rate, what is Kasha's minimum payment on a balance of $6428.74? Round the answer to the nearest dollar.

b) Find the amount of interest and the amount applied to reduce the principal in the minimum payment found in part (a).

c) If Kasha had transferred her balance to a card with an APR of 13.2%, how much of her payment would be interest and how much would be applied to reduce the principal?

d) Compare the amounts for 13.2% from part (c) with the amounts for 18.7% from part (b).

33. A fishing boat listed at $16,500 is on sale at 15% off. What is the sale price? [6.6c]

A. $14,025   B. $2475
C. 85%   D. $14,225

34. Find the amount in a money market account if $10,500 is invested at 6%, compounded semiannually, for $1\frac{1}{2}$ years. [6.7b]

A. $11,139.45   B. $12,505.67
C. $11,473.63   D. $10,976.03

Synthesis

35. Mike's Bike Shop reduces the price of a bicycle by 40% during a sale. By what percent must the store increase the sale price, after the sale, to get back to the original price? [6.6c]

36. A worker receives raises of 3%, 6%, and then 9%. By what percent has the original salary increased? [6.5a]

Understanding Through Discussion and Writing

1. Which is the better deal for a consumer and why: a discount of 40% or a discount of 20% followed by another of 22%? [6.6c]

2. Which is better for a wage earner, and why: a 10% raise followed by a 5% raise a year later, or a 5% raise followed by a 10% raise a year later? [6.5a]

3. Ollie buys a microwave oven during a 10%-off sale. The sale price that Ollie paid was $162. To find the original price, Ollie calculates 10% of $162 and adds that to $162. Is this correct? Why or why not? [6.6c]

4. You take 40% of 50% of a number. What percent of the number could you take to obtain the same result making only one multiplication? Explain your answer. [6.5a]

5. A firm must choose between borrowing $5000 at 10% for 30 days and borrowing $10,000 at 8% for 60 days. Give arguments in favor of and against each option. [6.7a]

6. On the basis of the mathematics presented in Section 6.7, discuss what you have learned about interest rates and credit cards. [6.7c]

CHAPTER 6

Test

For Extra Help

Step-by-step test solutions are found on the Chapter Test Prep Videos available via the Video Resources on DVD, in *MyMathLab*, and on You Tube (search "BittingerBasicMath" and click on "Channels").

1. *Land-Line Phone Service.* About 14.7% of U.S. households were projected to cut land-line phone service in 2009. Find decimal notation for 14.7%.
Source: Yankee Group 2006, *USA TODAY*, January 15, 2007

2. *Gravity.* The gravity of Mars is 0.38 as strong as Earth's. Find percent notation for 0.38.
Source: www.marsinstitute.info/epo/mermarsfacts.html

3. Find percent notation for $\frac{11}{8}$.

4. Find fraction notation for 65%.

5. Translate to a percent equation. Then solve.
What is 40% of 55?

6. Translate to a proportion. Then solve.
What percent of 80 is 65?

Solve.

7. *Organ Transplants.* In 2006, there were 28,291 organ transplants in the United States. The pie chart below shows the percentages for the main transplants. How many kidney transplants were there in 2006? liver transplants? heart transplants?

8. *Batting Average.* Garrett Atkins, third baseman for the Colorado Rockies, got 175 hits during the 2008 baseball season. This was about 28.64% of his at-bats. How many at-bats did he have?
Source: Major League Baseball

Organ Transplants in the United States, 2006

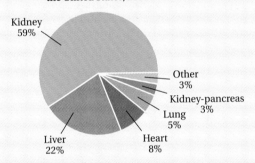

Kidney 59%
Other 3%
Kidney-pancreas 3%
Lung 5%
Heart 8%
Liver 22%

SOURCE: 2007 OPTN/SRTR Annual Report, Table 1.7; United Network for Organ Sharing

9. *Foreign Adoptions.* The number of foreign children adopted by Americans declined from 20,679 in 2006 to 19,292 in 2007. Find the percent of decrease.
Source: U.S. State Department, *USA TODAY*, August 13, 2008

10. There are about 6,603,000,000 people living in the world today, and approximately 4,002,000,000 live in Asia. What percent of people live in Asia?
Source: Population Division/International Programs Center, U.S. Census Bureau, U.S. Dept. of Commerce

11. *Oklahoma Sales Tax.* The sales tax rate in Oklahoma is 4.5%. How much tax is charged on a purchase of $560? What is the total price?

12. Noah's commission rate is 15%. What is the commission from the sale of $4200 worth of merchandise?

13. The marked price of a DVD player is $200 and the item is on sale at 20% off. What are the discount and the sale price?

14. What is the simple interest on a principal of $120 at the interest rate of 7.1% for 1 year?

15. A city orchestra invests $5200 at 6% simple interest. How much is in the account after $\frac{1}{2}$ year?

16. Find the amount in an account if $1000 is invested at $5\frac{3}{8}$% compounded annually, for 2 years.

17. The Suarez family invests $10,000 at an annual interest rate of 4.9%, compounded monthly. How much is in the account after 3 years?

18. *Job Opportunities.* The table below lists job opportunities in 2006 and projected increases for 2016. Complete the table by filling in the missing numbers.

OCCUPATION	TOTAL EMPLOYMENT IN 2006	PROJECTED EMPLOYMENT IN 2016	CHANGE	PERCENT OF INCREASE
Dental assistant	280,000	362,000	82,000	29.3%
Plumber	705,000		52,000	
Veterinary assistant	71,000	100,000		
Motorcycle repair technician		24,000	3000	
Fitness professional		298,000		26.8%

SOURCE: EarnMyDegree.com

19. Find the discount and the discount rate of the television in this ad.

19" LCD HDTV

$299⁹⁹

was $349⁹⁹

HDTV

20. *Credit Cards.* Jayden has a balance of $2704.27 on a credit card with an annual percentage rate of 16.3%. The minimum payment required on the current statement is $54. Find the amount of interest and the amount applied to reduce the principal in this payment and the balance after this payment.

21. 0.75% of what number is 300?

A. 2.25   **B.** 40,000   **C.** 400   **D.** 225

Synthesis

22. By selling a home without using a realtor, Juan and Marie can avoid paying a 7.5% commission. They receive an offer of $180,000 from a potential buyer. In order to give a comparable offer, for what price would a realtor need to sell the house? Round to the nearest hundred.

23. Karen's commission rate is 16%. She invests her commission from the sale of $15,000 worth of merchandise at an interest rate of 12%, compounded quarterly. How much is Karen's investment worth after 6 months?

Cumulative Review

1. *Passports.* In 2008, 44% of Canadians had passports while only 30% of Americans had passports. Find decimal notation for 44% and for 30%.

Sources: Government of Canada; U.S. Department of Homeland Security, U.S. Customs and Border Protection docket number USCBP 2007-0061

2. Find percent notation: 0.269.

3. Find percent notation: $\dfrac{9}{8}$.

4. Find decimal notation: $\dfrac{13}{6}$.

5. Write fraction notation for the ratio 5 to 0.5.

6. Find the rate in kilometers per hour: 350 km, 15 hr.

Use $<$, $>$, or $=$ for □ to write a true sentence.

7. $\dfrac{5}{7}$ □ $\dfrac{6}{8}$

8. $\dfrac{6}{14}$ □ $\dfrac{15}{25}$

Estimate the sum or the difference by first rounding to the nearest hundred.

9. $263{,}961 + 32{,}090 + 127.89$

10. $73{,}510 - 23{,}450$

Calculate.

11. $46 - [4(6 + 4 \div 2) + 2 \times 3 - 5]$

12. $[0.8(1.5 - 9.8 \div 49) + (1 + 0.1)^2] \div 1.5$

Compute and simplify.

13. $\dfrac{6}{5} + 1\dfrac{5}{6}$

14. $46.9 + 2.84$

15.
$$\begin{array}{r} 4\,8\,7{,}0\,9\,4 \\ 6{,}9\,3\,6 \\ +\ \ 2\,1{,}1\,2\,0 \\ \hline \end{array}$$

16. $35 - 34.98$

17. $3\dfrac{1}{3} - 2\dfrac{2}{3}$

18. $\dfrac{8}{9} - \dfrac{6}{7}$

19. $\dfrac{7}{9} \cdot \dfrac{3}{14}$

20.
$$\begin{array}{r} 2\,3\,6{,}9\,8\,4 \\ \times\ \ \ \ \ 3{,}6\,0\,0 \\ \hline \end{array}$$

21.
$$\begin{array}{r} 4\,6.0\,1\,2 \\ \times\ \ \ \ \ 0.0\,3 \\ \hline \end{array}$$

22. $6\dfrac{3}{5} \div 4\dfrac{2}{5}$

23. $431.2 \div 35.2$

24. $15 \overline{)1\,8\,5\,0}$

Solve.

25. $36 \cdot x = 3420$

26. $y + 142.87 = 151$

27. $\dfrac{2}{15} \cdot t = \dfrac{6}{5}$

28. $\dfrac{3}{4} + x = \dfrac{5}{6}$

29. $\dfrac{y}{25} = \dfrac{24}{15}$

30. $\dfrac{16}{n} = \dfrac{21}{11}$

31. *Museum Attendance.* Among leading art museums, the Indianapolis Museum of Art ranks 6th in the nation in the number of visitors as a percentage of the metropolitan population. In 2007, the number of visitors was 30.31% of the metropolitan population, which was 1,525,104. What was the total attendance?

Source: dashboard.imamuseum.org/series/2007+MSA+Attendance

32. *Salary of Medical Doctors.* In 2007, the starting salary for a doctor in radiology was $172,500 higher than the starting salary of a doctor in neurology. If the starting salary for a radiologist was $350,000, what was the starting salary of a neurologist?

Source: *Journal of the American Medical Association*

33. At one point during the 2008–2009 NBA season, the Cleveland Cavaliers had won 39 out of 49 games. At this rate, how many games would they win in the entire season of 82 games?

Source: National Basketball Association

34. *Shirts.* A total of $424.75 was paid for 5 shirts at an upscale men's store. How much did each shirt cost?

35. *Unit Price.* A 200-oz bottle of liquid laundry detergent costs $14.99. What is the unit price?

36. Patty walked $\frac{7}{10}$ mi to school and then $\frac{8}{10}$ mi to the library. How far did she walk?

37. On a map, 1 in. represents 80 mi. How much does $\frac{3}{4}$ in. represent?

38. *Compound Interest.* The Bakers invest $8500 in an investment account paying 8%, compounded monthly. How much is in the account after 5 years?

39. *Ribbons.* How many pieces of ribbon $1\frac{4}{5}$ yd long can be cut from a length of ribbon 9 yd long?

40. *Auto Repair Technicians.* In 2006, 773,000 people were employed as auto repair technicians. It is projected that by 2016, this number will increase by 110,000. Find the percent of increase in the number of auto repair technicians and the number of technicians in 2016.

Source: EarnMyDegree.com

41. Subtract and simplify: $\frac{14}{25} - \frac{3}{20}$.

A. $\frac{11}{500}$ **B.** $\frac{11}{5}$

C. $\frac{41}{100}$ **D.** $\frac{205}{500}$

42. The population of the state of Louisiana decreased from 4,468,976 in 2000 to 4,287,768 in 2006. What was the percent of decrease?

Source: U.S. Census Bureau

A. 4.1% **B.** 4.2%
C. 104% **D.** 95.9%

Synthesis

43. If a is 50% of b, then b is what percent of a?

Answers

CHAPTER 1

Exercise Set 1.1, p. 6

1. 5 thousands **3.** 5 hundreds **5.** 9 **7.** 7
9. 5 thousands + 7 hundreds + 0 tens + 2 ones, or
5 thousands + 7 hundreds + 2 ones
11. 9 ten thousands + 3 thousands + 9 hundreds + 8 tens +
6 ones **13.** 2 thousands + 0 hundreds + 5 tens + 8 ones, or
2 thousands + 5 tens + 8 ones **15.** 1 thousand +
5 hundreds + 7 tens + 6 ones **17.** 1 billion +
4 hundred millions + 2 ten millions + 4 millions +
1 hundred thousand + 6 ten thousands + 1 thousand +
9 hundreds + 4 tens + 8 ones **19.** 9 ten millions +
9 millions + 8 hundred thousands + 8 ten thousands +
6 thousands + 5 hundreds + 6 tens + 8 ones
21. 6 hundred thousands + 1 ten thousand + 7 thousands +
2 hundreds + 4 tens + 9 ones **23.** Eighty-five
25. Eighty-eight thousand **27.** One hundred twenty-three
thousand, seven hundred sixty-five **29.** Seven billion, seven
hundred fifty-four million, two hundred eleven thousand, five
hundred seventy-seven **31.** Seven hundred thousand, six
hundred thirty-four **33.** Three million, forty-eight thousand,
five **35.** 2,233,812 **37.** 8,000,000,000
39. 50,324 **41.** 632,896 **43.** 1,600,000,000
45. 64,186,000 **47.** 138

Calculator Corner, p. 10

1. 121 **2.** 1602 **3.** 1932 **4.** 864

Exercise Set 1.2, p. 12

1. 387 **3.** 164 **5.** 5198 **7.** 100 **9.** 8503 **11.** 5266
13. 4466 **15.** 6608 **17.** 34,432 **19.** 101,310 **21.** 230
23. 18,424 **25.** 31,685 **27.** 132 yd **29.** 1661 ft
31. 570 ft **33.** 8 ten thousands **34.** Nine billion, three
hundred forty-six million, three hundred ninety-nine
thousand, four hundred sixty-eight
35. $1 + 99 = 100, 2 + 98 = 100, \ldots, 49 + 51 = 100$. Then
49 100's = 4900 and $4900 + 50 + 100 = 5050$.

Calculator Corner, p. 15

1. 28 **2.** 47 **3.** 67 **4.** 119 **5.** 2128 **6.** 2593

Exercise Set 1.3, p. 17

1. 44 **3.** 533 **5.** 39 **7.** 14 **9.** 369 **11.** 26 **13.** 234
15. 417 **17.** 5382 **19.** 2778 **21.** 3069 **23.** 1089
25. 7748 **27.** 4144 **29.** 56 **31.** 454 **33.** 3831
35. 3749 **37.** 2191 **39.** 4418 **41.** 43,028 **43.** 95,974
45. 4206 **47.** 1305 **49.** 9989 **51.** 48,017 **53.** 1024

54. 12,732 **55.** 90,283 **56.** 29,364 **57.** 1345 **58.** 924
59. 22,692 **60.** 10,920 **61.** Six million, three hundred
seventy-five thousand, six hundred two **62.** 7 ten thousands
63. 3; 4

Calculator Corner, p. 23

1. 448 **2.** 21,970 **3.** 6380 **4.** 39,564 **5.** 180,480
6. 2,363,754

Exercise Set 1.4, p. 24

1. 520 **3.** 564 **5.** 1527 **7.** 64,603 **9.** 4770 **11.** 3995
13. 870 **15.** 1920 **17.** 46,296 **19.** 14,652 **21.** 258,312
23. 798,408 **25.** 20,723,872 **27.** 362,128 **29.** 302,220
31. 49,101,136 **33.** 25,236,000 **35.** 20,064,048
37. 529,984 sq mi **39.** 8100 sq ft **41.** 12,685 **42.** 10,834
43. 427,477 **44.** 111,110 **45.** 1241 **46.** 8889
47. 254,119 **48.** 66,444 **49.** 247,464 sq ft

Calculator Corner, p. 30

1. 28 **2.** 123 **3.** 323 **4.** 36

Exercise Set 1.5 p. 32

1. 12 **3.** 1 **5.** 22 **7.** 0 **9.** Not defined **11.** 6
13. 55 R 2 **15.** 108 **17.** 307 **19.** 753 R 3 **21.** 74 R 1
23. 92 R 2 **25.** 1703 **27.** 987 R 5 **29.** 12,700 **31.** 127
33. 52 R 52 **35.** 29 R 5 **37.** 40 R 12 **39.** 90 R 22
41. 29 **43.** 105 R 3 **45.** 1609 R 2 **47.** 1007 R 1 **49.** 23
51. 107 R 1 **53.** 370 **55.** 609 R 15 **57.** 304
59. 3508 R 219 **61.** 8070 **63.** Perimeter **64.** Minuend
65. Digits; periods **66.** Dividend **67.** Factors; product
68. Additive **69.** Associative **70.** Divisor; remainder;
dividend **71.** 54, 122; 33, 2772; 4, 8 **73.** 30 buses

Mid-Chapter Review: Chapter 1, p. 35

1. False **2.** True **3.** True **4.** False **5.** True **6.** False
7.

95,406,237

Ninety-five million,

four hundred six thousand,

two hundred thirty-seven

8.
$$\begin{array}{r} \overset{5}{}\overset{9}{}\overset{14}{} \\ \cancel{6}\,\cancel{0}\,4 \\ -\ 4\ 9\ 7 \\ \hline 1\ 0\ 7 \end{array}$$

9. 6 hundreds **10.** 6 ten thousands
11. 6 thousands **12.** 6 ones **13.** 2 **14.** 6 **15.** 5 **16.** 1
17. 5 thousands + 6 hundreds + 0 tens + 2 ones, or
5 thousands + 6 hundreds + 2 ones

18. 6 ten thousands + 9 thousands + 3 hundreds + 4 tens + 5 ones   **19.** One hundred thirty-six   **20.** Sixty-four thousand, three hundred twenty-five   **21.** 308,716
22. 4,567,216   **23.** 798   **24.** 1030   **25.** 7922   **26.** 7534
27. 465   **28.** 339   **29.** 1854   **30.** 4328   **31.** 216
32. 15,876   **33.** 132,275   **34.** 5,679,870   **35.** 253
36. 112 R 5   **37.** 23 R 19   **38.** 144 R 31   **39.** 25 m
40. 8 sq in.   **41.** When numbers are being added, it does not matter how they are grouped.   **42.** Subtraction is not commutative. For example, $5 - 2 = 3$, but $2 - 5 \neq 3$.
43. Answers will vary. Suppose one coat costs $150. Then the multiplication $4 \cdot \$150$ gives the cost of four coats. Or, suppose one ream of copy paper costs $4. Then the multiplication $\$4 \cdot 150$ gives the cost of 150 reams.   **44.** If we use the definition of division, $0 \div 0 = a$ such that $a \cdot 0 = 0$. We see that a could be *any* number since $a \cdot 0 = 0$ for any number a. Thus we cannot say that $0 \div 0 = 0$. This is why we agree not to allow division by 0.

Exercise Set 1.6, p. 43

1. 50   **3.** 460   **5.** 730   **7.** 900   **9.** 100   **11.** 1000
13. 9100   **15.** 32,800   **17.** 6000   **19.** 8000
21. 45,000   **23.** 373,000   **25.** $80 + 90 = 170$
27. $8070 - 2350 = 5720$ **29.** 220; incorrect
31. 890; incorrect   **33.** $7300 + 9200 = 16,500$
35. $6900 - 1700 = 5200$ **37.** 1600; correct   **39.** 1500; correct   **41.** $10,000 + 5000 + 9000 + 7000 = 31,000$
43. $92,000 - 23,000 = 69,000$ **45.** $50 \cdot 70 = 3500$
47. $30 \cdot 30 = 900$ **49.** $900 \cdot 300 = 270,000$
51. $400 \cdot 200 = 80,000$ **53.** $350 \div 70 = 5$
55. $8450 \div 50 = 169$ **57.** $1200 \div 200 = 6$
59. $8400 \div 300 = 28$ **61.** $11,200   **63.** $18,900; no
65. Answers will vary depending on the options chosen.
67. (a) $2,190,000; (b) $2,170,000   **69.** 90 people
71. <   **73.** >   **75.** <   **77.** >   **79.** >   **81.** >
83. $190,078 > 172,000$, or $172,000 < 190,078$
85. $1694 < 5249$, or $5249 > 1694$ **87.** 86,754   **88.** 13,589
89. 48,824   **90.** 4415   **91.** 1702   **92.** 17,748   **93.** 54 R 4
94. 208   **95.** Left to the student   **97.** Left to the student

Exercise Set 1.7, p. 52

1. 14   **3.** 0   **5.** 90,900   **7.** 450   **9.** 352   **11.** 25
13. 29   **15.** 0   **17.** 79   **19.** 45   **21.** 8   **23.** 14   **25.** 32
27. 143   **29.** 17,603   **31.** 37   **33.** 1035   **35.** 66
37. 324   **39.** 743   **41.** 175   **43.** 335   **45.** 18,252
47. 104   **49.** 45   **51.** 4056   **53.** 2847   **55.** 15   **57.** 205
59. 457   **61.** 142 R 5   **62.** 142   **63.** 334   **64.** 334 R 11
65. <   **66.** >   **67.** >   **68.** <   **69.** 6,376,000
70. 6,375,600   **71.** 347

Translating for Success, p. 63

1. E   **2.** M   **3.** D   **4.** G   **5.** A   **6.** O   **7.** F   **8.** K
9. J   **10.** H

Exercise Set 1.8, p. 64

1. 550 ft   **3.** 1250 ft   **5.** 95 milligrams   **7.** 18 rows
9. 1502 hr   **11.** 2054 mi   **13.** 2,073,600 pixels
15. 268,000 men   **17.** 168 hr   **19.** $23 per month
21. $233   **23.** $9276   **25.** 151,500   **27.** 1,190,000
motorcycles   **29.** $78   **31.** $40 per month   **33.** $24,456
35. 35 weeks; 2 episodes   **37.** 236 gal   **39.** 21 columns
41. (a) 4200 sq ft; (b) 268 ft   **43.** $247   **45.** 645 mi; 5 in.
47. 56 cartons   **49.** 32 $10 bills   **51.** $400   **53.** 525 min, or 8 hr 45 min   **55.** 700 min, or 11 hr 40 min   **57.** 104 seats
59. 106 bones   **61.** 234,600   **62.** 234,560   **63.** 235,000
64. 22,000   **65.** 16,000   **66.** 4000   **67.** 8000
68. 320,000   **69.** 720,000   **70.** 46,800,000
71. 792,000 mi; 1,386,000 mi

Calculator Corner, p. 72

1. 243   **2.** 15,625   **3.** 20,736   **4.** 2048

Calculator Corner, p. 74

1. 49   **2.** 85   **3.** 36   **4.** 0   **5.** 73   **6.** 49

Exercise Set 1.9, p. 77

1. 3^4 **3.** 5^2 **5.** 7^5 **7.** 10^3 **9.** 49   **11.** 729
13. 20,736   **15.** 243   **17.** 22   **19.** 20   **21.** 100   **23.** 1
25. 49   **27.** 5   **29.** 434   **31.** 41   **33.** 88   **35.** 4
37. 303   **39.** 20   **41.** 70   **43.** 295   **45.** 32   **47.** 906
49. 62   **51.** 102   **53.** 32   **55.** $94   **57.** 401   **59.** 110
61. 7   **63.** 544   **65.** 708   **67.** 27   **69.** 452   **70.** 835
71. 13   **72.** 37   **73.** 2342   **74.** 4898   **75.** 25   **76.** 100
77. 104,286 mi^2 **78.** 98 gal   **79.** 24; $1 + 5 \cdot (4 + 3) = 36$
81. 7; $12 \div (4 + 2) \cdot 3 - 2 = 4$

Summary and Review: Chapter 1, p. 80

Concept Reinforcement

1. True   **2.** True   **3.** False   **4.** False   **5.** True   **6.** False

Important Concepts

1. 2 thousands   **2.** 65,302   **3.** 3237   **4.** 225,036
5. 315 R 14   **6.** 36,500   **7.** 36,000   **8.** <   **9.** 36   **10.** 216

Review Exercises

1. 8 thousands   **2.** 3   **3.** 2 thousands + 7 hundreds + 9 tens + 3 ones   **4.** 5 ten thousands + 6 thousands + 0 hundreds + 7 tens + 8 ones, or 5 ten thousands + 6 thousands + 7 tens + 8 ones   **5.** 4 millions + 0 hundred thousands + 0 ten thousands + 7 thousands + 1 hundred + 0 tens + 1 one, or 4 millions + 7 thousands + 1 hundred + 1 one   **6.** Sixty-seven thousand, eight hundred nineteen   **7.** Two million, seven hundred eighty-one thousand, four hundred twenty-seven
8. Four million, eight hundred seventeen thousand, nine hundred forty-one   **9.** 476,588   **10.** 1,563,000,000
11. 14,272   **12.** 66,024   **13.** 21,788   **14.** 98,921
15. 5148   **16.** 1689   **17.** 2274   **18.** 17,757
19. 5,100,000   **20.** 6,276,800   **21.** 506,748   **22.** 27,589
23. 5,331,810   **24.** 12 R 3   **25.** 5   **26.** 913 R 3
27. 384 R 1   **28.** 4 R 46   **29.** 54   **30.** 452
31. 5008   **32.** 4389   **33.** 345,800   **34.** 345,760
35. 346,000   **36.** 300,000   **37.** >   **38.** <
39. $41,300 + 19,700 = 61,000$ **40.** $38,700 - 24,500 = 14,200$
41. $400 \cdot 700 = 280,000$ **42.** 8   **43.** 45   **44.** 58   **45.** 0
46. 4^3 **47.** 10,000   **48.** 36   **49.** 65   **50.** 233   **51.** 260
52. 165   **53.** $502   **54.** $484   **55.** 1982   **56.** 19 cartons
57. $13,585   **58.** 14 beehives   **59.** 98 sq ft; 42 ft   **60.** 137 beakers; 13 mL left over   **61.** $27,598   **62.** B   **63.** A
64. D   **65.** 8   **66.** $a = 8, b = 4$ **67.** 6 days

Understanding Through Discussion and Writing

1. No; if subtraction were associative, then $a - (b - c) = (a - b) - c$ for any a, b, and c. But, for example,

$$12 - (8 - 4) = 12 - 4 = 8,$$

whereas

$$(12 - 8) - 4 = 4 - 4 = 0.$$

Since $8 \neq 0$, this example shows that subtraction is not associative.   **2.** By rounding prices and estimating their sum, a shopper can estimate the total grocery bill while shopping. This is particularly useful if the shopper wants to spend no more than a certain amount.   **3.** Answers will vary. Anthony is driving from Kansas City to Minneapolis, a distance of 512 mi. He stops for gas after driving 183 mi. How much farther must he drive?

4. The parentheses are not necessary in the expression $9 - (4 \cdot 2)$. Using the rules for order of operations, the multiplication would be performed before the subtraction even if the parentheses were not present. The parentheses are necessary in the expression $(3 \cdot 4)^2$; $(3 \cdot 4)^2 = 12^2 = 144$, but $3 \cdot 4^2 = 3 \cdot 16 = 48$.

Test: Chapter 1, p. 85

1. [1.1a] 5　**2.** [1.1b] 8 thousands + 8 hundreds + 4 tens + 3 ones　**3.** [1.1c] Thirty-eight million, four hundred three thousand, two hundred seventy-seven
4. [1.2a] 9989　**5.** [1.2a] 63,791　**6.** [1.2a] 3165
7. [1.2a] 10,515　**8.** [1.3a] 3630　**9.** [1.3a] 1039
10. [1.3a] 6848　**11.** [1.3a] 5175　**12.** [1.4a] 41,112
13. [1.4a] 5,325,600　**14.** [1.4a] 2405　**15.** [1.4a] 534,264
16. [1.5a] 3 R 3　**17.** [1.5a] 70　**18.** [1.5a] 97
19. [1.5a] 805 R 8　**20.** [1.6a] 35,000　**21.** [1.6a] 34,530
22. [1.6a] 34,500　**23.** [1.6b] 23,600 + 54,700 = 78,300
24. [1.6b] 54,800 − 23,600 = 31,200
25. [1.6b] 800 · 500 = 400,000　**26.** [1.6c] >
27. [1.6c] <　**28.** [1.7b] 46　**29.** [1.7b] 13　**30.** [1.7b] 14
31. [1.7b] 381　**32.** [1.8a] 83 calories　**33.** [1.8a] 20 staplers
34. [1.8a] 1,256,615 sq mi　**35. (a)** [1.2b], [1.4b] 300 in., 5000 sq in.; 264 in., 3872 sq in.; 228 in., 2888 sq in.; **(b)** [1.8a] 2112 sq in.　**36.** [1.8a] 1852 12-packs; 7 cakes left over　**37.** [1.8a] $95　**38.** [1.9a] 12^4　**39.** [1.9b] 343
40. [1.9b] 100,000　**41.** [1.9c] 31　**42.** [1.9c] 98
43. [1.9c] 2　**44.** [1.9c] 18　**45.** [1.9d] 216　**46.** [1.9c] A
47. [1.4b], [1.8a] 336 sq in.　**48.** [1.9c] 9　**49.** [1.8a] 80 payments

CHAPTER 2

Calculator Corner, p. 90

1. No　**2.** Yes　**3.** Yes　**4.** No

Exercise Set 2.1, p. 93

1. No　**3.** Yes　**5.** 1, 2, 3, 6, 9, 18　**7.** 1, 2, 3, 6, 9, 18, 27, 54　**9.** 1, 2, 4　**11.** 1　**13.** 1, 2, 7, 14, 49, 98　**15.** 1, 3, 5, 15, 17, 51, 85, 255　**17.** 4, 8, 12, 16, 20, 24, 28, 32, 36, 40　**19.** 20, 40, 60, 80, 100, 120, 140, 160, 180, 200　**21.** 3, 6, 9, 12, 15, 18, 21, 24, 27, 30　**23.** 12, 24, 36, 48, 60, 72, 84, 96, 108, 120
25. 10, 20, 30, 40, 50, 60, 70, 80, 90, 100　**27.** 9, 18, 27, 36, 45, 54, 63, 72, 81, 90　**29.** No　**31.** Yes　**33.** Yes　**35.** No
37. No　**39.** Neither　**41.** Composite　**43.** Prime
45. Prime　**47.** 2 · 2 · 2　**49.** 2 · 7　**51.** 2 · 3 · 7　**53.** 5 · 5
55. 2 · 5 · 5　**57.** 13 · 13　**59.** 2 · 2 · 5 · 5　**61.** 5 · 7
63. 2 · 2 · 2 · 3 · 3　**65.** 7 · 11　**67.** 2 · 2 · 7 · 103
69. 3 · 17　**71.** 2 · 2 · 2 · 3 · 5 · 5　**73.** 3 · 7 · 13
75. 3 · 11 · 17　**77.** 26　**78.** 256　**79.** 425
80. 4200　**81.** 0　**82.** 22　**83.** 1　**84.** 3　**85.** $946
86. 201 min, or 3 hr 21 min　**87.** Row 1: 48, 90, 432, 63; row 2: 7, 2, 2, 10, 8, 6, 21, 10; row 3: 9, 18, 36, 14, 12, 11, 21; row 4: 29, 19, 42

Exercise Set 2.2, p. 99

1. 46, 224, 300, 36, 45,270, 4444, 256, 8064, 21,568
3. 224, 300, 36, 4444, 256, 8064, 21,568　**5.** 300, 36, 45,270, 8064　**7.** 36, 45,270, 711, 8064　**9.** 324, 42, 501, 3009, 75, 2001, 402, 111,111, 1005　**11.** 55,555, 200, 75, 2345, 35, 1005
13. 56, 784, 200　**15.** 200　**17.** 313,332, 7624, 111,126, 876, 1110, 5128, 64,000, 9990　**19.** 313,332, 111,126, 876, 1110, 9990　**21.** 9990　**23.** 1110, 64,000, 9990　**25.** 138
26. 139　**27.** 874　**28.** 56　**29.** 26　**30.** 13　**31.** 234
32. 4003　**33.** 45 gal　**34.** 4320 min
35. 2 · 2 · 2 · 3 · 5 · 5 · 13　**37.** 2 · 2 · 3 · 3 · 7 · 11
39. 95,238

Exercise Set 2.3, p. 106

1. Numerator: 3; denominator: 4　**3.** Numerator: 11; denominator: 2　**5.** Numerator: 0; denominator: 7　**7.** $\frac{6}{12}$
9. $\frac{1}{8}$　**11.** $\frac{4}{3}$　**13.** $\frac{9}{8}$　**15.** $\frac{7}{8}$　**17.** $\frac{3}{4}$　**19.** $\frac{12}{12}$　**21.** $\frac{4}{8}$
23. $\frac{5}{8}$　**25.** $\frac{4}{7}$　**27.** $\frac{12}{16}$　**29.** $\frac{38}{16}$　**31. (a)** $\frac{2}{3}$; **(b)** $\frac{6}{8}$
33. (a) $\frac{3}{8}$; **(b)** $\frac{5}{8}$　**35. (a)** $\frac{5}{8}$; **(b)** $\frac{5}{3}$; **(c)** $\frac{3}{8}$; **(d)** $\frac{3}{5}$
37. (a) $\frac{1014}{100,000}$; **(b)** $\frac{750}{100,000}$; **(c)** $\frac{812}{100,000}$; **(d)** $\frac{922}{100,000}$; **(e)** $\frac{865}{100,000}$
(f) $\frac{1379}{100,000}$　**39. (a)** $\frac{4}{15}$; **(b)** $\frac{4}{11}$; **(c)** $\frac{11}{15}$　**41. (a)** $\frac{16}{100}$; **(b)** $\frac{18.5}{100}$
43. 0　**45.** 7　**47.** 1　**49.** 1　**51.** 0　**53.** 1　**55.** 1
57. 1　**59.** 1　**61.** 18　**63.** Not defined　**65.** Not defined　**67.** 34,560　**68.** 34,600　**69.** 35,000
70. 30,000　**71.** 24,465 members　**72.** 71 gal
73. 2203　**74.** 848　**75.** 37,239　**76.** 11,851
77. $\frac{1}{6}$　**79.** $\frac{2}{16}$, or $\frac{1}{8}$　**81.** ⬡◯◯◼◼◼　**83.**

Exercise Set 2.4, p. 115

1. $\frac{4}{15}$　**3.** $\frac{70}{64}$　**5.** $\frac{49}{64}$　**7.** $\frac{2}{15}$　**9.** $\frac{40}{21}$　**11.** $\frac{6}{5}$　**13.** $\frac{1}{6}$　**15.** $\frac{85}{8}$
17. $\frac{7}{100}$　**19.** $\frac{2}{5}$　**21.** $\frac{14}{39}$　**23.** $\frac{5}{8}$　**25.** $\frac{1}{40}$　**27.** $\frac{160}{27}$　**29.** $\frac{182}{285}$
31. $\frac{9}{16}$　**33.** $\frac{8}{11}$　**35.** $\frac{40}{3}$ yd　**37.** $\frac{1}{2625}$　**39.** $\frac{1}{24}$　**41.** $\frac{9}{20}$
43. 204　**44.** 700　**45.** 3001　**46.** 204 R 8
47. 8 thousands　**48.** 8 millions　**49.** 8 ones
50. 8 hundreds　**51.** 50　**52.** 6399　**53.** $\frac{71,269}{180,433}$　**55.** $\frac{56}{1125}$

Calculator Corner, p. 120

1. $\frac{14}{15}$　**2.** $\frac{7}{8}$　**3.** $\frac{138}{167}$　**4.** $\frac{7}{25}$

Exercise Set 2.5, p. 122

1. $\frac{5}{10}$　**3.** $\frac{20}{32}$　**5.** $\frac{27}{30}$　**7.** $\frac{28}{32}$　**9.** $\frac{20}{48}$　**11.** $\frac{51}{54}$　**13.** $\frac{75}{45}$
15. $\frac{42}{132}$　**17.** $\frac{1}{2}$　**19.** $\frac{3}{4}$　**21.** $\frac{1}{5}$　**23.** 3　**25.** $\frac{3}{4}$　**27.** $\frac{7}{8}$
29. $\frac{6}{5}$　**31.** $\frac{1}{3}$　**33.** 6　**35.** $\frac{1}{3}$　**37.** $\frac{2}{3}$　**39.** $\frac{7}{90}$　**41.** =
43. ≠　**45.** =　**47.** ≠　**49.** =　**51.** ≠　**53.** =
55. ≠　**57.** $\frac{7}{25}$　**59.** $\frac{1}{4}$　**61.** 526,761 yd²; 3020 yd
62. $928　**63.** 186　**64.** 32　**65.** 3520　**66.** 89
67. No; $\frac{63}{82} \neq \frac{77}{100}$ because 63 · 100 ≠ 82 · 77.

Mid-Chapter Review: Chapter 2, p. 124

1. True　**2.** False　**3.** False　**4.** True　**5.** $\frac{25}{25} = 1$
6. $\frac{0}{9} = 0$　**7.** $\frac{8}{1} = 8$　**8.** $\frac{6}{13} = \frac{18}{39}$
9. $\frac{70}{225} = \frac{2 \cdot 5 \cdot 7}{3 \cdot 3 \cdot 5 \cdot 5} = \frac{5}{5} \cdot \frac{2 \cdot 7}{3 \cdot 3 \cdot 5} = 1 \cdot \frac{14}{45} = \frac{14}{45}$
10. 84, 17,576, 224, 132, 594, 504, 1632
11. 84, 300, 132, 500, 180　**12.** 17,576, 224, 500
13. 84, 300, 132, 120, 1632　**14.** 300, 180, 120　**15.** Prime
16. Prime　**17.** Composite　**18.** Neither
19. 1, 2, 4, 5, 8, 10, 16, 20, 32, 40, 80, 160; 2 · 2 · 2 · 2 · 2 · 5
20. 1, 2, 3, 6, 37, 74, 111, 222; 2 · 3 · 37
21. 1, 2, 7, 14, 49, 98; 2 · 7 · 7　**22.** 1, 3, 5, 7, 9, 15, 21, 35, 45, 63, 105, 315; 3 · 3 · 5 · 7　**23.** $\frac{8}{24}$, or $\frac{1}{3}$　**24.** $\frac{8}{6}$, or $\frac{4}{3}$　**25.** $\frac{7}{9}$
26. $\frac{8}{45}$　**27.** $\frac{40}{11}$　**28.** $\frac{2}{5}$　**29.** $\frac{11}{3}$　**30.** 1　**31.** 0
32. $\frac{9}{31}$　**33.** $\frac{9}{5}$　**34.** $\frac{5}{42}$　**35.** $\frac{21}{29}$　**36.** Not defined
37. =　**38.** ≠　**39.** $\frac{60}{500}$, or $\frac{3}{25}$　**40.** $\frac{21}{10,000}$ mi²
41. Find the product of two prime numbers.　**42.** If we use the divisibility tests, it is quickly clear that none of the even-numbered years is a prime number. In addition, the divisibility tests for 5 and 3 show that 2001, 2005, 2007, 2013, 2015, and 2019 are not prime numbers. Then the years 2003, 2009, 2011, and 2017 can be divided by prime numbers to determine whether they are prime. When we do this, we find that 2003, 2011, and 2017 are prime numbers. If the divisibility tests are not used, each of the numbers from 2000 to 2020 can be divided by prime numbers to determine if it is prime.

43. It is possible to cancel only when identical *factors* appear in the numerator and the denominator of a fraction. Situations in which it is not possible to cancel include the occurrence of identical *addends* or *digits* in the numerator and the denominator. **44.** No; since the only factors of a prime number are the number itself and 1, two different prime numbers cannot contain a common factor (other than 1).

Exercise Set 2.6, p. 128

1. $\frac{1}{3}$ **3.** $\frac{1}{8}$ **5.** $\frac{1}{10}$ **7.** $\frac{1}{6}$ **9.** $\frac{27}{10}$ **11.** $\frac{14}{9}$ **13.** 1 **15.** 1
17. 1 **19.** 1 **21.** 2 **23.** 4 **25.** 9 **27.** 9 **29.** $\frac{15}{2}$
31. $\frac{98}{5}$ **33.** 60 **35.** 30 **37.** $\frac{1}{5}$ **39.** $\frac{9}{25}$ **41.** $\frac{11}{40}$ **43.** $\frac{5}{14}$
45. $\frac{5}{8}$ in. **47.** \$48,280 **49.** 625 addresses **51.** $\frac{1}{3}$ cup
53. \$115,500 **55.** 160 mi **57.** Food: \$8400; housing:
\$10,500; clothing: \$4200; savings: \$3000; taxes: \$8400; other
expenses: \$7500 **59.** 35 **60.** 85 **61.** 125 **62.** 120
63. 4989 **64.** 8546 **65.** 6498 **66.** 6407 **67.** 4673
68. 5338 **69.** $\frac{129}{485}$ **71.** $\frac{1}{12}$ **73.** $\frac{1}{168}$

Translating for Success, p. 136

1. C **2.** H **3.** A **4.** N **5.** O **6.** F **7.** I **8.** L
9. D **10.** M

Exercise Set 2.7, p. 137

1. $\frac{6}{5}$ **3.** $\frac{1}{6}$ **5.** 6 **7.** $\frac{3}{10}$ **9.** $\frac{4}{5}$ **11.** $\frac{4}{15}$ **13.** 4 **15.** 2
17. $\frac{1}{8}$ **19.** $\frac{3}{7}$ **21.** 8 **23.** 35 **25.** 1 **27.** $\frac{2}{3}$ **29.** $\frac{9}{4}$
31. 144 **33.** 75 **35.** 2 **37.** $\frac{3}{5}$ **39.** 315
41. 960 extension cords **43.** 32 pairs **45.** 24 bowls
47. 16 L **49.** 288 km; 108 km **51.** $\frac{1}{16}$ in.
53. Associative **54.** Factors **55.** Prime
56. Denominator **57.** Additive **58.** Reciprocals
59. Whole **60.** Equation **61.** $\frac{9}{19}$ **63.** 36 **65.** $\frac{3}{8}$

Summary and Review: Chapter 2, p. 140

Concept Reinforcement

1. True **2.** False **3.** True **4.** True

Important Concepts

1. 1, 2, 4, 8, 13, 26, 52, 104 **2.** $2 \cdot 2 \cdot 2 \cdot 13$ **3.** 0, 1, 18
4. $\frac{56}{96}$ **5.** $\frac{5}{14}$ **6.** $\neq$ **7.** $\frac{70}{9}$ **8.** $\frac{7}{10}$ **9.** $\frac{7}{3}$ cups

Review Exercises

1. 1, 2, 3, 4, 5, 6, 10, 12, 15, 20, 30, 60 **2.** 1, 2, 4, 8, 11, 16, 22, 44, 88, 176 **3.** 8, 16, 24, 32, 40, 48, 56, 64, 72, 80 **4.** Yes
5. No **6.** Prime **7.** Neither **8.** Composite **9.** $2 \cdot 5 \cdot 7$
10. $2 \cdot 3 \cdot 5$ **11.** $3 \cdot 3 \cdot 5$ **12.** $2 \cdot 3 \cdot 5 \cdot 5$
13. $2 \cdot 2 \cdot 2 \cdot 3 \cdot 3 \cdot 3 \cdot 3$ **14.** $2 \cdot 3 \cdot 5 \cdot 5 \cdot 5 \cdot 7$
15. 4344, 600, 93, 330, 255,555, 780, 2802, 711
16. 140, 182, 716, 2432, 4344, 600, 330, 780, 2802
17. 140, 716, 2432, 4344, 600, 780 **18.** 2432, 4344, 600
19. 140, 95, 475, 600, 330, 255,555, 780 **20.** 4344, 600, 330, 780, 2802 **21.** 255,555, 711 **22.** 140, 600, 330, 780
23. Numerator: 2; denominator: 7 **24.** $\frac{3}{5}$ **25.** $\frac{7}{6}$ **26.** $\frac{2}{7}$
27. (a) $\frac{3}{5}$; **(b)** $\frac{5}{3}$; **(c)** $\frac{3}{5}$ **28.** $\frac{2}{5}$ **29.** $\frac{1}{4}$ **30.** 1 **31.** 0
32. $\frac{39}{40}$ **33.** 18 **34.** $\frac{1}{3}$ **35.** $\frac{11}{23}$ **36.** Not defined **37.** 6
38. $\frac{2}{7}$ **39.** $\frac{32}{225}$ **40.** $\frac{15}{100} = \frac{3}{20}, \frac{38}{100} = \frac{19}{50}, \frac{23}{100}, \frac{24}{100} = \frac{6}{25}$
41. $\neq$ **42.** $=$ **43.** $\neq$ **44.** $=$ **45.** $\frac{3}{2}$ **46.** 56
47. $\frac{5}{2}$ **48.** 24 **49.** $\frac{2}{3}$ **50.** $\frac{1}{14}$ **51.** $\frac{2}{3}$ **52.** $\frac{1}{22}$ **53.** $\frac{3}{20}$
54. $\frac{10}{7}$ **55.** $\frac{5}{4}$ **56.** $\frac{1}{3}$ **57.** 9 **58.** $\frac{36}{47}$ **59.** $\frac{9}{2}$ **60.** 2
61. $\frac{11}{6}$ **62.** $\frac{1}{4}$ **63.** $\frac{9}{4}$ **64.** 300 **65.** 1 **66.** $\frac{4}{9}$ **67.** $\frac{3}{10}$
68. 240 **69.** 9 days **70.** About 22,800,000 metric tons
71. 1000 km **72.** $\frac{1}{3}$ cup; 2 cups **73.** \$15 **74.** 60 bags
75. D **76.** B **77.** $a = 11,176; b = 9887$ **78.** 13, 11, 101, 37

Understanding Through Discussion and Writing

1. The student is probably multiplying the divisor by the reciprocal of the dividend rather than multiplying the dividend by the reciprocal of the divisor.
2. $9432 = 9 \cdot 1000 + 4 \cdot 100 + 3 \cdot 10 + 2 \cdot 1 =$
$9(999 + 1) + 4(99 + 1) + 3(9 + 1) + 2 \cdot 1 =$
$9 \cdot 999 + 9 \cdot 1 + 4 \cdot 99 + 4 \cdot 1 + 3 \cdot 9 + 3 \cdot 1 + 2 \cdot 1$. Since 999, 99, and 9 are each a multiple of 9, $9 \cdot 999$, $4 \cdot 99$, and $3 \cdot 9$ are multiples of 9. This leaves $9 \cdot 1 + 4 \cdot 1 + 3 \cdot 1 + 2 \cdot 1$, or $9 + 4 + 3 + 2$. If $9 + 4 + 3 + 2$, the sum of the digits, is divisible by 9, then 9432 is divisible by 9. **3.** Taking $\frac{1}{2}$ of a number is equivalent to multiplying the number by $\frac{1}{2}$. Dividing by $\frac{1}{2}$ is equivalent to multiplying by the reciprocal of $\frac{1}{2}$, or 2. Thus taking $\frac{1}{2}$ of a number is not the same as dividing by $\frac{1}{2}$.
4. We first consider an object, and take $\frac{4}{7}$ of it. We divide the object into 7 parts and take 4 of them, as shown by the shading below.

Next, we take $\frac{2}{3}$ of the shaded area above. We divide it into 3 parts and take two of them, as shown below.

The entire object has been divided into 21 parts, 8 of which have been shaded twice. Thus, $\frac{2}{3} \cdot \frac{4}{7} = \frac{8}{21}$. **5.** Since $\frac{1}{7}$ is a smaller number than $\frac{2}{3}$, there are more $\frac{1}{7}$'s in 5 than $\frac{2}{3}$'s. Thus, $5 \div \frac{1}{7}$ is a greater number than $5 \div \frac{2}{3}$. **6.** No; in order to simplify a fraction, we must be able to remove a factor of the type $\frac{n}{n}$, $n \neq 0$, where n is a factor that the numerator and the denominator have in common.

Test: Chapter 2, p. 145

1. [2.1a] 1, 2, 3, 4, 5, 6, 10, 12, 15, 20, 25, 30, 50, 60, 75, 100, 150, 300 **2.** [2.1c] Prime **3.** [2.1c] Composite
4. [2.1d] $2 \cdot 3 \cdot 3$ **5.** [2.1d] $2 \cdot 2 \cdot 3 \cdot 5$ **6.** [2.2a] Yes
7. [2.2a] No **8.** [2.2a] No **9.** [2.2a] Yes **10.** [2.3a] Numerator: 4; denominator: 5 **11.** [2.3a] $\frac{3}{4}$ **12.** [2.3a] $\frac{3}{7}$
13. [2.3a] **(a)** $\frac{371}{555}$; **(b)** $\frac{184}{555}$ **14.** [2.3b] 26 **15.** [2.3b] 1
16. [2.3b] 0 **17.** [2.5b] $\frac{1}{2}$ **18.** [2.5b] 6 **19.** [2.3b] Not defined **20.** [2.3b] Not defined **21.** [2.5b] $\frac{2}{3}$ **22.** [2.5c] $=$
23. [2.5c] $\neq$ **24.** [2.6a] 32 **25.** [2.6a] $\frac{3}{2}$ **26.** [2.6a] $\frac{5}{2}$
27. [2.6a] $\frac{2}{9}$ **28.** [2.7a] $\frac{8}{5}$ **29.** [2.7a] 4 **30.** [2.7a] $\frac{1}{18}$
31. [2.7b] $\frac{8}{5}$ **32.** [2.7b] 18 **33.** [2.7b] $\frac{18}{7}$ **34.** [2.7c] 64
35. [2.7c] $\frac{7}{4}$ **36.** [2.6b] 4375 students **37.** [2.7d] $\frac{3}{40}$ m
38. [2.7d] 5 qt **39.** [2.6b] $\frac{3}{4}$ in. **40.** [2.3a] C
41. [2.6b] $\frac{7}{48}$ acre **42.** [2.6a], [2.7b] $\frac{7}{960}$

CHAPTER 3

Exercise Set 3.1, p. 153

1. 4 **3.** 50 **5.** 40 **7.** 54 **9.** 150 **11.** 120 **13.** 72
15. 420 **17.** 144 **19.** 288 **21.** 30 **23.** 90 **25.** 72
27. 60 **29.** 36 **31.** 900 **33.** 48 **35.** 50 **37.** 143
39. 420 **41.** 378 **43.** 810 **45.** 2160 **47.** 9828
49. 6000 **51.** Every 60 yr **53.** Every 420 yr

55. 659 tornadoes   **56.** 940,000,000   **57.** $\frac{8}{7}$ **58.** 33,135
59. 6,356,118   **60.** 77,699   **61.** 5 in. by 24 in.

Exercise Set 3.2, p. 159
1. 1   **3.** $\frac{3}{4}$ **5.** $\frac{3}{2}$ **7.** $\frac{7}{24}$ **9.** $\frac{3}{2}$ **11.** $\frac{19}{24}$ **13.** $\frac{9}{10}$ **15.** $\frac{29}{18}$
17. $\frac{31}{100}$ **19.** $\frac{41}{60}$ **21.** $\frac{189}{100}$ **23.** $\frac{7}{8}$ **25.** $\frac{13}{24}$ **27.** $\frac{17}{24}$ **29.** $\frac{3}{4}$
31. $\frac{437}{500}$ **33.** $\frac{53}{40}$ **35.** $\frac{391}{144}$ **37.** $\frac{37}{12}$ mi   **39.** $\frac{13}{16}$ lb
41. $\frac{4}{5}$ qt; $\frac{8}{5}$ qt; $\frac{2}{5}$ qt   **43.** $\frac{27}{32}$ **45.** $\frac{51}{32}$ in.   **47.** 210,528
48. 4,194,000   **49.** 3,387,807   **50.** 352,350   **51.** 544,683
votes   **52.** 5 votes   **53.** 84 votes   **54.** 510,314 votes
55. 21,484,549 votes   **56.** 33,128,217 votes   **57.** 6,140,000 lb
58. $\frac{3}{64}$ acre   **59.** $\frac{4}{15}$; $320

Translating for Success, p. 166
1. J   **2.** E   **3.** D   **4.** B   **5.** I   **6.** N   **7.** A   **8.** C
9. L   **10.** F

Exercise Set 3.3, p. 167
1. $\frac{2}{3}$ **3.** $\frac{3}{4}$ **5.** $\frac{5}{8}$ **7.** $\frac{1}{24}$ **9.** $\frac{1}{2}$ **11.** $\frac{9}{14}$ **13.** $\frac{3}{5}$ **15.** $\frac{7}{10}$
17. $\frac{17}{60}$ **19.** $\frac{53}{100}$ **21.** $\frac{26}{75}$ **23.** $\frac{9}{100}$ **25.** $\frac{13}{24}$ **27.** $\frac{1}{10}$ **29.** $\frac{1}{24}$
31. $\frac{13}{16}$ **33.** $\frac{31}{75}$ **35.** $\frac{31}{75}$ **37.** <   **39.** >   **41.** <   **43.** <
45. >   **47.** >   **49.** <   **51.** $\frac{1}{15}$ **53.** $\frac{2}{15}$ **55.** $\frac{1}{2}$
57. $\frac{5}{12}$ hr   **59.** $\frac{1}{32}$ in.   **61.** $\frac{1}{4}$ tub   **63.** $\frac{11}{20}$ lb   **65.** 1
66. Not defined   **67.** Not defined   **68.** 4   **69.** $\frac{4}{21}$ **70.** $\frac{3}{2}$
71. 21   **72.** $\frac{1}{32}$ **73.** About 12.8 billion, or 12,800,000,000,
crayons   **74.** 9 cups   **75.** $\frac{14}{3553}$ **77.** $\frac{21}{40}$ km   **79.** $\frac{19}{24}$
81. $\frac{145}{144}$ **83.** >   **85.** *Day 1:* Cut off $\frac{1}{7}$ of bar and pay him.
Day 2: Cut off $\frac{2}{7}$ of bar. Trade him for the $\frac{1}{7}$. *Day 3:* Give him back
the $\frac{1}{7}$. *Day 4:* Trade him the $\frac{4}{7}$ for his $\frac{3}{7}$. *Day 5:* Give him the $\frac{1}{7}$
again. *Day 6:* Trade him the $\frac{2}{7}$ for the $\frac{1}{7}$. *Day 7:* Give him the $\frac{1}{7}$
again. This assumes that he does not spend parts of the gold bar
immediately.

Exercise Set 3.4, p. 173
1. $\frac{65}{2}, \frac{125}{6}, \frac{47}{4}$ **3.** $4\frac{1}{4}, 3\frac{1}{3}, 1\frac{1}{8}$ **5.** $\frac{17}{3}$ **7.** $\frac{13}{4}$ **9.** $\frac{81}{8}$ **11.** $\frac{51}{10}$
13. $\frac{103}{5}$ **15.** $\frac{59}{6}$ **17.** $\frac{73}{10}$ **19.** $\frac{13}{8}$ **21.** $\frac{51}{4}$ **23.** $\frac{43}{10}$ **25.** $\frac{203}{100}$
27. $\frac{200}{9}$ **29.** $\frac{279}{50}$ **31.** $\frac{1621}{16}$ **33.** $3\frac{3}{5}$ **35.** $4\frac{2}{3}$ **37.** $4\frac{1}{2}$
39. $5\frac{7}{9}$ **41.** $7\frac{4}{7}$ **43.** $7\frac{1}{2}$ **45.** $11\frac{1}{5}$ **47.** $1\frac{1}{2}$ **49.** $7\frac{57}{100}$
51. $43\frac{1}{8}$ **53.** $108\frac{5}{9}$ **55.** $618\frac{1}{5}$ **57.** $40\frac{4}{7}$ **59.** $55\frac{1}{51}$
61. $2292\frac{23}{35}$ **63.** 45,800   **64.** 45,770   **65.** $\frac{8}{15}$ **66.** $\frac{21}{25}$
67. $\frac{16}{27}$ **68.** $\frac{583}{669}$ **69.** 18   **70.** $\frac{5}{2}$ **71.** $\frac{1}{4}$ **72.** $\frac{2}{5}$ **73.** 24
74. 49   **75.** $\frac{2560}{3}$ **76.** $\frac{4}{3}$ **77.** $237\frac{19}{541}$ **79.** $8\frac{2}{3}$ **81.** $52\frac{2}{7}$

Mid-Chapter Review: Chapter 3, p. 176
1. True   **2.** True   **3.** False   **4.** False
5.
$$\frac{11}{42} - \frac{3}{35} = \frac{11}{2 \cdot 3 \cdot 7} - \frac{3}{5 \cdot 7}$$
$$= \frac{11}{2 \cdot 3 \cdot 7} \cdot \left(\frac{5}{5}\right) - \frac{3}{5 \cdot 7} \cdot \left(\frac{2 \cdot 3}{2 \cdot 3}\right)$$
$$= \frac{11 \cdot 5}{2 \cdot 3 \cdot 7 \cdot 5} - \frac{3 \cdot 2 \cdot 3}{5 \cdot 7 \cdot 2 \cdot 3}$$
$$= \frac{55}{2 \cdot 3 \cdot 5 \cdot 7} - \frac{18}{2 \cdot 3 \cdot 5 \cdot 7}$$
$$= \frac{55 - 18}{2 \cdot 3 \cdot 5 \cdot 7} = \frac{37}{210}$$

6.
$$x + \frac{1}{8} = \frac{2}{3}$$
$$x + \frac{1}{8} - \frac{1}{8} = \frac{2}{3} - \frac{1}{8}$$
$$x + 0 = \frac{2}{3} \cdot \frac{8}{8} - \frac{1}{8} \cdot \frac{3}{3}$$
$$x = \frac{16}{24} - \frac{3}{24}$$
$$x = \frac{13}{24}$$

7.

45 and 50	120
50 and 80	720
30 and 24	400
18, 24, and 80	450
30, 45, and 50	

8. $\frac{16}{45}$ **9.** $\frac{25}{12}$ **10.** $\frac{1}{18}$ **11.** $\frac{19}{90}$ **12.** $\frac{7}{240}$ **13.** $\frac{156}{119}$ **14.** $\frac{79}{720}$
15. $\frac{6}{91}$ **16.** $\frac{22}{15}$ mi   **17.** $\frac{101}{20}$ hr   **18.** $\frac{1}{5}, \frac{7}{2}; \frac{3}{10}, \frac{4}{9}$ **19.** $\frac{13}{80}$

20. $17\frac{8}{15}$ **21.** C   **22.** C   **23.** No; if one number is a
multiple of the other, for example, the LCM is the larger of the
numbers.   **24.** We multiply by 1, using the notation n/n,
to express each fraction in terms of the least common
denominator.   **25.** Write $\frac{8}{5}$ as $\frac{16}{10}$ and $\frac{8}{2}$ as $\frac{40}{10}$ and since
taking 40 tenths away from 16 tenths would give a result less
than 0, it cannot possibly be $\frac{8}{5}$. You could also find the sum
$\frac{8}{3} + \frac{8}{2}$ and show that it is not $\frac{8}{5}$.   **26.** No; $2\frac{1}{3} = \frac{7}{3}$ but $2 \cdot \frac{1}{3} = \frac{2}{3}$.

Exercise Set 3.5, p. 183
1. $28\frac{3}{4}$ **3.** $185\frac{7}{8}$ **5.** $6\frac{1}{2}$ **7.** $2\frac{11}{12}$ **9.** $14\frac{7}{12}$ **11.** $12\frac{1}{10}$
13. $16\frac{5}{24}$ **15.** $21\frac{1}{2}$ **17.** $27\frac{7}{8}$ **19.** $27\frac{13}{24}$ **21.** $1\frac{3}{5}$
23. $4\frac{1}{10}$ **25.** $21\frac{17}{24}$ **27.** $12\frac{1}{4}$ **29.** $15\frac{3}{8}$ **31.** $7\frac{5}{12}$
33. $13\frac{3}{8}$ **35.** $11\frac{5}{18}$ **37.** $14\frac{13}{24}$ flats   **39.** $7\frac{5}{12}$ lb   **41.** $6\frac{5}{12}$ in.
43. $20\frac{1}{12}$ yd   **45.** $5\frac{7}{8}$ in.   **47.** $95\frac{5}{11}$ mi   **49.** $134\frac{1}{4}''$
51. $20\frac{1}{8}$ in.   **53.** $78\frac{1}{2}$ in.   **55.** $3\frac{4}{5}$ hr   **57.** $28\frac{3}{4}$ yd
59. $7\frac{3}{8}$ ft   **61.** $5\frac{3}{8}$ yd   **63.** $4\frac{5}{6}$ ft $\times$ $10\frac{7}{12}$ ft   **65.** 16 packages
66. 286 cartons; 2 oz left over   **67.** Yes   **68.** No   **69.** No
70. Yes   **71.** No   **72.** Yes   **73.** Yes   **74.** Yes   **75.** $\frac{10}{13}$
76. $\frac{1}{10}$ **77.** $8568\frac{786}{1189}$ **79.** $5\frac{3}{4}$ ft

Calculator Corner, p. 192
1. $\frac{7}{12}$ **2.** $\frac{11}{10}$ **3.** $\frac{35}{16}$ **4.** $\frac{3}{10}$ **5.** $10\frac{2}{15}$ **6.** $1\frac{1}{28}$ **7.** $10\frac{11}{15}$
8. $2\frac{91}{115}$

Translating for Success, p. 193
1. O   **2.** K   **3.** F   **4.** D   **5.** H   **6.** G   **7.** L   **8.** E
9. M   **10.** J

Exercise Set 3.6, p. 194
1. $22\frac{2}{3}$ **3.** $2\frac{5}{12}$ **5.** $8\frac{1}{6}$ **7.** $9\frac{31}{40}$ **9.** $24\frac{91}{100}$ **11.** $975\frac{4}{5}$
13. $6\frac{1}{4}$ **15.** $1\frac{1}{5}$ **17.** $3\frac{9}{16}$ **19.** $1\frac{1}{8}$ **21.** $1\frac{8}{43}$ **23.** $\frac{9}{40}$
25. 45,000 beagles   **27.** About 690,000   **29.** $62\frac{1}{2}$ ft^2
31. $13\frac{1}{3}$ tsp   **33.** $343\frac{3}{4}$ lb   **35.** 68°F   **37.** About 1,800,750
39. About 301,360,000   **41.** $5\frac{1}{2}$ cups of flour, $2\frac{2}{3}$ cups of sugar
43. 15 mpg   **45.** 400 cu ft   **47.** $16\frac{1}{2}$ servings   **49.** $35\frac{115}{256}$ sq in.
51. $59,538\frac{1}{8}$ sq ft   **53.** Divisor; quotient; dividend
54. Common   **55.** Composite   **56.** Divisible; divisible
57. Multiplications; divisions; additions; subtractions
58. Addends   **59.** Numerator   **60.** Reciprocal   **61.** $360\frac{60}{473}$
63. $35\frac{57}{64}$ **65.** $\frac{4}{9}$ **67.** $1\frac{4}{5}$

Exercise Set 3.7, p. 202
1. $\frac{59}{30}$, or $1\frac{29}{30}$ **3.** $\frac{3}{20}$ **5.** $\frac{211}{8}$, or $26\frac{3}{8}$ **7.** $\frac{7}{16}$ **9.** $\frac{1}{36}$ **11.** $\frac{3}{8}$
13. $\frac{17}{6}$, or $2\frac{5}{6}$ **15.** $\frac{8395}{84}$, or $99\frac{79}{84}$ **17.** $\frac{37}{48}$ **19.** $\frac{25}{72}$
21. $\frac{103}{16}$, or $6\frac{7}{16}$ **23.** $16\frac{7}{96}$ mi   **25.** $9\frac{19}{40}$ lb   **27.** 0   **29.** 1
31. $\frac{1}{2}$ **33.** $\frac{1}{2}$ **35.** 0   **37.** 1   **39.** 3   **41.** 13   **43.** 2
45. $1\frac{1}{2}$ **47.** $\frac{1}{2}$ **49.** $271\frac{1}{2}$ **51.** 3   **53.** 100   **55.** $29\frac{1}{2}$
57. $\frac{8}{3}$ **58.** $\frac{8}{3}$ **59.** Prime: 5, 7, 23, 43; composite: 9, 14;
neither: 1   **60.** 59 R 77   **61.** 16 people   **62.** 43 mg
63. $a = 2, b = 8$ **65.** The largest is $\frac{4}{3} + \frac{5}{2} = \frac{23}{6}$.

Summary and Review: Chapter 3, p. 205
Concept Reinforcement
1. True   **2.** True   **3.** False   **4.** True

Important Concepts
1. 156   **2.** $\frac{112}{180}$, or $\frac{28}{45}$ **3.** $\frac{4}{35}$ **4.** <   **5.** $\frac{59}{99}$ **6.** $\frac{26}{3}$ **7.** $7\frac{5}{6}$
8. $7\frac{27}{28}$ **9.** $14\frac{14}{25}$ **10.** About 4,500,000   **11.** $\frac{9}{2}$, or $4\frac{1}{2}$ **12.** $4\frac{1}{2}$

Review Exercises
1. 36   **2.** 90   **3.** 30   **4.** 1404   **5.** $\frac{63}{40}$ **6.** $\frac{19}{48}$ **7.** $\frac{25}{12}$
8. $\frac{891}{1000}$ **9.** $\frac{1}{3}$ **10.** $\frac{1}{8}$ **11.** $\frac{5}{27}$ **12.** $\frac{11}{18}$ **13.** >   **14.** >
15. $\frac{19}{40}$ **16.** $\frac{2}{5}$ **17.** $\frac{15}{2}$ **18.** $\frac{67}{8}$ **19.** $\frac{13}{3}$ **20.** $\frac{75}{7}$ **21.** $2\frac{1}{3}$

22. $6\frac{3}{4}$ **23.** $12\frac{3}{5}$ **24.** $3\frac{1}{2}$ **25.** $877\frac{1}{3}$ **26.** $456\frac{5}{23}$
27. $10\frac{2}{5}$ **28.** $11\frac{11}{15}$ **29.** $10\frac{2}{3}$ **30.** $8\frac{1}{4}$ **31.** $7\frac{7}{9}$ **32.** $4\frac{11}{15}$
33. $4\frac{3}{20}$ **34.** $13\frac{3}{8}$ **35.** 16 **36.** $3\frac{1}{2}$ **37.** $2\frac{21}{50}$ **38.** 6
39. 12 **40.** $1\frac{7}{17}$ **41.** $\frac{1}{8}$ **42.** $\frac{9}{10}$ **43.** $4\frac{1}{4}$ yd **44.** $177\frac{3}{4}$ in^2
45. $50\frac{1}{4}$ in^2 **46.** $1\frac{73}{100}$ in. **47.** 24 lb **48.** $\$850$
49. $8\frac{3}{8}$ cups **50.** $13\frac{1}{4}$ in. $\times$ $13\frac{1}{4}$ in.: perimeter $= 53$ in.,
area $= 175\frac{9}{16}$ sq in.; $13\frac{1}{4}$ in. $\times$ $3\frac{1}{4}$ in.: perimeter $= 33$ in.,
area $= 43\frac{1}{16}$ sq in. **51.** $63\frac{2}{3}$ pies; $19\frac{1}{3}$ pies **52.** 1 **53.** $\frac{7}{40}$
54. 3 **55.** $\frac{77}{240}$ **56.** $\frac{1}{2}$ **57.** 0 **58.** 1 **59.** 7 **60.** 10
61. $5\frac{1}{2}$ **62.** 0 **63.** $2\frac{1}{2}$ **64.** $28\frac{1}{2}$ **65.** A **66.** D
67. 12 min **68.** $\frac{6}{3} + \frac{5}{4} = 3\frac{1}{4}$

Understanding Through Discussion and Writing

1. No; if the sum of the fractional parts of the mixed numerals is n/n, then the sum of the mixed numerals is an integer. For example, $1\frac{1}{5} + 6\frac{4}{5} = 7\frac{5}{5} = 8$. **2.** A wheel makes $33\frac{1}{3}$ revolutions per minute. It rotates for $4\frac{1}{2}$ min. How many revolutions does it make? Answers may vary. **3.** The student is multiplying the whole numbers to get the whole-number portion of the answer and multiplying fractions to get the fraction part of the answer. The student should have converted each mixed numeral to fraction notation, multiplied, simplified, and then converted back to a mixed numeral. The correct answer is $4\frac{6}{7}$. **4.** It might be necessary to find the least common denominator before adding or subtracting. The least common denominator is the least common multiple of the denominators. **5.** Suppose that a room has dimensions $15\frac{3}{4}$ ft by $28\frac{5}{8}$ ft. The equation $2 \cdot 15\frac{3}{4} + 2 \cdot 28\frac{5}{8} = 88\frac{3}{4}$ gives the perimeter of the room, in feet. Answers may vary. **6.** Note that $5 \cdot 3\frac{2}{7} = 5\left(3 + \frac{2}{7}\right) = 5 \cdot 3 + 5 \cdot \frac{2}{7}$. The products $5 \cdot 3$ and $5 \cdot \frac{2}{7}$ should be added rather than multiplied together. The student could also have converted $3\frac{2}{7}$ to fraction notation, multiplied, simplified, and converted back to a mixed numeral. The correct answer is $16\frac{3}{7}$.

Test: Chapter 3, p. 211

1. [3.1a] 48 **2.** [3.1a] 600 **3.** [3.2a] 3 **4.** [3.2a] $\frac{37}{24}$
5. [3.2a] $\frac{921}{1000}$ **6.** [3.3a] $\frac{1}{3}$ **7.** [3.3a] $\frac{1}{12}$ **8.** [3.3a] $\frac{77}{120}$
9. [3.3c] $\frac{15}{4}$ **10.** [3.3c] $\frac{1}{4}$ **11.** [3.3b] $>$ **12.** [3.4a] $\frac{7}{4}$
13. [3.4a] $\frac{79}{8}$ **14.** [3.4a] $4\frac{1}{2}$ **15.** [3.4a] $8\frac{2}{9}$ **16.** [3.4b] $162\frac{7}{11}$
17. [3.5a] $14\frac{1}{5}$ **18.** [3.5a] $14\frac{5}{12}$ **19.** [3.5b] $4\frac{7}{24}$ **20.** [3.5b] $6\frac{1}{6}$
21. [3.6a] 39 **22.** [3.6a] $4\frac{1}{2}$ **23.** [3.6b] 2 **24.** [3.6b] $\frac{1}{36}$
25. [3.6c] About 105 kg **26.** [3.6c] 80 books **27.** [3.5c]
(a) 3 in.; **(b)** $4\frac{1}{2}$ in. **28.** [3.3d] $\frac{1}{16}$ in. **29.** [3.7a] $6\frac{11}{36}$ ft
30. [3.7a] $3\frac{1}{2}$ **31.** [3.7a] $\frac{3}{4}$ **32.** [3.7b] 0 **33.** [3.7b] 1
34. [3.7b] 16 **35.** [3.7b] $1214\frac{1}{2}$ **36.** [3.1a] D
37. [3.1a] **(a)** $24, 48, 72$; **(b)** 24 **38.** [3.3b], [3.5c] Rebecca walks $\frac{17}{56}$ mi farther.

Cumulative Review: Chapters 1-3, p. 213

1. (a) [3.5c] $14\frac{13}{24}$ mi; **(b)** [3.7a] $4\frac{61}{72}$ mi **2.** [1.8a] 31 people
3. [2.6b] $\frac{2}{5}$ tsp; 4 tsp **4.** [3.6c] 16 pieces **5.** [1.8a] $\$108$
6. [3.2b] $\frac{33}{20}$ mi **7.** [2.3a] $\frac{5}{16}$ **8.** [2.3a] $\frac{4}{3}$ **9.** [1.2a] 8982
10. [1.3a] 4518 **11.** [1.4a] 5004 **12.** [1.4a] $293,232$
13. [3.2a] $\frac{5}{12}$ **14.** [3.5a] $8\frac{1}{4}$ **15.** [3.3a] $\frac{5}{12}$ **16.** [3.5b] $1\frac{1}{6}$
17. [2.6a] $\frac{3}{2}$ **18.** [2.6a] 15 **19.** [2.7b] $\frac{4}{7}$ **20.** [3.6b] $7\frac{1}{3}$
21. [1.5a] 715 **22.** [1.5a] 56 R 11 **23.** [3.4b] $56\frac{11}{100}$
24. [1.1a] 5 **25. (a)** [3.6c] $142\frac{1}{4}$ ft^2; **(b)** [3.5c] 54 ft
26. [1.6a] $38,500$ **27.** [3.1a] 72 **28.** [3.7a] $\frac{1377}{100}$, or $13\frac{77}{100}$
29. [3.3b] $>$ **30.** [2.5c] $=$ **31.** [3.3b] $<$ **32.** [3.7b] 1
33. [3.7b] $\frac{1}{2}$ **34.** [3.7b] 0 **35.** [2.5b] $\frac{4}{5}$ **36.** [2.3b] 0
37. [2.5b] 32 **38.** [3.4a] $\frac{37}{8}$ **39.** [3.4a] $5\frac{2}{3}$ **40.** [1.7b] 93
41. [3.3c] $\frac{5}{9}$ **42.** [2.7c] $\frac{12}{7}$ **43.** [1.7b] 905
44. [2.1a, c, d], [2.2a]

Factors of 68: 1, 2, 4, 17, 34, 68
Factorization of 68: $2 \cdot 2 \cdot 17$, or $2 \cdot 34$
Prime factorization of 68: $2 \cdot 2 \cdot 17$
Numbers divisible by 6: 12, 54, 72, 300
Numbers divisible by 8: 8, 16, 24, 32, 40, 48, 64, 864
Numbers divisible by 5: 70, 95, 215
Prime numbers: 2, 3, 17, 19, 23, 31, 47, 101
45. [2.1c] 2003

CHAPTER 4

Exercise Set 4.1, p. 223

1. Four hundred eighty-six and thirty-four hundredths **3.** One hundred forty-six thousandths **5.** Two hundred forty-nine and eighty-nine hundredths **7.** Three and seven hundred eighty-five thousandths **9.** Twenty-seven and one thousand, two hundred forty-five ten-thousandths **11.** $\frac{83}{10}$
13. $\frac{356}{100}$ **15.** $\frac{20,003}{1000}$ **17.** $\frac{10,008}{10,000}$ **19.** $\frac{372}{10}$ **21.** $\frac{13}{100,000}$
23. 0.8 **25.** 3.798 **27.** 8.89 **29.** 0.00019 **31.** 0.0078
33. 0.376193 **35.** 2.9 **37.** 3.098 **39.** 99.44 **41.** 2.1739
43. 0.58 **45.** 0.91 **47.** 0.001 **49.** 235.07 **51.** $\frac{4}{100}$
53. 0.4325 **55.** 0.1 **57.** 0.5 **59.** 2.7 **61.** 123.7
63. 0.89 **65.** 0.67 **67.** 1.00 **69.** 0.09 **71.** 0.325
73. 17.001 **75.** 10.101 **77.** 9.999 **79.** 800 **81.** 809.573
83. 810 **85.** 34.5439 **87.** 34.54 **89.** 35 **91.** 6170
92. 6200 **93.** 6000 **94.** $2 \cdot 2 \cdot 2 \cdot 2 \cdot 5 \cdot 5 \cdot 5$, or $2^4 \cdot 5^3$
95. $2 \cdot 3 \cdot 3 \cdot 5 \cdot 17$, or $2 \cdot 3^2 \cdot 5 \cdot 17$ **96.** $2 \cdot 7 \cdot 11 \cdot 13$
97. $2 \cdot 2 \cdot 2 \cdot 7 \cdot 7 \cdot 11$, or $2^3 \cdot 7^2 \cdot 11$ **99.** $2.000001, 2.0119,$
$2.018, 2.0302, 2.1, 2.108, 2.109$ **101.** 6.78346 **103.** 0.03030

Calculator Corner, p. 228

1. 317.645 **2.** 506.553 **3.** 17.15 **4.** 49.08 **5.** 4.4
6. 33.83 **7.** 454.74 **8.** 0.99

Exercise Set 4.2, p. 231

1. 334.37 **3.** 1576.215 **5.** 132.560 **7.** 50.0248 **9.** 40.007
11. 977.955 **13.** 771.967 **15.** 8754.8221 **17.** 49.02
19. 85.921 **21.** 2.4975 **23.** 3.397 **25.** 8.85 **27.** 3.37
29. 1.045 **31.** 3.703 **33.** 0.9902 **35.** 99.66 **37.** 4.88
39. 0.994 **41.** 17.802 **43.** 51.13 **45.** 32.7386 **47.** 4.0622
49. 11.65 **51.** 384.68 **53.** 582.97 **55.** $15,335.3$
57. The balance forward should read:
$\quad$ \$ 9704.56
$\qquad$ 9677.12
$\qquad$ 10,677.12
$\qquad$ 10,553.17
$\qquad$ 10,429.15
$\qquad$ 10,416.72
$\qquad$ 12,916.72
$\qquad$ 12,778.94
$\qquad$ 12,797.82
$\qquad$ 9997.82
59. $35,000$ **60.** $34,000$ **61.** $\frac{1}{6}$ **62.** $\frac{34}{45}$ **63.** 6166
64. 5366 **65.** $16\frac{1}{2}$ servings **66.** $60\frac{1}{5}$ mi **67.** 345.8

Calculator Corner, p. 237

1. 48.6 **2.** 6930.5 **3.** 142.803 **4.** 0.5076 **5.** 7916.4
6. 20.4153

Exercise Set 4.3, p. 239

1. 60.2 **3.** 6.72 **5.** 0.252 **7.** 0.522 **9.** 237.6
11. $583,686.852$ **13.** 780 **15.** 8.923 **17.** 0.09768
19. 0.782 **21.** 521.6 **23.** 3.2472 **25.** 897.6 **27.** 322.07
29. 55.68 **31.** 3487.5 **33.** 50.0004 **35.** 114.42902
37. 13.284 **39.** 90.72 **41.** 0.0028728 **43.** 0.72523
45. 1.872115 **47.** $45,678$ **49.** 2888¢ **51.** 66¢
53. $\$0.34$ **55.** $\$34.45$ **57.** $47,300,000,000$ **59.** $9,300,000$
61. $23,400,000,000$ **63.** $11\frac{1}{5}$ **64.** $\frac{35}{72}$ **65.** $2\frac{7}{15}$ **66.** $7\frac{2}{15}$

67. 342 **68.** 87 **69.** 4566 **70.** 1257 **71.** 87
72. 1176 R 14 **73.** 10^{21} = 1 sextillion **75.** 10^{24} = 1 septillion

Calculator Corner, p. 243

1. 14.3 **2.** 2.56 **3.** 0.064 **4.** 75.8

Exercise Set 4.4, p. 249

1. 2.99 **3.** 23.78 **5.** 7.48 **7.** 7.2 **9.** 1.143 **11.** 4.041
13. 56 **15.** 70 **17.** 20 **19.** 0.4 **21.** 0.41 **23.** 8.5
25. 9.3 **27.** 0.625 **29.** 0.26 **31.** 15.625 **33.** 2.34
35. 0.47 **37.** 0.2134567 **39.** 2.359 **41.** 4.26487
43. 169.4 **45.** 1023.7 **47.** 4256.1 **49.** 9.3 **51.** 0.0090678
53. 45.6 **55.** 2107 **57.** 303.003 **59.** 446.208 **61.** 24.14
63. 13.0072 **65.** 19.3204 **67.** 473.188278 **69.** 10.49
71. 911.13 **73.** 205 **75.** \$1288.36 **77.** 5.42 million stays
79. $\frac{6}{7}$ **80.** $\frac{7}{8}$ **81.** $\frac{19}{73}$ **82.** $\frac{23}{31}$ **83.** $2 \cdot 2 \cdot 3 \cdot 3 \cdot 19$, or
$2^2 \cdot 3^2 \cdot 19$ **84.** $2 \cdot 3 \cdot 3 \cdot 3 \cdot 3$, or $2 \cdot 3^4$ **85.** $3 \cdot 3 \cdot 223$, or
$3^2 \cdot 223$ **86.** $5 \cdot 401$ **87.** $15\frac{1}{8}$ **88.** $5\frac{7}{8}$ **89.** 6.254194585
91. 1000 **93.** 100

Mid-Chapter Review: Chapter 4, p. 253

1. False **2.** True **3.** True
4.
$$y + 12.8 = 23.35$$
$$y + 12.8 - 12.8 = 23.35 - 12.8$$
$$y + 0 = 10.55$$
$$y = 10.55$$
5. $5.6 + 4.3 \times (6.5 - 0.25)^2 = 5.6 + 4.3 \times (6.25)^2$
$$= 5.6 + 4.3 \times 39.0625$$
$$= 5.6 + 167.96875$$
$$= 173.56875$$
6. Nine and sixty-nine hundredths **7.** 1,050,000 **8.** $\frac{453}{100}$
9. $\frac{287}{1000}$ **10.** 0.13 **11.** 5.2 **12.** 0.7 **13.** 6.39 **14.** 35.67
15. 8.002 **16.** 28.462 **17.** 28.46 **18.** 28.5 **19.** 28
20. 50.095 **21.** 1214.862 **22.** 5.228 **23.** 18.24 **24.** 272.19
25. 5.593 **26.** 15.55 **27.** 39.37 **28.** 4.14 **29.** 92.871
30. 8123.6 **31.** 2.937 **32.** 5.06 **33.** 3.2 **34.** 763.4
35. 0.914036 **36.** 2045¢ **37.** \$1.47 **38.** 12.7 **39.** 8.4
40. 59.774 **41.** 33.33 **42.** The student probably rounded
over successively from the thousandths place as follows:
$236.448 \approx 236.45 \approx 236.5 \approx 237$. The student should have
considered only the tenths place and rounded down.
43. The decimal points were not lined up before the subtraction
was carried out. **44.** $10 \div 0.2 = \frac{10}{0.2} = \frac{10}{0.2} \cdot \frac{10}{10} = \frac{100}{2} = 100 \div 2$.
45. $0.247 \div 0.1 = \frac{247}{1000} \div \frac{1}{10} = \frac{247}{1000} \cdot \frac{10}{1} = \frac{247 \cdot 10}{10 \cdot 100} = \frac{247}{100} =$
$2.47 \neq 0.0247$;
$0.247 \div 10 = \frac{247}{1000} \div 10 = \frac{247}{1000} \cdot \frac{1}{10} = \frac{247}{10,000} = 0.0247 \neq 2.47$

Exercise Set 4.5, p. 260

1. 0.23 **3.** 0.6 **5.** 0.325 **7.** 0.2 **9.** 0.85 **11.** 0.375
13. 0.975 **15.** 0.52 **17.** 20.016 **19.** 0.25 **21.** 1.16
23. 1.1875 **25.** $0.2\overline{6}$ **27.** $0.\overline{3}$ **29.** $1.\overline{3}$ **31.** $1.1\overline{6}$
33. $0.\overline{571428}$ **35.** $0.91\overline{6}$ **37.** 0.3; 0.27; 0.267 **39.** 0.3; 0.33;
0.333 **41.** 1.3; 1.33; 1.333 **43.** 1.2; 1.17; 1.167 **45.** 0.6; 0.57;
0.571 **47.** 0.9; 0.92; 0.917 **49.** 0.2; 0.18; 0.182 **51.** 0.3; 0.28;
0.278 **53.** (a) 0.429; (b) 0.75; (c) 0.571; (d) 1.333
55. 15.8 mpg **57.** 17.8 mpg **59.** 15.2 mph **61.** \$29.5625;
\$29.56 **63.** \$27.875; \$27.88 **65.** \$31.484375; \$31.48
67. 11.06 **69.** 8.4 **71.** $417.51\overline{6}$ **73.** 0 **75.** 2.8125
77. 0.20425 **79.** 317.14 **81.** 0.1825 **83.** 18 **85.** 2.736
87. 21 **88.** $238\frac{7}{8}$ **89.** 10 **90.** $\frac{43}{52}$ **91.** $50\frac{5}{24}$ **92.** $30\frac{7}{10}$
93. $1\frac{1}{2}$ **94.** $14\frac{13}{24}$ **95.** $1\frac{1}{24}$ cups **96.** $1\frac{33}{100}$ in. **97.** $0.\overline{142857}$
99. $0.\overline{428571}$ **101.** $0.\overline{714285}$ **103.** $0.\overline{1}$ **105.** $0.\overline{001}$

Exercise Set 4.6, p. 267

1. (d) **3.** (c) **5.** (a) **7.** (c) **9.** 1.6 **11.** 6 **13.** 60
15. 2.3 **17.** 180 **19.** (a) **21.** (c) **23.** (b) **25.** (b)
27. $1800 \div 9 = 200$ posts; answers may vary

29. $\$2 \cdot 12 = \$24$; answers may vary **31.** Repeating
32. Multiple **33.** Distributive **34.** Solution
35. Multiplicative **36.** Commutative **37.** Denominator;
multiple **38.** Divisible; divisible **39.** Yes **41.** No
43. (a) $+, \times$; (b) $+, \times, -$

Translating for Success, p. 277

1. I **2.** C **3.** N **4.** A **5.** G **6.** B **7.** D **8.** O **9.** F
10. M

Exercise Set 4.7, p. 278

1. \$43.1 billion **3.** 6.29 million passengers **5.** \$151.1 million
7. \$0.51 **9.** 102.8°F **11.** \$64,333,333.33 **13.** Area: 8.125 sq cm;
perimeter: 11.5 cm **15.** 22,691.5 mi **17.** 20.2 mpg
19. 11.9752 cu ft **21.** 78.1 cm **23.** 28.5 cm **25.** \$24.33
27. 2.31 cm **29.** 876 calories **31.** \$1171.74 **33.** 227.75 sq ft
35. 0.364 **37.** 2152.56 sq yd **39.** 10.8¢ **41.** \$906.50
43. 6.052 billion **45.** 1.4°F **47.** \$262,153 **49.** \$83,782
51. \$1,401,429 **53.** \$29,133 **55.** \$53.04 **57.** \$3745.41
59. \$1406.75 **61.** 6335 **62.** $\frac{31}{24}$ **63.** $6\frac{5}{6}$ **64.** $\frac{23}{15}$ **65.** $\frac{1}{24}$
66. 2803 **67.** $\frac{2}{15}$ **68.** $1\frac{5}{6}$ **69.** $\frac{129}{251}$ **70.** $\frac{5}{16}$ **71.** $\frac{13}{25}$ **72.** $\frac{25}{19}$
73. 28 min **74.** $7\frac{1}{5}$ min **75.** 186 calories **76.** 30 calories
77. \$17.28

Summary and Review: Chapter 4, p. 285

Concept Reinforcement

1. True **2.** False **3.** True **4.** False **5.** True

Important Concepts

1. $\frac{5093}{100}$ **2.** 81.7 **3.** 42.159 **4.** 153.35 **5.** 38.611
6. 207.848 **7.** 19.11 **8.** 0.176 **9.** 60,437 **10.** 7.4
11. 0.047 **12.** 15,690

Review Exercises

1. 6,590,000 **2.** 3,100,000,000 **3.** Three and forty-seven
hundredths **4.** Thirty-one thousandths **5.** Twenty-seven and
one ten-thousandth **6.** Nine tenths **7.** $\frac{9}{100}$ **8.** $\frac{4561}{1000}$
9. $\frac{89}{1000}$ **10.** $\frac{30,227}{10,000}$ **11.** 0.034 **12.** 4.2603 **13.** 27.91
14. 867.006 **15.** 0.034 **16.** 0.91 **17.** 0.741 **18.** 1.041
19. 17.4 **20.** 17.43 **21.** 17.429 **22.** 17 **23.** 574.519
24. 0.6838 **25.** 229.1 **26.** 45.551 **27.** 29.2092 **28.** 790.29
29. 29.148 **30.** 70.7891 **31.** 12.96 **32.** 0.14442 **33.** 4.3
34. 0.02468 **35.** 7.5 **36.** 0.45 **37.** 45.2 **38.** 1.022
39. 0.2763 **40.** 1389.2 **41.** 496.2795 **42.** 6.95 **43.** 42.54
44. 4.9911 **45.** \$15.52 **46.** 1.9 lb **47.** \$784.47 **48.** \$171.24
49. 14.5 mpg **50.** (a) 106.2 lb; (b) 15.2 lb **51.** 272 **52.** 216
53. \$125 **54.** 0.52 **55.** 0.45 **56.** 2.75 **57.** 3.25 **58.** 1.16
59. $1.5\overline{4}$ **60.** 1.5 **61.** 1.55 **62.** 1.545 **63.** \$82.73
64. \$4.87 **65.** 2493¢ **66.** 986¢ **67.** 1.8045 **68.** 57.1449
69. 15.6375 **70.** D **71.** B **72.** (a) $2.56 \times 6.4 \div 51.2 -$
$17.4 + 89.7 = 72.62$; (b) $(11.12 - 0.29) \times 3^4 = 877.23$
73. $1 = 3 \cdot \frac{1}{3} = 3(0.33333333\ldots) = 0.99999999\ldots$, or $0.\overline{9}$

Understanding Through Discussion and Writing

1. Count the number of decimal places. Move the decimal point
that many places to the right and write the result over a denominator of 1 followed by that many zeros.
2. $346.708 \times 0.1 = \frac{346,708}{1000} \times \frac{1}{10} = \frac{346,708}{10,000} = 34.6708 \neq 3467.08$
3. When the denominator of a fraction is a multiple of 10, long
division is not the fastest way to convert the fraction to decimal
notation. Many times when the denominator is a factor of some
multiple of 10, this is also the case. The latter situation occurs when
the denominator has only 2's or 5's or both as factors. **4.** Multiply
by 1 to get a denominator that is a power of 10:

$$\frac{44}{125} = \frac{44}{125} \cdot \frac{8}{8} = \frac{352}{1000} = 0.352.$$

We can also divide to find that $\frac{44}{125} = 0.352$.

Test: Chapter 4, p. 290

1. [4.3b] 18,400,000   **2.** [4.3b] 13,100,000,000   **3.** [4.1a] Two and thirty-four hundredths   **4.** [4.1a] One hundred five and five ten-thousandths   **5.** [4.1b] $\frac{91}{100}$ **6.** [4.1b] $\frac{2769}{1000}$ **7.** [4.1b] 0.074   **8.** [4.1b] 3.7047   **9.** [4.1b] 756.09   **10.** [4.1b] 91.703   **11.** [4.1c] 0.162   **12.** [4.1c] 0.078   **13.** [4.1c] 0.9   **14.** [4.1d] 6   **15.** [4.1d] 5.68   **16.** [4.1d] 5.678   **17.** [4.1d] 5.7   **18.** [4.2a] 0.7902   **19.** [4.2a] 186.5   **20.** [4.2a] 1033.23   **21.** [4.2b] 48.357   **22.** [4.2b] 19.0901   **23.** [4.2b] 152.8934   **24.** [4.3a] 0.03   **25.** [4.3a] 0.21345   **26.** [4.3a] 73,962   **27.** [4.4a] 4.75   **28.** [4.4a] 30.4   **29.** [4.4a] 0.19   **30.** [4.4a] 0.34689   **31.** [4.4a] 34,689   **32.** [4.4b] 84.26   **33.** [4.2c] 8.982   **34.** [4.7a] $133.99   **35.** [4.7a] 28.3 mpg   **36.** [4.7a] $592.45   **37.** [4.7a] $293.93   **38.** [4.7a] 67.44 million passengers   **39.** [4.6a] 198   **40.** [4.6a] 4   **41.** [4.5a] 0.35   **42.** [4.5a] 0.88   **43.** [4.5a] 5.25   **44.** [4.5a] 0.75   **45.** [4.5a] 1.$\overline{2}$ **46.** [4.5a] 2.$\overline{142857}$ **47.** [4.5b] 2.1   **48.** [4.5b] 2.14   **49.** [4.5b] 2.143   **50.** [4.4c] 40.0065   **51.** [4.4c] 384.8464   **52.** [4.5c] 302.4   **53.** [4.3b] B   **54.** [4.7a] $35   **55.** [4.1b, c] $\frac{2}{3}, \frac{5}{7}, \frac{15}{19}, \frac{11}{13}, \frac{17}{20}, \frac{13}{15}$

Cumulative Review: Chapters 1-4, p. 293

1. [3.4a] $\frac{20}{9}$ **2.** [4.1b] $\frac{3051}{1000}$ **3.** [4.5a] 1.4   **4.** [4.5a] 0.$\overline{54}$ **5.** [2.1c] Prime   **6.** [2.2a] Yes   **7.** [1.9c] 1754   **8.** [4.4c] 4.364   **9.** [4.1d] 584.97   **10.** [4.5b] 218.56   **11.** [4.6a] 160   **12.** [4.6a] 4   **13.** [1.6b] 12,800,000   **14.** [4.6a] 6   **15.** [3.5a] $6\frac{1}{20}$ **16.** [1.2a] 139,116   **17.** [3.2a] $\frac{31}{18}$ **18.** [4.2a] 145.953   **19.** [1.3a] 710,137   **20.** [4.2b] 13.097   **21.** [3.5b] $\frac{5}{7}$ **22.** [3.3a] $\frac{1}{110}$ **23.** [2.6a] $\frac{1}{6}$ **24.** [1.4a] 5,317,200   **25.** [4.3a] 4.78   **26.** [4.3a] 0.0279431   **27.** [4.4a] 2.122   **28.** [1.5a] 1843   **29.** [4.4a] 13,862.1   **30.** [2.7b] $\frac{5}{6}$ **31.** [4.2c] 0.78   **32.** [1.7b] 28   **33.** [4.4b] 8.62   **34.** [1.7b] 367,251   **35.** [3.3c] $\frac{1}{18}$ **36.** [2.7c] $\frac{1}{2}$ **37.** [1.8a] 24,929 transplants   **38.** [1.8a] $177 billion   **39.** [2.7d] $1500   **40.** [2.6b] $2400   **41.** [4.7a] $258.77   **42.** [3.5c] $6\frac{1}{2}$ lb   **43.** [3.2b] 2 lb   **44.** [4.7a] 467.28 sq ft   **45.** [3.7a] $\frac{9}{32}$ **46.** [4.4c] 527.04   **47.** [4.7a] $2.39   **48.** [3.6c] 144 packages

CHAPTER 5

Exercise Set 5.1, p. 300

1. $\frac{4}{5}$ **3.** $\frac{178}{572}$ **5.** $\frac{0.4}{12}$ **7.** $\frac{3.8}{7.4}$ **9.** $\frac{56.78}{98.35}$ **11.** $\frac{8\frac{3}{4}}{9\frac{5}{6}}$ **13.** $\frac{21}{4}, \frac{4}{21}$ **15.** $\frac{100}{84.7}, \frac{84.7}{100}$ **17.** $\frac{113}{366}$ **19.** $\frac{1000}{126}, \frac{126}{1000}$ **21.** $\frac{60}{100}, \frac{100}{60}$ **23.** $\frac{2}{3}$ **25.** $\frac{3}{4}$ **27.** $\frac{12}{25}$ **29.** $\frac{7}{9}$ **31.** $\frac{2}{3}$ **33.** $\frac{14}{25}$ **35.** $\frac{1}{2}$ **37.** $\frac{3}{4}$ **39.** $\frac{478}{213}, \frac{213}{478}$ **41.** =   **42.** ≠   **43.** ≠   **44.** =   **45.** 50   **46.** 9.5   **47.** 14.5   **48.** 152   **49.** $6\frac{7}{20}$ cm   **50.** $17\frac{11}{20}$ cm   **51.** $\frac{30}{47}$ **53.** 1:2:3

Exercise Set 5.2, p. 306

1. 40 km/h   **3.** 7.48 mi/sec   **5.** 25 mpg   **7.** 32 mpg   **9.** 43,728 people/sq mi   **11.** 25 mph; 0.04 hr/mi   **13.** About 24.6 points/game   **15.** 0.623 gal/ft^2 **17.** 186,000 mi/sec   **19.** 124 km/h   **21.** 25 beats/min   **23.** 26.188¢/oz; 26.450¢/oz; 16 oz   **25.** 8.867¢/oz; 7.320¢/oz; 75 oz   **27.** 13.889¢/oz; 17.464¢/oz; 18 oz   **29.** 12.991¢/oz; 10.346¢/oz; 26 oz   **31.** B: 18.719¢/oz; E: 14.563¢/oz; Brand E   **33.** A: 10.375¢/oz; B: 9.139¢/oz; H: 8.022¢/oz; Brand H   **35.** 4063 theaters   **36.** $37\frac{1}{2}$ servings   **37.** $9.1 billion   **38.** 281.4 million tracks   **39.** 109.608   **40.** 67,819   **41.** 5833.56   **42.** 466,190.4   **43.** 6-oz; 10.833¢/oz; 5.5-oz; 10.909¢/oz

Exercise Set 5.3, p. 314

1. No   **3.** Yes   **5.** Yes   **7.** No   **9.** 45   **11.** 12   **13.** 10   **15.** 20   **17.** 5   **19.** 18   **21.** 22   **23.** 28   **25.** $\frac{28}{3}$, or $9\frac{1}{3}$

27. $\frac{26}{9}$, or $2\frac{8}{9}$ **29.** 5   **31.** 5   **33.** 0   **35.** 14   **37.** 2.7   **39.** 1.8   **41.** 0.06   **43.** 0.7   **45.** 12.5725   **47.** 1   **49.** $\frac{1}{20}$ **51.** $\frac{3}{8}$ **53.** $\frac{16}{75}$ **55.** $\frac{51}{16}$, or $3\frac{3}{16}$ **57.** $\frac{546}{185}$, or $2\frac{176}{185}$ **59.** Quotient   **60.** Sum   **61.** Average   **62.** Dollars; cents   **63.** Subtrahend   **64.** Terminating   **65.** Commutative   **66.** Cross products   **67.** Approximately 2731.4   **69.** Ruth: 1.863 strikeouts per home run; Schmidt: 3.436 strikeouts per home run

Mid-Chapter Review: Chapter 5, p. 317

1. True   **2.** True   **3.** False   **4.** False
5. $\frac{120 \text{ mi}}{2 \text{ hr}} = \frac{120}{2}\,\frac{\text{mi}}{\text{hr}} = 60$ mi/hr
6.
$$\frac{x}{4} = \frac{3}{6}$$
$$x \cdot 6 = 4 \cdot 3$$
$$\frac{x \cdot 6}{6} = \frac{4 \cdot 3}{6}$$
$$x = 2$$
7. $\frac{4}{7}$ **8.** $\frac{313}{199}$ **9.** $\frac{35}{17}$ **10.** $\frac{59}{101}$ **11.** $\frac{2}{3}$ **12.** $\frac{1}{3}$ **13.** $\frac{8}{7}$ **14.** $\frac{25}{19}$ **15.** $\frac{2}{1}$ **16.** $\frac{5}{1}$ **17.** $\frac{2}{7}$ **18.** $\frac{3}{5}$ **19.** 60.75 mi/hr, or 60.75 mph   **20.** 48.67 km/h   **21.** 13 m/sec   **22.** 16.17 ft/sec   **23.** 27 in./day   **24.** About 0.929 free throw made/attempt   **25.** 11.611¢/oz   **26.** 49.917¢/oz   **27.** Yes   **28.** No   **29.** No   **30.** Yes   **31.** 12   **32.** 40   **33.** 9   **34.** 35   **35.** 2.2   **36.** 4.32   **37.** $\frac{1}{2}$ **38.** $\frac{65}{4}$, or $16\frac{1}{4}$ **39.** Yes;

every ratio $\frac{a}{b}$ can be written as $\frac{\frac{a}{b}}{1}$.   **40.** By making some sketches, we see that the rectangle's length must be twice the width.   **41.** The student's approach will work. However, when we use the approach of equating cross products, we eliminate the need to find the least common denominator.   **42.** The instructor thinks that the longer a student studies, the higher his or her grade will be. An example is the situation in which one student gets a test grade of 96 after studying for 8 hr while another student gets a score of 78 after studying for $6\frac{1}{2}$ hr. This is represented by the proportion $\frac{96}{8} = \frac{78}{6\frac{1}{2}}$.

Translating for Success, p. 324

1. N   **2.** I   **3.** A   **4.** K   **5.** J   **6.** F   **7.** M   **8.** B   **9.** G   **10.** E

Exercise Set 5.4, p. 325

1. 11.04 hr   **3.** 60 students   **5.** (a) About 122 gal; (b) 3080 mi   **7.** 212.52 million, or 212,520,000   **9.** 175 bulbs   **11.** 2975 ft^2 **13.** 450 pages   **15.** (a) 28.065 British pounds; (b) $15,392.84   **17.** (a) 13,399.625 Japanese yen; (b) $47.76   **19.** 13,500 mi   **21.** 880 calories   **23.** 120 lb   **25.** 64 gal   **27.** 100 oz   **29.** 954 deer   **31.** 58.1 mi   **33.** 9.75 gal   **35.** (a) 171 games; (b) about 47 home runs   **37.** Neither   **38.** Prime   **39.** Composite   **40.** Composite   **41.** Prime   **42.** $2 \cdot 2 \cdot 2 \cdot 101$, or $2^3 \cdot 101$ **43.** $2 \cdot 2 \cdot 2 \cdot 7$, or $2^3 \cdot 7$ **44.** $2 \cdot 433$ **45.** $3 \cdot 31$ **46.** $2 \cdot 2 \cdot 5 \cdot 101$, or $2^2 \cdot 5 \cdot 101$ **47.** 17 positions   **49.** 2150 earned runs   **51.** CD player: $150; receiver: $450; speakers: $300

Exercise Set 5.5, p. 334

1. 25   **3.** $\frac{4}{3}$, or $1\frac{1}{3}$ **5.** $x = \frac{27}{4}$, or $6\frac{3}{4}$; $y = 9$ **7.** $x = 7.5$; $y = 7.2$ **9.** 1.25 m   **11.** 36 ft   **13.** 7 ft   **15.** 100 ft   **17.** 4   **19.** $10\frac{1}{2}$ **21.** $x = 6$; $y = 5.25$; $z = 3$ **23.** $x = 5\frac{1}{3}$, or $5.\overline{3}$; $y = 4\frac{2}{3}$, or $4.\overline{6}$; $z = 5\frac{1}{3}$, or $5.\overline{3}$ **25.** 20 ft   **27.** 152 ft   **29.** $59.81   **30.** 9.63   **31.** 679.4928   **32.** 2.74568   **33.** 27,456.8   **34.** 0.549136   **35.** 0.85   **36.** 1.825   **37.** 0.909   **38.** 0.843   **39.** 13.75 ft   **41.** 1.25 cm   **43.** 3681.437   **45.** $x = 0.4$; $y \approx 0.35$

Summary and Review: Chapter 5, p. 338

Concept Reinforcement

1. True   **2.** True   **3.** False   **4.** True

Important Concepts

1. $\frac{17}{3}$ **2.** $\frac{8}{7}$ **3.** \$7.50/hr **4.** A: 9.964¢/oz; B: 10.281¢/oz; Brand A **5.** Yes **6.** $\frac{27}{8}$ **7.** 175 mi **8.** 21

Review Exercises

1. $\frac{47}{84}$ **2.** $\frac{46}{1.27}$ **3.** $\frac{83}{100}$ **4.** $\frac{0.72}{197}$ **5. (a)** $\frac{12,480}{16,640}$, or $\frac{3}{4}$; **(b)** $\frac{16,640}{29,120}$, or $\frac{4}{7}$ **6.** $\frac{3}{4}$ **7.** $\frac{9}{16}$ **8.** 26 mpg **9.** 6300 revolutions/min
10. 0.638 gal/ft² **11.** 6.33¢/tablet **12.** 11.208¢/oz
13. 14.969¢/oz; 12.479¢/oz; 15.609¢/oz; 48 oz **14.** Yes
15. No **16.** 32 **17.** 7 **18.** $\frac{1}{40}$ **19.** 24 **20.** 27 circuits
21. (a) 267 Canadian dollars; **(b)** 46.82 U.S. dollars
22. 832 mi **23.** 27 acres **24.** About 3,418,140 lb
25. 6 in. **26.** About 13,644 lawyers **27.** $x = \frac{14}{3}$, or $4\frac{2}{3}$
28. $x = \frac{56}{5}$, or $11\frac{1}{5}$; $y = \frac{63}{5}$, or $12\frac{3}{5}$ **29.** 40 ft
30. $x = 3$; $y = \frac{21}{2}$, or $10\frac{1}{2}$; $z = \frac{15}{2}$, or $7\frac{1}{2}$ **31.** B **32.** C
33. 1.329¢/sheet; 1.554¢/sheet; 1.110¢/sheet; 6 big rolls
34. $x = 4258.5$; $z \approx 10{,}094.3$ **35.** Finishing paint: 11 gal; primer: 16.5 gal

Understanding Through Discussion and Writing

1. In terms of cost, a low faculty-to-student ratio is less expensive than a high faculty-to-student ratio. In terms of quality of education and student satisfaction, a high faculty-to-student ratio is more desirable. A college president must balance the cost and quality issues. **2.** Yes; unit prices can be used to solve proportions involving money. In Example 3 of Section 5.4, for instance, we could have divided \$90 by the unit price, or the price per ticket, to find the number of tickets that could be purchased for \$90. **3.** Leslie used 4 gal of gasoline to drive 92 mi. At the same rate, how many gallons would be needed to travel 368 mi? **4.** Yes; consider the following pair of triangles.

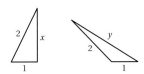

Two pairs of sides are proportional, but we can see that x is shorter than y, so the ratio of x to y is clearly not the same as the ratio of 1 to 1 (or 2 to 2).

Test: Chapter 5, p. 343

1. [5.1a] $\frac{85}{97}$ **2.** [5.1a] $\frac{0.34}{124}$ **3.** [5.1b] $\frac{9}{10}$ **4.** [5.1b] $\frac{25}{32}$
5. [5.2a] 0.625 ft/sec **6.** [5.2a] $1\frac{1}{3}$ servings/lb
7. [5.2a] 32 mpg **8.** [5.2b] About 15.563¢/oz
9. [5.2b] 16.475¢/oz; 13.980¢/oz; 11.490¢/oz; 16.660¢/oz; 100 oz **10.** [5.3a] Yes **11.** [5.3a] No **12.** [5.3b] 12
13. [5.3b] 360 **14.** [5.3b] 42.1875 **15.** [5.3b] 100
16. [5.4a] 1512 km **17.** [5.4a] 4.8 min **18.** [5.4a] 525 mi
19. [5.5a] 66 m **20.** [5.4a] **(a)** 3501.45 Hong Kong dollars; **(b)** \$102.17 **21.** [5.4a] About \$59.17 **22.** [5.5a] $x = 8$; $y = 8.8$ **23.** [5.5b] $x = \frac{24}{5}$, or 4.8; $y = \frac{32}{5}$, or 6.4; $z = 12$
24. [5.2a] C **25.** [5.4a] 5888 marbles

Cumulative Review: Chapters 1–5, p. 345

1. [4.2a] 513.996 **2.** [3.5a] $6\frac{3}{4}$ **3.** [3.2a] $\frac{7}{20}$
4. [4.2b] 30.491 **5.** [4.2b] 72.912 **6.** [3.3a] $\frac{7}{60}$
7. [4.3a] 222.076 **8.** [4.3a] 567.8 **9.** [3.6a] 3 **10.** [4.4a] 43
11. [1.5a] 899 **12.** [2.7b] $\frac{3}{2}$ **13.** [1.1b] 3 ten thousands + 0 thousands + 0 hundreds + 7 tens + 4 ones, or 3 ten thousands + 7 tens + 4 ones **14.** [4.1a] One hundred twenty and seven hundredths **15.** [4.1c] 0.7 **16.** [4.1c] 0.8
17. [2.1d] $2 \cdot 2 \cdot 2 \cdot 2 \cdot 3 \cdot 3$, or $2^4 \cdot 3^2$ **18.** [3.1a] 90
19. [2.3a] $\frac{5}{8}$ **20.** [2.5b] $\frac{5}{8}$ **21.** [4.5c] 5.718 **22.** [4.5c] 0.179
23. [5.1a] $\frac{0.3}{15}$ **24.** [5.3a] Yes **25.** [5.2a] 55 m/sec
26. [5.2b] 8-oz can **27.** [5.3b] 30.24 **28.** [4.4b] 26.4375

29. [2.7c] $\frac{8}{9}$ **30.** [5.3b] 128 **31.** [4.2c] 33.34 **32.** [3.3c] $\frac{76}{175}$
33. [2.6b] 390 cal **34.** [5.4a] 7 min **35.** [3.5c] $2\frac{1}{4}$ cups
36. [2.7d] 12 doors **37.** [4.7a], [5.2a] 42.2025 mi
38. [4.7a] 132 orbits **39.** [2.1c] D **40.** [5.2a] B
41. [5.5a] $10\frac{1}{2}$ ft

CHAPTER 6

Calculator Corner, p. 350

1. 0.14 **2.** 0.00069 **3.** 0.438 **4.** 1.25

Exercise Set 6.1, p. 352

1. $\frac{90}{100}$; $90 \times \frac{1}{100}$; 90×0.01 **3.** $\frac{12.5}{100}$; $12.5 \times \frac{1}{100}$; 12.5×0.01
5. 0.67 **7.** 0.456 **9.** 0.5901 **11.** 0.1 **13.** 0.01 **15.** 2
17. 0.001 **19.** 0.0009 **21.** 0.0018 **23.** 0.2319
25. 0.14875 **27.** 0.565 **29.** 0.97 **31.** 0.07; 0.08
33. 0.548 **35.** 47% **37.** 3% **39.** 870% **41.** 33.4%
43. 75% **45.** 40% **47.** 0.6% **49.** 1.7% **51.** 27.18%
53. 2.39% **55.** 27% **57.** 5.7%; 17.6% **59.** 90.6%; 88%
61. 0.64; 0.3; 0.04; 0.02 **63.** $33\frac{1}{3}$ **64.** $37\frac{1}{2}$ **65.** $9\frac{3}{8}$
66. $18\frac{9}{16}$ **67.** $5\frac{11}{14}$ **68.** $111\frac{2}{3}$ **69.** $0.\overline{6}$ **70.** $0.\overline{3}$
71. $0.8\overline{3}$ **72.** $1.41\overline{6}$ **73.** $2.\overline{6}$ **74.** 0.9375 **75.** 50%
77. 70% **79.** 20%

Calculator Corner, p. 355

1. 52% **2.** 38.46% **3.** 110.26% **4.** 171.43% **5.** 59.62%
6. 28.31%

Exercise Set 6.2, p. 359

1. 41% **3.** 5% **5.** 20% **7.** 30% **9.** 50% **11.** 87.5%, or $87\frac{1}{2}$% **13.** 80% **15.** $66.\overline{6}$%, or $66\frac{2}{3}$% **17.** $16.\overline{6}$%, or $16\frac{2}{3}$% **19.** 18.75%, or $18\frac{3}{4}$% **21.** 81.25%, or $81\frac{1}{4}$%
23. 16% **25.** 5% **27.** 34% **29.** 8%; 59% **31.** 22%
33. 12% **35.** 15% **37.** $\frac{17}{20}$ **39.** $\frac{5}{8}$ **41.** $\frac{1}{3}$ **43.** $\frac{1}{6}$ **45.** $\frac{29}{400}$
47. $\frac{1}{125}$ **49.** $\frac{203}{800}$ **51.** $\frac{176}{225}$ **53.** $\frac{711}{1100}$ **55.** $\frac{3}{2}$ **57.** $\frac{13}{40,000}$
59. $\frac{1}{3}$ **61.** $\frac{3}{50}$ **63.** $\frac{3}{25}$ **65.** $\frac{3}{4}$ **67.** $\frac{3}{20}$ **69.** $\frac{209}{1000}$
71.

Fraction Notation	Decimal Notation	Percent Notation
$\frac{1}{8}$	0.125	12.5%, or $12\frac{1}{2}$%
$\frac{1}{6}$	$0.1\overline{6}$	$16.\overline{6}$%, or $16\frac{2}{3}$%
$\frac{1}{5}$	0.2	20%
$\frac{1}{4}$	0.25	25%
$\frac{1}{3}$	$0.\overline{3}$	$33.\overline{3}$%, or $33\frac{1}{3}$%
$\frac{3}{8}$	0.375	37.5%, or $37\frac{1}{2}$%
$\frac{2}{5}$	0.4	40%
$\frac{1}{2}$	0.5	50%

73.

Fraction Notation	Decimal Notation	Percent Notation
$\frac{1}{2}$	0.5	50%
$\frac{1}{3}$	$0.\overline{3}$	$33.\overline{3}\%$, or $33\frac{1}{3}\%$
$\frac{1}{4}$	0.25	25%
$\frac{1}{6}$	$0.1\overline{6}$	$16.\overline{6}\%$, or $16\frac{2}{3}\%$
$\frac{1}{8}$	0.125	12.5%, or $12\frac{1}{2}\%$
$\frac{3}{4}$	0.75	75%
$\frac{5}{6}$	$0.8\overline{3}$	$83.\overline{3}\%$, or $83\frac{1}{3}\%$
$\frac{3}{8}$	0.375	37.5%, or $37\frac{1}{2}\%$

75. 70 **76.** 5 **77.** 400 **78.** 18.75 **79.** 23.125
80. 25.5 **81.** 4.5 **82.** 8.75 **83.** $33\frac{1}{3}$ **84.** $37\frac{1}{2}$ **85.** $83\frac{1}{3}$
86. $20\frac{1}{2}$ **87.** $43\frac{1}{8}$ **88.** $62\frac{1}{6}$ **89.** $18\frac{3}{4}$ **90.** $7\frac{4}{9}$ **91.** $\frac{18}{17}$
92. $\frac{209}{10}$ **93.** $\frac{203}{8}$ **94.** $\frac{259}{8}$ **95.** $11.\overline{1}\%$ **97.** $257.\overline{46317}\%$
99. $0.01\overline{5}$ **101.** $1.04\overline{142857}$ **103.** $\frac{1}{6}\%, \frac{2}{7}\%, 0.5\%, 1\frac{1}{6}\%, 1.6\%,$ $16\frac{1}{6}\%, 0.2, \frac{1}{2}, 0.\overline{54}, 1.6$

Calculator Corner, p. 366

1. $5.04 **2.** 0.0112 **3.** 450 **4.** $1000 **5.** 2.5% **6.** 12%

Exercise Set 6.3, p. 367

1. $a = 32\% \times 78$ **3.** $89 = p \times 99$ **5.** $13 = 25\% \times b$
7. 234.6 **9.** 45 **11.** $18 **13.** 1.9 **15.** 78% **17.** 200%
19. 50% **21.** 125% **23.** 40 **25.** $40 **27.** 88 **29.** 20
31. 6.25 **33.** $846.60 **35.** 1216 **37.** $\frac{9}{100}$ **38.** $\frac{179}{100}$
39. $\frac{875}{1000}$, or $\frac{7}{8}$ **40.** $\frac{125}{1000}$, or $\frac{1}{8}$ **41.** $\frac{9375}{10,000}$, or $\frac{15}{16}$ **42.** $\frac{6875}{10,000}$, or $\frac{11}{16}$
43. 0.89 **44.** 0.07 **45.** 0.3 **46.** 0.017 **47.** $800 (can vary); $843.20 **49.** $10,000 (can vary); $10,400 **51.** $1875

Exercise Set 6.4, p. 373

1. $\frac{37}{100} = \frac{a}{74}$ **3.** $\frac{N}{100} = \frac{4.3}{5.9}$ **5.** $\frac{25}{100} = \frac{14}{b}$ **7.** 68.4 **9.** 462
11. 40 **13.** 2.88 **15.** 25% **17.** 102% **19.** 25%
21. 93.75%, or $93\frac{3}{4}\%$ **23.** $72 **25.** 90 **27.** 88 **29.** 20
31. 25 **33.** $780.20 **35.** 200 **37.** 8 **38.** 4000 **39.** 8
40. 2074 **41.** 100 **42.** 15 **43.** 8.04 **44.** $\frac{3}{16}$, or 0.1875
45. $\frac{43}{48}$ qt **46.** $\frac{1}{8}$ T **47.** $1170 (can vary); $1118.64

Mid-Chapter Review: Chapter 6, p. 375

1. True **2.** False **3.** True **4.** $\frac{1}{2}\% = \frac{1}{2} \cdot \frac{1}{100} = \frac{1}{200}$
5. $\frac{80}{1000} = \frac{8}{100} = 8\%$ **6.** $5.5\% = \frac{5.5}{100} = \frac{55}{1000} = \frac{11}{200}$
7. $0.375 = \frac{375}{1000} = \frac{37.5}{100} = 37.5\%$.
8. $15 = p \times 80$
$\frac{15}{80} = \frac{p \times 80}{80}$
$\frac{15}{80} = p$
$0.1875 = p$
$18.75\% = p$
9. 0.28 **10.** 0.0015

11. 0.05375 **12.** 2.4 **13.** 71% **14.** 9% **15.** 38.91%
16. 18.75%, or $18\frac{3}{4}\%$ **17.** 0.5% **18.** 74% **19.** 600%
20. $83.\overline{3}\%$, or $83\frac{1}{3}\%$ **21.** $\frac{17}{20}$ **22.** $\frac{3}{6250}$ **23.** $\frac{91}{400}$ **24.** $\frac{1}{6}$
25. 62.5%, or $62\frac{1}{2}\%$ **26.** 45% **27.** 58 **28.** $16.\overline{6}\%$, or $16\frac{2}{3}\%$
29. 2560 **30.** $50 **31.** 20% **32.** $455 **33.** 0.05%, 0.1%, $\frac{1}{2}\%$, 1%, 5%, 10%, $\frac{13}{100}$, 0.275, $\frac{3}{10}$, $\frac{7}{20}$ **34.** D **35.** B **36.** Some will say that the conversion will be done most accurately by first finding decimal notation. Others will say that it is more efficient to become familiar with some or all of the fraction and percent equivalents that appear inside the back cover and to make the conversion by going directly from fraction notation to percent notation. **37.** Since $40\% \div 10 = 4\%$, we can divide 36.8 by 10, obtaining 3.68. Since $400\% = 40\% \times 10$, we can multiply 36.8 by 10, obtaining 368. **38.** Answers may vary. Some will say this is a good idea since it makes the computations in the solution easier. Others will say it is a poor idea since it adds an extra step to the solution. **39.** They all represent the same number.

Translating for Success, p. 384

1. J **2.** M **3.** N **4.** E **5.** G **6.** H **7.** O **8.** C
9. D **10.** B

Exercise Set 6.5, p. 385

1. South Korea: 62,381 students; Japan: 35,272 students
3. $46,656 **5.** 140 items **7.** 940,000,000 acres **9.** $36,400
11. 74.4 items correct; 5.6 items incorrect **13.** Egypt: 24,596,374; United States: 62,315,472 **15.** About 17.3%
17. $230.10 **19.** About 612,000 fast-food cooks **21.** Alcohol: 43.2 mL; water: 496.8 mL **23.** Air Force: 25.2%; Army: 36.5%; Navy: 25.3%; Marines: 13.0% **25.** 5% **27.** 15% **29.** About 34.7% **31.** $33\frac{1}{3}\%$ **33.** About 49.5% **35.** About 59%
37. About 9.5% **39.** 34.375%, or $34\frac{3}{8}\%$ **41.** 15,081; 2.5%
43. 5,130,632; 20.2% **45.** 1,466,465; 13.3% **47.** 1.0%
49. $2.\overline{27}$ **50.** 0.44 **51.** 3.375 **52.** $4.\overline{7}$ **53.** 0.92 **54.** $0.8\overline{3}$
55. 0.4375 **56.** 2.317 **57.** 3.4809 **58.** 0.675 **59.** $42

Exercise Set 6.6, p. 396

1. $9.56 **3.** $1.62 **5.** $11.59; $171.39 **7.** 4% **9.** 2%
11. $5600 **13.** $116.72 **15.** $276.28 **17.** $194.08
19. $2625 **21.** 12% **23.** $185,000 **25.** $5440 **27.** 15%
29. $355 **31.** $30; $270 **33.** $2.55; $14.45 **35.** $125; $112.50 **37.** 40%; $360 **39.** $50; 30% **41.** $849; 21.2%
43. 18 **44.** $\frac{22}{7}$ **45.** 265.625 **46.** 1.15 **47.** $0.\overline{5}$ **48.** $2.\overline{09}$
49. $0.91\overline{6}$ **50.** $1.\overline{857142}$ **51.** $2.\overline{142857}$ **52.** $1.58\overline{3}$
53. 4,030,000,000,000 **54.** 5,800,000 **55.** 42,700,000
56. 6,090,000,000,000 **57.** $4.99 **59.** $17,700

Calculator Corner, p. 403

1. $16,357.18 **2.** $12,764.72

Exercise Set 6.7, p. 406

1. $8 **3.** $113.52 **5.** $925 **7.** $671.88 **9. (a)** $147.95; **(b)** $10,147.95 **11. (a)** $84.14; **(b)** $6584.14
13. (a) $46.03; **(b)** $5646.03 **15.** $441 **17.** $2802.50
19. $7853.38 **21.** $99,427.40 **23.** $4243.60
25. $28,225.00 **27.** $9270.87 **29.** $129,871.09
31. $4101.01 **33.** $1324.58 **35.** Interest: $20.88; amount applied to principal: $4.69; balance after the payment: $1273.87
37. (a) $98; **(b)** interest: $86.56; amount applied to principal: $11.44; **(c)** interest: $51.20; amount applied to principal: $46.80; **(d)** At 12.6%, the principal is reduced by $35.36 more than at the 21.3% rate. The interest at 12.6% is $35.36 less than at 21.3%. **39.** Reciprocals **40.** Divisible by 6 **41.** Additive
42. Unit price **43.** Perimeter **44.** Divisible by 3
45. Prime **46.** Proportional **47.** 9.38%

Summary and Review: Chapter 6, p. 409

Concept Reinforcement

1. True   **2.** False   **3.** True

Important Concepts

1. 0.62625   **2.** $63.\overline{63}$%, or $63\frac{7}{11}$%   **3.** $\frac{17}{250}$ **4.** $4.1\overline{6}$%, or $4\frac{1}{6}$%
5. 10,000   **6.** About 15.3%   **7.** 6%   **8.** $185,000
9. Simple interest: $22.60; total amount due: $2522.60
10. $6594.26

Review Exercises

1. 0.04; 0.144   **2.** 0.621; 0.842   **3.** 37.5%, or $37\frac{1}{2}$%
4. $33.\overline{3}$%, or $33\frac{1}{3}$%   **5.** 170%   **6.** 6.5%   **7.** $\frac{6}{25}$ **8.** $\frac{63}{1000}$
9. $30.6 = p \times 90$; 34%   **10.** $63 = 84\% \times b$; 75
11. $a = 38\frac{1}{2}\% \times 168$; 64.68   **12.** $\frac{24}{100} = \frac{16.8}{b}$; 70
13. $\frac{42}{30} = \frac{N}{100}$; 140%   **14.** $\frac{10.5}{100} = \frac{a}{84}$; 8.82   **15.** 178 students;
84 students   **16.** 46%   **17.** 2500 mL   **18.** 12%   **19.** 92
20. $24   **21.** 6%   **22.** 11%   **23.** $42; $308   **24.** 14%
25. $2940   **26.** About 18.3%   **27.** $36   **28. (a)** $394.52;
(b) $24,394.52   **29.** $121   **30.** $7575.25   **31.** $9504.80
32. (a) $129; **(b)** interest: $100.18; amount applied to principal:
$28.82; **(c)** interest: $70.72; amount applied to principal: $58.28;
(d) At 13.2%, the principal is decreased by $29.46 more than at
the 18.7% rate. The interest at 13.2% is $29.46 less than at 18.7%.
33. A   **34.** C   **35.** $66.\overline{6}$%, or $66\frac{2}{3}$% **36.** About 19%

Understanding Through Discussion and Writing

1. A 40% discount is better. When successive discounts are taken,
each is based on the previous discounted price rather than on the
original price. A 20% discount followed by a 22% discount is the
same as a 37.6% discount off the original price.   **2.** Let $S =$ the
original salary. After both raises have been given, the two
situations yield the same salary: $1.05 \cdot 1.1S = 1.1 \cdot 1.05S$.
However, the first situation is better for the wage earner, because
$1.1S$ is earned the first year when a 10% raise is given while in
the second situation $1.05S$ is earned that year.   **3.** No; the 10%
discount was based on the original price rather than on the sale
price.   **4.** For a number n, 40% of 50% of n is $0.4(0.5n)$, or
$0.2n$, or 20% of n. Thus taking 40% of 50% of a number is the
same as taking 20% of the number.   **5.** The interest due on the
30-day loan will be $41.10 while that due on the 60-day loan will
be $131.51. This could be an argument in favor of the 30-day
loan. On the other hand, the 60-day loan puts twice as much
cash at the firm's disposal for twice as long as the 30-day loan
does. This could be an argument in favor of the 60-day loan.
6. Answers will vary.

Test: Chapter 6, p. 415

1. [6.1b] 0.147   **2.** [6.1b] 38%   **3.** [6.2a] 137.5%   **4.** [6.2b] $\frac{13}{20}$
5. [6.3a, b] $a = 40\% \cdot 55$; 22   **6.** [6.4a, b] $\frac{N}{100} = \frac{65}{80}$; 81.25%
7. [6.5a] 16,692 kidney transplants; 6224 liver transplants; 2263
heart transplants   **8.** [6.5a] About 611 at-bats
9. [6.5b] 6.7%   **10.** [6.5a] 60.6%   **11.** [6.6a] $25.20; $585.20
12. [6.6b] $630   **13.** [6.6c] $40; $160   **14.** [6.7a] $8.52
15. [6.7a] $5356   **16.** [6.7b] $1110.39   **17.** [6.7b] $11,580.07
18. [6.5b] Plumber: 757,000, 7.4%; veterinary assistant: 29,000,
40.8%; motorcycle repair technician: 21,000, 14.3%; fitness
professional: 235,000, 63,000   **19.** [6.6c] $50; about 14.3%
20. [6.7c] Interest: $36.73; amount applied to the principal:
$17.27; balance after payment: $2687   **21.** [6.3a, b], [6.4a, b] B
22. [6.6b] $194,600   **23.** [6.6b], [6.7b] $2546.16

Cumulative Review: Chapters 1-6, p. 417

1. [6.1b] 0.44; 0.3   **2.** [6.1b] 26.9%   **3.** [6.2a] 112.5%
4. [4.5a] $2.1\overline{6}$ **5.** [5.1a] $\frac{5}{0.5}$, or $\frac{10}{1}$ **6.** [5.2a] $\frac{70\,km}{3\,hr}$, or
$23.\overline{3}$ km/hr, or $23\frac{1}{3}$ km/hr   **7.** [2.5c], [3.3b] $<$ **8.** [2.5c], [3.3b] $<$
9. [1.6b], [4.6a] 296,200   **10.** [1.6b], [4.6a] 50,000
11. [1.9d] 13   **12.** [4.4c] 1.5   **13.** [3.5a] $3\frac{1}{30}$
14. [4.2a] 49.74   **15.** [1.2a] 515,150   **16.** [4.2b] 0.02
17. [3.5b] $\frac{2}{3}$ **18.** [3.3a] $\frac{2}{63}$ **19.** [2.6a] $\frac{1}{6}$ **20.** [1.4a] 853,142,400
21. [4.3a] 1.38036   **22.** [3.6b] $\frac{3}{2}$, or $1\frac{1}{2}$ **23.** [4.4a] 12.25
24. [1.5a] 123 R5   **25.** [1.7b] 95   **26.** [4.2c] 8.13   **27.** [2.7c] 9
28. [3.3c] $\frac{1}{12}$ **29.** [5.3b] 40   **30.** [5.3b] $\frac{176}{21}$ **31.** [6.5a] 462,259
visitors   **32.** [1.8a] $177,500   **33.** [5.4a] About 65 games
34. [4.7a] $84.95   **35.** [5.2b] 7.495 cents/oz
36. [3.2b], [3.5c] $\frac{3}{2}$ mi, or $1\frac{1}{2}$ mi   **37.** [5.4a] 60 mi
38. [6.7b] $12,663.69   **39.** [3.6c] 5 pieces
40. [6.5b] About 14.2%; 883,000 technicians   **41.** [3.3a] C
42. [6.5b] A   **43.** [6.3b], [6.4b] 200%

Photo Credits

Glossary

A

Addend   In addition, a number being added

Additive identity   The number 0

Area   The number of square units that fill a plane region

Arithmetic numbers   The whole numbers and the positive fractions

Associative law of addition   The statement that when three numbers are added, regrouping the addends gives the same sum

Associative law of multiplication   The statement that when three numbers are multiplied, regrouping the factors gives the same product

Average   A center point of a set of numbers found by adding the numbers and dividing by the number of items of data; also called the arithmetic mean or mean

B

Base   In exponential notation, the number being raised to a power

C

Commission   A percent of total sales paid to a salesperson

Commutative law of addition   The statement that when two numbers are added, changing the order in which the numbers are added does not affect the sum

Commutative law of multiplication   The statement that when two numbers are multiplied, changing the order in which the numbers are multiplied does not affect the product

Composite number   A natural number, other than 1, that is not prime

Compound interest   Interest paid on interest

Cross products   Given an equation with a single fraction on each side, the products formed by multiplying the left numerator and the right denominator, and the left denominator and the right numerator

D

Decimal notation   A representation of a number containing a decimal point

Denominator   The bottom number in a fraction

Difference   The result of subtracting one number from another

Digit   A number 0, 1, 2, 3, 4, 5, 6, 7, 8, or 9 that names a place-value location

Discount   The amount subtracted from the original price of an item to find the sale price

Distributive law of multiplication over addition   The statement that multiplying a factor by the sum of two numbers gives the same result as multiplying the factor by each of the two numbers and then adding

Distributive law of multiplication over subtraction   The statement that multiplying a factor by the difference of two numbers gives the same result as multiplying the factor by each of the two numbers and then subtracting

Dividend   In division, the number being divided

Divisible   The number a is divisible by another number b if there exists a number c such that $a = b \cdot c$

Divisor   In division, the number dividing another number

E

Equation   A number sentence that says that the expressions on either side of the equals sign, $=$, represent the same number

Equivalent fractions   Fractions that represent the same number

Even number   A number that is divisible by 2; that is, it has an even ones digit

Exponent   In expressions of the form a^n, the number n; for n a natural number, a^n represents n factors of a

Exponential notation   A representation of a number using a base raised to a power

F

Factor   A multiplier

Factorization   A number expressed as a product of natural numbers

Fraction notation   A number written using a numerator and a denominator

I

Inequality   A mathematical sentence using $<$, $>$, $\leq$, $\geq$, or $\neq$

Interest   A percentage of an amount invested or borrowed

L

Least common denominator (LCD)   The least common multiple of the denominators of two or more fractions

Least common multiple (LCM)   The smallest number that is a multiple of two or more numbers

M

Marked price   The original price of an item

Minuend   The number from which another number is being subtracted

Mixed numeral   A number represented by a whole number and a fraction less than 1

Multiple   A product of a number and some natural number

Multiplicative identity   The number 1

N

Natural numbers   The numbers 1, 2, 3, 4, 5, ...

Numerator   The top number in a fraction

O

Original price   The price of an item before a discount is deducted

P

Percent notation   A representation of a number as parts per 100

Perimeter   The distance around an object, or the sum of the lengths of its sides

Prime factorization   A factorization of a composite number as a product of prime numbers

Prime number   A natural number that has exactly two *different* factors: only itself and 1

Principal   The amount invested

Product   The result in multiplication

Proportion   An equation that states that two ratios are equal

Purchase price   The price of an item before sales tax is added

Q

Quotient   The result when one number is divided by another

R

Rate   A ratio used to compare two different kinds of measure

Ratio   The quotient of two quantities

Rational numbers   All numbers that can be named in the form a/b, where a and b are integers and $b \neq 0$

Reciprocals   Two numbers whose product is 1

Repeating decimal   A decimal that cannot be written using a finite number of decimal places

Rounding   Approximating the value of a number; used when estimating

S

Sale price   The price of an item after a discount has been deducted

Sales tax   A tax added to the purchase price of an item

Similar triangles   Triangles in which corresponding sides are proportional

Simple interest   A percentage of an amount P invested or borrowed for t years, computed by calculating principal $\times$ interest rate $\times$ time

Solution of an equation   A number that makes an equation true

Subtrahend   In subtraction, the number being subtracted

Sum   The result in addition

T

Terminating decimal   A decimal that can be written using a finite number of decimal places

Total price   The sum of the purchase price of an item and the sales tax on the item

U

Unit price   The ratio of price to the number of units; also called unit rate

Unit rate   The ratio of price to the number of units; also called unit price

V

Variable   A letter that represents an unknown number

W

Whole numbers   The natural numbers and 0: 0, 1, 2, 3, ...

Index